Introduction to
Physical Chemistry

By the same author:
Introduction to Inorganic Chemistry
A New Introduction to Organic Chemistry

Development of atomic structure

1672	Newton	corpuscular theory
1690	Huygens	wave theory
1800	Volta	Voltaic pile
1807	Dalton	atomic theory
1811	Avogadro	molecular theory
1834	Faraday	Laws of electrolysis
1864	Newlands	Law of octaves
1869	Mendeléef	periodic table
1876	Goldstein	cathode rays
1885	Balmer	hydrogen spectrum
1886	Goldstein	positive rays
1887	Hertz	photo-electric effect
1894	Rayleigh and Ramsay	inert gases
1895	Röntgen	X-rays
1896	Becquerel	radioactivity
1897	J. J. Thomson	e/m ratio
1898	M. and P. Curie	isolation of radium
1899	Thomson and Townsend	charge on electron
1900	Planck	quantum theory
1902	Rutherford and Soddy	α-, β-, and γ-rays
1909	Millikan	charge on electron
1910	Soddy	isotopes
1911	Rutherford	the nuclear atom
1912	Wilson	cloud chamber
1912	J. J. Thomson	positive rays
1912	von Laue and Bragg	X-ray crystallography
1913	Moseley	X-ray spectra
1914	Bohr	atomic orbits
1919	Aston	mass spectrograph
1919	Rutherford	transmutation
1926	Schrödinger	wave mechanics
1927	Heisenberg	uncertainty principle
1927	Davisson and Germer	diffraction
1932	Chadwick	neutron
1932	Anderson	positron
1932	Washburn and Urey	deuterium
1933	I. and J-F. Joliot-Curie	artificial radioactivity
1934	Fermi	fission of uranium
1939	Hahn, Strassman, Frisch and Meitner	nuclear fission
1940	McMillan and Abelson	Np and Pu
1942	Fermi and others	first atomic pile
1944–55	Seaborg and others	transuranium elements (Am–Fm)
1945		atomic bomb
1952		thermonuclear test
1955		anti-proton
1956		Calder Hall reactor
1958	Ghiorso and others	nobelium
1959		first AGR (Dounreay)
1961–68	Ghiorso and others	Lr, Ku and Ha
1963	Gell-Mann and Zweig	idea of quarks
1976		AGR (Hinckley Point)
1977		gas centrifuge (Capenhurst)

Introduction to Physical Chemistry

Third Edition

G I Brown BA, BSc

Formerly Assistant Master, Eton College

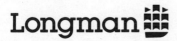

LONGMAN GROUP UK LIMITED
Longman House,
Burnt Mill, Harlow, Essex CM20 2JE, England
and Associated Companies throughout the World.

© Longman Group Limited 1964, 1972, 1982

First published 1964
Second edition 1972
Third edition 1983
Third impression 1986

Set in Monophoto 10/11pt Times New Roman

Produced by Longman Group (FE) Ltd
Printed in Hong Kong

ISBN 0-582-35365-3

Preface

It is almost twenty years since this book was first published, and this new edition has been prepared to bring it up-to-date. As in the earlier editions, the intention is to provide a thorough introduction to physical chemistry and to arouse a student's interest in the subject.

Its contents are based on the requirements of such examinations as the GCE (A and S levels) and university entrance and scholarship examinations. At points of special interest or significance, however, the discussion is taken beyond intermediate examination requirements. The aim has been to try to give, at a reasonably elementary level, and in a not too mathematical form, as balanced and complete a picture of modern physical chemistry as possible.

The presentation is in numbered sections so that particular topics can be omitted or rearranged to suit individual tastes. It is hoped that the book, by selective use, may be of value to the rather weak A-level candidate, to the open scholarship winner, to first-year undergraduates, and to students in technical colleges.

Much of the material from the first edition has been retained, but some older topics, no longer relevant or fashionable, have been omitted, and a certain amount of new material has been added. Some chapters and sections have been rewritten or rearranged, the emphasis has been changed in places, and many detailed alterations and improvements have been made throughout. A wide range of questions, totalling more than 725, is provided; some require the use of a data book. The whole book has been completely modernised, and slightly shortened, without, it is hoped, losing its original flavour.

Contents

Acknowledgements

We are grateful to the following for permission to reproduce past examination questions:
The Associated Examining Board, Cambridge Tutorial Representatives, Cambridge University Press, The Clarendon Press, Her Majesty's Stationery Office, Joint Matriculation Board, Oxford and Cambridge Schools Examination Board, Oxford Delegacy of Local Examinations, Oxford Joint Scholarship and Entrance, Southern Universities' Joint Board for School Examinations, University of London, and Welsh Joint Education Committee.

We would like to take this opportunity to accept full responsibility for the answers supplied, and the conversion of units to the new system, together with minor alterations to avoid unnecessary arithmetic.

1
Gases

1 Physical quantities

Physical chemistry is mainly concerned with the measurement and study of the physical properties, e.g. the density, freezing point, vapour pressure, solubility, energy content and electrical conductivity, of chemicals and mixtures of chemicals. The measurements are expressed by a number and a unit in what is known as a physical quantity. The length of a piece of lead, for example, might be $10\,m$; its mass, $50\,kg$; its density, $11.3\,g\,cm^{-3}$; its temperature, $25\,K$; and its resistivity $19 \times 10^{-8}\,\Omega\,m$. Each physical quantity has its own symbol and unit; it is essential to become familiar with these quantities and the units used. An outline of the system used is given here; and further details are given in Appendix I (p. 470).

a Basic physical quantities and basic SI units Seven physical quantities have been chosen, internationally, as independent, basic quantities. Names and symbols have been allotted to these quantities and the Système Internationale d'Unités has defined the basic units (SI units) to be used:

physical quantity	symbol for quantity	basic SI unit	symbol for unit
length	l	metre	m
mass	m	kilogram	kg
time	t	second	s
electric current	I	ampere	A
thermodynamic temperature	T	kelvin	K
amount of substance	n	mole	mol
luminous intensity	I_v	candela	cd

Notice that when the unit is named after a person it is not given a capital initial letter, but the symbol for the unit is. The candela is of little importance in chemistry.

b Derived physical quantities and derived units All other physical quantities can be derived from the seven basic quantities by multiplication, division, differentiation and/or integration. The units used for the derived quantities are obtained from the basic units by multiplication or division without the introduction of any numerical factors. The SI system is said to be *coherent*.

1

Area, for example, is length × length, with units of m^2; volume is length × length × length, with units of m^3; density is mass divided by volume, with units of $kg\,m^{-3}$; velocity is length divided by time, with units of $m\,s^{-1}$. Other derived units are given special names and symbols as, for example, below:

derived physical quantity	symbol for quantity	derived SI unit	definition of unit
force	F	newton (N)	$1\,N = 1\,kg\,m\,s^{-2}$
pressure	p	pascal (Pa)	$1\,Pa = 1\,N\,m^{-2}$
			$= 1\,kg\,m^{-1}\,s^{-2}$
energy	E	joule (J)	$1\,J = 1\,kg\,m^2\,s^{-2}$

c Prefixes for SI units Decimal fractions or multiples of the basic SI units, or of the specially named derived units, are indicated by use of the following prefixes:

fraction	prefix	symbol	multiple	prefix	symbol
10^{-1}	deci-	d	10	deka-	da
10^{-2}	centi-	c	10^2	hecto-	h
10^{-3}	milli-	m	10^3	kilo-	k
10^{-6}	micro-	μ	10^6	mega-	M
10^{-9}	nano-	n	10^9	giga-	G
10^{-12}	pico-	p	10^{12}	tera-	T

These prefixes are generally attached to the SI base unit. Small lengths, for example, are often quoted in nanometres (nm); larger lengths in kilometres (km). For masses, however, the prefix is attached to g and not to kg, which is the base SI unit. Thus $10^3\,kg$ is $1\,Mg$ and not $1\,kkg$; $10^{-6}\,kg$ is $1\,mg$ and not $1\,\mu kg$.

Prefixes differing in step by 10^3 are preferred but not always used. It is a matter of choosing a unit that is a convenient size.

d Using the units When feeding data into a mathematical equation the units used for the quantities on both sides of the equation must be coherent. It is advisable to get into the habit of using only the basic SI units, and this is essential if using any physical constants (p. 474) quoted in such units.

2 Laws, hypotheses, theories

Many of the experimental measurements made on matter in bulk can be summarised as laws, and, once these laws are firmly established, attempts are made to explain them. This is done by putting forward a *hypothesis*, which is an idea or a collection of ideas, capable of accounting for the facts.

These first ideas are generally somewhat tentative, but if they become widely accepted as true, after consideration, discussion and modification, they are then restated in what is called a *theory*. Finally, if a theory can be built up which effectively accounts for a variety of facts, it is often possible to use it to predict new facts.

The development of chemistry has depended very heavily on the process of discovering experimental laws, and then devising hypotheses and theories to account for them. The atomic, molecular, kinetic and ionic theories have played a very large part.

It is one of the major themes of chemistry that the properties of matter in bulk, and the associated laws, can be accounted for in terms of the nature and structure of individual atoms, molecules and ions, of how they can pack together, and of the forces within and between them.

3 The states of matter

Solids, liquids and gases are known as the three states of matter; alternatively, the terms solid, liquid and gas phase are used. At very high temperatures, matter exists as plasma with its own rather peculiar characteristics; it is the subject of much modern research (p. 33).

So far as gases are concerned, the early gas laws led to the ideas of the kinetic theory. A gas is envisaged as a random arrangement of atoms or molecules, not closely packed, and with very weak forces between them so that they can move about quite freely. That is why gases have low densities, why they expand to occupy any container, and why they are easily compressed.

In contrast, crystalline solids are rigid and incompressible. They are made up of atoms, molecules or ions tightly packed within a lattice and with strong, sometimes very strong, forces between them.

Liquids are intermediate between gases and solids. They lack rigidity, but they are relatively incompressible.

4 Boyle's law, 1662

This law states that *the volume of a given mass of gas is inversely proportional to the pressure, if the temperature remains constant.* Expressed mathematically

$$p \propto 1/V \qquad \text{or} \qquad pV = \text{a constant value}$$

The original data on which Boyle based his law was scanty, and later work showed that real gases did not fully obey the law. The deviations were particularly marked at high pressures, at low temperatures, and for gases which could be liquefied easily (p. 28).

It is now known that the deviations are caused by interactions between the molecules in a gas. If such interactions did not exist, a gas would obey Boyle's law. Such a gas is referred to as an *ideal* or *perfect* gas, but there are no known examples and the concept is a theoretical one.

Real gases approach the ideal or perfect gas at low pressures and high temperatures, and deviations from Boyle's law are not very large for gases at normal temperatures and pressures.

5 Units of pressure

Pressure is defined as force per unit area and, as force (p. 471) is measured in newtons ($1\,N = 1\,kg\,m\,s^{-2}$) and area in square metres (m^2), the main unit of pressure is newtons per square metre ($N\,m^{-2}$). $1\,N\,m^{-2}$ is called a pascal (Pa).

Other units of pressure are also in common use. Normal atmospheric pressure (1 atmosphere or 101.325 kPa) will support a column of mercury 760 mm high so that pressure can be expressed in millimetres of mercury, with 1 mm of mercury being known as a torr. Thus

$$1\,atm = 101.325\,kPa = 760\,mm\ of\ mercury = 760\,torr$$

10^5 Pa, approximately 1 atm, is also known as a bar, and kilobars are used in high pressure work.

6 Charles's law, 1787

This law relates the volume of a given mass of gas to the temperature, at a constant pressure. The volume of a fixed mass of any gas can be measured at different temperatures, in °C, i.e. on the Celsius scale of temperature, and, when plotted (Fig. 1), extrapolation shows that the volume of any gas would be zero at -273.15°C. This value was, therefore, taken as the zero of a new scale of temperature known as the absolute, Kelvin or thermodynamic scale.

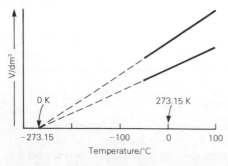

Fig. 1 Variation in the volume of a given mass of gas with temperature, at a fixed pressure.

The SI unit of temperature, known as the kelvin (1 K), is defined as 1/273.16 of the thermodynamic temperature of the triple point of water (273.16 K, see p. 216), the triple point of water being chosen because of its reproducibility. 1 K means a temperature of 1 kelvin above absolute zero, or the temperature interval between n K and $(n+1)$ K. This temperature interval is equal to 1 °C, so that

temperature in K = temperature in °C + 273.15

Charles's law, which is sometimes known as Gay-Lussac's law, was, originally, expressed in terms of °C. In modern terms it is best expressed as: *the volume of a given mass of gas is proportional to the temperature in K, if the pressure remains constant*, i.e.

$$V \propto T \qquad \text{or} \qquad V = kT$$

where T is the temperature in K and k is a constant.

Because real gases are not perfect, they deviate from Charles's law as they do from Boyle's law.

7 The gas equation

Charles's and Boyle's laws can be combined together into one general expression which is known as the gas equation.

If a given mass of gas has a volume of V_1 at a pressure of p_1, it will have a volume, V_x, at a pressure of p_2, and, if the temperature remains constant,

$$V_1 p_1 = V_x p_2 \qquad \text{(by Boyle's law)}$$

If, now, the temperature at which V_x is measured is changed from T_1 K to T_2 K, then the new volume of the gas, V_2, will be given by

$$V_x/T_1 = V_2/T_2 \qquad \text{(by Charles's law)}$$

Substituting from one of these equations into the other gives

$$p_1 V_1/T_1 = p_2 V_2/T_2$$

and, in more general form, this can be expressed as

$$pV/T = \text{a constant} \qquad \text{or} \qquad pV = kT$$

which is known as the gas equation. k is a numerical constant, and the temperature is expressed in K.

8 The molar gas constant

The value of k in the gas equation depends on the mass of gas and on the units in which p and V are measured. If 1 mol (p. 16) of gas is considered, V is the molar volume, (V_m), k is written as R, and

$$pV_m = RT \qquad \text{(for 1 mol of gas)}.$$

R is known as the *molar gas constant*. In SI units, pressure is measured in Pa, and volume in m^3; the corresponding value of the molar gas constant is $8.31441\,J\,K^{-1}\,mol^{-1}$.

Knowing this value the generalised relationship for n mol,

$$pV = nRT \qquad \text{(for } n \text{ mol)}$$

enables any of the four quantities p, V, n or T to be calculated if the other three of them are known.

The equation is known as the *ideal gas equation*; it is an *equation of state* for the temperature, pressure and volume and any ideal gas sample must conform with it. Alternatively, any gas that obeys the equation must be an ideal gas.

It is often convenient to quote gas volumes at an arbitrarily chosen standard temperature and pressure, s.t.p. 101.325 kPa (1 atm) and 0 °C (273.15 K) are chosen.

9 Gaseous diffusion

Diffusion means 'spreading out' and one of the most striking characteristics of gases is their ability to diffuse. One of the essential differences between a gas and a solid or a liquid, for instance, is that a gas will always spread out or diffuse so as to occupy fully any container. This diffusion takes place in all directions, even against gravity, and enables gases to pass through porous substances, or through very small apertures, though the escape of a gas through a single hole is often referred to as *effusion*. Gases, in simple language, leak well.

Gaseous diffusion may be demonstrated very readily by placing a gas jar full of air over one full of nitrogen dioxide. Though nitrogen dioxide has a greater density than air, it will diffuse upwards and its spreading out can be observed because of its colour.

The fact that all gases are completely miscible, provided there is no chemical reaction between them, also indicates the readiness with which they diffuse, as does the short time-lag between an escape of, say, ammonia or hydrogen sulphide and detection of its smell, at some distance.

10 Graham's Law of gaseous diffusion

The apparatus shown in Fig. 2 can be used to demonstrate that gases diffuse at different rates. If a beaker full of hydrogen is inverted over the porous pot, a jet of water is pushed out of the tube, A. This is because the hydrogen diffuses into the porous pot more quickly than the air inside it diffuses out of the pot. If carbon dioxide is used instead of hydrogen, the pressure inside the pot is reduced, and air is forced in through A. Carbon dioxide must, therefore, diffuse less rapidly than air.

Graham, in 1846, compared the rates at which various gases diffused

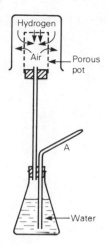

Fig. 2

through porous pots, and also the rates of effusion through a small aperture. His results were summarised in his law of diffusion which states that *the rate of diffusion (or effusion) of a gas, at a constant temperature and pressure, is inversely proportional to the square root of its density*:

$$\text{rate of diffusion or effusion} \propto \sqrt{1/\text{density}}$$

By comparing the rate of diffusion of two gases, A and B, it is possible to measure the density of one if that of the other is known, because

$$\frac{\text{rate of diffusion of A}}{\text{rate of diffusion of B}} = \sqrt{\frac{\text{density of B}}{\text{density of A}}}$$

As hydrogen has the lowest density of any gas it diffuses most rapidly.

The density and relative molecular mass (M_r) of a gas are proportional (p. 18) so that Graham's law can also be expressed in terms of M_r, i.e.

$$\frac{\text{rate of diffusion of A}}{\text{rate of diffusion of B}} = \sqrt{\frac{M_r \text{ of B}}{M_r \text{ of A}}}$$

11 Comparison of rates of effusion

Fairly accurate results can be obtained for relative rates of effusion of gases using a gas syringe (Fig. 3). The syringe is connected, via a 3-way tap, to a tube closed at the end by metal foil with a small hole in it.

The syringe is filled to a predetermined point with a dry gas, and the time taken for a particular volume of gas to effuse through the hole is measured. The experiment is then repeated using a second dry gas. If necessary, a weight can be placed on the plunger to give suitable time periods.

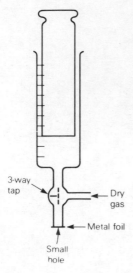

3-way tap

Dry gas

Metal foil

Small hole

Fig. 3

If the times are t_1 for gas 1 and t_2 for gas 2, then, by Graham's law

$$\frac{t_2}{t_1} = \sqrt{\frac{\text{density of gas 2}}{\text{density of gas 1}}} = \sqrt{\frac{M_r \text{ of gas 2}}{M_r \text{ of gas 1}}}$$

If the density or relative molecular mass of one gas is known, the values for the other gas can be found.

12 The density of a gas

There are two ways of expressing the density of a gas or vapour. It can be expressed as a *density*, in g dm^{-3} or kg m^{-3}, or, as a *relative density*, by comparing the mass of any volume of the gas or vapour with that of an equal volume of hydrogen. The relative density of a gas or vapour is also known as the *vapour density* and, by definition,

$$\frac{\text{relative (vapour) density}}{\text{of a gas or vapour}} = \frac{\text{mass of any volume of gas or vapour}}{\text{mass of an equal volume of hydrogen}}$$

the volumes being measured at the same temperature and pressure.

The density of a gas can be determined by measuring the mass of an evacuated globe, filling it with gas at a known temperature and pressure, and measuring the new mass. This gives the mass of the gas required to fill the globe. The volume of the globe is then measured by filling it with water, and finding the mass of the water. The density is obtained by dividing the mass of the gas by its volume. Measuring the mass of gases, however, is always difficult, for a reasonable volume of gas is not very heavy. Moreover, evacuation of vessels is necessary for, normally, a vessel is full of air though it may be spoken of as being empty.

The method was first used by Regnault (1845) but it was developed by Lord Rayleigh between 1885 and 1892. It was in this work that he discovered the discrepancy between the density of atmospheric nitrogen $(1.2572\,\mathrm{g\,dm^{-3}})$ and nitrogen prepared from chemical sources $(1.2560\,\mathrm{g\,dm^{-3}})$, an observation that led to the discovery of the noble gases.

When relative densities of gases, and not densities, are required to be measured by this direct method, the density of hydrogen must also be obtained. The relative density of any gas, X, is then given by

$$\frac{\text{relative density}}{\text{of gas X}} = \frac{\text{density of X}}{\text{density of hydrogen}}$$

Measurement of rates of diffusion or effusion (p. 7) can be used to obtain either densities or relative densities, so long as the value for one gas is known.

13 Measurement of gas density by buoyancy method

Gas densities can be measured by comparing the buoyancy of a small, evacuated quartz bulb in gases at different pressures. The apparatus used is known as a *microbalance*. A quartz bulb is attached to one end of a quartz balance beam, the beam being supported on a knife-edge, or, better, by a torsion fibre. At the other end of the beam there is a counterweight and a pointer, the whole being enclosed in a glass case (Fig. 4).

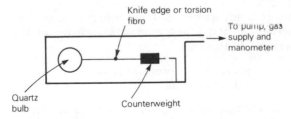

Fig. 4 Diagrammatic arrangement of a microbalance.

The glass container is first evacuated and then filled with the gas under test at such a pressure that the beam balances with the pointer at the zero mark. The pressure, p_g, is measured. After a second evacuation, oxygen, or some other reference gas, is introduced and the pressure, p_0, necessary to maintain the pointer on the zero mark, is recorded.

If the density of the gas under test is ρ_g at s.t.p. it will be $\rho_g p_g$ at a pressure of p_g. The upthrust on the quartz bulb will, therefore, be $\rho_g p_g$

multiplied by the volume of the bulb. When supported in oxygen (with density ρ_0 at s.t.p.), at a pressure of p_0 the upthrust will be $\rho_0 p_0$ multiplied by the volume of the bulb.

As the balance beam rests in the same position in both cases, the upthrust on the bulb must be the same whichever gas is being used, so that $\rho_g p_g$ is equal to $\rho_0 p_0$. By measuring p_g and p_0, and knowing the density of oxygen, ρ_g can be calculated.

The measurement is a relative one against a standard gas of known density. Only pressure measurements have to be made, and very small quantities of gas can be used.

14 Dalton's law of partial pressures

Dalton found, in 1801, that *the total pressure exerted by a mixture of gases, which do not interact, is the sum of the pressures which each gas would exert if it were present alone in the entire volume occupied by the mixture*. The pressures exerted by each gas separately are known as their partial pressures, and the above statement is called Dalton's law of partial pressures.

Dalton arrived at the law experimentally, but it can be deduced from the gas equation. If a mixture of gases consisting of n_1 mol of A and n_2 mol of B occupy a volume, V, then, if p_1 and p_2 are the partial pressures of A and B,

$$p_1 V = n_1 RT \quad \text{and} \quad p_2 V = n_2 RT$$

If the total pressure exerted by the mixture of gases is p, then,

$$pV = (n_1 + n_2)RT$$

and, therefore, p must be equal to $p_1 + p_2$.

The law of partial pressure is not widely used, but it may be necessary to find the pressure exerted by a dry gas from a knowledge of the pressure exerted by the wet gas. In this case,

$$\frac{\text{total pressure of dry}}{\text{gas plus water vapour}} = \frac{\text{pressure}}{\text{of dry gas}} + \frac{\text{pressure of}}{\text{water vapour}}$$

The pressure of the water vapour, at any given temperature, can be obtained from tables, so that the pressure of the dry gas is easily obtained.

The partial pressure of a gas in a gaseous mixture is also used as a convenient measure of its concentration (p. 287).

15 Gay-Lussac's law of combining volumes

Cavendish had found, in 1781, that water was formed by sparking a mixture of hydrogen and air, and he had estimated that two volumes of hydrogen combined with one volume of oxygen. In 1805, Gay-Lussac and Humboldt repeated the determination, and again obtained the ratio 2:1. Struck by the apparent simplicity of this ratio, Gay-Lussac examined other data relating to the volumes in which gases reacted together, and made further investigations.

His results led, in 1808, to the law of combining volumes which states that *when gases react they do so in volumes which bear a simple ratio to each other and to the volume of any gaseous product, all volumes being measured under the same conditions of temperature and pressure.* For example,

oxygen + hydrogen $\longrightarrow$ water vapour
1 vol.   2 vols.   2 vols.

nitrogen + hydrogen $\longrightarrow$ ammonia
1 vol.   3 vols.   2 vols.

As with other laws relating to gases Gay-Lussac's law has been shown, by modern work, to be not strictly accurate, simply because real gases are not perfect. But the impact of the law, coming, as it did, very shortly after Dalton had put forward his atomic theory, was of immense importance, for it led to Avogadro's hypothesis and molecular theory (p. 15).

Questions on Chapter 1

1 Use the gas equation to calculate   (a) the volume occupied by 200 g of chlorine at 15 °C and 54.71 kPa   (b) the volume occcupied by 200 g of carbon dioxide at 98.65 kPa and 30 °C.

2 If the gas constant is expressed per molecule instead of per mole it is sometimes called the Boltzmann constant. What is the value of this constant?

3 Derive the relationship between the density of a gas in $g\,dm^{-3}$ (d), the gas pressure in atmospheres (p), the relative molecular mass of the gas (M_r), the absolute temperature (T) and the gas constant (R). If dry air is regarded as 21 per cent by volume of oxygen, 1 per cent of argon and 78 per cent of nitrogen, what will its density ($g\,dm^{-3}$) be at 20 °C and 98.65 kPa pressure?

4 The density of solid carbon dioxide is $1.53\,g\,cm^{-3}$. What volume of gas is obtained, at 15 °C and 101.325 kPa, from $1\,cm^3$ of solid carbon dioxide?

5 State Graham's law of diffusion of gases, and give a qualitative explanation of it in terms of the kinetic theory. Describe an experiment to demonstrate gaseous diffusion. A certain volume of ethanoic acid vapour was found to diffuse in 580.8 seconds, whilst the same volume of oxygen, with the same experimental conditions, took 300 seconds to diffuse. What deductions can be made as a result of this experiment? (JMB)

6 Discuss the diffusion of gases in terms of the kinetic theory. Illustrate the use of this phenomenon for (a) the determination of relative molecular mass (b) the separation of gaseous mixtures. (OS)

7 Describe some simple experiments which illustrate gaseous diffusion. A given volume of a certain gas was found to diffuse in two-thirds of the time taken by an equal volume of hydrogen chloride under the same physical conditions. What is the relative molecular mass of this gas?

8 A compound gave, on analysis, X = 90.35 per cent and H = 9.65 per cent, and was found to diffuse through a porous plug at two-thirds of the rate of nitrogen. What is the molecular formula of the compound? (X = 28.1; N = 14.0.)

9 Calculate the ratio of the rate of diffusion of a gas at 91 °C and 0 °C at a constant pressure.

10 If the density of hydrogen at s.t.p. is $x \, g \, dm^{-3}$ what will it be at 10 °C and 100 kPa pressure?

11 Compare the rates of diffusion of (a) ammonia and hydrogen (b) hydrogen and deuterium (c) oxygen and ozone and (d) $^{238}UF_6$ and $^{235}UF_6$. Why is the result obtained in (d) of practical importance?

12 Sketch the graph which would be obtained by plotting the rates of diffusion of different gases (under the same conditions) against the reciprocal of the square root of their relative molecular masses.

13 Devise an experimental arrangement which could be used to demonstrate the truth of Dalton's law of partial pressures.

14 Write an account of plasmas.

15 A mixture of gases at s.t.p. contains 65 per cent of nitrogen, 15 per cent of carbon dioxide and 20 per cent of oxygen by volume. What is the partial pressure of each gas in kPa?

16 The partial pressures of the components of a mixture of gases are 26.64 kPa (oxygen), 34 kPa (nitrogen) and 42.66 kPa (hydrogen). What is the percentage by volume of oxygen in the mixture?

17 Prove that the fraction of the total gas pressure exerted by one component of a gaseous mixture is equal to the fraction of the total number of molecules provided by that component.

18* Record the numerical values of the gas constant and the Boltzmann constant. How are they related?

19* If the volume of a gas saturated with water vapour is $V \, dm^3$ at a pressure, $P \, kPa$, and a temperature of 25 °C, what will the volume of the dry gas be under the same conditions?

---

* Questions marked with an asterisk require the use of a book of data.

2
Early atomic and molecular theory

1 Dalton's atomic theory

The hypothesis that matter is made up of atoms was first put forward in a speculative way by many Greek philosophers such as Democritus and Leucippus; and, later, by Boyle and Newton. Dalton, however, was the first person to restate and extend the older notions, and to show how they could be used to account for the experimental laws of chemical combination. He did this about 1807.

Dalton's ideas, many of which have had to be modified with the passage of time, can be summarised as follows.

a All matter is composed of atoms.

b Atoms are indivisible, indestructible and cannot be created.

c All the atoms of any one element are the same, and they are different from those of any other element.

d Compounds are formed by the chemical combination of atoms in small, whole numbers. The result is a small group of atoms chemically combined together. Dalton called such a group a 'compound atom': we now call it a molecule.

2 The laws of chemical combination

Four experimental laws formulated between 1774 and 1803 could be accounted for by Dalton's atomic theory.

a The law of conservation of mass (Lavoisier, 1774) *Matter is neither created nor destroyed in the course of a chemical reaction.* A reaction is simply a rearrangement of atoms and, as there will be the same number of atoms both before and after the reaction, there will be no change in mass.

As it is now known that mass can be converted into energy (p. 96) the law must strictly be expressed as a law of conservation of mass plus energy, i.e. *the sum of the quantity of matter and energy in an isolated system is always constant.*

Any change in mass due to energy changes is of no significance when the ordinary gravimetric (mass) aspects of a reaction are being studied, or when the ordinary methods of chemical analysis are being applied. The validity of the law of conservation of mass is assumed whenever a qantitative chemical experiment is carried out. But conversion of matter into energy is of great importance in atomic energy considerations (p. 101).

b The law of constant composition (Proust, 1799) *All pure samples of the same chemical compound contain the same elements combined together in the same proportions by mass.* In a quantity of a compound between elements A and B there will be, say, x atoms of A and y of B. The fraction of the mass which is A will be equal to the mass of x atoms of A divided by the mass of x atoms of A plus the mass of y atoms of B. Because all A atoms and all B atoms are alike, it follows that the fraction by mass of A in any pure sample of the compound must always be the same, i.e. the compound must have a constant composition.

c The law of multiple proportions (Dalton, 1803) *When two elements, A and B, combine together to form more than one compound, the different masses of A which combine with a fixed mass of B are in a simple ratio.* Some of the various compounds that might be formed between A and B are represented by AB, A_2B and AB_2. The masses of A combining with a fixed mass of B are in the simple ratios $2:4:1$.

d The law of reciprocal proportions and the law of equivalents (Richter, 1792) *The masses of elements, A, B and C, etc., which combine separately with some fixed mass of another element, are the masses in which A, B and C, etc., will combine with each other, or some simple multiple of them.* If each element is allotted a number, or numbers, known as its combining or equivalent mass, then *elements will always combine together in the ratio of their combining or equivalent masses.* The equivalent of an element was, historically, a very important quantity to measure and to know, but it plays little part in modern chemistry.

If, for example, A, B and C form compounds represented by AB, A_2C and BC_2 (as in HCl, H_2O and ClO_2), 2 atoms of B or 1 atom of C will combine with 2 atoms of A. The ratio of B to C $(2:1)$ is four times the ratio of $1:2$ which is the way in which B and C combine.

3 Relative atomic mass

Dalton's idea that all the atoms of any one element are alike, but different from those of any other element, enabled him to represent any atom by a symbol. He used geometrical shapes, but these were replaced by letters, first used by Berzelius in 1818.

Because the atomic theory had followed the formulation of gravimetric laws, it was realised that the masses of the atoms of the elements were of particular importance. Dalton recognised that, in his day, the direct measurement of the absolute mass of such an extremely small particle as a single atom was not possible, and he was more concerned with relative atomic mass. Historically the hydrogen atom, the lightest atom known, was chosen as the standard of reference so that relative atomic mass was originally defined as

$$\frac{\text{relative atomic mass}}{\text{of an element, } A_r} = \frac{\text{mass of 1 atom of the element}}{\text{mass of 1 atom of hydrogen}}$$

Later, oxygen, and particularly the ^{16}O isotope, was used as standard of reference. But, because of the modern way in which relative atomic masses are measured using mass spectrometry (p. 89), the existing definition is in terms of the ^{12}C isotope, i.e.

$$\frac{\text{relative atomic mass}}{\text{of an element, } A_r} = \frac{\text{mass of 1 atom of the element}}{1/12 \times \text{mass of 1 atom of } ^{12}C}$$

In the nineteenth century, however, the immediate problem was how to measure accurate values of relative atomic masses. Great ingenuity was used and, slowly, tables of values were built up enabling, for instance, the suggestion of the periodic table classification (p. 70) by Mendeléef (1869).

As 12 g of ^{12}C atoms (1 mol) contains $6.022\,045 \times 10^{23}$ atoms (the Avogadro constant), it follows that one twelfth of the mass of 1 atom of ^{12}C is $1.660\,565\,5 \times 10^{-27}$ kg. This is known as the *atomic mass unit* (amu) or *constant*.

4 Avogadro's hypothesis

Just as Dalton's atomic theory accounted for the laws of chemical combination, so Avogadro's hypothesis accounted for Gay-Lussac's law of combining volumes (p. 10). Avogadro made a bold and imaginative suggestion that gaseous elements might sometimes exist as groups of atoms (molecules), and not as single atoms, which had been Dalton's view. In modern terms the idea of *atomicity*, i.e. the number of atoms in 1 molecule of an element, was born.

Avogadro also suggested that *equal volumes of all gases under the same conditions of temperature and pressure contain the same number of molecules*. This is known as Avogadro's hypothesis.

The two ideas, taken together, can account for Gay-Lussac's law as, for example, in the formation of two volumes of hydrogen chloride from one of hydrogen and one of chlorine:

reaction	hydrogen	+ chlorine	→ hydrogen chloride
experimental result	1 vol.	+ 1 vol.	→ 2 vol.
by Avogadro's hypothesis	n molecules 1 molecule	+ n molecules + 1 molecule	→ 2n molecules → 2 molecules
equation	$H_2(g)$	+ $Cl_2(g)$	→ $2HCl(g)$

Dalton had attempted to explain this in terms of reaction between single atoms of hydrogen and chlorine, but the resulting equation,

$$H + Cl \longrightarrow HCl$$

cannot account, satisfactorily, for the volume measurements.

5 Relative molecular mass, M_r

The relative molecular mass of an element or compound is defined on the same basis as the relative atomic mass of an element. This makes it possible to calculate the relative molecular mass of a substance by adding up the contributory relative atomic masses.

Relative molecular mass is defined, then, in terms of ^{12}C, i.e.

$$\begin{array}{l} M_r \text{ of element or} \\ \text{compound} \end{array} = \frac{\text{mass of 1 molecule of element or compound}}{1/12 \times \text{mass of 1 atom of } ^{12}C}$$

6 The mole

The number of atoms in 1 mol of ^{12}C is obtained by dividing 12 g by the mass of an individual atom of ^{12}C. The answer, known as the *Avogadro constant* (L), has a value of $6.022\,045 \times 10^{23}\,mol^{-1}$ and this is used in defining an amount of substance known as a mole. This is one of the seven basic physical quantities (p. 1).

A mole is the amount of a substance containing as many elementary particles as there are in 12 g of ^{12}C. Or, alternatively, a mole is the amount of substance which contains $6.022\,045 \times 10^{23}$ elementary particles.

The particles concerned may be molecules, atoms, ions, radicals, electrons, or any other specified particle. As a mole is an amount it can be expressed in a variety of different units of mass or volume. So far as mass is concerned, the recommended units, in the SI system, are kilograms. Thus the mass of:

1 mol of hydrogen, H_2, is $2.0158 \times 10^{-3}\,kg$
2 mol of hydrogen, H_2, is $4.0316 \times 10^{-3}\,kg$
1 mol of hydrogen ion, H^+, is $1.0079 \times 10^{-3}\,kg$
1 mol of sulphate ion, $SO_4{}^{2-}$, is $96.06 \times 10^{-3}\,kg$
1 mol of electrons, e^-, is $0.548\,58 \times 10^{-6}\,kg$

The particles concerned need not be existing particles. It is possible, for example, to refer to 1 mol of $(2H_2)$ or 6 mol of $(1/3\ H_2)$; their masses will be equal to that of 2 mol of H_2.

There are two main advantages in expressing quantities in moles. In the first place, the number of moles of a substance is directly proportional to the number of elementary particles. 64 kg of oxygen does not, at first sight, appear to be twice as much as 2 kg of hydrogen. On a mass basis it is not twice as much; but on a molar basis it is, for there are twice as many molecules in 64 kg of oxygen as there are in 2 kg of hydrogen.

Secondly, as is explained in the following section, 1 mol of any gas or vapour occupies the same volume under the same conditions.

7 The molar volume

Since 1 mol of any gas contains, by definition, the same number of molecules it must also, by Avogadro's hypothesis, occupy the same volume at s.t.p. It is, therefore, only necessary to find the volume occupied by 1 mol of any gas to know the volume occupied by 1 mol of any other.

The density of hydrogen at s.t.p has been measured and found to be $0.089\,87\,\mathrm{kg\,m^{-3}}$. Taking 1 mol of hydrogen as $2.0158 \times 10^{-3}\,\mathrm{kg}$, it follows that 1 mol of hydrogen must occupy $22.430\,\mathrm{dm^3}$ at s.t.p. A more accurate figure is $22.4138\,\mathrm{dm^3\,mol^{-1}}$ or $2.241\,38 \times 10^{-2}\,\mathrm{m^3\,mol^{-1}}$, and this value is known as the molar volume of an ideal gas at s.t.p.

8 The Avogadro constant

The Avogadro constant can be measured in a variety of ways.

a From measurement of the charge on the electron A charge of $96\,484.56n$ coulomb (C) is required to deposit 1 mol of a metal from its ions, M^{n+}, during electrolysis (p. 436). This involves the discharge of L ions each with a charge of ne coulomb, where L is the Avogadro constant and e is the charge on the electron. The value of e is $1.602\,189\,2 \times 10^{-19}\,\mathrm{C}$, so that

$$L = \frac{96\,484.56}{1.602\,189\,2 \times 10^{-19}} = 6.022\,045 \times 10^{23}\,\mathrm{mol^{-1}}$$

b From measurement of the lattice spacing in a crystal In this method, the measured density of a crystalline substance is equated to the density of the crystal calculated from the measured lattice spacing.

The interionic distance in the crystal of potassium chloride, for instance, is found to be $0.314\,54\,\mathrm{nm}$ by X-ray analysis (p. 165). The volume of the small cube formed by four K^+ and four Cl^- ions will, therefore, be $(0.314\,54)^3\,\mathrm{nm^3}$.

Four ion pairs of potassium chloride will be found within this small cube, but each ion will be shared equally with seven other similar and surrounding, cubes. The mass of the material in the small cube can, therefore, be regarded as $4m/8$, i.e. $m/2$, where m is the actual mass of an ion pair of potassium chloride. m will be equal to M/L, where M is the molar mass of potassium chloride and L is the Avogadro constant, so that the mass of the small cube will be $M/2L$ g. Its calculated density will be $M/2L(0.314\,54)^3$ g nm^{-3}.

As the measured density of the crystal is 1.9893×10^{-21} g nm^{-3}:

$$\frac{M}{2L(0.314\,54)^3} = 1.9893 \times 10^{-21}$$

which leads to a value for L of 6.022×10^{23} mol^{-1}.

c Measurement of the volume of helium gas from decay of radium Radium decays (p. 79) by emitting α-particles and the number emitted from a given amount of radium in a given time can be measured using a Geiger counter. It is 11.6×10^{17} g^{-1} year^{-1}.

If the α-particles are allowed to pass into a surrounding container they will form helium gas, and the volume of gas formed in a given time can be measured. It is 0.043 cm^3 year^{-1}.

0.043 cm^3 of helium gas must, therefore, contain 11.6×10^{17} atoms, so that $22\,414$ cm^3 will contain 6.05×10^{23} atoms. This gives a value for the Avogadro constant.

9 $M_r = 2 \times$ relative density (d)

This important, approximate relationship was derived by Cannizzaro by a straightforward application of Avogadro's hypothesis. Using modern definitions,

$$\frac{\text{relative density of}}{\text{a gas or vapour } (d)} = \frac{\text{mass of a volume of gas or vapour}}{\text{mass of an equal volume of hydrogen}}$$

all volumes being measured at the same temperature and pressure. This definition can be rewritten, by applying Avogadro's hypothesis, as

$$d = \frac{\text{mass of 1 molecule of gas or vapour}}{\text{mass of 1 molecule of hydrogen}}$$

or
$$d = \frac{\text{mass of 1 molecule of gas or vapour}}{2.0158 \times 1/12 \times \text{mass of 1 atom of } ^{12}\text{C}}$$

But
$$M_r = \frac{\text{mass of 1 molecule of gas or vapour}}{1/12 \times \text{mass of 1 atom of } ^{12}\text{C}}$$

and, therefore, $\quad M_r = 2.0158 \times d$

which is conveniently remembered as $M_r = 2d$.

The importance of the relationship is that it enables values of relative molecular masses to be obtained from relative density values which can be measured easily (p. 8).

10 Measurement of relative molecular mass for gases

The relative density or the density of a gas can be obtained either by direct weighing (p. 8) with a microbalance (p. 9), or by comparing rates of effusion (p. 7).

Once the relative density of a gas is known, its relative molecular mass can be found by multiplying by 2.0158, or 2 if only an approximate result is required. If the density of the gas is known, the relative molecular mass is equal to the numerical value of the molar mass in $g\,mol^{-1}$, i.e. the mass, in g, of $22.414\,dm^3$ of the gas at s.t.p.

Such relative molecular mass values are quite accurate enough for many purposes, but they are not absolutely accurate as the relationships used in their calculation are only really true for ideal or perfect gases.

11 Measurement of relative molecular mass for volatile liquids

A number of different methods have been used to measure the volume occupied by the vapour obtained from a known mass of a volatile liquid. A modern, elementary method uses a gas syringe surrounded by a steam jacket or immersed in a furnace (Fig. 5).

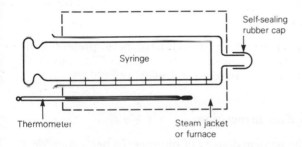

Fig. 5 Use of a gas syringe to measure the relative molecular mass of a volatile liquid.

The nozzle of the syringe is fitted with a self-sealing rubber cap. Once thermal equilibrium is established, i.e. when the piston no longer moves and the thermometer reading is constant, a measured mass of a volatile liquid is injected, through the rubber cap, using a hypodermic syringe. The liquid vaporises and the volume of vapour it produces can be measured (at atmospheric pressure and the temperature of the gas syringe).

The measured volume can be converted to s.t.p. and the mass of liquid necessary to produce $22.414\,dm^3$ of vapour at s.t.p. is equal to the molar mass of the liquid.

12 Effect of dissociation on relative density values

When heating causes a solid, liquid or gas to split up reversibly into other molecules or atoms, it is said to undergo *thermal dissociation*. Examples of substances which dissociate in this way are phosphorus pentachloride, dinitrogen tetroxide and iodine. The nature of the dissociation is shown by the equations:

$$PCl_5(g) \rightleftharpoons PCl_3(g) + Cl_2(g) \qquad N_2O_4(g) \rightleftharpoons 2NO_2(g)$$

$$I_2(g) \rightleftharpoons 2I(g)$$

The extent of the dissociation is measured by the *degree of dissociation,* usually represented by the symbol α, which is *the fraction, or percentage, of the original undissociated molecules which have dissociated.* The degree of dissociation is constant at any one temperature and pressure, for an equilibrium mixture is set up, but the degree of dissociation may change with temperature and pressure. When the dissociation is complete, α is equal to 1 or 100 per cent.

When dissociation leads to a change in the number of molecules present, it affects relative density measurements. If the degree of dissociation of phosphorus pentachloride, for example, is α, at a certain temperature, and if 1 mol of phosphorus pentachloride be considered, the state of affairs as a result of dissociation can be summarised in the equation,

$$\begin{array}{ccc} PCl_5(g) & \rightleftharpoons PCl_3(g) + & Cl_2(g) \\ (1-\alpha)\,mol & \alpha\,mol & \alpha\,mol \end{array}$$

This means that the 1 mol which would have been present if no dissociation had taken place is replaced by $(1+\alpha)$ mol. This causes an increase in volume and a *lowering* of relative density, i.e.,

$$\frac{\text{relative density as a result of dissociation}}{\text{relative density if no dissociation}} = \frac{1}{1+\alpha}$$

In numerical terms, the relative density of phosphorus pentachloride, if it did not dissociate, would be half its relative molecular mass, i.e. 208.5/2 or 104.25. The measured relative density of phosphorus pentachloride at 200 °C is, in fact, found to be 70. It follows that

$$70/104.25 = 1/(1 + \alpha)$$

so that α is equal to 0.489 or 48.9 per cent at 200 °C. Different values would be found at different temperatures.

Notice that the effect of thermal dissociation on relative density measurements only becomes apparent when the dissociation produces a change in the number of molecules. Hydrogen iodide dissociates,

$$2HI(g) \rightleftharpoons H_2(g) + I_2(g)$$

but, as there is no change in the number of molecules, the measurement of relative density is not affected. Notice, too, that the formula derived above applies only to dissociation in which 1 molecule splits up into 2. In the more general case of 1 molecule splitting up into n molecules

$$\begin{array}{cc} A_n & \rightleftharpoons & nA \\ (1-\alpha)\,\text{mol} & & n\alpha\,\text{mol} \end{array}$$

$$\frac{\text{relative density as a result of dissociation}}{\text{relative density if no dissociation}} = \frac{1}{1-\alpha+n\alpha}$$

13 Effect of association on relative density values

Association is the joining together of two or more like molecules to form a single molecule of higher relative molecular mass.

Association in the vapour state causes a decrease in the number of molecules and a corresponding increase in the measured relative density. The relative density of ethanoic acid, for instance, at certain temperatures, has a measured value of 60, giving a relative molecular mass of 120. The expected relative molecular mass on the basis of the formula, CH_3COOH, would be 60. The measured relative molecular mass of 120 comes about because the acid is fully associated into double molecules in the vapour state,

$$2CH_3COOH(g) \longrightarrow (CH_3COOH)_2(g)$$

Other examples of two-fold association (dimerisation) are found in iron(III) chloride, Fe_2Cl_6, aluminium chloride, Al_2Cl_6, phosphorus(III) oxide, P_4O_6, and phosphorus(V) oxide, P_4O_{10}.

Questions on Chapter 2

1 What are the claims of Bacon, Boyle, Hooke and Newton to be 'the fathers of chemical philosophy'?
2 'Chemistry is a French science. Its founder was Lavoisier of immortal memory.' What was Lavoisier's contribution to chemical development?
3 The word 'law' is used in many different ways. Illustrate the different meanings it may have.
4 Explain the importance of (a) the law of conservation of mass (b) the law of constant composition, to someone who has no knowledge of science.
5 Give a definition of a chemical compound. Criticise the definition you give.
6 A definite compound always contains the same elements in the same proportions. Is this statement valid?

7 If you were provided with supplies of hydrogen and of chlorine explain, in detail, what you would do to illustrate the truth of Gay-Lussac's law.

8 Explain fully, with varied examples, the meaning of the terms atom, molecule and ion.

9 Why do you think that (a) hydrogen (b) nitrogen (c) oxygen are diatomic? What is the evidence for the monatomicity of neon? How would you attempt to discover the atomicity of mercury in the vapour state?

10 Write short biographical sketches on (a) Avogadro (b) Dalton (c) Gay-Lussac and (d) Cannizzaro.

11 Explain fully why Avogadro's hypothesis was first put forward and why it eventually became so important.

12 Chlorine is said to be a diatomic gas. Is this true at all temperatures? Is I_2 a reasonable formula for iodine over a wide range of temperature?

13 How many gas molecules will there be in an X-ray tube of volume $2 \, dm^3$ if the temperature is $17\,°C$ and the pressure inside the tube is $1.333\,22 \times 10^{-3} \, Pa$?

14 A vacuum tube at $27\,°C$ was evacuated until it contained only 4×10^{13} molecules. At this stage the pressure was $1.6 \times 10^{-3} \, Pa$. What was the volume of the tube?

15 What type of particle, and what number of particles, would be found in (a) 1 mol of methane (b) 0.5 mol of iodine at a temperature high enough for dissociation to be complete (c) 1 mol of fully ionised barium chloride (d) 1 mol of fully dissociated phosphorus pentachloride?

16 State Avogadro's hypothesis and explain its importance in chemistry. Outline one method of determining the Avogadro constant. The density of mercury is $1.36 \times 10^4 \, kg\,m^{-3}$ and its relative atomic mass is 200.6. Given that the Avogadro constant is 6.02×10^{23} calculate the mass of a single mercury atom and an approximate value for its radius. Comment on any approximation used. (OJSE)

17 What is the mass of (a) 1 mol of electrons (b) 1 mol of protons (c) 1 mol of sucrose (d) a mixture of 0.2 mol of H_2 and 0.8 mol of O_2?

18 To what extent do you believe in molecules? Summarise the evidence for your views.

19 Explain what is meant by the terms density and relative density, and describe, briefly, how you would measure the two values for (a) nitrogen and (b) propanone. What conclusions do you draw from the observation that the relative density of sulphuric acid at $450\,°C$ is 24.5?

20 Discuss the various methods that have been adopted to find the numerical value of the Avogadro constant.

21 0.05 g of a volatile liquid, X, was injected into a gas syringe in a furnace at $100\,°C$. The volume of vapour produced was $33.3 \, cm^3$. Calculate the relative molecular mass of X from the data provided. Discuss the main sources of error in the method.

22 What volume, at $80\,°C$ and $100 \, kPa$ pressure, would be occupied by 0.1 g of a volatile liquid (b.p. $= 35\,°C$) with $M_r = 80$?

23 The rate of diffusion of a volatile metallic fluoride containing 32.39 per cent of fluorine is 13.27 times as slow as that of hydrogen. Establish the formula of this fluoride and the relative atomic mass of the metal. (W)

24 Explain the reasoning which leads to the conclusion that the relative molecular mass of a gas is twice its relative density. The chloride of a certain element has a relative density of 69 and contains 77.5 per cent of chlorine. Comment on these figures.

25 Iodine is 10 per cent dissociated at 900 °C at atmospheric pressure. If 0.254 g of iodine were dropped into a bulb of a Victor Meyer apparatus at 900 °C, what volume of air would be collected at s.t.p. and what would be the apparent relative molecular mass? (W)

26 Explain what is meant by the molar volume of a substance. At 50 °C and 98.65 kPa pressure 1 dm³ of partially dissociated dinitrogen tetroxide (N_2O_4) was found to have a mass of 2.5 g. What is the degree of dissociation of the compound under these conditions?

27 The apparent relative molecular mass of iodine at 55 °C was found to be 165; calculate the degree of dissociation of iodine into atoms at this temperature.

28 The relative density of steam at 2000 °C is 8.9. What is the percentage dissociation into hydrogen and oxygen?

29 What evidence, other than the abnormal vapour density results, is there for the thermal dissociation of   (a) N_2O_4 (b) PCl_5 (c) NH_4Cl?

30 Write equations representing as many examples of thermal dissociation as you can think of.

31 If a gas of density d_1 has a degree of dissociation, x, and gives, on dissociation, a gaseous mixture of density d_2, prove that $x = (d_1 - d_2)/(n-1)d_2$, where n is the number of molecules obtained from the complete dissociation of 1 molecule of the original gas.

3
Kinetic theory

The kinetic theory was developed, between 1860 and 1890, mainly by Clausius, Clerk Maxwell and Boltzmann. The theory is mainly useful in accounting for the known properties of gases, but it also clarifies many problems concerned with liquids and solids.

1 Outline of the theory

The ideas underlying the theory may be summarised as follows.

a Matter is made up of particles. These particles may be small groups of atoms (molecules) or, in monatomic gases, single atoms.

b The particles in a gas are in continual, rapid, random motion in straight lines in every direction. They continually collide with each other and with the walls of any containing vessel. The pressure exerted by the gas on the walls of the vessel is due to bombardment by the moving particles. Though the same idea of random motion applies to the particles in a liquid, the motion is greatly decreased, and it is still further decreased in a solid.

c The particles in a gas are separated from each other by distances which are large compared with the size of the particles. In a liquid, the particles are closer together, and they are still closer in a solid. In simple treatments the particles are regarded as points.

d The particles are regarded as being perfectly elastic, so that the collisions they undergo in a gas do not result in any change in the total amount of kinetic energy of the gas. In perfectly elastic collisions, loss of kinetic energy by one molecule is balanced by an equal gain by another. In inelastic collisions, some of the kinetic energy is converted into internal energy, e.g. vibrational or rotational energy (p. 176).

The *average* kinetic energy of the individual particles remains constant, at a constant temperature, though the kinetic energy of one particular particle may vary enormously depending on the nature of the collisions it undergoes.

e Increase in temperature causes the motion of the particles to increase, the average kinetic energy of the particles in a gas being proportional to the absolute temperature of the gas.

24

2 The fundamental kinetic equation

The quantitative applications of the kinetic theory are based on the fundamental equation,

$$pV = \tfrac{1}{3}mnu^2$$

where p is the pressure exerted by the gas, V is its volume, m is the mass of one molecule, n is the total number of molecules of gas present, and u is the root mean square velocity.

The root mean square is a rather unusual way of expressing an average value. The average of such quantities as 1, 2, 3, 4 and 6 would, most commonly, be taken as the sum of the quantities divided by five, i.e. 16/5 or 3.2. This is correctly described as the *arithmetic mean* of the quantities.

The root mean square of the quantities is given by

$$\sqrt{(1^2 + 2^2 + 3^2 + 4^2 + 5^2)/5} = \sqrt{66/5} = \sqrt{13.2} = 3.633$$

The root mean square velocity of gas molecules, u, is, therefore, obtained from the expression,

$$u = \sqrt{(u_1{}^2 + u_2{}^2 + u_3{}^2 + u_4{}^2 \ldots + u_n{}^2)/n}$$

where n is the total number of molecules concerned, and u_1, u_2, u_3, $u_4 \ldots u_n$ are their individual velocities. The root mean square of the velocities of the gas molecules is used in the fundamental equation so that the expression $\tfrac{1}{2}mu^2$ will accurately give the average kinetic energy per molecule.

3 Derivation of $pV = \tfrac{1}{3}mnu^2$

The equation is derived by considering n molecules, each of mass $m\,\mathrm{g}$ contained in a cube of side $l\,\mathrm{cm}$.

The velocity of any one single molecule, u_1, can be resolved into components, x, y and z acting in directions parallel to the three sides of the cube. If this is done, then,

$$u_1{}^2 = x^2 + y^2 + z^2$$

The pressure on one face of the cube is due to all the components in the same direction as x; that on the other faces is due to the components in the same directions as y and z.

In the direction of x, the molecule with total velocity u_1 travels $x\,\mathrm{cm}$ in 1 s. It can, however, only travel $l\,\mathrm{cm}$ before colliding with a wall, and for each $l\,\mathrm{cm}$ it travels it will undergo one collision with a wall. In one second, therefore, it will undergo x/l collisions.

Before collision, the momentum of the single particle, in the direction of x, will be mx. After a collision, which is perfectly elastic, the momentum will have the same value, but opposite sign. The change of momentum for every single collision is, therefore, $mx - (-mx)$, i.e. $2mx$. As there are x/l collisions per second, the change of momentum per second, in the direction of x, will be $2mx^2/l$. In the direction of y, the change of momentum will, similarly, be $2my^2/l$ and in the direction of z, it will be $2mz^2/l$.

The total change of momentum per second, caused by one single molecule will, therefore, be

$$2mx^2/l + 2my^2/l + 2mz^2/l \quad \text{or} \quad 2mu_1^2\ l$$

For n molecules the total change of momentum per second will be

$$2m(u_1^2 + u_2^2 + u_3^2 \ldots + u_n^2)/l = 2mnu^2/l$$

Now, by Newton's law, the force exerted on a surface by bombarding particles is equal to the rate of change of momentum of the particles. The force exerted on the faces of the cube by molecular bombardment is, therefore, $2mnu^2/l$.

Because pressure is force per unit area, and because the area of the six faces of the cube is $6l^2$, the pressure exerted by the gas will be

$$2mnu^2/6l^3 \quad \text{or} \quad mnu^2/3l^3$$

But l^3 is the volume of the cube, i.e. the volume of gas under consideration and, therefore,

$$p = mnu^2/3V \quad \text{or} \quad pV = \tfrac{1}{3}mnu^2$$

If u is in ms^{-1}, m in kg and V in m^3, then p will be in Pa.

4 The kinetic energy of a gas

The total kinetic energy, E, of all the molecules in a gas is given by $mnu^2/2$; the average kinetic energy of each molecule by $mu^2/2$. On the basis of the kinetic theory, both these quantities are proportional to the absolute temperature, i.e.

$$\tfrac{1}{2}mnu^2 = E = kT$$

where k is a constant. It follows that a gas at $0\,\text{K}$ has no kinetic energy; the molecules are at rest*.

The fundamental equation of kinetic theory can, then, be written as

$$pV = \tfrac{1}{3}mnu^2 = \tfrac{2}{3}E = k'T$$

---

* This result of classical mechanics is an oversimplification. Quantum mechanics demands that a particle, even at absolute zero, contains a small, irremovable amount of energy referred to as zero-point energy.

or, for 1 mol of gas,

$$pV_m = \tfrac{1}{3}mLu^2 = \tfrac{2}{3}E_m = RT$$

where L is the Avogadro constant and R the gas constant.

The value of R is $8.314\,\mathrm{J\,K^{-1}\,mol^{-1}}$ so that the kinetic energy of the molecules in 1 mol of a gas is approximately $12.5T$ joules.

5 Accounting for the gas laws

The equations for pV in the preceding section can be used to show how the kinetic theory accounts for the gas laws.

a Boyle's law For a given mass of gas at a given temperature the total energy, E, is constant so that pV must also be constant.

b Charles's law As pV equals $k'T$ then, at a given pressure, V equals $k''T$, which is Charles's law.

c Dalton's law of partial pressures The total energy of a mixture of two gases, A and B, will be equal to the sum of the separate energies of A and B, i.e.

$$E = E_A + E_B$$

It follows that $\quad pV = 2(E_A + E_B)/3 = p_A V + p_B V$

where p_A and p_B are the partial pressures of A and B. Thus

$$p = p_A + p_B$$

which is the statement of the law of partial pressures.

d Graham's law of diffusion As the density of a gas, ρ, is given by mn/V, it follows that

$$u = \sqrt{3pV/mn} = \sqrt{3p/\rho}$$

At constant pressure, u will be inversely proportional to the square root of the gas density, and Graham's law follows if it is assumed that the rate of diffusion is proportional to u.

6 Deviations from the gas laws

Experimental measurements by Regnault and Amagat showed that Boyle's law was not valid over a wide range of pressures, and work by Andrews showed that deviations from the law occurred at low temperatures. Such deviations are due to the fact that real gases are not ideal, and the simple ideas of the kinetic theory have to be modified to account for the deviations.

Introduction to Physical Chemistry

a The effect of pressure An ideal gas would obey Boyle's law so that pV_m/RT, known as the compression or compressibility factor (Z), would equal one at any pressure. For real gases, however, Z only approaches the value of one at very low pressures (Fig. 6). At other pressures, Z may be greater than one (meaning that the gas is more difficult to compress than an ideal gas), or less than one.

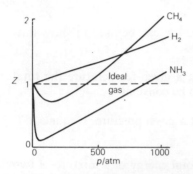

Fig. 6 Plot of compression factor ($Z = PV_m/RT$) against pressure.

At high pressures, the molecules of a gas are close together, and there are strong repulsive forces between the molecules which hinder compression. At moderate pressures, the forces between molecules are attractive and favour compression. At very low pressures, neither repulsive nor attractive forces are significant, which is why real gases approach ideal behaviour at low pressures.

b The effect of temperature A plot of p against V_m for a fixed mass of an ideal gas at a fixed temperature would be expected to give a rectangular hyperbola. This is so, or very nearly so, at high temperatures, but there are deviations at lower temperatures. Typical results, for carbon dioxide, are summarised in Fig. 7.

At low temperatures, the isothermals are discontinuous, splitting up into three parts. If the curve ABCD is considered, the part AB represents the effect of increasing pressure in decreasing the volume of a gas. Between B and C, however, there is a very large volume change with no change in pressure. This represents the liquefaction of the gas. At C, liquefaction is complete, and the curve CD represents the effect of pressure on the volume of liquid produced.

Point X is the *critical point*, with corresponding critical temperature (31.2 °C), critical pressure (7.27 MPa) and critical molar volume (94 cm^3).

At temperatures above the critical temperature, it is not possible to liquefy a gas however high the pressure may be.

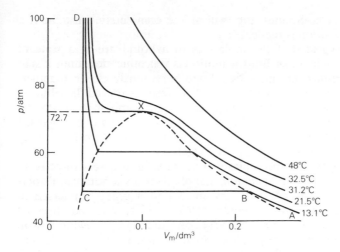

Fig. 7 Plot of pressure against molar volume at different temperatures.

c The van der Waals' equation, 1873 Van der Waals replaced the simple gas equation by

$$(V_m - b)(p + a/V_m^2) = RT$$

where a and b are numerical constants.

He argued that the actual size of gas molecules means that the volume in which they are free to move is less than the total volume which a gas occupies. That is why he replaced V_m by $(V_m - b)$.

He also argued that the attractive forces between gas molecules lowers the pressure that they can exert by lowering the frequency of impact with the walls of the containing vessel, and lowering the force of the impact when it does take place. As both these effects will be proportional to the gas density, and inversely proportional to the volume, van der Waals replaced p by $(p + a/V_m^2)$.

The van der Waals' equation fits the experimental data better than the simple gas equation, and values of a and b for different gases can be obtained by fitting the equation to known p, V and T values, or by using their relationships to the critical constants of a gas.

More complicated equations give still better agreement between theory and experiment. They are empirical, virial equations of the type

$$pV_m = RT + Bp + Cp^2 + Dp^3 + \ldots$$

where B, C, D, etc. are small coefficients dependent on temperature.

7 Kinetic theory and liquids

Most liquids, like all gases, are miscible. Liquids can also be compressed, though to a much smaller extent than gases and not in a regular way covered by any laws. Solutes, too, will diffuse completely throughout a solvent when a solution is made.

These facts suggest that the molecules in a liquid are in a state of random motion, as in a gas, but the motion is very much less than it is in a gas and the molecules in a liquid are very much closer together. Liquids, in fact, lie midway between the disorderly, scattered distribution found in a gas and the orderly, compact design found in a crystalline solid (p. 160). On the one hand, the structure of a liquid may be regarded as being something like that of a very imperfect gas; on the other, it may be viewed as that of a very disordered crystal.

The random motion of molecules in a liquid is shown, very directly, in the phenomenon of *Brownian motion*, first observed by a botanist, Robert Brown, in 1827. He found that very small pollen grains immersed in water underwent a curiously irregular motion which could be observed under a microscope. It can be seen very conveniently by viewing a drop of Indian ink on a microscope slide. Such motion is due to molecular bombardment of the small grains by molecules of the liquid in which the grains are suspended, and this simple observation provides one of the most direct pieces of evidence that the general ideas of the kinetic theory are correct.

8 Surface tension

Cohesive forces between molecules in a gas have to be taken into account in modifying the simple gas laws, but they are not very strong forces. In a liquid, however, the molecules are much closer together and the cohesive forces are much stronger.

Molecules on the surface of a liquid are subjected to a net inward force, which accounts for the surface tension of liquids. This causes liquids to behave as though their surfaces consisted of an elastic skin, and one sign of this is that liquid surfaces usually assume the smallest possible area. This means that freely suspended liquids form into spherical drops, for a sphere has the minimum surface-to-volume ratio. It also accounts for the fact that bubbles of gas in a liquid are spherical.

The surface tension of a liquid is defined as the force acting, in the surface, upon a line of length $1\,m$; the units are $N\,m^{-1}$. The values for water and mercury at $20\,°C$ are 72.75×10^{-3} and $470 \times 10^{-3}\,N\,m^{-1}$ respectively. Values usually decrease with increasing temperature.

A clearer picture is provided by the idea of surface energy. To move molecules from the interior of a liquid into the surface, i.e. to increase the surface area, must involve doing work against the internal forces in the liquid. In other words, energy must be supplied to increase a surface area. For a surface tension of $x\,N\,m^{-1}$, a surface increase of $1\,m^2$ would require $x\,J$.

The pent up energy in a surface can be seen by the release of energy on pricking a soap bubble.

9 Vapour pressure

Although the inward forces on the molecules at the surface of a liquid may be quite strong, some molecules with kinetic energy higher than the average will have sufficient energy to escape from the surface. This is the phenomenon of evaporation. The escaping molecules constitute the vapour above the liquid.

If evaporation takes place into a closed space, the concentration of vapour will increase, and the molecules in the vapour will exert a vapour pressure. Molecules will escape from the liquid surface to form vapour, but some vapour molecules will also be attracted by the liquid and will condense, i.e. pass back into the liquid. Eventually, as many molecules will be leaving the liquid as will be passing back into it, and the rate of evaporation will be equal to the rate of condensation; a dynamic or kinetic equilibrium will be set up. At this stage, the vapour reaches its maximum concentration and exerts its maximum pressure. The vapour is said to be saturated, and the pressure it exerts is known as the saturated vapour pressure (p. 215).

10 Vapour pressure and temperature

The saturated vapour pressure exerted by the vapour above a liquid is independent of the amount of liquid present, so long as there is enough to form a saturated vapour, but it increases with temperature. Clearly, at a higher temperature, more molecules will have sufficient energy to escape from the liquid surface. When the temperature reaches a high enough value and the saturated vapour pressure becomes equal to the external pressure, the liquid changes rapidly and completely into vapour and is said to boil.

The boiling point of a liquid is defined as the temperature at which its saturated vapour pressure becomes equal to the external pressure on the liquid. The boiling point of a liquid depends, then, on the external pressure. Water, for instance, boils at 100 °C under a pressure of 101.325 kPa, i.e. at 100 °C the saturated vapour pressure of water is 101.325 kPa but, under an external pressure of 1600 Pa, water will boil at 14 °C. The change of vapour pressure with temperature for a typical liquid is shown graphically on p. 217.

11 Latent heat of evaporation

The escape of molecules of higher than average kinetic energy from the surface of a liquid reduces the mean kinetic energy of the liquid which, therefore, becomes cooler. Evaporation, in fact, causes cooling, and this is the basis of refrigeration.

If the temperature of a liquid is to be maintained, when it evaporates, heat must be supplied, and this is the latent heat of evaporation. This is a measure of the energy which has to be supplied to cause molecules to escape from a liquid surface, i.e. to evaporate.

12 Kinetic theory and solids

The molecular motion in a liquid is much less than it is in a gas, but there is no really orderly array of molecules in either a gas or a liquid. When a liquid is cooled, its molecules lose energy and molecular motion decreases until a point is reached where the cohesive forces between molecules draw the molecules together into a solid structure with a definite, orderly array. This is the kinetic picture of crystallisation from a liquid.

In a solid crystal, molecular motion is limited to vibration or oscillation about a fixed position, and the molecules, atoms or ions are close together. This close-packing of the particles is shown by the very great difficulties encountered in compressing a solid.

The slight motion in a solid, at ordinary temperatures, is shown by the fact that two solids when placed in close contact may diffuse very slightly into each other, and also by the fact that solids can exert a vapour pressure. The vapour pressure of a solid, at normal temperatures, is generally so small as to be negligible, but some solids, e.g. iodine and naphthalene, exert considerable vapour pressures at temperatures below their melting points. On heating, such solids are converted directly into vapours, and, on cooling the vapour formed it condenses directly into a solid. Such a change from solid to vapour or from vapour to solid, without an intermediate liquid stage, is known as *sublimation* (p. 216).

The majority of solids melt into a liquid on heating. In the process of melting, it is thought that the vibrational energy of a solid increases as it is heated and that the particles of the solid vibrate more and more until the cohesive forces can no longer hold them together. The particles break loose, and the solid melts.

In non-crystalline, or amorphous, solids, e.g. glass and pitch, there is no regular pattern of molecules, and they have no fixed melting point. They are, in fact, super-cooled liquids (p. 216). Many so-called amorphous solids, e.g. amorphous sulphur and copper(II) oxide, appear to have no crystalline structure but are, in fact, made up of random mixtures of micro-crystalline aggregates.

12 Liquid crystals

Some solids, e.g. ammonium *cis*-octadec-9-enoate (oleate) and cholesteryl esters, with long and/or flat molecules, can be melted to give what are known as liquid crystals. They are cloudy liquids, which are said to be in the mesomorphic state. They are, in fact, intermediate between the liquid and solid state, having some regions within the liquid in which the molecules are aligned in a regular crystal-like way. They resemble liquids in that they are fluid, but they resemble crystals in their optical properties, e.g. they are doubly refracting. On heating to a higher temperature, they pass into normal liquids. Liquid crystals are used in instrument displays, e.g. in calculators, watches and thermometers.

14 Plasma

As a gas is heated to about $10\,000\,K$, the intermolecular or interatomic collisions become more and more violent until, eventually, the molecules or atoms disintegrate into electrons and positively charged particles; what is known as a plasma has been formed. Plasmas have overall neutrality, but the strong forces within them, between the charged particles, are very different from those between the neutral molecules or atoms in a gas. That is why plasmas are very different from gases. In particular, external electric or magnetic fields affect the charged particles in a plasma, but have little or no effect on the neutral particles in a gas. Plasmas are also at such high temperatures that they emit a lot of light and other electromagnetic radiation.

Much research is taking place to try to understand the characteristics of plasmas more fully. One of the reasons is that the material involved in nuclear fusion (p. 103) is in the plasma state.

15 Molecular velocities

The root mean square velocity, u, can be calculated from the fundamental equation of the kinetic theory. For 1 mol of gas,

$$pV_m = mLu^2/3 = RT$$

root mean square velocity, u $= \sqrt{3pV_m/mL} = \sqrt{3p/\rho} = \sqrt{3RT/M}$

where M is the molar mass of the gas. M and m must be expressed in kg.

Molecular velocities can also be expressed[†] as *mean velocities*, c, or *most probable velocities*, c^* (p. 34). The three values are not equal, for

$$c = 0.921u \quad \text{and} \quad c^* = 0.816u$$

[†] Mean velocity, $c = \sqrt{8RT/\pi M}$. Most probable velocity, $c^* = \sqrt{2RT/M}$. The general difference between u, c and c^* can be appreciated in terms of a simple example of six molecules, one of which has a velocity of $x\,m\,s^{-1}$, two have velocities of $y\,m\,s^{-1}$ and three have velocities of $z\,m\,s^{-1}$. The most probable velocity, c^*, is z (as three molecules have that velocity); the mean velocity, c, is $(x + 2y + 3z)/6$; and the root mean square velocity, u, is the square root of $(x^2 + 2y^2 + 3z^2)/6$.

Values for hydrogen and carbon dioxide, at s.t.p., are

	H_2	CO_2
root mean square velocity, u/m s^{-1}	1838	393
mean velocity, c/m s^{-1}	1693	362
most probable velocity, c^*/m s^{-1}	1500	321

All three velocities are proportional to the square root of the absolute temperature so that a rise in temperature from 0 °C to 1000 °C just about doubles the velocities. They are inversely proportional to the relative molecular mass of the molecules concerned so that a 25-fold increase in that causes a 5-fold decrease in velocity.

Hydrogen, with the lowest relative molecular mass and density, has the highest root mean square velocity of any gas. At s.t.p. it is about that of a rifle bullet and over one and a half kilometres per second.

16 Distribution of molecular velocities

The result of the numerous collisions (p. 36) within a gas is that the velocities of the individual molecules vary enormously. Most molecules have a velocity close to the mean, but some may acquire considerably higher or lower values as a result of a series of favourable or unfavourable collisions.

a Theoretical treatment The distribution of velocities amongst molecules in a gas was calculated, by Maxwell and Boltzmann, from the laws of probability. Some typical results are shown graphically in Fig. 8. At higher temperatures, the number of molecules with high velocity increases, whilst the number with the most probable velocity falls.

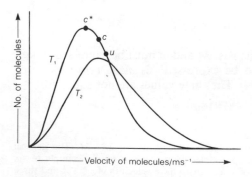

Fig. 8 The distribution of molecular velocities in a gas at a temperature, T_1, and a higher temperature, T_2. At T_1, the most probable velocity (c^*), the mean velocity (c) and the root mean square velocity (u) are all shown.

The distribution of kinetic energy amongst molecules follows a similar pattern. It can be expressed, in a simplified form*, as

$$n = n_0 e^{-E/RT}$$

where n_0 is the total number of molecules and n is the number having an energy greater than the value E (p. 319).

b Experimental measurement The velocities of molecules can be measured by using a *velocity selector*. Gas molecules from a source are passed into an evacuated chamber containing two, or more, rotating discs (Fig. 9). Each disc has a slit cut in it, and, for any particular speed of rotation, only molecules with a certain velocity will be able to pass through both slits onto the detector. From the dimensions of the apparatus, and the speed of rotation of the discs, it is possible to calculate the velocity of the molecules passing through. By altering the speed of rotation of the discs, molecules with different velocities can be detected, and their relative numbers measured.

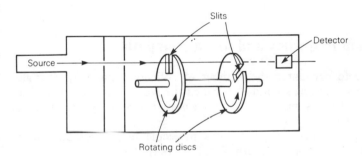

Fig. 9 A velocity selector.

In Zartmann's apparatus (Fig. 10), vapour from a boiling metal, such as tin or silver, is passed into an evacuated chamber containing a cylindrical drum with a slit in. The drum is rotated at a known speed. Atoms enter the drum through the slit, pass across the diameter and condense on a curved glass plate on the opposite side of the drum. The

---

* That the expression should take this form can be seen quite readily. If the chance of one favourable collision is $1/x$, the chance, P, of n favourable collisions in a row will be $(1/x)^n$. Thus

$$P = (1/x)^n \quad \therefore \ \ln P = n\ln 1/x \quad \text{or} \quad \ln P = yn \quad \text{or} \quad P = e^{-yn}$$

The chance of a molecule having an energy greater than E will be proportional to e^{-zE}, for E will be proportional to n.

e, equal to $2.71828\ldots$, is the base of Napierian logarithms, just as 10 is the base of ordinary logarithms. If $y = 10^x$, then $x = \log_{10} y$; if $y = e^x$, then $x = \log_e y$. $\log_{10} y$ is, nowadays, written as $\lg y$; $\log_e y$ as $\ln y$. ($\ln y = 2.3026\lg y$.)

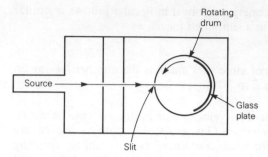

Fig. 10 Zartmann's apparatus.

exact position at which they condense on the plate will depend on their velocity. When sufficient metal has condensed, the thickness can be measured at various points and the number of atoms with each particular velocity can be calculated.

17 Collision frequency and mean free path

It is possible to calculate* the number of collisions made by a single molecule in a gas in one second; it is called the *collision frequency, z*. The total number of collisions in $1 \, cm^3$ of a gas in 1 second, Z, can also be calculated (p. 320).

The average distance travelled by a molecule before colliding with another is called its mean free path, λ. It follows that

$$\text{mean free path, } \lambda = \frac{\text{mean velocity, } c}{\text{collision frequency, } z}$$

The mean velocity for hydrogen at s.t.p. is $1693 \, m \, s^{-1}$, and the collision frequency is $9.6 \times 10^9 \, s^{-1}$. The mean free path is 176 nm.

Values for mean free paths can also be obtained from viscosity, thermal conductivity and diffusion measurements. For example

$$\lambda = \eta \sqrt{3/p\rho}$$

where η is the coefficient of viscosity, p the pressure, and ρ the density.

---

* $z = \sqrt{2}\pi d^2 cn$ where d is the 'diameter' of the molecule and n is the number of molecules per unit volume. $\pi d^2(\delta)$ is called the *collision cross-section*; it is the cross-sectional area of an imaginary sphere surrounding the molecule into which the centre of another molecule cannot penetrate.

Questions on Chapter 3

1 What assumptions are made in the kinetic theory of gases? How does the theory explain the following facts: (a) that equal volumes of all gases at the same temperature and pressure contain the same number of molecules (b) that the vapour pressure of a liquid depends only on its temperature (c) that gases diffuse at rates inversely proportional to the square roots of their densities.

2 Derive the fundamental kinetic theory equation by considering a gas in (a) a cylinder (b) a sphere, rather than a gas in a cube.

3 Both temperature and pressure are manifestations of molecular motion. Explain what this means in non-technical language.

4 Elemental iodine, at room temperature, is a black crystalline solid, consisting of an array of iodine molecules in a regular pattern. Show by the kinetic theory what changes occur when iodine is heated from room temperature to about 2000 °C, always at atmospheric pressure. You are recommended to refer, in your description, to the essential difference between a solid and a liquid, a liquid and a gas, and between a perfect and an imperfect gas. Indicate as far as you can what forces of cohesion are involved in each change the iodine undergoes.

5 Calculate the root mean square velocities of (a) chlorine (b) deuterium (c) methane, at s.t.p.

6 Explain what is meant by the gas constant. If the volume of a mole of gas at s.t.p. is 22.4 dm^3, calculate the gas constant in $J K^{-1} mol^{-1}$. Calculate the root mean square velocity of carbon dioxide at 23 °C.

7 The equation $pV_m = RT$ represents the behaviour of a perfect gas. Give some account of the deviations from this equation which are observed experimentally. How are these deviations accounted for by van der Waals' equation?

8 Write short notes on the following: (a) root mean square (b) critical temperature (c) the reduced equation of state (d) Brownian motion.

9 If 1 cm^3 of oxygen at s.t.p. weighs 1.43 mg, calculate the root mean square velocity of oxygen at s.t.p.

10 Give an account of the methods which have been used for liquefying gases.

11 The molar volume of hydrogen, at s.t.p., is 22 430 cm^3, but the comparative figure for ammonia is 22 094 cm^3. Comment.

12 If the densities of nitrogen in the solid, liquid and gaseous states be taken as 1.03, 0.80 and 0.001 25 g cm^{-3} respectively, calculate the molar volumes of nitrogen in each state. Comment on your results.

13 Sketch, on the same scale, the graphs of number of molecules against molecular velocity of CO_2, O_2, N_2 and H_2, as in Fig. 8.

14 The word 'gas' is derived from 'chaos'. To what extent are gases chaotic; to what extent orderly? Why is it that solids are more orderly than gases?

15 Describe how it is possible to measure molecular velocities, and in what different ways can the term be used.

4
Outlines of atomic structure

The early successes in applying simple atomic and molecular theory have been followed by so many more that the evidence for the existence of atoms and molecules is overwhelming. It is possible, indeed, to photograph them by field-ionisation microscopy, and electron or X-ray diffraction.

In the 170 years since Dalton, more and more has been discovered about the sizes, shapes and structures of atoms and molecules, and the knowledge gained can be used to provide ever more satisfying explanations of chemical phenomena.

It was the elucidation of the structure of atoms at the beginning of this century that led the way. It is a long story of great scientific achievement. A concise summary of the main features will be given first, with more details in the following chapters. The chronological tables on the front endpaper may help the reader to keep his or her bearings.

1 Fundamental particles

Atoms are made up, essentially, of three fundamental particles, which differ in mass and electric charge as follows:

	electron	proton	neutron
symbol	e or e^-	p	n
approximate relative mass	1/1836	1	1
approximate relative charge	-1	$+1$	not charged
actual mass /kg	$9.109\,534 \times 10^{-31}$	$1.672\,648\,5 \times 10^{-27}$	$1.674\,954\,3 \times 10^{-27}$
/amu	$5.485\,802\,6 \times 10^{-4}$	$1.007\,276\,471$	$1.008\,665\,012$
actual charge/C	$1.602\,189\,2 \times 10^{-19}$	$1.602\,189\,2 \times 10^{-19}$	0

The atomic mass unit (amu) is 1/12th of the mass of an individual atom of ^{12}C, i.e. $1.660\,565\,5 \times 10^{-27}\,kg$ (p. 15). It is convenient to remember that the neutron and proton have approximately equal masses of 1 amu and that the electron is about 1836 times lighter; its mass can sometimes be neglected as an approximation. The electron and proton have equal, but opposite, electrical charges; the neutron is not charged.

The existence of electrons in atoms was first suggested, by J. J.

Thomson, as a result of experimental work on the conduction of electricity through gases at low pressures, which produces cathode rays and X-rays, and a study of radioactivity. The term electron had, however, been introduced earlier by Johnstone Stoney (1881) in a rather different sense; it had resulted from considerations of Faraday's laws (p. 44) governing the passage of electricity through solutions. Accurate measurement of the charge and mass of an electron has been very important (p. 46).

An atom is electrically neutral, and if it contains negatively charged electrons it must also contain some positively charged particles. Protons are positively charged, and the supposition that they existed within atoms came about as a result of Lord Rutherford's experiments in which he bombarded elements with the α- and β-rays given off by radioactive elements. The neutron was discovered in 1932 by Sir James Chadwick by bombarding beryllium with α-rays.

Electrons, protons and neutrons were the first small particles to be discovered, and many of the properties of matter can be explained in terms of them. That is why they came to be called fundamental, basic or elementary particles. More recent research has shown, however, that many other small particles exist (p. 100).

2 The nuclear atom

Rutherford's bombardment experiments indicated that an atom consisted of a heavy, positively charged central nucleus with electrons distributed around it. The nucleus contains protons and neutrons (except that there are no neutrons in a hydrogen atom), and its positive charge is neutralised by the negative charge of the electrons round the nucleus so that the atom as a whole is electrically neutral.

Loss of electrons gives rise to positively charged ions (cations); gain of electrons gives negatively charged ions (anions).

In certain conditions the nucleus of an atom can be split so that one element can be transmuted into another, and energy, called nuclear energy, may be released in the process.

3 Atomic number and relative atomic mass

The value of the positive charge on the nucleus, or of the number of electrons, in the atom of any element is equal to the ordinal number of that element in the periodic table arrangement. This number is called the atomic number, Z.

In passing from one atom to the next in the periodic table there is a unit increase of positive charge on the nucleus of the atoms concerned and a consequent addition of one electron. Such an arrangement was supported by the results of Moseley's investigation of X-ray spectra (p. 50).

The approximate relative atomic mass, A_r, of an element is obtained by adding up the number of protons and neutrons in the atom concerned for, by comparison, the mass of the electrons in an atom is very small.

Typical atomic structures may be represented as follows:

hydrogen $\begin{array}{l} A_r = 1 \\ Z = 1 \end{array}$ helium $\begin{array}{l} A_r = 4 \\ Z = 2 \end{array}$ lithium $\begin{array}{l} A_r = 7 \\ Z = 3 \end{array}$

$\left(\begin{array}{c} 1p \end{array} \right)$ 1e $\left(\begin{array}{c} 2p \\ 2n \end{array} \right)$ 2e $\left(\begin{array}{c} 3p \\ 4n \end{array} \right)$ 3e

4 Electrons in atoms

The chemical properties of an atom depend, very largely, on the arrangement of the electrons around the nucleus, for chemical combination depends on interaction between the outer electrons of the atoms combining together.

The working out of the detailed arrangements of electrons rested mainly on the ideas of the quantum theory, first put forward by Planck in 1900, and on the interpretation of a mass of spectroscopic data originally undertaken by Bohr.

Bohr regarded the electron as a negatively charged particle and was able to account for many spectroscopic results by allotting the various electrons in an atom to different orbits arranged round the nucleus. Electrons in different orbits had different energies, and transfer of electrons, from orbit to orbit, explained the absorption or emission of energy, which could be related, in simple cases, to spectroscopic results.

Such an orbital distribution of electrons also accounted for the chemical properties of elements in relation to their position in the periodic table. Each element in Group 1 is found, for example, to have one electron in the outermost orbit. The elements in Group 2 have two electrons in the outermost orbit, and so on. Further details about the arrangement of electrons are known, and provide greater insight into the properties of different elements.

5 Isotopes

Because the chemical properties depend, in the main, on the arrangement of their electrons, it is possible to have atoms with the same chemical properties, i.e. the same arrangements of electrons, but with different relative atomic masses, i.e. different nuclear structures. Such atoms are known as isotopes, e.g.

hydrogen, or
protium, $_1^1H$

heavy hydrogen,
or deuterium, $_1^2H$ or D

tritium,
$_1^3H$ or T

(1p) 1e

(1p / 1n) 1e

(1p / 2n) 1e

uranium-238, $_{92}^{238}U$ uranium-235, $_{92}^{235}U$ uranium-234, $_{92}^{234}U$

(92p / 146n) 92e

(92p / 143n) 92e

(92p / 142n) 92e

In the symbols given to each isotope the superscript shows what is known as the mass number (A), whilst the subscript gives the atomic number (Z).

The existence of isotopes was first discovered when the way in which naturally occurring radioactive elements disintegrated was worked out (p. 80) by Rutherford, Soddy and Russell, and when positive rays were analysed.

Most naturally occurring elements have been found to consist of several isotopes, and mass spectrometers have been of particular importance in detecting these isotopes. Isotopes which do not occur naturally can be made by atomic transmutations, and some of them, especially the radioactive ones, are very useful.

6 The wave nature of the electron

Davisson and Germer observed, in 1927, that a stream of moving electrons could be diffracted by crystals acting as diffraction gratings. Since it is only possible to account for diffraction phenomena in terms of waves, we must assume that electrons, like light, have a dual nature. They may appear to be small, negatively charged particles, as envisaged by Thomson, Rutherford and Bohr; but they also have a wave-like nature, similar to that of light or X-rays.

The waves associated with moving electrons can be treated mathematically by a specialised technique known as wave-mechanics, developed, mainly, by Schrödinger. The idea of tiny, negatively charged particles existing in fixed orbits round the nucleus of an atom is replaced by the idea of 'clouds' of electrical charge, of varying charge density, round the nucleus. The electron in a specific, limited orbit is replaced by a more diffuse region of electric charge referred to as an atomic orbital. The shape of such orbitals, and the distribution of charge within them, is of major importance.

7 The atomic nucleus

When an atom is split, it is the nucleus of the atom which splits into two, or more, smaller nuclei. In naturally occurring radioactive elements, this nuclear splitting is going on all the time, and cannot be stopped. Rays, known as α-, β- and γ-rays are given off.

It is now possible to bring about a great number of nuclear reactions in which one nucleus is changed into another, so that the alchemist's dream of converting, say, lead into gold can now be achieved, though not commercially. This has made possible the preparation of many isotopes not occurring naturally, and radioactive isotopes of naturally occurring elements, which are not themselves radioactive, can be made. Such isotopes are said to be artificially radioactive. It has also been possible to make trans-uranium elements with heavier nuclei than that of uranium, which has the heaviest nucleus of all naturally occurring elements.

Moreover, by splitting some big nuclei, particularly those of uranium-235 and plutonium, into smaller nuclei, nuclear energy can be released by nuclear fission. Similarly, energy can be obtained from nuclear fusion by uniting the nuclei of two light isotopes. In such matters, consideration of nuclear structure and nuclear stability plays a big part.

5
The electron

The term electron was first introduced, by Johnstone Stoney, to describe a certain quantity of electricity, following a consideration of the application of Faraday's laws (p. 44) to the passage of electricity through solutions.

Further experimental work on the passage of electricity through gases at reduced pressures led to the discovery of cathode rays (p. 44) consisting of very small, negatively charged particles, and it is these particles which are now known as electrons. Similar particles were also found to be emitted, as β-rays (p. 53) by radioactive substances, or by heated wires as in thermionic valves, or when light of appropriate wavelengths falls on to metals, as in the photoelectric effect.

The possibility of obtaining electrons from such widely different sources led J. J. Thomson to suggest that electrons must be a component part of atoms, and the part played by electrons in atomic structure is one of the major themes of the following chapters. In developing this theme it will be seen that the electron has to be regarded as a minute, negatively charged particle, and that the wave nature of the electron has also to be taken into account (p. 72).

1 Conduction of electricity through solutions. The Faraday

It has been known for a very long time that amber rubbed with fur, and glass rubbed with silk, will attract small pieces of dry paper. The word electric, originating from the Greek for amber, was first used by Gilbert in 1600. It was also realised, very early, that amber rubbed with fur would attract glass rubbed with silk. Two varieties of electricity, known as resinous and vitreous, were supposed to exist. Ingenious frictional machines were developed to build up high electric charges.

But electric currents of any magnitude were only obtained following the chance observation by Galvani, in 1791, of the twitching of a partially dissected frog's leg when the nerve and leg were connected through a metal dipping into the body fluids. This observation was developed by Volta, in 1800, into the Voltaic pile, and a new era in science began.

In 1800, Nicholson and Carlisle decomposed water into hydrogen and oxygen by passing an electric current through it, and in the first decade of the nineteenth century Davy isolated sodium and potassium, and other similar metals, by passing electric currents through their molten hydroxides. Such activities gave a great fillip to electrical experimentation, and to the view that combination between atoms was electrical in nature.

The results of much measurement on the decomposition of solutions of salts, acids and alkalis by electricity were eventually summarised, in 1834, by Faraday who began his scientific work as Davy's laboratory assistant.

He discovered that the mass of a substance liberated at an electrode during electrolysis was proportional to the quantity of electricity passed. Quantitatively, in modern terms, he found that $96\,485\,C$ would discharge 1 mol of an X^+ ion at the cathode and 1 mol of a Y^- ion at the anode. To discharge 1 mol of X^{n+} and Y^{n-} ions required n times as many coulombs, 1 C being provided by a current of 1 A passing for 1 s. $96\,485\,C$ is called 1 faraday (1 F), or $96\,485\,C\,mol^{-1}$ is referred to as the Faraday constant. The more accurate value is $9.648\,456 \times 10^4\,C\,mol^{-1}$.

Because Faraday did not express his results in such a modern form, it was not immediately realised that every atom of a monovalent element requires a definite small quantity of electricity to liberate it during electrolysis. A divalent atom requires twice the quantity of electricity; a trivalent atom, three times the quantity.

Such conclusions, published about 1881 by Helmholtz and Johnstone Stoney, strongly suggested that electricity, like matter, was not continuous but was made up of definite small units. Johnstone Stoney called the quantity of electricity required to liberate one atom of a monovalent element an electron. Its value, using modern figures, was $(9.648\,456 \times 10^4) \div (6.022\,045 \times 10^{23})$, i.e. $1.602\,189\,2 \times 10^{-19}\,C$.

This was the origin of the word electron, and little did Helmholtz or Johnstone Stoney realise what an incredibly large part this remarkably small quantity of electricity was destined to play. In modern terms,

$$\frac{\text{the Faraday constant}}{\text{the Avogadro constant}} = \text{the charge on an electron}$$

Once any two of these quantities are known, the third is obtainable.

2 Conduction of electricity through gases. Cathode rays

At normal pressures, gases are good insulators but, if a sufficiently high voltage is used, a current can be passed between two electrodes separated by a gas under reduced pressure. The effect observed depends on the pressure, and a remarkable series of changes takes place as the pressure is progressively lowered by pumping more and more gas out of the discharge tube. It is interesting to reflect on the part played by the development of pumps in scientific progress.

At a pressure of about $0.25\,Pa$ the tube is occupied by a dark space (known as the Crookes dark space after the man who first studied the effects), whilst the glass of the tube fluoresces. The nature of the dark space within the discharge tube was originally in much doubt, but it is now known to consist of cathode rays (a name first used by Goldstein in

1876) made up of a stream of electrons passing from the cathode to the anode.

Such a conclusion is supported by the following experimental evidence.

a When a thick metallic shape is placed in the discharge tube between the cathode and the anode, a well-defined shadow of the shape is formed (Hittorf, 1869). This shows that the cathode rays travel in straight lines.

b The cathode rays will penetrate a thin sheet of metal.

c A small, light paddle wheel placed between the cathode and anode is caused to rotate by the cathode rays, and it can be rotated in the opposite direction by reversing the current. This shows that cathode rays can cause mechanical movement, and that they move in a direction away from the cathode.

d Cathode rays cause fluorescence when they impinge on most substances. In particular, they cause very clear fluorescence on a screen coated with zinc sulphide. Such a screen is, therefore, useful in observing cathode rays.

e A beam of cathode rays is deflected by an external magnetic field in a direction at right angles to both the direction of the beam and the direction of the lines of force of the magnetic field (Plücker, 1858). Application of Fleming's left-hand rule shows that the cathode rays behave like a flow of negative charge.

f A beam of cathode rays is also deflected by an electrical field, again in a direction indicating negative charge.

g If the cathode rays are led into a metal cylinder connected to an electroscope or electrometer outside the discharge tube, the cylinder is found to become negatively charged (Perrin, 1895). This again shows that the cathode rays are negatively charged.

h Cathode rays passed through a supersaturated vapour cause the formation of a fog or vapour trail in their path by producing nuclei for condensation of the vapour.

3 Measurement of charge, mass and velocity of electrons

J. J. Thomson, in 1897, measured the velocity of cathode rays, and the value of the charge/mass ratio for the electrons of which they are composed. He did this by subjecting a beam of cathode rays, obtained by passing them through a perforated anode and a slit, to the effect of both magnetic and electric fields so arranged as to deflect them in opposite directions. The deflection caused to the beam of electrons could be observed on a zinc sulphide strip at the end of the tube away from the cathode (Fig. 11).

Application of the magnetic field caused the beam to move in an arc of a circle whilst passing through the field so that the beam was deflected from X to Y. If the magnetic field is of strength, B, and the electrons in the beam are carrying a charge, e, and moving with velocity, v, the

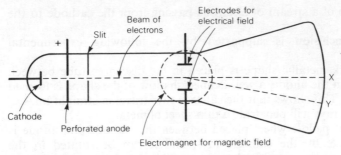

Fig. 11 Thomson's method of measuring e/m and v for electrons.

magnetic force exerted on each electron will be Bev. This will be balanced by a centrifugal force of mv^2/r, where m is the mass of an electron and r the radius of the circle in which the electrons move. Thus

magnetic force on electrons = centrifugal force on electrons

$$Bev = mv^2/r \quad \text{or} \quad e/m = v/rB$$

The value of r can be obtained from the dimensions of the apparatus and the distance XY.

By applying an electric field of the required strength, at right angles to the magnetic field, the deflected beam can be brought back on to its original path. When this is done, the magnetic force (Bev) on the electrons must be balanced by the electrical force, which is given by Ee, where E is the strength of the electrical field. Thus

magnetic force = electrical force

$$Bev = Ee \quad \text{or} \quad v = E/B$$

Knowing the values of E and B, v can be calculated, and substitution into the previous equation gives the value of e/m for an electron.

With different gases, different pressures and different potential differences between cathode and anode, v varied, though it was of the order of one-tenth of the velocity of light. But the value of e/m was constant under all conditions, and this suggested that the electron was a quite definite, individual particle which was probably a component of all matter.

This experiment could only give a value for the ratio e/m. If the charge on the electron was assumed to be the same as that on a univalent ion (p. 44), the mass of an electron must be approximately 1/1840 that of a hydrogen atom. No smaller particle has ever had a bigger future.

4 Determination of the charge on an electron

Thomson's experiment, described in the preceding section, gives only a value of the ratio e/m. This determination was followed by a direct

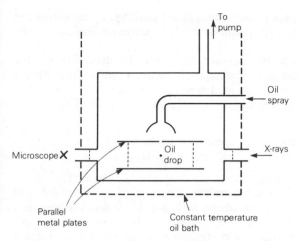

Fig. 12 Diagrammatic representation of Millikan's oil-drop apparatus.

measurement of e by Townsend, and by Thomson himself, and, later (1909), by a more accurate method, known as the oil-drop method, by Millikan.

Millikan observed the motion of a charged drop of non-evaporating oil between two circular plates of metal set accurately parallel by glass, insulating separators. Oil drops, from a spray, were allowed to pass through a small hole in the upper plate (Fig. 12). The drops became charged, by friction, in the spraying process, and could be further charged, as required, by passing a beam of X-rays (p. 50) between the plates. The drops were illuminated from one side and viewed through a microscope at right angles to the direction of illumination. Single drops, appearing as a bright spot, could be observed in this way.

With the plates earthed, an oil drop falls under the influence of gravity. Measurement of the terminal velocity of the drop, and application of Stokes's law, enables the weight of the drop to be calculated*.

The gravitational fall of a charged drop, between the plates, can be stopped by applying a potential difference of several thousand volts

---

* Stokes's law states that the retarding force, F, on a spherical particle of radius, r, falling through a gas of viscosity, η, with a velocity of v is given by,

$$F = 6\pi\eta rv$$

If v is the terminal velocity of an oil drop, the corresponding retarding force will be equal to the weight of the drop minus the upthrust on it, i.e. $\frac{4}{3}\pi r^3 g(\rho_1 - \rho_2)$ where g is the acceleration due to gravity and ρ_1 and ρ_2 are the densities of the oil and the gas respectively. Thus, for terminal velocity,

$$6\pi\eta rv = \frac{4}{3}\pi r^3 g(\rho_1 - \rho_2)$$

If the radius of an oil drop be calculated from experimental measurements, its weight can be readily obtained from its volume and its density. For small drops, however, such as those used in Millikan's experiment, a slightly modified form of Stokes's law must be used for real accuracy.

between the plates, the upper plate being made positive. By adjusting the potential difference a single drop can be made to move up or down, or be held stationary.

For a stationary drop, the upward force on the drop due to the electrical field must be equal to the mass of the drop minus the upthrust on it. The upward force on the drop, due to the electrical field, is equal to the electrical field multiplied by the charge on the drop. Thus, for a stationary drop,

$$\begin{matrix} \text{electrical} \\ \text{field} \end{matrix} \times \begin{matrix} \text{charge on} \\ \text{drop} \end{matrix} = \begin{matrix} \text{mass of} \\ \text{drop} \end{matrix} - \begin{matrix} \text{upthrust on} \\ \text{drop} \end{matrix}$$

from which the value of the charge on the drop can be obtained.

Millikan found that the charge on any single drop of oil varied, and that it could be changed by treatment with X-rays. Whatever charge a drop might have, however, it was always found to be an integral multiple of 1.60×10^{-19} C. Change of gas or pressure in the apparatus did not affect this basic value, which was taken to be the charge on a single electron.

The presently accepted value for the charge on an electron is $1.6021892 \times 10^{-19}$ C, which is, of course, equal to the charge on a univalent ion (p. 000). In conjunction with the most accurate e/m value, this gives a value of 9.109534×10^{-31} kg for the mass of an electron.

Questions on Chapter 5

(Questions on Faraday's laws will be found on p. 448.)

1 Write a short biographical sketch on one of the following: J. J. Thomson, Davy, Millikan, Crookes, Helmholtz.

2 Give an account on either (a) thermionic valves, or (b) the photo-electric effect.

3 Write short notes on the following terms: electron volt, electron spin, electron alloy, electron diffraction, electron affinity.

4 Describe, with diagrams, what is observed when the pressure is slowly reduced inside a discharge tube connected to a source of high potential difference.

5 Use Fleming's left-hand rule to show the direction of the force acting on an electron when it moves at right angles to a magnetic field. Draw a diagram, and give an explanation of it.

6 All electrons have the same charge and the same mass. In what other ways might they differ from each other?

7 Outline some modern developments which would probably not have been possible without the discovery of the electron.

8 If a stream of electrons each of mass m, charge e, and velocity $3 \times 10^7 \,\mathrm{m\,s^{-1}}$ is deflected 2 mm in passing for 100 mm through an electrostatic field of $1.8 \,\mathrm{V\,mm^{-1}}$ perpendicular to their path, find the value of e/m in $\mathrm{C\,kg^{-1}}$.

9 A charged particle is found to follow a circular track of radius 200 mm when it enters a perpendicular magnetic field of $7.5 \times 10^{-4}\,Wb\,m^{-2}$ at a velocity of $26.4 \times 10^{6}\,m\,s^{-1}$. What is the charge/mass ratio for the particle in $C\,kg^{-1}$?

10 Why is it that Faraday's laws might be said to have led to the discovery of the electron?

11 What charge, in C, is carried by (a) 1 mol of Al^{3+} ions (b) 2 mol of $PO_4{}^{3-}$ ions (c) 2 mol of Na^+ ions?

6
X-rays, radioactivity and the nuclear atom

1 X-rays

Röntgen discovered, in 1895, that wrapped photographic plates became fogged when left near a discharge tube, and this is now known to be due to the emission of a very penetrating radiation, known as X-rays, from solids bombarded by cathode rays.

Besides being very penetrating and affecting photographic plates, X-rays travel in straight lines and cause luminescence of, for example, zinc sulphide. The rays can, moreover, ionise a gas when they are passed through it, thus causing it to conduct electricity. X-rays are not, however, deflected by magnetic or electrical fields.

The real nature of X-rays was not discovered until 1912. In particular, it was not known whether they consisted of a stream of particles, like cathode rays, or whether they were wave-like, similar to light.

If like light, then X-rays ought to be capable of being diffracted if a suitable diffraction grating were available, but early attempts to diffract X-rays, using ruled gratings, were not successful. In 1912, von Laue suggested that the wavelength of X-rays might be too small to give diffraction patterns with a ruled grating, but that the regular, close-spaced array of planes of atoms within a crystal might serve as a diffraction grating for such short wavelengths. Friedrich tested this suggestion experimentally, and found that a copper(II) sulphate crystal would, in fact, diffract X-rays. This result established the nature of X-rays as light-like radiation of very short wavelength.

This was the beginning of X-ray crystallography (p. 165), which enables crystal structures to be investigated by X-rays, and which, using crystals of known structure, also enables the wavelengths of X-rays and other similar radiation to be determined.

2 X-ray spectra

X-ray spectra were studied by Moseley (1913) and this led to the elucidation of the meaning of atomic numbers. When a solid is bombarded by cathode rays to produce X-rays, the X-rays given out are of varying wavelength. Bombardment of every element gives a general, or white, X-radiation which shows up in any X-ray spectrum as a continuous background of low intensity. Superimposed on this background are a number of lines of higher intensity characteristic of the particular element being used as the anode. These lines occur in groups known as the K, L, M, N . . . groups.

By measuring the wavelengths of corresponding lines in the X-ray spectra of as many elements as possible, Moseley discovered that the square root of the frequency of corresponding lines for different elements gave almost a straight line when plotted against the atomic numbers of the elements concerned (Fig. 13).

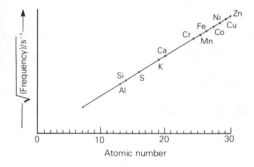

Fig. 13 The relationship between the atomic numbers of some simple elements and the square root of the frequency of the K_α lines in their X-ray spectra.

This suggested that the atomic number of an element is really of more significance than the relative atomic mass in the periodic table arrangement (p. 69), and on this basis, Moseley was able to make some adjustments in the older periodic table orders. Cobalt, for example, has a higher relative atomic mass than nickel and ought to follow it in the periodic table so far as relative atomic mass is concerned. Moseley found, however, that the X-ray spectral lines of nickel had higher frequencies than those of cobalt, and he therefore placed nickel after cobalt.

The lines of high intensity in any X-ray spectrum are caused by the bombarding electrons knocking electrons out of the inner shells of the bombarded atoms. Displaced electrons are then replaced by electrons from higher shells with a corresponding emission of radiation. Electrons returning to the K-shell give rise to the K-lines, and so on. The low intensity background X-radiation arises from the deceleration of the bombarding electrons as they enter the metal and the emission of their energy as electromagnetic radiation.

3 Radioactive elements

In the year following the discovery of X-rays (1896), Becquerel found, whilst investigating various fluorescent substances, that uranium and uranium compounds would also emit a penetrating radiation capable of affecting wrapped photographic plates, and he called this phenomenon radioactivity.

The discovery was followed, in 1898, by the isolation of two more strongly radioactive elements, polonium and radium, by M. and Mme Curie. The Curies examined the radioactivity of a uranium mineral called pitchblende and found that it was much more radioactive than would be expected from its uranium content. A prolonged and tedious extraction process led to the isolation of polonium and radium from pitchblende. 1000 kg of the mineral gave only about 0.2 g of radium, but this element was about 2 million times more radioactive than uranium. Mme Curie also found, in 1898, that thorium was radioactive, and in 1900, actinium was found to be radioactive.

The naturally occurring elements now known to be radioactive are polonium, radon, radium, actinium, thorium, protactinium and uranium. All the transuranium elements (p. 70) are also radioactive, and artificial radioactivity can be induced in elements not naturally radioactive (p. 82).

4 α-, β- and γ-rays

Radioactive substances emit three different types of ray, which can be separated and investigated because they are affected differently by magnetic and electric fields. If the radiation from a radioactive substance is passed through a magnetic field it is split up into what are now known as α-, β- and γ-rays.

The α-rays are deflected in a direction indicating that they are positively charged; the β-rays are deflected in the opposite direction, and must be negatively charged; the γ-rays are not deflected at all (Fig. 14).

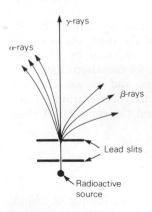

Fig. 14 Effect of a magnetic field, acting normally to the plane of the paper, on α-, β- and γ-rays.

Furthermore, the β-rays are deflected to a much greater extent than the α-rays, which indicates that the β-rays are much lighter than the α-rays, and the β-ray beam is dispersed more than the α-ray beam, which indicates that the β-rays consist of particles of more widely varying velocity than the α-rays. The γ-rays are the most penetrating, and the α-rays the least.

a α-rays Rutherford measured the charge/mass ratio for α-rays by the method of applying magnetic and electric fields similar to that used by Thomson in measuring e/m for cathode rays (p. 45). The result agreed with the supposition that α-rays were made up of helium atoms bearing two positive charges, i.e. of helium nuclei, and this was proved in two ways.

First, some radon was sealed in a glass tube thin enough to allow the α-rays emitted to escape through the glass walls into a surrounding, evacuated tube. After some time, an electric discharge was passed through the outer tube and analysis of the resulting spectrum showed that helium was present.

Secondly, Rutherford made a direct measurement of the charge on an alpha particle. α-rays cause a glow when they fall on a zinc sulphide screen, and, if the intensity of radiation is not too great, each α-particle causes a definite flash which can be seen through an eyepiece. In this way, in what is known as a spinthariscope, α-particles can be counted by counting the number of flashes.

Rutherford measured the number of α-particles falling on a given area in a given time, and then measured the charge imparted to a metal screen of the same area in the same time by the same radioactive source. He was, in this way, able to calculate the charge carried by each α-particle and this came out to be twice the charge on a hydrogen ion. In conjunction with the measured charge/mass ratio for an α-particle this showed that the mass of an α-particle was four times that of a hydrogen atom. The α-particle is, therefore, a helium atom bearing two positive charges, i.e. a helium nucleus (p. 40).

The velocity of α-particles emitted by radioactive substances varies, but it is about 1/20 of the velocity of light. α-rays can penetrate about 70 mm of air at atmospheric pressure, or a thin sheet of aluminium a small fraction of a millimetre thick.

b β-rays β-rays are found to be very similar to cathode rays in their general properties and this, coupled with measurement of their charge/mass ratio, shows that they consist of a stream of electrons.

The velocity of β-particles from radioactive sources varies from 3 per cent to 99 per cent the velocity of light, and β-rays will penetrate several millimetres of aluminium or about 3 mm of lead.

c γ-rays γ-rays are not affected by electric or magnetic fields and they are extremely penetrating electromagnetic radiation similar to X-rays but of much shorter wavelength. They will penetrate about 150 mm of lead, and travel with the speed of light.

The nature of α-, β- and γ-rays is summarised below.

	nature	electrical charge	mass of particle	velocity	relative penetration
α-rays	helium nuclei	+2 units	4 units	c. 1/20 velocity of light	1
β-rays	electrons	−1 unit	1/1836 unit	3–99% velocity of light	100
γ-rays	electro-magnetic radiation	no charge		velocity of light	10 000

5 Bombardment of matter by α- and β-rays

The existence of cathode rays and β-rays, obtained from such different sources, and yet both consisting of electrons, provides evidence that electrons are a component part of atoms. This suggestion is also supported by the fact that electrons are given off from a hot metallic wire, as in a thermionic valve, and when light of appropriate wavelength falls on to metals, as in the photoelectric effect first observed by Hertz in 1887.

But, if an atom contains electrons (negatively charged), it must also contain an equal number of some positively charged particles to give it electrical neutrality. Rutherford proved that it was the nucleus of an atom that was positively charged as a result of experiments he directed on the scattering of α- and β-rays by metallic foil.

a Bombardment by β-rays When a parallel beam of β-rays is directed towards a thin sheet of metal, the β-rays pass through, but the beam diverges. This observation is explained as being due to the repulsion of the electrons in the β-rays by the electrons in the atoms of the foil (Fig. 15).

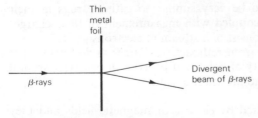

Fig. 15 The divergence of a beam of β-rays on passing through thin metal foil.

The measured divergence shows, moreover, that the number of electrons in an atom necessary to cause this divergence must be about half the relative atomic mass of the atom concerned. Magnesium, with relative atomic mass of 24, is shown, for example, to have about 12 electrons in its atom.

As the mass of the electron is very small, the electrons in an atom cannot contribute very much to the total mass of the atom, and the main mass of an atom must be situated in the positively charged part of the atom. This viewpoint was firmly supported by the results of experiments in which α-rays were used as bombarding particles instead of β-rays.

b Bombardment by α-rays In these experiments, carried out by Geiger and Marsden, a parallel beam of α-particles, from a radioactive source, was directed towards a platinum or gold foil. The effect was observed by picking up the α-particles on a zinc sulphide screen, and it was found that the α-particles passed right through the metallic foil but that they were deflected from their course in many directions. The average deflection was less than one degree, but some α-particles were deflected through a much greater angle, and much to everyone's surprise at that time, a very few (about 1 in 20 000) were deflected back through angles greater than 90° (Fig. 16). This back-deflection was not simply a reflection from the front surface of the foil, for the number of back-deflections increased as thicker pieces of foil were used.

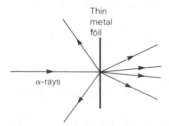

Fig. 16 The scattering of α-particles by a sheet of thin metal.

As the α-particle is heavy, it could only be deflected through large angles by a near collision with something of its own mass and charge. Rutherford was able to show that the observations of the deflections of α-particles could be accounted for if it was assumed that an atom had a positively charged nucleus, of diameter about 10^{-12} mm, carrying a positive charge equal in number to about half the relative atomic mass of the atom concerned. Thus it was shown that the charge necessary on the positive portion of the atom was equal in magnitude to the total negative charge carried by the electrons.

6 The nucleus

The general idea of a nuclear atom was widely accepted by 1914. A positively charged nucleus, contributing almost all the mass of an atom, was surrounded by electrons, the net positive charge on the nucleus being equal to the total negative charge of the surrounding electrons.

The nucleus of a hydrogen atom, with a mass of 1 unit and a charge of +1 unit, became known as a proton. It was, originally, thought that larger nuclei contained both protons and electrons, but this was not a very satisfactory suggestion so far as electrical forces and energy considerations were concerned (p. 97). The difficulties were overcome after the discovery by Chadwick, in 1932, of a neutral particle known as the neutron. Chadwick bombarded beryllium with α-rays, from a radioactive substance, and found that particles were emitted which were not affected by magnetic or electric fields. These particles are neutrons, with zero charge and unit mass.

With the discovery of the neutron it was no longer necessary to postulate the existence of protons and electrons together in an atomic nucleus. Instead, the nucleus was thought to consist of protons and neutrons, as shown in the typical atomic structures on page 41.

Further details of the atomic nucleus are given on pages 95–104. They are of importance in considerations of atom splitting for, when an atom splits, either naturally in radioactive elements or artificially by bombardment, it is the nucleus that splits. From the point of view of chemical combination between atoms, it is the arrangement of the electrons around the nucleus that is of major importance (pps. 58–75).

Questions on Chapter 6

1 Write an account on the production and uses of X-rays.

2 How was it shown that (a) β-rays consisted of electrons, and (b) α-rays consisted of helium nuclei?

3 Draw a diagram showing how Fleming's left-hand rule can be used to predict the effect of a magnetic field on (a) an electron (b) an α-particle (c) a proton.

4 Explain the difference between (a) a β-particle and an α-particle (b) γ-rays and X-rays (c) a proton and a deuteron.

5 What were the major contributions to the elucidation of atomic structure of (a) Rutherford (b) Thomson (c) Moseley (d) Röntgen?

6 Explain how it is possible to get an approximate value for the Avogadro constant by measuring the rate of α-particle emission from a known mass of radium using a spinthariscope and, also, the rate of helium formation from a known mass of radium.

7 By placing a minute, but known, mass of radium close to a zinc sulphide screen and counting the flashes on the screen it was estimated that 1 g of radium emitted 34 000 million alpha particles per second. The volume of helium obtained from decaying radium was found to be $0.156\,cm^3$ at s.t.p. per gram of radium per year. What value do these figures give for the Avogadro constant?

8 How many protons and how many neutrons are there in the nuclei of the following isotopes: ^{19}F, ^{75}As, ^{70}Ge, ^{16}O, ^{1}H, ^{2}H, ^{235}U, ^{238}U and ^{239}Np?

9 Write down the nuclear structures of the isotopes formed when (a) $^{239}_{92}U$ loses first one and then a second β-particle, and (b) $^{239}_{92}U$ loses first one and then a second α-particle.

10 Give the numbers of protons, neutrons and electrons which will be found in the following atoms: $^{27}_{13}Al$, $^{235}_{92}U$, $^{209}_{83}Bi$, $^{4}_{2}He$, $^{3}_{2}He$, $^{1}_{1}H$, $^{2}_{1}H$, $^{3}_{1}H$, $^{16}_{8}O$, $^{17}_{8}O$, $^{18}_{8}O$, $^{35}_{17}Cl$, $^{37}_{17}Cl$.

11 Write down the nuclear structures of all the naturally radioactive elements.

12 Why was the supposition that an atomic nucleus contained protons and electrons unsatisfactory, and how was this difficulty overcome?

7
Electrons in atoms

1 Introduction

The chemical properties of an atom are largely controlled by its electrons, for it is interaction between the electrons of two or more atoms that leads to chemical combination between the atoms. The detailed arrangement of electrons within an atom is, therefore, of fundamental importance. The working out of these arrangements is an amazing feat similar to the fitting together of a jig-saw puzzle or the solution of a cipher.

The solution of the problems involved comes from interpretation of spectroscopic data and the application of the ideas of the quantum theory. It also involves the relative position of the various elements in the periodic table, and a consideration of the wave-like nature of electrons.

2 Spectral series

As opposed to particle-like rays, wave-like radiation can cause interference patterns, can be diffracted if a suitable diffraction grating can be found (p. 50), and can be 'sorted out' into its component wavelengths by using a spectrometer.

In this way, visible light is readily shown to be made up of different coloured lights, each colour corresponding to a group of waves of different wavelengths and frequencies. Violet light, at one end of the visible spectrum, has a shorter wavelength (420 nm) than red light, at the other end (750 nm). Frequencies are expressed in vibrations per second, s^{-1}, or hertz (Hz). Wavenumbers (the reciprocal of wavelength) are also used, generally in cm^{-1}.

Visible light represents only a small part of all radiation, and a more complete representation of the possible types of radiation is given in Fig. 17.

Characteristic spectra can be obtained from substances by causing them to emit radiation (p. 178). This can be done in a variety of ways, e.g., by heating a substance or by subjecting it to electrical stimulation or excitation by using an electric arc, spark or discharge; and a variety of *emission spectra* can be obtained.

The emission spectra from the vapours of elements are known as *atomic* or *line spectra*. They consist of a series of lines, each corresponding to a particular wavelength, and indicating that the number of possible energy changes that can take place within the atom of the element is, in some way, limited.

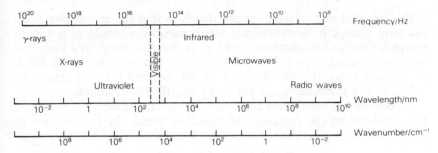

Fig. 17 The electromagnetic spectrum. The boundaries between one type of radiation and another are not very sharply defined. See also Fig. 96 on page 178.

So far as the historical development of atomic structure is concerned, a study of the line spectrum of hydrogen is of great importance.

This spectrum consists of lines corresponding to widely different frequencies, but over a period of time, starting in 1885, it was found that many of the numerous lines could be fitted into series. These series, known after their discoverers as the Balmer (1885), Paschen (1896), Lyman (1915), Brackett (1922) and Pfund (1925) series, can be expressed in one overall formula:

$$1/\lambda = R_\infty(1/n^2 - 1/m^2) \qquad \text{or} \qquad v = R_\infty c(1/n^2 - 1/m^2)$$

where λ is the wavelength, v the frequency, c the velocity of light, R_∞ a constant, known as the Rydberg constant ($109\,737\,\text{cm}^{-1}$), and n and m have integral values as follows:

series	n	m	main spectral region
Lyman	1	2, 3, 4, etc.	ultra-violet
Balmer	2	3, 4, 5, etc.	visible
Paschen	3	4, 5, 6, etc.	infra-red
Brackett	4	5, 6, 7, etc.	infra-red
Pfund	5	6, 7, 8, etc.	infra-red

It is a remarkable experimental fact that so many apparently, at first sight, unrelated lines in a spectrum can be expressed by a simple formula. Line spectra of the alkali metals are also made up of similar series of lines known as the sharp, principal, diffuse and fundamental series. The lines in these series can be related in a single formula as for the hydrogen spectrum.

3 Outline of quantum theory

Progress in working out the arrangement of electrons in an atom came first when Bohr, in 1914, applied the ideas of the quantum theory, put forward by Planck in 1900, to the interpretation of spectroscopic data.

The essential idea of the quantum theory is that the energy of a body can only change by some definite whole-number multiple of a unit of energy known as the quantum. This means that the energy of a body can increase or decrease by 1, 2, 3, 4 ... n quanta, but never by $1\frac{1}{2}$, $2\frac{3}{4}$, 107.3, etc. quanta. It is rather like the fact that our currency can only change by 1, 2, 3, 4 ... n halfpennies, but not by $1\frac{1}{2}$, $2\frac{3}{4}$, 107.3 halfpennies.

Unlike the halfpenny, however, the value of the quantum is not fixed, but is related to the frequency of radiation which, by its emission or absorption, causes the change in energy. This relationship is expressed as

$$E = h\nu \qquad \text{or} \qquad E = hc/\lambda$$

where E is the value of the quantum (J), h is the Planck constant $(6.626\,176 \times 10^{-34}\,\text{J s})$, ν is the frequency $(\text{s}^{-1}$ or Hz), λ is the wavelength (m) and c is the velocity of light (m s^{-1}). It is, then, a simple matter to calculate the value of the quantum corresponding to any known frequency.

When radiation of frequency, ν, or wavelength, λ, is absorbed by a body, there will be an energy increase, from E_2 to E_1, where

$$E_1 - E_2 = nh\nu = nhc/\lambda$$

and n is an integer. Emission of similar radiation would cause an equal decrease in energy.

4 Bohr's interpretation of spectral series

Rutherford assumed that the electrons circulated round the nucleus of an atom in orbits, rather as planets circulate round the sun, and atoms were pictured as minute solar systems. The electrical forces of attraction between the negatively charged electrons and the positively charged nucleus were just counterbalanced by centrifugal forces.

Bohr pointed out, however, that charged particles could not circulate in an orbit without having an acceleration towards the centre of the orbit, and according to the accepted electrodynamic theory of the time, an electric charge must radiate energy when it is accelerated.

If Rutherford's idea is correct, then an atom would radiate energy continuously; in doing so it would undergo spontaneous destruction. Moreover, the continuous emission of radiation would not account for the formation of line spectra.

To deal with these difficulties, Bohr put forward two suggestions which, in effect, deny the truth of older electrodynamic theories as applied to the motion of electrons. First, the electrons in an atom could only rotate in certain selected orbits and that they did not, in these orbits, radiate energy. Such orbits were called *stationary states*. Secondly, each stationary state corresponds to a certain energy level and emission of radiant energy was caused by the movement of an electron from one stationary state to another of lower energy. Conversely, absorption of energy took place by an electron moving into a stationary state of higher energy.

According to the quantum theory an electron passing from a higher stationary state, with energy E_1, to a lower one, with energy E_2, would cause an emission of radiation of frequency, v, or wavelength, λ, i.e.

$$E_1 - E_2 = hv = hc/\lambda$$

Similarly, absorption of radiation of this frequency would cause an electron to pass from energy level E_2 to the higher energy level E_1.

On this view, the series observed in the line spectrum of hydrogen are explained by the various limited energy changes which an electron can undergo in moving between the various stationary states. The general idea is made clear by a study of Figs. 18 and 19.

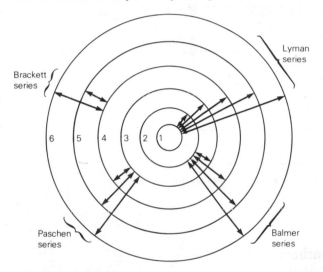

Fig. 18 The energy changes which an electron can undergo in moving from one energy level to another in an atom. The diagram is not to scale, for the radii of the various stationary states shown are, in fact, proportional to the squares of the principle quantum numbers allotted to them.

The atom is in its normal or ground state when the electron is in the stationary state of least energy; when in any other state the atom is said to be excited. On excitation, the electron moves into stationary states of higher energy content, and it is the return of the electron to stationary states of lower energy which results in the emission of radiant energy and the formation of spectral lines. It is like lifting a ball up and letting it fall again, both the lifting and falling being done in definite stages.

For hydrogen, and hydrogen-like spectra, the Bohr theory is able to account for the observed spectral series in detail and with accuracy. Values for the radii and the energy levels of the various stationary states can be calculated, and these fit in well with the spectroscopic results.

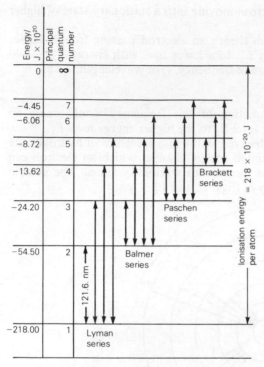

Fig. 19 Some of the energy levels in the hydrogen atom.

5 Quantum numbers

The extension of Bohr's ideas came about as a result of a more detailed investigation of spectra. In particular, to account for the greater number of lines observed, it was found necessary to increase the number of possible orbits in which an electron could exist within an atom. In other words, it was necessary to allow more energy changes within an atom to explain the formation of all the observed spectral lines.

The term quantum number is used to label the various energy levels or orbits. The number allotted to Bohr's original stationary states, visualised as circular orbits, is called the principal quantum number. The first orbit, nearest the nucleus, has a principal quantum number of 1; the second has a principal quantum number of 2, and so on. Alternatively, letters are used to characterise the orbits, the first being referred to as the K orbit, the second as the L, the third as the M, and so on. The choice of letters originates from Moseley's work on X-ray spectra (p. 50).

A more detailed study of spectra showed, however, that for each principal quantum number there were several associated orbits, so that the principal quantum number now represents a group or shell of orbits.

a Principal quantum number This represents a group or shell of orbits, but the total number of electrons that can occupy any shell, i.e., that can have the same principal quantum number, is limited and is given by $2n^2$ where n is the principal quantum number concerned, i.e.

shell	K	L	M	N
principal quantum number (n)	1	2	3	4
maximum number of electrons	2	8	18	32

b Subsidiary quantum number This represents the various subsidiary orbits within a shell; they may be visualised as elliptical orbits. Thus in any one shell there are various subsidiary orbits denoted as the 1, 2, 3, 4 . . . or the s, p, d, f . . . orbits. The number of orbits in any one shell is, however, limited, and in dealing with the structures of atoms in their normal states, it is only necessary to consider the orbits listed on page 66. The 1s orbit may be referred to as the K1 orbit, but the former is the commoner usage.

c Spin quantum number The Pauli or exclusion principle states that all the electrons in any atom must be distinguishable, or that no two electrons in a single atom can have all their quantum numbers alike.

To agree with the principle it is assumed that electrons spin on their axes; this assumption also accounts for the splitting of spectral lines observed with a spectroscope of good resolving power. Because of electron spin, a spin quantum number is allotted to an electron to characterise it.

The shell of principal quantum number 1 can hold two electrons. As these cannot be exactly alike, because of the Pauli principle, it is assumed that they have different spins. In general, if two electrons occupy the same orbit they must have different spins, and, as electrons in the same orbit can only differ by having different spins, it follows that no orbit can contain more than two electrons.

d Magnetic quantum number For electrons in orbits with a principal quantum number greater than 1, a further complication arises. The shell with principal quantum number 2 can contain a maximum of eight electrons. Of these, two, with opposed spins, will be in the 2s level. The remaining six will be in the 2p level, but for all six to be different it is necessary to subdivide the 2p level still further, into three, so that each of the three subdivisions may contain two electrons. These three 2p orbits may be envisaged as in different planes and can be denoted as $2p_x$, $2p_y$ and $2p_z$ orbits (Fig. 20).

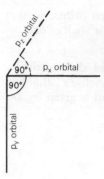

Fig. 20 The directional arrangement of p-orbitals.

This subdivision of p orbits into three, and a similar subdivision of d orbits into five and of f orbits into seven, necessitates a fourth quantum number known as the magnetic quantum number. For simple purposes this is not of great importance since the energy of an electron in any of the three p orbits is the same unless the atom is placed in a strong magnetic field. When this happens the three p orbits take up different positions with respect to the lines of force of the field and attain slightly different energy levels. This accounts for the splitting of spectral lines when the source of emission is placed in a magnetic field (the Zeeman effect) or in an electric field (the Stark effect).

e Summary To sum up, four quantum numbers are required to characterise completely any individual electron in a particular orbit. In a simple way this corresponds to a normal post office address. To characterise a particular Mr X it is necessary to allot a particular address to him, e.g. Mr X, 114 High Street, Wimbledon, England. The country corresponds to the principal quantum number, the town to the subsidiary quantum number, the street to the spin quantum number, and the street number to the magnetic quantum number.

The idea of quantum numbers has been presented from a pictorial point of view. Solution of the Schrödinger equation (p. 73) determines which quantum numbers are allowed. Those quantum numbers which are allowed can be summarised as follows.

i n = principal quantum number. The allowed values are 1, 2, 3, 4 . . . .
ii l = subsidiary quantum number. The allowed values of l depend on the value of n. When n is 1, l is 0, i.e. there are only s electrons in the shell with principal quantum number 1.
     When n is 2, l can be 0 or 1, i.e. the shell with principal quantum number of 2 contains both s and p electrons.
     When n is 3, l can be 0, 1 or 2, i.e. s, p and d electrons. When n is 4, l may be 0, 1, 2 or 3, i.e. s, p, d and f electrons.

It will be seen that $l = 0$ for an s electron, 1 for a p electron, 2 for a d electron and 3 for an f electron.

iii m = magnetic quantum number. For an s electron with $l = 0$, m is 0. For a p electron, with $l = 1$, m can be -1, 0 or $+1$. For a d electron with $l = 2$, m can be -2, -1, 0, $+1$ and $+2$, and so on. In general, m can have $(2l + 1)$ different values.

iv s = spin quantum number. The allowed values are $+\frac{1}{2}$ and $-\frac{1}{2}$.

The maximum number of electrons with the same principal quantum number can be summarised as follows, bearing in mind the fact that no two electrons in the same atom can have the same values for the four quantum numbers.

n	1	2				3								
l	0	0	1			0	1			2				
m	0	0	-1	0	$+1$	0	-1	0	$+1$	-2	-1	0	$+1$	$+2$
s	$+\frac{1}{2}$ $-\frac{1}{2}$	$+\frac{1}{2}$ $-\frac{1}{2}$	$+\frac{1}{2}$ $-\frac{1}{2}$	$+\frac{1}{2}$ $-\frac{1}{2}$	$+\frac{1}{2}$ $-\frac{1}{2}$	$+\frac{1}{2}$ $-\frac{1}{2}$	$+\frac{1}{2}$ $-\frac{1}{2}$	$+\frac{1}{2}$ $-\frac{1}{2}$	$+\frac{1}{2}$ $-\frac{1}{2}$	$+\frac{1}{2}$ $-\frac{1}{2}$	$+\frac{1}{2}$ $-\frac{1}{2}$	$+\frac{1}{2}$ $-\frac{1}{2}$	$+\frac{1}{2}$ $-\frac{1}{2}$	$+\frac{1}{2}$ $-\frac{1}{2}$

    2   8   18

6 The arrangement of electrons in orbits

The arrangement of electrons in an atom in its normal state is that which makes its energy a minimum, i.e. it is the most stable arrangement, within the limitations already mentioned.

The energy levels of the various orbits can be obtained from spectroscopic data, and the detailed arrangements of electrons can be built up from a knowledge of these energy levels, from the position of the atom concerned in the periodic table, and from the allowable quantum numbers.

The relative energy levels of orbits in an atom, as obtained from spectroscopic measurements, are shown in Fig. 21. The precise positioning of one orbit in relation to another depends on the atomic number of the element concerned. The order given is that for elements of low atomic number, and it changes to some extent with elements of higher atomic number. This is because the attraction between the nucleus and the electrons changes as the positive charge on the nucleus changes.

In Fig. 21 each circle represents an orbit which can be occupied either by a single electron or by two electrons with different spins. The circles enclosed within a rectangle represent orbits of equal energy. In passing along the periodic table the electrons occupy the orbits in energy order, as shown by the arrows in Fig. 21. The order is conveniently

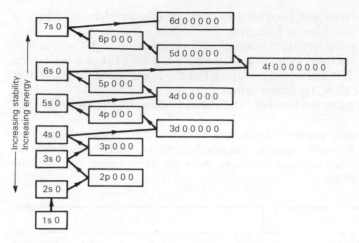

Fig. 21 The approximate comparative stabilities of the various orbits in an atom.

remembered, as in Fig. 22, though there are a few exceptions to this oversimplified mnemonic.

The one electron in a hydrogen atom will occupy the most stable orbit, i.e. the 1s orbit, and the second electron in the helium atom will occupy the same orbit but will have a different spin. The 1s orbit is now full; a third electron will occupy the next most stable orbit, i.e. the 2s orbit, and so on. The best 'seats' are occupied first. This building-up process was called the *aufbau prinzip* (building principle) by Pauli.

On this basis, but not differentiating between the three p, five d or seven f levels (p. 71), the arrangement of electrons in the atoms of the elements in their normal states is shown in the tables on pages 68 and 69.

These electronic arrangements are commonly written in the form $1s^2$, $2s^2$, $2p^6$, $3s^2$, $3p^6$, which is the argon arrangement.

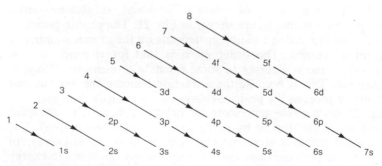

Fig. 22 The order of filling of orbitals in an atom (simplified).

7 Electronic structures and the periodic table

The periodic table of the elements was first drawn up, by Mendeléef in 1869, by listing the elements in the order of their relative atomic masses; a full, modern version is given on page 70. At first, there was no understanding of why such an arrangement should fit elements into groups with similar chemical properties. The deeper significance of the arrangement came, first, from a study of X-ray spectra (p. 50), and more completely, from a knowledge of the arrangements of electrons in different atoms.

In particular, all the atoms in any one group of the table were often found to have the same number of electrons in their outermost orbits. Moreover, five general types of element were revealed, as follows.

a The noble gases The arrangement of electrons in the noble gases is as follows.

| | 1 | 2 | 3 | 4 | 5 | 6 |
	s	s p	s p d	s p d f	s p d	s p
He	2					
Ne	2	2 6				
Ar	2	2 6	2 6			
Kr	2	2 6	2 6 10	2 6		
Xe	2	2 6	2 6 10	2 6 10	2 6	
Rn	2	2 6	2 6 10	2 6 10 14	2 6 10	2 6

Every orbit which is occupied at all is fully occupied and it is this unique arrangement of electrons which accounts for the chemical inactivity of the noble gases. Helium can be regarded as slightly different, for its atom only has 2 electrons in the outer orbit, whereas all the others have 8.

b s-block elements These elements, found in groups 1A and 2A, have atoms containing either 1 or 2 electrons in the outermost s orbit. All the other orbits which contain electrons are fully occupied. The structures of the atoms concerned may be summarised as follows:

Li	2.1	Be	2.2
Na	2.8.1	Mg	2.8.2
K	2.8.8.1	Ca	2.8.8.2
Rb	2.8.18.8.1	Sr	2.8.18.8.2
Cs	2.8.18.18.8.1	Ba	2.8.18.18.8.2
Fr	2.8.18.32.18.8.1	Ra	2.8.18.32.18.8.2

Both hydrogen and helium can also be regarded as s block elements, but it is best to link helium with the noble gases and the chemical functioning of hydrogen is, in many ways, unique.

The arrangement of electrons in the atoms of the elements in their normal states.

	1s	2s	2p	3s	3p	3d	4s	4p	4d	4f	5s	5p	5d	5f	6s	6p	6d	7s	
1 H	1																		
2 He	2																		1s full
3 Li	2	1																	
4 Be	2	2																	2s full
5 B	2	2	1																
6 C	2	2	2																
7 N	2	2	3																
8 O	2	2	4																
9 F	2	2	5																
10 Ne	2	2	6																2p full
11 Na	2	8		1															
12 Mg	2	8		2															3s full
13 Al	2	8		2	1														
14 Si	2	8		2	2														
15 P	2	8		2	3														
16 S	2	8		2	4														
17 Cl	2	8		2	5														
18 Ar	2	8		2	6														3p full
19 K	2	8		8			1												
20 Ca	2	8		8			2												4s full
21 Sc	2	8		8		1	2												
22 Ti	2	8		8		2	2												
23 V	2	8		8		3	2												
24 Cr	2	8		8		5	1												
25 Mn	2	8		8		5	2												
26 Fe	2	8		8		6	2												
27 Co	2	8		8		7	2												
28 Ni	2	8		8		8	2												
29 Cu	2	8		8		10	1												
30 Zn	2	8		8		10	2												3d full
31 Ga	2	8			18		2	1											
32 Ge	2	8			18		2	2											
33 As	2	8			18		2	3											
34 Se	2	8			18		2	4											
35 Br	2	8			18		2	5											
36 Kr	2	8			18		2	6											4p full
37 Rb	2	8			18		8				1								
38 Sr	2	8			18		8				2								5s full
39 Y	2	8			18		8		1		2								
40 Zr	2	8			18		8		2		2								
41 Nb	2	8			18		8		4		1								
42 Mo	2	8			18		8		5		1								
43 Tc	2	8			18		8		6		1								
44 Ru	2	8			18		8		7		1								
45 Rh	2	8			18		8		8		1								
46 Pd	2	8			18		8		10										
47 Ag	2	8			18		8		10		1								
48 Cd	2	8			18		8		10		2								4d full
49 In	2	8			18			18			2	1							
50 Sn	2	8			18			18			2	2							

transition elements: 21 Sc – 28 Ni

transition elements: 39 Y – 46 Pd

68

			1s	2s	2p	3s	3p	3d	4s	4p	4d	4f	5s	5p	5d	5f	6s	6p	6d	7s	
	51	Sb	2	8			18			18			2	3							
	52	Te	2	8			18			18			2	4							
	53	I	2	8			18			18			2	5							
	54	Xe	2	8			18			18			2	6							5p full
	55	Cs	2	8			18			18			8				1				
	56	Ba	2	8			18			18			8				2				6s full
transition elements	57	La	2	8			18			18			8		1		2				
rare earths or lanthanons	58	Ce	2	8			18			18		2	8				2				
	59	Pr	2	8			18			18		3	8				2				
	60	Nd	2	8			18			18		4	8				2				
	61	Pm	2	8			18			18		5	8				2				
	62	Sm	2	8			18			18		6	8				2				
	63	Eu	2	8			18			18		7	8				2				
	64	Gd	2	8			18			18		7	8		1		2				
	65	Tb	2	8			18			18		9	8				2				
	66	Dy	2	8			18			18		10	8				2				
	67	Ho	2	8			18			18		11	8				2				
	68	Er	2	8			18			18		12	8				2				
	69	Tm	2	8			18			18		13	8				2				
	70	Yb	2	8			18			18		14	8				2				
	71	Lu	2	8			18			18		14	8		1		2				4f full
	72	Hf	2	8			18				32		8		2		2				
	73	Ta	2	8			18				32		8		3		2				
	74	W	2	8			18				32		8		4		2				
	75	Re	2	8			18				32		8		5		2				
	76	Os	2	8			18				32		8		6		2				
	77	Ir	2	8			18				32		8		9						
	78	Pt	2	8			18				32		8		9		1				
	79	Au	2	8			18				32		8		10		1				
	80	Hg	2	8			18				32		8		10		2				5d full
	81	Tl	2	8			18				32			18			2	1			
	82	Pb	2	8			18				32			18			2	2			
	83	Bi	2	8			18				32			18			2	3			
	84	Po	2	8			18				32			18			2	4			
	85	At	2	8			18				32			18			2	5			
	86	Rn	2	8			18				32			18			2	6			6p full
	87	Fr	2	8			18				32			18			8			1	
	88	Ra	2	8			18				32			18			8			2	7s full
	89	Ac	2	8			18				32			18			8		1	2	
actinons	90	Th	2	8			18				32			18			8		2	2	
	91	Pa	2	8			18				32			18		2	8		1	2	
	92	U	2	8			18				32			18		3	8		1	2	
	93	Np	2	8			18				32			18		5	8			2	
	94	Pu	2	8			18				32			18		6	8			2	
	95	Am	2	8			18				32			18		7	8			2	
	96	Cm	2	8			18				32			18		7	8		1	2	
	97	Bk	2	8			18				32			18		9	8			2	
	98	Cf	2	8			18				32			18		10	8			2	
	99	Es	2	8			18				32			18		11	8			2	
	100	Fm	2	8			18				32			18		12	8			2	
	101	Md	2	8			18				32			18		13	8			2	
	102	No	2	8			18				32			18		14	8			2	
	103	Lr	2	8			18				32			18		14	8		1	2	

c p-block elements These occur in Groups 3 to 7B. They have atoms in which the outermost p orbit is filling up from 1 to 5 electrons, the total number of electrons in the outermost shell rising from 3 for Group 3 elements to 7 for Group 7 elements.

d d-block elements In these elements an inner 3d, 4d or 5d orbit fills up from one to ten electrons so that there are three series of ten elements each: Sc to Zn, Y to Cd and La to Hg (omitting the rare earths). In the first eight of each of these series, the 3d, 4d or 5d orbit contains one to eight electrons, i.e. it is not fully occupied. These elements are the *transition elements*, forming the first, second and third transition series.

The periodic table

s-block

1A	2A															0
										p-block						2
1																He
H										3B	4B	5B	6B	7B		
3	4									5	6	7	8	9		10
Li	Be	d-block								B	C	N	O	F		Ne
11	12		← transition elements →							13	14	15	16	17		18
Na	Mg									Al	Si	P	S	Cl		Ar

d-block

3A	4A	5A	6A	7A		8		1B	2B
21	22	23	24	25	26	27	28	29	30
Sc	Ti	V	Cr	Mn	Fe	Co	Ni	Cu	Zn
39	40	41	42	43	44	45	46	47	48
Y	Zr	Nb	Mo	Tc	Ru	Rh	Pd	Ag	Cd
57*	72	73	74	75	76	77	78	79	80
La	Hf	Ta	W	Re	Os	Ir	Pt	Au	Hg

19	20										31	32	33	34	35		36
K	Ca										Ga	Ge	As	Se	Br		Kr
37	38										49	50	51	52	53		54
Rb	Sr										In	Sn	Sb	Te	I		Xe
55	56										81	82	83	84	85		86
Cs	Ba										Tl	Pb	Bi	Po	At		Rn
87	88	89†															
Fr	Ra	Ac															

*the rare earths or lanthanons

f-block

58	59	60	61	62	63	64	65	66	67	68	69	70	71
Ce	Pr	Nd	Pm	Sm	Eu	Gd	Tb	Dy	Ho	Er	Tm	Yb	Lu

†the actinons

90	91	92	93	94	95	96	97	98	99	100	101	102	103
Th	Pa	U	Np	Pu	Am	Cm	Bk	Cf	Es	Fm	Md	No	Lr

e f-block elements There are two series of f-block elements in which the 4f and 5f orbits fill up from 1 to 14. The first series contains the *rare earths* or *lanthanons*, and the second series the *actinons*. There is remarkable chemical similarity within each series because each element has the same number of electrons in the two outermost shells.

8 Electron spin

A full interpretation of the doublets that occur in the fine structure of many spectra requires the assumption that electrons can spin on their axes and, in this way, act as tiny magnets. Such an assumption was demonstrated in the Stern–Gerlach experiment (1921) in which a beam of atoms of an alkali metal, or silver, was passed through a powerful non-uniform magnetic field. This field was found to split the beam of atoms into two, corresponding to two different spins of the outermost electrons in the atoms.

A detailed consideration of the arrangements of electrons in many atoms must take electron spin into account. The arrangement can readily be predicted by application of the rule of maximum multiplicity. This empirical rule, suggested by Hund, states that the distribution of electrons in a free atom between the three p, five d and seven f orbits is such that as many of the available orbits as possible are occupied by single (unpaired) electrons before any pairing or coupling of electrons with opposed spins takes place. Thus, if three electrons are to occupy the three p orbits in any one shell, one will go into each of the three available orbits. This can be interpreted as meaning that electrons repel each other and keep as far away from each other as is possible. Alternatively, the pairing of electrons requires an input of energy.

	1s	2s	$2p_x$	$2p_y$	$2p_z$
1 H	↓				
2 He	↓↑				
3 Li	↓↑	↓			
4 Be	↓↑	↓↑			
5 B	↓↑	↓↑	↓		
6 C	↓↑	↓↑	↓	↓	
7 N	↓↑	↓↑	↓	↓	↓
8 O	↓↑	↓↑	↓↑	↓	↓
9 F	↓↑	↓↑	↓↑	↓↑	↓
10 Ne	↓↑	↓↑	↓↑	↓↑	↓↑

Fig. 23 The distribution of electrons in the elements from H to Ne.

The electronic arrangements in the atoms H to Ne, and in those of the first transition series, with relevant spins shown by ↑ or ↓ arrows, are given in Figs. 23 and 24.

	1s	2s	2p	3s	3p	3d	4s
Sc	2	2	6	2	6	↓	2
Ti	2	2	6	2	6	↓ ↓	2
V	2	2	6	2	6	↓ ↓ ↓	2
Cr	2	2	6	2	6	↓ ↓ ↓ ↓ ↓	1
Mn	2	2	6	2	6	↓ ↓ ↓ ↓ ↓	2
Fe	2	2	6	2	6	↑↓ ↓ ↓ ↓ ↓	2
Co	2	2	6	2	6	↑↓ ↑↓ ↓ ↓ ↓	2
Ni	2	2	6	2	6	↑↓ ↑↓ ↑↓ ↓ ↓	2
Cu	2	2	6	2	6	↑↓ ↑↓ ↑↓ ↑↓ ↑↓	1
Zn	2	2	6	2	6	↑↓ ↑↓ ↑↓ ↑↓ ↑↓	2

Fig. 24 The distribution of electrons in the first transition series showing the electron spin of the 3d electrons.

9 Dual nature of light

The word 'ray' is used somewhat indiscriminately to describe both a stream of particles and wave-like radiation. The usage is, perhaps, valid when it is realised that particles may have a wave-like aspect whilst waves may have a particle-like aspect. Newton originally regarded light as fitting a corpuscular theory, but Huyghens introduced the wave theory of light. The wave theory is essential to explain interference and diffraction phenomena. The corpuscular theory is necessary to explain, for instance, the photoelectric effect.

In this effect, metals give off electrons when illuminated with light of appropriate wavelengths. Moreover, there is a simple relationship between the energy of the emitted electrons and the frequency of the incident radiation. This result was interpreted, by Einstein in 1905, as meaning that radiation could be regarded as made up of small 'packets' of energy, known as photons. The energy of a photon was dependent on the frequency, according to the basic equation of the quantum theory (p. 60), which explained why the energy of the emitted electrons in the photoelectric effect was related to the frequency of the radiation.

10 The wave nature of an electron

So far, the electron has been regarded as a tiny, negatively charged particle, but very important results come from a consideration of the wave nature of an electron. In 1924, de Broglie suggested that moving electrons had waves of definite wavelength associated with them, and this theoretical prediction was demonstrated experimentally when Davisson and Germer showed (1927) that a stream of electrons could be diffracted by crystals acting as simple diffraction gratings, just as light- or X-rays can. As it is only possible to account for diffraction in terms of waves, it is necessary to assume that a stream of electrons behaves as a wave-like radiation.

The de Broglie relationship between moving electrons and the waves associated with them can be expressed mathematically as

$$\lambda = h/mv$$

where λ is the wavelength of the waves, h is the Planck constant, m is the mass of the electron, and v is the velocity of the electron. If this expression is written as

$$\text{momentum} \times \lambda = h$$

it relates the particle-like aspect of an electron, i.e. momentum, to the wave-like aspect, i.e. wavelength.

The expression is true for all particles, but it is only with very small particles that the wave-like aspect is of any significance. The wavelength of a large particle will be so small that its wave-like properties will not be measurable or observable. An electron of energy 1.6×10^{-19} J or 1 eV (p. 97) has an associated wavelength of 1.2 nm; an α-particle from radium, a wavelength of 6.6×10^{-6} nm; and a golf ball, travelling at 30 m s^{-1}, a wavelength of 4.9×10^{-25} nm.

It is not easy to obtain a pictorial idea of this concept of an electron, but it is possible to treat the wave nature of an electron mathematically. This is done by a specialised technique known as wave mechanics or quantum mechanics.

So far as atomic structure is concerned, the idea of tiny, negatively charged particles existing in fixed orbits around the nucleus of an atom is replaced by the idea of charge clouds of varying charge density existing in a wave pattern around the nucleus. The wave pattern for an electron, or a group of electrons, can be expressed in the form of a mathematical equation, which involves a wave function (ψ), the total energy (W) and the potential energy (V) of the system, the mass of the electron (m), the Planck constant (h) and the co-ordinates of the system. The equation, developed by Schrödinger, is as follows, for a single-electron system,

$$\frac{\partial^2 \psi}{\partial x^2} + \frac{\partial^2 \psi}{\partial y^2} + \frac{\partial^2 \psi}{\partial z^2} + \frac{8\pi^2 m}{h^2}(W - V)\psi = 0$$

The equation can only be solved satisfactorily when the total energy of the system has certain definite values, and these values correspond to those of the more precise stationary states in the Bohr model of the atom. The allowable wave functions corresponding to the energy values which allow the Schrödinger equations to be solved are known as characteristic functions or eigenfunctions. From these eigenfunctions, the probability of finding an electron in the region around a nucleus can be calculated, and the region in which the electron might have any probability of existing is referred to as an *orbital*.

An electron can no longer be said to occupy a specific, limited orbit.

Instead, it must be thought of as existing in a much more diffuse region known as an atomic orbital. The relation between this new idea and the older one of stationary states is that the position of greatest probability for finding the electron closely corresponds to that of the more definite orbits previously envisaged.

11 Representation of atomic orbitals

As there is a probability of finding an electron in a diffuse region round a nucleus, there is a distribution of charge round the nucleus, and atomic orbitals are represented as charge clouds having varying charge density. On this basis, the single electron in the 1s orbital of a hydrogen atom may be represented as shown in Fig. 25(a), which tries to depict the variation in charge density. A rather simpler method, however, is more commonly used, and this maps out a boundary surface within which an electron might be said to exist; the boundary surface for a 1s orbital is shown in Fig. 25(b).

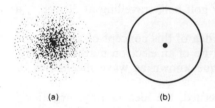

(a) (b)

Fig. 25 Two methods of representing a 1s atomic orbital. (a) Showing the variation of charge density by shading. (b) The boundary surface for a 1s atomic orbital.

This means that the electron exists somewhere within a sphere with the atomic nucleus at its centre. All electrons in s orbitals have similar spherical boundary surfaces, and s orbitals are said to be spherically symmetrical. The charge cloud is not concentrated in any particular direction.

p orbitals, however, are dumb-bell shaped and have a marked directional character depending on whether a p_x, a p_y or a p_z orbital is being considered. The boundary surfaces of p orbitals are shown in Fig. 26. These p orbitals possess a nodal plane, i.e. a plane in which the probability of finding the electron is zero.

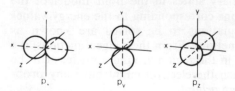

p_x p_y p_z

Fig. 26 The boundary surfaces of p_x, p_y and p_z orbitals.

The shape of the boundary surfaces of s and p orbitals, and of other orbitals obtained by combination of s and p orbitals, is of very great significance in considering chemical combination between atoms, as will be seen in the next chapter.

12 The uncertainty principle

One of the fundamental differences between an electron and a larger particle is that the wave-like aspect of the electron is of much greater significance than the wave-like aspect of the larger particle (p. 73).

There is a further difference. Both the position and the velocity of a large particle, e.g. a planet, at any one time can be measured with reasonable accuracy, but this cannot be done for anything so small as an electron. There is simply no way of measuring the velocity of an electron exactly and of locating it exactly at any one time. This is because any method of measurement affects the electron being measured. An electron might be detected, for example, by using very short wavelength X-rays, or γ-rays, if the electron would cause scattering of the rays, but the speed and direction of the electron would be affected by the rays, and electrical or mechanical methods would have the same result.

The information which can be obtained about an individual electron is, therefore, far from precise, and this is one example of the application of the uncertainty principle put forward by Heisenberg in 1927. This states that the more accurately the position of a particle is defined the less accurately is its velocity known, and the more accurately the velocity is defined the less accurately is its position known.

De Broglie's relation (p. 73) indicates why this is so. A long wavelength can be measured with greater fractional accuracy than a short one. If, therefore, a particle has a small momentum, and a correspondingly large wavelength, the wavelength can be measured with some accuracy, but this is at the expense of a relatively inaccurate determination of the small momentum. Alternatively, if the momentum is large and can be measured with some accuracy, the wavelength will be small and will not be known with any accuracy.

Because it is impossible to known the position and the velocity of an electron, the best that can be done is to try to determine the probable position and the probable velocity. Fortunately, wave mechanics enables the various probabilities to be calculated, but the simple physical picture of a single, particle-like electron moving with known velocity in a definite and precise orbit is lost.

Questions on Chapter 7

1 How would you try to convince a person with little or no scientific knowledge that light is a form of energy?
2 Describe any experiment which shows that light is a form of energy.

3 What different types of spectra are there, and how could one example of each type be obtained?

4 What was the origin of the symbols K, L, M and s, p, d, f?

5 The value of the Rydberg constant is $1.097373 \times 10^7 \, m^{-1}$. Calculate the wavelength of the first lines in each of the five spectral series of hydrogen.

6 The value of the Planck constant is $6.626 \times 10^{-34} \, J \, s$. Express it in any three other sets of units.

7 If the energy levels in a hydrogen atom corresponding with principal quantum numbers 1, 2, 3, 4, 5 and 6 are -217.9, -54.48, -24.2, -13.62, -8.7 and $-6.05 \times 10^{20} \, J$ respectively, calculate the wavelengths of lines you would expect to observe in a hydrogen spectrum. What would be the ionisation energy for hydrogen?

8 The relationship $\lambda = h/mv$ shows that an electron of mass $9.1 \times 10^{-31} \, kg$ and velocity $5.9 \times 10^5 \, m \, s^{-1}$ has a wavelength of $1.234 \, nm$. Calculate the wavelength associated with a golf ball of mass $45 \, g$ travelling at $30 \, m \, s^{-1}$.

9 What is the importance of the Pauli principle?

10 Write short notes on the following: (a) stationary states (b) Planck's constant (c) electron spin (d) J. J. Thomson.

11 What part did X-ray spectra play in the development of chemistry?

12 Explain the relationship between the position of an element in the periodic table and the atomic structure of its atoms.

13 Give an account of the experimental observations which led to the discovery of the noble gases, and describe a chemical method of preparing crude argon from air. Comment on the importance of the inert gas group in the study of the periodic classification of the elements and the understanding of atomic structures. The atomic numbers of the noble gases are He, 2; Ne, 10; Ar, 18; Kr, 36; Xe, 54. (W)

14 Explain concisely what is meant by the periodic classification of the elements, indicating, with one example of each, the significance of the terms 'group', 'short period' and 'transitional element'. How is the electronic structure of an element related to (a) its position in the periodic classification (b) its valency?

Give the electronic structures of the following atoms: argon, carbon, chlorine, sodium. How is it that iron (atomic number 26) can have two valencies? (JMB)

15 Justify the statement: 'The atomic number of an element gives more information regarding its atomic structure and is of greater usefulness in its classification than its relative atomic mass.' (JMB)

16 What do you understand by the term quantum number? What quantum numbers are needed to define the state of an electron in a polyelectronic atom, and what values can these quantum numbers adopt? In the atomic spectrum of hydrogen a line of wavelength $397 \, nm$ is observed in the Balmer series. This transition is between electronic levels with principal quantum numbers 2 and 7. Calculate the ionisation energy of the hydrogen atom. ($h = 6.626 \times 10^{-34} \, J \, s$; $c = 2.998 \times 10^8 \, m \, s^{-1}$; $1 \, nm = 10^{-9} \, m$). (CC)

17 Discuss the atomic spectrum of hydrogen and its relation to our understanding of the electronic structure of atoms. Suggest explanations for the following observations: (a) The atomic spectrum of hydrogen contains lines in the radio frequency region of the electromagnetic spectrum (b) A line in the spectrum of atomic hydrogen on a distant object in the universe occurs at a wavelength of 300 nm though it is known to occur in the laboratory at 121.6 nm (OJSE)

18 What type of information can be obtained from a study of atomic spectra? Why does the spectrum of the hydrogen atom contain so few lines compared with the spectra of all other atoms? In the spectrum of the hydrogen atom Lyman observed a transition from the $n = 1$ to the $n = 2$ level at 121.56 nm. Where would you expect the first line of the Brackett series ($n = 4$ to $n = 5$) to occur? (1 nm = 10^{-9} m). (CC)

8
Radioactive decay

1 Rate of decay

The natural decay of a radioactive element cannot be retarded or speeded up by any physical or chemical means, and the rate of a radioactive change is a characteristic of the atom of the element concerned. There are various ways in which the rate can be expressed.

a The half-life This is the time required for one half of the element concerned to decay, or the time for the radioactivity of an element to be reduced to half its initial value. The half-life period of radium, for example, is 1620 years. This means that 1 g of radium will be reduced to 0.5 g in 1620 years, to 0.25 g in a further 1620 years, and so on. The original amount makes no difference; 1 tonne would be reduced to 0.5 tonne in the same time as 1 g is reduced to 0.5 g. For other elements the half-life period may be a fraction of a second, or millions of years. It is convenient to remember that a time of 10 half-lives reduces the activity of a radioactive element about 1000 times, i.e. by about 99.9 per cent.

b The decay constant The rate of decay is directly proportional to the number of atoms present, and the rate gets slower and slower as more and more atoms break up. The rate can be expressed in a relationship similar to that used for unimolecular reactions (p. 310),

$$-\frac{dN}{dt} = \lambda N \qquad \text{or} \qquad -\frac{dN}{N} = \lambda dt$$

where N is the number of undecayed atoms, t is the time in seconds and λ is a constant. The rate of decay (or the activity) is generally expressed as the number of atoms disintegrating per second. If so, the constant, known as the *radioactive decay constant*, has units of s^{-1}.

On integration,

$$-\ln N = \lambda t + X$$

and, if N_0 is the number of atoms present at zero time, then

$$X = -\ln N_0 \qquad \ln(N/N_0) = -\lambda t \qquad 2.303 \lg(N/N_0) = -\lambda t$$

At the half-life time, N will equal $0.5 N_0$, so that

$$t_{\frac{1}{2}} = \ln 2/\lambda = 0.693/\lambda$$

where $t_{\frac{1}{2}}$ is the half-life period. A large value of λ gives a small half-life; and vice versa.

c Graphical representation The decay of a radioactive substance can be shown graphically by plotting the activity in number of disintegrations

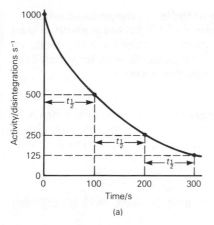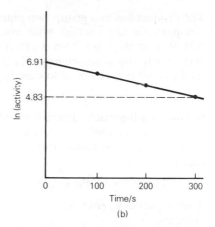

Fig. 27 Radioactive decay. **(a)** The activity falls off with time in relation to the half-life. **(b)** The plot of ln(activity) against time is a straight line, the slope of which gives the value of $-\lambda$. Compare Fig. 146 on page 311.

per second (which is proportional to N, the number of undecayed atoms) against the time. If a radioactive sample has a half-life of 100 seconds, for example, such a plot would be as shown in Fig. 27(a). Fig. 27(b) shows that the plot of ln(activity) against time is a straight line.

The slope of the line, $-(6.91 - 4.83)/300$, gives the value $(-0.006\,93)$ of $-\lambda$ in s^{-1}. This agrees with the value for λ calculated from $0.693/t_{\frac{1}{2}}$.

d The curie This unit was originally the rate of decay of 1 g of radium, but it is now defined as a rate of decay of 3.7×10^{10} disintegrations per second. The amount of a substance which undergoes 3.7×10^{10} disintegrations per second is also called a curie.

1 mol of an element with relative atomic mass, A_r, will weigh A_r g and contain 6.022×10^{23} atoms. The rate of decay per gram will be $6.022\lambda \times 10^{23}/A_r$ disintegrations per second, so that

$$\text{Number of curies per gram} = \frac{1.628\lambda \times 10^{13}}{A_r} = \frac{1.128 \times 10^{13}}{A_r \times t_{\frac{1}{2}}}$$

2 Common decay processes

Naturally-occurring radioactive elements generally decay by emission of α-particles or β-particles or γ-rays. Sometimes the processes take place together or one after the other. In artificial radioactive elements (p. 82), other processes may also be involved.

a Loss of an α-particle The α-particle is a helium nucleus so that its loss causes a decrease of 2 in the atomic number and 4 in the mass number, e.g.

$$^{238}_{92}U \longrightarrow {}^{234}_{90}Th + {}^{4}_{2}He$$

The product lies in a group two places to the left in the periodic table.

α-particles are emitted with energies (p. 97) between about 2 and 9 MeV, and all those from a given nucleus generally have the same, or very nearly the same, energy. Radioactive elements with long half-lives tend to emit α-particles of low energy; and vice versa.

b Loss of a β-particle There is no change in the mass number when a β-particle (an electron*) is emitted, but the atomic number increases by 1, with a consequent move of one group to the right in the periodic table, e.g.

$$^{234}_{90}\text{Th} \longrightarrow {}^{234}_{91}\text{Pa} + {}^{0}_{-1}\text{e}$$

The β-particles emitted by any one nucleus have widely differing energies.

The net effect is the conversion of a neutron into a proton, but it is not absolutely clear how this happens, and to account for the energy and mass changes in detail it is necessary to introduce the antineutrino (p. 100), a particle of zero charge and almost zero mass, into the process. Thus

$$\text{neutron} \longrightarrow \text{proton} + \beta\text{-particle} + \text{antineutrino}$$

c γ-rays The loss of particles from a nucleus often leaves it in an excited state. Any excess energy is generally emitted as γ-rays, which cause no change in either the mass or the atomic number.

3 Radioactive series

The final outcome of much research into the naturally-occurring radioactive elements could be summarised in three radioactive series, known as the uranium, thorium and actinium series.

The thorium series is typical, beginning with $^{232}_{90}\text{Th}$ and ending with $^{208}_{82}\text{Pb}$, which is stable. Historically, special names were used for the particular isotopes. $^{228}_{88}\text{Ra}$, for example, was called mesothorium I; $^{224}_{88}\text{Ra}$ was thorium X. Each radioactive change has its own half-life, and the changes in atomic number and/or mass number are summarised in Fig. 28.

The actinium and uranium series are similar. A fourth series, the neptunium series, includes trans-uranium elements and the end product is an isotope of bismuth.

---

*In simple work, a β-particle is, invariably, an electron, and it is symbolised by β or $_{-1}^{0}\text{e}$. But the term is, sometimes, used in more advanced work, for the positron (a particle with the same mass as an electron but opposite, equal, charge). It may be symbolised as β⁺ or $^{0}_{1}\text{e}$; if β⁺ is used, then the electron is β⁻; it can be called a negatron to distinguish it from the positron.

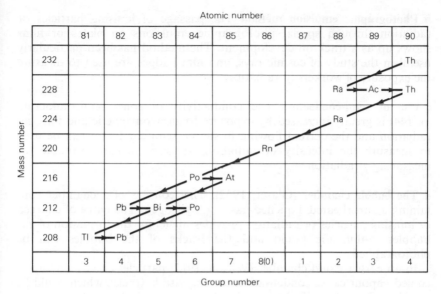

Fig. 28 The thorium decay series. The horizontal changes represent a loss of a β-particle and an increase in atomic number of 1. The diagonal changes represent a loss of an α-particle and decrease in atomic number of 2 and in mass number of 4.

4 Detection and measurement of radiation

X-rays and radiation from radioactive substances can be detected in a variety of ways, the best method in any situation depending on the nature and energy of the radiation. The methods depend, in principle, on the detection of charged particles. α- and β-rays are directly charged, but X-rays and γ-rays only produce charged particles when passed through gases or directed onto certain chemicals. Neutrons, similarly, give rise to charged particles by collision with hydrogen atoms for fast neutrons, or by collision with ^{10}B or ^{6}Li for slow neutrons.

a Scintillation counters Visual observation of the flashes of light from a zinc sulphide screen provided one of the earliest methods of detecting α-rays (p. 53) and the method is used in modern scintillation counters. Flashes of light, from a material known as a phosphor or scintillator, are allowed to fall onto a cathode with a photo-sensitive coating. The electrons released by the photoelectric effect are accelerated from one cathode to another, releasing, at each cathode, an increasing number of electrons until a pulse of current suitable for amplification is obtained. The device is known as a photo-multiplier.

By using suitable phosphors, neutrons, α-rays, β-rays and γ-rays can all be detected and measured.

b Photographic emulsion method The passage of ionising particles or radiation through special photographic emulsions on plates or films shows up as a track on development. The method has been particularly useful in the study of cosmic rays, and film badges are used to measure the exposure of workers to radiation.

c Use of semi-conductors The conductivity of some semiconductors (p. 144) is greatly increased by exposure to electromagnetic and ionising radiation and the current flowing in such a semi-conductor can be used to measure the intensity of radiation. It is particularly suitable for measuring γ-radiation.

d The bubble chamber (Glaser, 1952). This consists of a chamber containing a superheated, liquefied gas, e.g. hydrogen. The entry of charged or ionising particles or radiation provides nuclei for the formation of gas bubbles within the liquid and the tracks of the bubbles can be photographed.

In an earlier cloud chamber (Wilson, 1912) particles entering a super-cooled vapour cause condensation leaving visible tracks, which could be photographed, along the line of particles.

e Geiger counters A Geiger-Müller counter consists, essentially, of a tube containing a central anode wire and a surrounding, cylindrical cathode. The tube is filled with gases at low pressure and a potential of 1–2.5 kV is maintained between the anode and cathode. Passage of charged, or ionising, particles through the gas enables a pulse of current to flow between the two electrodes. Such a pulse can be amplified so that it can be recorded on a loudspeaker as a sharp click or on an automatic recorder which counts the number of pulses.

5 Artificial radioactivity

Elements with atomic number greater than 83 are naturally radioactive, but other naturally-occurring isotopes are not. It is possible, however, to make isotopes which are radioactive by nuclear reactions (p. 96). The radioactivity is referred to as artificial or induced. Very many radioactive isotopes of many different elements have been made, and many of them are very useful (p. 83).

The first radioactive isotopes to be produced were obtained by M and Mme Curie-Joliot in 1933. They bombarded aluminium, beryllium and magnesium with α-particles and obtained radioactive isotopes of phosphorus, nitrogen and silicon respectively, e.g.

$$^{27}_{13}Al + {}^{4}_{2}He \longrightarrow {}^{30}_{15}P + {}^{1}_{0}n$$

The $^{30}_{15}P$ isotope decayed by the loss of a positron, a particle with the same mass as an electron and a positive charge of the same magnitude as that on an electron (p. 100), i.e.

$$^{30}_{15}P \longrightarrow {}^{30}_{14}Si + {}^{0}_{1}e$$

Other nuclear reactions giving radioactive isotopes of sodium, carbon and sulphur respectively are represented by the following equations:

$$^{23}_{11}Na + {}^{1}_{0}n \longrightarrow {}^{24}_{11}Na$$

$$^{10}_{5}B + {}^{2}_{1}H \longrightarrow {}^{11}_{6}C + {}^{1}_{0}n$$

$$^{35}_{17}Cl + {}^{1}_{0}n \longrightarrow {}^{35}_{16}S + {}^{1}_{1}H$$

6 Use of radioactive isotopes

Naturally occurring radium has been used for some time in radiotherapy, particularly in the treatment of cancer, and the γ-rays from radium or radon have also been used in taking radiographs. The present commercial availability of artificial radioactive isotopes has, however, greatly increased the research and industrial applications.

a Tracer techniques An object with a very small amount of a radioactive isotope attached to it, or incorporated in it, can readily be detected by picking up the radiation from it. A joint in a buried pipe or cable can be marked, for example, in this way, or the efficiency of a mixing process can be followed by adding a radioactive isotope to one of the ingredients before mixing and observing the level of radiation throughout the mixture.

The wear inside an engine can be measured by making the various moving parts radioactive and measuring the resulting radioactivity in the circulating oil. The flow of a material, e.g. molten glass through a furnace, fertiliser through a drier, gases in ventilation systems, or leaking gases or liquids, can be investigated by tracer techniques.

There are, too, many biological and medical uses. The uptake of phosphorus by a plant from a phosphate fertiliser can be traced by using a fertiliser containing ^{32}P, and radioactive tracer studies, using ^{14}C, have been very valuable in elucidating the nature of photosynthesis. Introduction of radioactive ^{59}Fe into the blood stream has enabled the part played by iron in blood formation and use to be studied, and ^{131}I has been used both in the diagnosis of thyroid diseases and in research on the functioning of the thyroid gland and the kidneys.

b Direct use of radiation The β- and γ-rays emitted by radioactive isotopes can be used to measure or control thicknesses of materials. The amount of radiation passing through a material will decrease as the material gets thicker. If a source of β- or γ-rays is placed on one side of the material and a detector on the other side, the scale reading on the detector will give a measure of the thickness.

Typical uses are in controlling the thickness of paper, plastic sheeting and sheet metal, in checking the packing of tobacco in cigarettes, in measuring or monitoring the thickness of coating of one metal on another, and in checking the level to which a container is filled with a liquid or solid.

γ-rays are also used in radiography, instead of X-rays. A radioactive isotope which gives γ-rays is much cheaper, much smaller, and more portable and manoeuvrable than X-ray equipment, but γ-ray photography is much slower than that using X-rays.

c **Radiodating** Radioactive measurements can be used to get good estimates of the age of ancient materials. Carbon-dating is the best known technique. The radioactive isotope, ^{14}C, with a half-life of 5720 years, is produced in the upper atmosphere when neutrons released by cosmic rays bombard nitrogen atoms.

$$^{14}_{7}N + ^{1}_{0}n \longrightarrow ^{14}_{6}C + ^{1}_{1}H$$

The resulting $^{14}CO_2$ is absorbed by plants so that ^{14}C atoms get into plants, and then into animals. Throughout the universe there is a constant exchange between ^{14}C atoms and non-radioactive ^{12}C atoms. This has gone on for so long that the ratio of ^{14}C to ^{12}C in plants, animals and the atmosphere will be constant. All living matter, then, will contain the same proportion of ^{14}C as occurs in the atmosphere. The resulting radioactivity of living matter is 15.3 disintegrations per minute per gram, and this is found in living samples.

When a plant or animal dies, however, the natural exchange between ^{14}C and ^{12}C stops and the existing ^{14}C atoms decay. 5720 years after death the activity would be halved. By measuring the actual activity of the ^{14}C content of old, dead samples it is, therefore, possible to date the sample. The method has been used for bones, coffins, ancient manuscripts such as the Dead Sea scrolls, etc. Within the range of about 600 to 10 000 years, ages can be estimated with an accuracy of 1–200 years.

The possible age of rocks between 50 and 4 000 million years old can also be estimated by measuring the relative amounts of two isotopes. A rock containing ^{238}U, for example, will also contain ^{206}Pb (p. 80). The older the rock is, the more ^{206}Pb and the less ^{238}U there will be. The ratio between the two can, therefore, be used to estimate the age of the rock.

(Questions on this chapter will be found on p. 92.)

9
Isotopes and mass spectrometry

1 Positive rays

Investigation of radioactive disintegration series shows that more than one element can occupy the same position in the periodic table. Such elements must have the same atomic numbers, but they are not identical for they have different relative atomic masses. Soddy called such elements isotopes. They have the same arrangement of electrons and the same number of protons, but different numbers of neutrons, in the nucleus.

The existence of isotopes of non-radioactive elements was shown by positive ray analysis. Positive rays, first discovered by Goldstein in 1886, are formed in a discharge tube as well as cathode rays, and may be observed by using a tube with a perforated cathode. If this is done, the radiation passing through the holes in the cathode (Fig. 29) is found, by the effect of electric and magnetic fields, to consist of positively charged particles much heavier than the electrons of cathode rays.

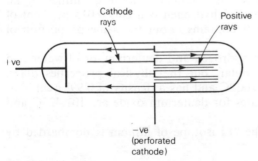

Fig. 29 Positive rays.

Measurement of the charge/mass ratio for positive rays obtained in tubes containing different gases indicates that the rays are made up of positively charged atoms or molecules formed by loss of electrons from the atoms or molecules of the gas in the tube.

Application of magnetic and electric fields enables the various particles in the positive rays to be 'sorted out' according to their mass. Thomson first investigated positive rays from a tube containing neon, of relative atomic mass 20.183, and he found that neon was, in fact, made up of two isotopes, one of mass 20 and one of mass 22. Later work has shown that there is, also, a third isotope of mass 21.

The 'sorting out' of isotopes is, nowadays, done in a mass spectrometer (p. 88).

2 Examples of isotopes

Almost all elements have now been found to have stable, naturally-occurring isotopes, and many have a wide variety. Carbon consists of 98.9 per cent of $^{12}_{6}C$ and 1.1 per cent of $^{13}_{6}C$; chlorine is 75.77 per cent of $^{35}_{17}Cl$ and 24.23 per cent of $^{37}_{17}Cl$; tin has ten isotopes. Fluorine, sodium, aluminium and phosphorus are some of the few elements that have only one isotope.

a Isotopes of hydrogen Hydrogen has two main isotopes, hydrogen or protium ($^{1}_{1}H$) and heavy hydrogen or deuterium ($^{2}_{1}H$ or D). Tritium ($^{3}_{1}H$ or T) can also be made, but it is doubtful whether it occurs naturally in ordinary hydrogen.

The existence of deuterium was discovered by Washburn and Urey, in 1931, by a careful examination of the line spectrum (p. 58) of the hydrogen remaining after several litres of liquid hydrogen had been evaporated.

This was followed, in 1932, by the discovery of deuterium oxide, or heavy water, D_2O or ^2H_2O, also by Washburn and Urey. They found that the residue remaining, after water had been almost completely decomposed by electrolysis, was rich in deuterium oxide, and almost pure deuterium oxide can now be obtained in large quantities by an extension of this method. Normal hydrogen contains 0.015 per cent of deuterium, and ordinary water contains about the same proportion of heavy water.

The chemical properties of hydrogen and deuterium, or of water and deuterium oxide, are very similar, but the physical properties differ. Water, for example, boils at 100 °C and has a density of $0.998 \, \text{g cm}^{-3}$ at 20 °C; the corresponding figures for deuterium oxide are 101.42 °C and $1.106 \, \text{g cm}^{-3}$.

Tritium is formed when the ^6Li isotope of lithium is bombarded by neutrons (p. 96).

$$^{6}_{3}Li + ^{1}_{0}n \longrightarrow \, ^{4}_{2}He + ^{3}_{1}H$$

b Isotopes of uranium These isotopes are of particular significance in atomic energy work (p. 101). Natural uranium consists almost entirely of ^{238}U, with 0.72 per cent (1 part in 140) of ^{235}U and 0.0056 per cent of ^{234}U.

c Isotopes of oxygen Oxygen has isotopes ^{16}O (99.8 per cent), ^{17}O (0.037 per cent) and ^{18}O (0.20 per cent).

In conjunction with the three possible isotopes of hydrogen it is evident that there are nine possible sorts of water formulated as $^1H_2{}^{16}O$, $^1H_2{}^{17}O$, $^1H_2{}^{18}O$, $^2H_2{}^{16}O$, $^2H_2{}^{17}O$, $^2H_2{}^{18}O$, $^3H_2{}^{16}O$, $^3H_2{}^{17}O$, and $^3H_2{}^{18}O$.

3 Separation of isotopes

As isotopes have the same chemical properties, their separation must depend on physical methods making use of the slight mass difference between different isotopes.

a Gaseous diffusion The rate of diffusion of a gas is inversely proportional to its density (Graham's law, p. 6). Thus, when a gas containing a mixture of two isotopes is allowed to diffuse through a porous partition, the lighter isotope passes through more rapidly than the heavier one. A single stage diffusion through one partition gives only a very partial separation, and, in practice, several thousand stages, in a cascade arrangement, may be necessary to bring about a good separation. The lighter gas mixture passing through a porous partition is taken on to the next stage, whilst the heavier gas mixture which has not diffused is returned to an earlier stage.

This method was adopted for separating ^{235}U and ^{238}U, using a mixture of $^{235}UF_6$ and $^{238}UF_6$, which are gases. The hexafluoride is one of the few gaseous compounds of uranium, and has the advantage that fluorine has only one isotope. It is, however, very corrosive.

b Thermal diffusion A long, vertical, cylindrical tube, with an electrically heated wire running down its axis, is used in this method. If a gaseous mixture of isotopes is introduced into the tube, the lighter isotopes diffuse more rapidly towards the central, hot region where they are carried upwards by convection currents. The heavier isotopes, near the cold, outer surface of the tube, are carried downwards. Thus the lighter isotope collects at the top of the tube, and the heavier one at the bottom.

c Centrifugal method A mixture of isotopes is rotated in a high speed centrifuge, the heavier isotope being drawn off from the periphery and the lighter one from the centre. By using a cascade arrangement of a lot of centrifuges, it is possible to achieve good results.

A commercial plant was opened at Capenhurst in 1977. It is used for making enriched uranium containing about five times more ^{235}U than natural uranium.

d Use of exchange reactions Compounds containing different isotopes of the same element may not have the same reactivity so that the equilibrium position for the reaction, for example,

$$^{15}NH_3(g) + {}^{14}NH_4{}^+(aq) \rightleftharpoons {}^{14}NH_3(g) + {}^{15}NH_4{}^+(aq)$$

lies slightly to the right. Ordinary nitrogen contains only 0.37 per cent of ^{15}N, but if this is converted into ammonia which is passed up a tower down which ammonium nitrate solution is dripping, the ^{15}N becomes concentrated in the ammonium nitrate solution from which it can be recovered.

e Electromagnetic separation This method adopts the principle of the mass spectrometer (Fig. 30). Groups of different isotopes are deflected differently and can be collected in containers placed in appropriate positions. The method is expensive if used for large quantities of material.

4 The mass spectrometer

Positive-ray analysis and the identification of isotopes was extended by Aston using an apparatus known as a mass spectrograph. Particles of differing mass showed up as lines on a photographic plate. As it looked not unlike a line spectrum it became known as a *mass spectrum*. But Aston's equipment had certain limitations and it is the development of the mass spectrometer, which records the particles of different mass electrically, that has been of greatest importance.

The instrument (Fig. 30) is operated at very low pressure. A stream of vapour of the substance under investigation is passed, through a pin-hole leak, from a reservoir into the main apparatus. The vapour then passes through a beam of electrons from a hot filament. Collisions with these electrons knock electrons out of the atoms or molecules in the vapour and form a beam of positively charged ions.

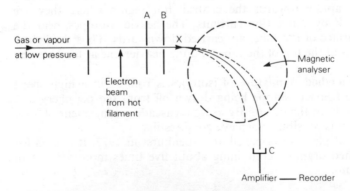

Fig. 30 The principle of a simple mass spectrometer. In a more accurate, double-focusing instrument, an electrostatic analyser is used before the magnetic analyser.

The beam is now passed through an electrical field, between plates A and B, and this produces a beam (X) of particles all having about the same kinetic energy but with different charge/mass ratios. On passing the beam through a magnetic field, particles with the same charge/mass ratio are focused on to a collector, C. The charge that builds up on C, which can be measured via an amplifier and a recorder, measures the number of particles with that particular charge/mass ratio. Changing the magnetic field, or the electrical field, between A and B brings another group of particles with a different charge/mass ratio into focus on C, and the abundance of that group can be recorded. The record of mass against abundance is known as the mass spectrum.

The accuracy of the mass spectrometer depends on its *resolving power*. An instrument with a resolving power of 1 in 200, for example, can distinguish between particles of mass 200 and 201. Instruments with resolving powers of 1 in 10^5 can be made. This high accuracy is achieved by what is known as double-focusing. An electrostatic field (the electrostatic analyser) is introduced (Fig. 30) before the magnetic field (the magnetic analyser) and this can be used to ensure that all the particles entering the magnetic field have precisely the same kinetic energy. The magnetic field then 'sorts them out' according to their charge/mass ratio.

5 Measurement of isotopic mass and A_r values

The A_r values for elements can be measured very accurately using a mass spectrometer. Values relative to ^{12}C are used, this being chosen as the basis for definition (p. 15) because it is easy to obtain ions containing many carbon atoms in a mass spectrometer.

For elements with only one isotope, the measurement is straightforward. If isotopes occur it is necessary to find the masses of the individual isotopes and their abundance; the A_r value is, then, the weighted mean. Chlorine, for example, consists of ^{35}Cl and ^{37}Cl isotopes as follows:

isotope	mass number	isotopic mass	abundance (per cent)
^{35}Cl	35	34.9694	75.77
^{37}Cl	37	36.9660	24.23

The A_r for chlorine is, therefore,

$$\frac{(34.9694 \times 75.77) + (36.9660 \times 24.23)}{100} = 35.453$$

The mass number of an isotope is the number of protons plus the number of neutrons in the nucleus.

6 Measurement of M_r by mass spectrometer

The first use of the mass spectrometer was in measuring relative atomic masses, but it is now widely used for measuring relative molecular masses, particularly for carbon compounds. The substance concerned, M, is introduced into the spectrometer. If the bombarding electrons are of low energy (8–14 eV) the molecules of M will only lose one electron to form a positively charged, molecular ion, M^+, i.e.

$$M + e^- \longrightarrow M^+ + 2e^-$$

M^+ will then show up as a single peak on the recorder and its mass (and, hence, the mass of M) can be found. Very accurate results can be obtained from small amounts of material.

If M contains atoms which have heavier isotopes there will be small peaks at masses above that of M. They are known as the $(M+1)$, $(M+2)\ldots$ peaks. Methane, for example, would give a main peak, corresponding to CH_4^+, at a mass of 16.0312. But some molecules would contain ^{13}C and ^2H and these would give higher peaks.

If bombarding electrons of higher energy (50–70 eV) are used, they will split any covalent bonds in M so that a wide variety of ions will be formed giving many peaks in the mass spectrum (see 7b below).

7 Other uses of the mass spectrometer

a Identification of compounds The M_r of a compound can be measured so accurately that it may be possible to identify it by comparing its measured M_r value with tabulated lists of relative molecular masses of a wide range of possible compounds. A compound with M_r of 134.0368 known to contain only ^{12}C, ^1H and ^{16}O, for example, would be $C_8H_6O_2$.

A good indication of the number of C atoms in a molecule can, also, be obtained by comparing the intensity of the $(M+1)$ peak with that of the M peak. The percentage abundance of ^{13}C is 1.108, so that, for a compound containing x atoms of carbon, there will be a $1.1x$ per cent chance that it will contain one atom of ^{13}C. The intensity of the $(M+1)$ peak to that of the M peak would, therefore, be $1.1x$ to 100.

The percentage abundances of ^2H and ^{17}O are only 0.015 and 0.037 per cent respectively so that their contribution to an $(M+1)$ peak is small. For compounds with one atom of bromine, however, the $(M+2)$ and the M peaks are of almost equal intensity because the ^{79}Br and ^{81}Br isotopes occur almost equally.

b Fragmentation patterns If methane is bombarded by high energy electrons all its bonds will be split so that CH_4^+, CH_3^+, CH_2^+, CH^+, C^+ and H^+ ions will all be formed and will be recorded by the mass spectrometer, together with peaks for the corresponding isotope-containing ions. The recorded plot is known as a fragmentation pattern or a stick diagram.

The fragmentation pattern for ethanol is shown in Fig. 31. The abundance of each ionic species is represented by a vertical line as a

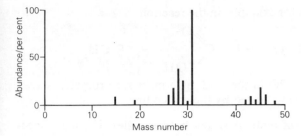

Fig. 31 The mass spectrum of ethanol showing the fragmentation pattern.

percentage of the most abundant ion, which, for ethanol, is CH_2OH^+. This is formed by cleavage of the C—C bond in the original molecular ion, $C_2H_5OH^+$

$$C_2H_5OH + e^- \longrightarrow C_2H_5OH^+ + 2e^-$$
$$C_2H_5OH^+ \longrightarrow \cdot CH_3 + CH_2OH^+$$

The other main ions present are summarised below:

mass number	15	26	27	28	29	31	45	46
ion	CH_3^+	$C_2H_2^+$	$C_2H_3^+$	$C_2H_4^+$	$C_2H_5^+$	CH_2OH^+	$C_2H_5O^+$	$C_2H_5OH^+$

This particular fragmentation pattern is unique for ethanol, and every other substance has its own pattern.

A study of the fragmentation pattern of an unknown compound can, therefore, throw considerable light on its likely structure, particularly as some bonds in molecules are split more easily than others. The possible weak and strong bonds in an unknown structure might be identified by comparing the fragmentation pattern from bombardment by low energy electrons with that using high energy electrons. Those fragments whose formation requires the splitting of strong bonds, will not show up under low energy electron bombardment.

c Analysis of mixtures As each chemical has its own unique fragmentation pattern, a mixture of chemicals gives a composite pattern based on those of its components. A mixture of chemicals can, therefore, be analysed by studying the composite pattern. Small quantities of mixtures containing many components can be analysed, particularly if a computer is available to interpret the overall fragmentation pattern in terms of the individual ones.

In another application, gas chromatography is used to separate the components of a mixture and each component is then identified by a mass spectrometer.

d The detection of labelled atoms Radioactive atoms are commonly used as tracers (p. 83), but non-radioactive isotopes can be used in the same way. The use of $H_2{}^{18}O$, for example, in the reaction

$$R-C{\overset{\displaystyle /\!/O}{\underset{\displaystyle \diagdown OR'}{}}} + H_2{}^{18}O \longrightarrow R-C{\overset{\displaystyle /\!/O}{\underset{\displaystyle \diagdown {}^{18}OH}{}}} + R'OH$$

and the detection of the ^{18}O in the acid, by a mass spectrometer, shows that the bond fission is as indicated by the dotted line.

e The ionisation energy of molecules, and bond energies The mass spectrum of a compound changes as the energy of the bombarding electrons is increased. The minimum energies required to form M^+, M^{2+}, etc. can be measured and they represent the first, second, etc. ionisation energies of M.

The breaking of a bond in M will also show up as new peaks in the mass spectrum and the energy required to bring this about will be the bond energy of the bond concerned.

Questions on Chapters 8 and 9

1 If the half-life of radium is 1620 years, what is its radioactive decay constant? What will be the rate of disintegration of 1 g of radium in terms of atoms disintegrating per second? How many curies is this? How many grams of radium represent 1 curie?

2 If the decay constant for radium is $1.356 \times 10^{-11}\,s^{-1}$, calculate the time required for (a) 10 per cent (b) 90 per cent of a sample of radium to disintegrate.

3 If 1 curie of radium of isotopic mass 226 and a half-life 1620 years is 1.02 g, how many grams of cobalt-60, of half-life 5.2 years, will represent 1 curie? If 1 curie of potassium-40 is 146 kg, what is its half-life?

4 Radon has a half-life of 3.8 days. Plot a graph of the percentage of a sample of radon which has decayed against the time in days.

5 If a sample of a radioactive isotope of half-life 3.11 hours has an activity of $1000\,s^{-1}$ at a certain time, what will be its activity one hour later?

6 What is the percentage loss in the activity of a radioactive element after 1, 10 and 100 half-life periods?

7 The rates of decay for two isotopes, X and Y, are given below in disintegrations per second. Use a graphical method to find the half-lives and the decay constants of X and Y.

time/min	0	10	20	30	50
X/s^{-1}	200	161	135	110	76
Y/s^{-1}	11.3	8.3	6.0	4.4	2.35

8 ^{238}U emits an α-particle to form UX_1. This, in turn, emits a β-particle to give UX_2. What are the atomic and mass numbers of UX_1 and UX_2?

9 $^{239}_{93}Np$ emits a β-particle to give an isotope which then decays to ^{235}U. What particle is emitted in the final decay?

10 ^{238}U decomposes by emitting A and forming ^{234}Th. This isotope undergoes decomposition with β-emission forming B. What are A and B?

11 The thorium, uranium and actinium disintegration series are sometimes known as the $4n$, $4n+2$, and $4n+3$ series respectively. Why is this? What isotopes might exist in the $4n+1$ series?

12 Write symbols showing the mass number and the atomic number for an electron, a proton, a neutron, a positron, a deuteron and an α-particle.

13 Write symbols showing the mass number and the atomic number of the atoms formed when   (a) $^{226}_{88}$Ra loses an α-particle   (b) $^{210}_{83}$Bi loses a β-particle   (c) $^{234}_{92}$U loses five α-particles   (d) $^{231}_{91}$Pa loses an α- and a β-particle   (e) $^{226}_{88}$Ra loses two β-particles.

14 Explain why it is that the charge/mass ratio for cathode rays is independent of the nature of the gas in the discharge tube, whilst the charge/mass ratio for positive rays is smaller than that for cathode rays and depends on the nature of the gas in the tube.

15 If natural oxygen consists of 99.758 per cent of an isotope of mass 16, 0.0373 per cent of an isotope of mass 17.004 534 and 0.2039 per cent of an isotope of mass 18.004 855, calculate the average mass of an atom in natural oxygen.

16   Compare the functioning of a mass spectrometer with that of an achromatic prism.

17 You have discovered a new element. How would you determine its relative atomic mass and investigate its radiochemical properties? Could you distinguish different isotopes in it? Suggest experiments by which you might determine its atomic number. (OJSE)

18 Write an essay on the use of isotopes in chemistry. (OJSE)

19 Discuss the differences between radioactive decay and chemical reactions. When meteorites travel through space they are bombarded by cosmic radiation which generates in them small amounts of radioactive nuclei. These generative processes cease when a meteorite strikes the earth. A recent stony meteorite was found to have a tritium, ^{3}H, to argon, ^{39}Ar, β-ray activity ratio of 40:1, compared with a ratio of 1:1 for an older meteorite of similar composition. Calculate the number of years that have elapsed since the older meteorite fell on the surface of the earth. State what assumptions you have made in working out your answer. (Decay constant of ^{3}H $= 5.8 \times 10^{-2}$ years^{-1}. Decay constant of ^{39}Ar $= 2.1 \times 10^{-3}$ years^{-1}.) (OLS)

20 Give one example in each case of nuclear reactions brought about by bombardment with   (a) protons   (b) alpha-particles. Write equations for these reactions which show the identity, mass number and atomic number of each nucleus involved. Explain concisely the principles involved in the technique of radiocarbon dating. Indicate in outline how a mass spectrometer can be used   (c) to show that bromine has two (stable) isotopes   (d) to obtain an accurate value for the relative atomic mass of this element. (OL)

21 State what you understand by each of the following terms: atomic number, isotope, atomic mass, mole. Draw a diagram of a mass spectrometer, labelling the parts. Explain how the instrument may be used to determine (a) molecular mass (b) isotopic ratio. What limitations are there in the determination of molecular mass by mass spectrometry? (W)

22 Some water was analysed in a mass spectrometer and peaks occurred corresponding to relative masses of 1, 2, 3, 4, 17, 18, 19 and 20. Give the formula of one ion in each case which might be responsible for these peaks.

23 What is meant by a fragmentation pattern? The pattern for ethanol shows peaks at 15, 26, 27, 28, 29, 31, 45 and 46. Suggest how these peaks are formed. What peaks would you expect for methanol?

24 Sketch the fragmentation patterns you would expect to get from high energy electron bombardment of (a) air (b) sulphur and (c) hydrogen chloride.

25* Potassium contains three isotopes, one of which is radioactive. If there is 140 g of potassium in the human body, what level of radioactivity, in disintegrations per second, will there be in the body?

26* What will be the ratio of the radioactivity emanating from equal masses of ^{14}C, ^{24}Na and ^{90}Sr?

27* What will be the intensity of the radioactivity, in curies, of a 100 g sample of calcium carbonate containing 0.001 per cent of $^{90}SrCO_3$?

28* Record the isotopic masses and the relative abundances of the isotopes of (a) bromine (b) neon and (c) sulphur. Use the value to calculate the relative atomic masses of the naturally occurring elements.

29* Examine the positions of (a) $_{52}Te$ and $_{53}I$ (b) $_{18}Ar$ and $_{19}K$ in the periodic table in relation to the relative atomic masses of the naturally occurring elements and the relative abundance of their various isotopes.

30* Three different compounds with molecular formula $C_xH_yO_z$ give peaks in a mass spectrometer at 60.0574, 60.0210 and 30.0105 respectively. What values might x, y and z have in the three compounds?

31* The fragmentation pattern of CH_3Cl shows peaks at 50 and 52, and that of CH_3Br peaks at 94 and 96. How do these arise? What would you expect the ratio of the intensities of (a) the peaks at 50 and 52 (b) the peaks at 94 and 96 to be? What peaks would you expect to find for CH_3F and CH_3I?

10
The atomic nucleus

The concept of a nuclear atom arose initially from Rutherford's attempts to account for the scattering of α-particles by atoms (p. 55). The nucleus, now known to contain protons and neutrons (nucleons), is very small yet it contains most of the mass of an atom. Consideration of the extent of the scattering by nucleus gives estimates of the diameter of the nuclei of an atom of $1.4 \times 10^{-15} A_r^{0.33}$ m, which corresponds with a density of nuclear material of about 10^{17} kg m^{-3}. A drop of nuclear material just large enough to see would weigh about 10^{10} kg.

Such high densities are found nowhere else, and they show how unusual the forces within a nucleus must be. Splitting atoms involves splitting nuclei and the breaking down of such forces.

1 Splitting the atom

In naturally occurring radioactive elements the nuclei of the atoms concerned are constantly splitting, and the process cannot be stopped. The nuclei of other atoms are more stable, but they can often be split by bombardment with particles such as α-particles, protons or neutrons.

The first artificial disintegration of a nucleus was observed by Rutherford, in 1919, when he passed α-particles through nitrogen gas and discovered that protons were ejected. In 1925 Blackett demonstrated that this ejection of a proton was accompanied by the formation of ^{17}O. The nuclear reaction was represented as,

$$^{14}_{7}\text{N} + {}^{4}_{2}\text{He} \longrightarrow {}^{1}_{1}\text{H} + {}^{17}_{8}\text{O}$$

At this time, α-particles were the most convenient particles available for use as projectiles in bombarding nuclei. β-particles (electrons) could have been used, equally simply, but they are far too light to be effective. It would be like shooting peas at elephants. The α-particle is heavy, in comparison, but it is not a perfect bombarding particle because of its positive charge. The nucleus being bombarded is positively charged so that another positively charged particle must have sufficient energy to overcome the electrostatic repulsion if it is to 'score a hit'.

The developments in nuclear disintegration have centred round the introduction of new bombarding particles and of new methods of increasing their energy. In 1932, Cockcroft and Walton used an electrostatic accelerator to provide high-speed protons. By this means they converted atoms of lithium into atoms of helium,

$$^{1}_{1}\text{H} + {}^{7}_{3}\text{Li} \longrightarrow 2\,{}^{4}_{2}\text{He}$$

Deuterons (deuterium nuclei) were also used; they have twice the mass of a proton but still suffer from being positively charged. Electrostatic accelerators were followed by cyclotrons, synchrotons and cosmotrons, all designed to provide charged particles with greater and greater energy.

Charged particles have their limitations, however, no matter what their energy may be, and the discovery of the neutron (p. 39) provided one of the best bombarding particles, for it is quite heavy and has no electrical charge. It can, therefore, approach a nucleus without being repelled.

2 Types of nuclear reaction

It is conventional to represent a nuclear reaction by writing the target element, the bombarding particle, the emitted particle or radiation, and the new isotope formed. Thus the reaction,

$$\ce{^{35}_{17}Cl} + \ce{^{1}_{0}n} \longrightarrow \ce{^{35}_{16}S} + \ce{^{1}_{1}H}$$

is written as $\ce{^{35}_{17}Cl}$ (n, p) $\ce{^{35}_{16}S}$, where n and p stand for neutron and proton respectively.

On this basis, some possible nuclear reactions, with examples, may be summarised as follows.

reaction type	example
n, p	$\ce{^{14}_{7}N} + \ce{^{1}_{0}n} \longrightarrow \ce{^{14}_{6}C} + \ce{^{1}_{1}H}$
n, α	$\ce{^{6}_{3}Li} + \ce{^{1}_{0}n} \longrightarrow \ce{^{3}_{1}H} + \ce{^{4}_{2}He}$
n capture	$\ce{^{107}_{47}Ag} + \ce{^{1}_{0}n} \longrightarrow \ce{^{108}_{47}Ag}$
p, n	$\ce{^{63}_{29}Cu} + \ce{^{1}_{1}H} \longrightarrow \ce{^{63}_{30}Zn} + \ce{^{1}_{0}n}$
p, α	$\ce{^{19}_{9}F} + \ce{^{1}_{1}H} \longrightarrow \ce{^{16}_{8}O} + \ce{^{4}_{2}He}$
d, n	$\ce{^{56}_{26}Fe} + \ce{^{2}_{1}H} \longrightarrow \ce{^{57}_{27}Co} + \ce{^{1}_{0}n}$

In any balanced equation for a nuclear reaction the sum of the superscripts (the atomic masses), and the sum of the subscripts (the atomic numbers), on both sides of the equation, must balance.

3 Energy changes in nuclear reactions

Einstein predicted, as part of his theory of relativity, that matter and energy could be interconverted according to the equation

$$E = mc^2$$

where E is the energy in joules, m the mass in kilograms and c the velocity of light in $\mathrm{m\,s^{-1}}$.

The relationship between mass and energy can therefore be summarised as follows:

$$1 \, \text{kg of mass} = 8.988 \times 10^{16} \, \text{J} = 5.610 \times 10^{29} \, \text{MeV}$$
$$1 \, \text{a.m.u. (p. 15)} = 1.493 \times 10^{-10} \, \text{J} = 931.7 \, \text{MeV}$$

where one electron volt (1 eV), which is the energy acquired or lost by a particle of unit electronic charge on passing through a potential difference of 1 volt, is equal to 1.602×10^{-19} joule. Very small amounts of mass would give very large amounts of energy if the change could be made effectively.

The energy liberated in any ordinary chemical reaction requires a conversion of only about 100×10^{-12} kg of matter. Such a small change of mass cannot be detected on any chemical balance, but the Einstein equation can be verified experimentally, by measuring the mass and energy changes in nuclear reactions.

In the bombardment of lithium by high-speed protons (p. 95) the range of the α-particles produced is such that their energy must be 8.6 MeV, and, as two α-particles are formed,

$$^1_1\text{H} + {}^7_3\text{Li} \longrightarrow 2\,{}^4_2\text{He}$$

the energy liberated in the reaction must be 17.2 MeV. This is equivalent to 0.0185 a.m.u.

The atomic masses participating in the reaction are

^1_1H	+	^7_3Li	$\longrightarrow$	$2\,^4_2\text{He}$
1.007 825		7.016 004		$2 \times 4.002\,603$

giving a decrease in mass of 0.0186 a.m.u.

It will be seen that the decrease in mass expected from the amount of energy liberated is in good agreement with the actual decrease in mass. The agreement between decrease in mass and energy released in nuclear reactions is, in fact, so close that measurement of energy release can be used for finding mass values.

4 Nuclear stability and binding energy

An atomic nucleus is made up of a number of protons equal to its atomic number, plus a number of neutrons which make up the mass number.

Careful measurements, by mass spectrometry (p. 89), show, however, that relative atomic masses are not exactly whole numbers, except for ^{12}C. For example,

^1H	1.007 825	^4He	4.002 603	^{12}C	12.000 000
^2H	2.014 102	^7Li	7.016 004	^{56}Fe	55.934 937

The nearest whole number is the mass number of the isotope; the actual mass is the relative isotopic mass multiplied by the atomic mass unit (p. 15).

For any nucleus, this actual mass is slightly less than the sum of the masses of the protons and neutrons in it; the difference is known as the *mass defect* or *deficit*. It arises because some of the mass of the protons and neutrons is converted into energy. This is the *binding energy* that holds the nucleus together; the greater it is, the more stable the nucleus.

For a nucleus containing Z protons, with a mass number of A, the number of neutrons will be $(A - Z)$. If m_p is the mass of the proton, m_n the mass of the neutron, and m the mass of the nucleus, then

$$\text{mass defect} = Zm_p + (A - Z)m_n - m$$

and the corresponding binding energy, in MeV, will be 931.7 times the mass defect, in a.m.u., i.e.

$$\text{binding energy} = 931.7\{Zm_p + (A - Z)m_n - m\}$$

For deuterium, with a simple nucleus containing 1 proton and 1 neutron, the actual mass of the nucleus will be the isotopic mass less the mass of 1 electron, i.e.

$$\begin{aligned}\text{mass of D nucleus} &= 2.014\,102 - (5.485\,802 \times 10^{-4})\,\text{a.m.u.} \\ &= 2.013\,553\,\text{a.m.u.}\end{aligned}$$

The binding energy will be

$$931.7(1.007\,276 + 1.008\,665 - 2.013\,553) = 2.225\,\text{MeV}$$

5 Nuclear structure

The binding energy per nucleon (the *binding fraction*) for different isotopes depends on the mass number (Fig. 32), with maximum values for elements of mass number around 60 (Cr to Zn). They are, therefore, the most stable elements. If heavier, less stable elements could be converted, by fission, into these more stable elements there would be a release of energy. Similarly, a fusion of lighter elements into more stable ones would release energy.

Light isotopes show irregularities with successive maxima for ^{4}He, ^{8}Be, ^{12}C, ^{16}O, ^{20}Ne and ^{24}Mg. These isotopes all have nuclei containing an equal number of protons and neutrons, and a pairing of two protons and two neutrons (as in the He nucleus or α-particle) causes an increase in binding energy. Such a pairing can be regarded as a pre-formed α-particle within the nucleus, which may explain why such particles are produced in radioactive decay.

As a pairing of protons and neutrons, in twos, causes increased binding energy, it might be expected that all stable nuclei would contain even, and equal, numbers of protons and neutrons. But this is not so. The great majority of stable nuclei do have even numbers of protons and neutrons; indeed, only four stable nuclei, ^{2}H, ^{6}Li, ^{10}B and ^{14}N, have odd

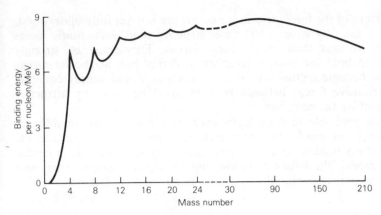

Fig. 32 Binding energy per nucleon plotted against mass number. (Note the change of scale.)

numbers of both protons and neutrons. For light isotopes, too, the numbers of protons and neutrons are approximately equal but, as the isotopic mass increases, the number of neutrons becomes gradually larger than the number of protons (Fig. 33). To counterbalance the repulsive forces between an increasing number of protons requires more and more neutrons. It is, however, very significant that the proton-neutron combinations that exist in stable nuclei are very limited. Nuclei outside the narrow band in Fig. 33 are radioactive. When they decay the numbers of protons and/or neutrons in the nuclei are changed, generally by loss of α- or β-particles or positrons (p. 100); more stable nuclei are formed.

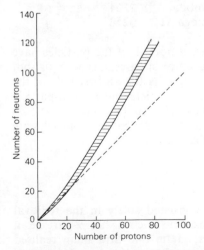

Fig. 33 Variation of number of neutrons with number of protons in stable, non-radioactive nuclei. All the known nuclei lie within the shaded area.

The nature of the forces within a nucleus are not yet fully understood. The forces are very strong; 100 times stronger than electrostatic forces and 10^{38} stronger than gravitational forces. They must be strongly attractive to hold the nucleus together so firmly but, at shorter range, they must become repulsive or else the nucleus would simply collapse. Strong repulsive forces between protons must be offset by attractive forces involving the neutrons.

It seems probable that nucleons exist within a nucleus in different energy levels in much the same way as electrons exist outside the nucleus. Many models have been postulated, e.g. the liquid drop model, the shell model, the adiabatic model and the optical model. They may also involve particles other than protons, neutrons and electrons.

6 Subatomic particles

A study of cosmic rays and the nature of nuclear changes, particularly in particle accelerators, has led to the prediction and/or discovery of well over 200 subatomic particles. These are classified, nowadays, according to the strength of the forces between them, their mass and their spin.

Some of the commoner particles are summarised below, with their masses relative to that of the electron as 1.

leptons		hadrons			
		mesons		baryons	
neutrino, ν	0	pions, π^+	318	proton, p^+	1836
electron, e^-	1	π^0	264	neutron, n	1839
muon, μ^-	207	kaons, K^+	967	lambda, Λ^0	2164
		K^0	967	omega, Ω^-	3252

Many baryons have been omitted. Most, if not all, of the particles have an associated *anti-particle*, which, in the simplest cases, has an equal but opposite charge. A positron, e^+, for example, is the anti-particle of the electron, e^-; p^- is an anti-proton. When a particle and its anti-particle meet, radiation is produced; it is known as pair annihilation.

It may, or may not, be that some or all of these subatomic particles are made up of still smaller particles known as *quarks*.

7 Nuclear fission

Before 1939 all nuclear reactions had consisted solely in the removal from a nucleus of a relatively small particle such as an α-particle, a proton or a neutron. In 1939, however, Hahn and Strassman realised that isotopes of medium mass number were obtained when uranium was irradiated with neutrons, and that this must represent a splitting (fission) of an atom into two, more or less similar, halves. Similar fission of thorium was also discovered about the same time.

To determine which isotope of uranium was undergoing fission, Nier separated ^{235}U and ^{238}U (p. 87), whereupon it was found that ^{235}U generally splits when bombarded by slow (thermal) or fast neutrons, whereas ^{238}U tends to absorb slow neutrons, with subsequent formation of neptunium and plutonium, and is only split by fast neutrons with high energy.

In the fission of ^{235}U by slow neutrons, the mass change, summarised below,

$$\underbrace{\underset{235.0439}{^{235}_{92}\text{U}} \quad + \quad \underset{1.0087}{^{1}_{0}\text{n}}}_{236.0526} \longrightarrow \underbrace{\underset{143.881}{^{144}_{56}\text{Ba}} \quad + \quad \underset{89.947}{^{90}_{36}\text{Kr}} \quad + \quad \underset{2.0174}{2\,^{1}_{0}\text{n}}}_{235.8454}$$

is 0.2072 a.m.u. This represents a release of about 200 MeV of energy for every one atom of ^{235}U which undergoes fission.

8 Nuclear weapons

When ^{235}U nuclei are split, two or three neutrons (referred to as secondary neutrons) are released, as shown in the equation above. Some of these may cause other ^{235}U atoms to split and a chain reaction may develop; one fission leading to three, three leading to nine, and so on (Fig. 34). What is known as the *multiplication factor* is, in this example, three. In a bomb the chain is encouraged; in a nuclear reactor it is controlled.

Both ^{235}U and ^{239}Pu are split up by neutrons and they can both be used for making nuclear bombs. The latter is more common, nowadays, because of the difficulty of extracting ^{235}U from naturally occurring uranium.

There can be no explosion unless the multiplication factor is greater than one. It is also necessary to have a piece of fissionable material greater than a certain *critical mass*. With too small a piece, too many of the secondary neutrons will escape without causing fission (Fig. 35).

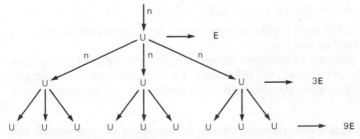

Fig. 34 Multiplication factor. The build-up of a chain reaction.

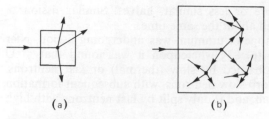

Fig. 35 Critical size. **(a)** The secondary neutrons escape without causing fission. **(b)** The secondary neutrons cause fission and a chain reaction can be set up.

There is, then, a critical size below which the material is safe but above which it will explode once the chain reaction is initiated. In a nuclear bomb an explosion is obtained, at the right moment, by bringing together two pieces of fissionable material, each below the critical size, to form one piece above that size.

9 Nuclear reactors

These involve a controlled chain reaction which is allowed to liberate heat energy steadily. The heat is used to boil water to drive a steam turbine. Nuclear energy is being increasingly used to provide electricity and to drive large ships.

The overall amount of fuel (some form of uranium, uranium oxides or plutonium) must be greater than the critical mass or else it would never be possible to achieve a chain reaction. The fuel is arranged in rods within a central core, and control rods of some neutron-absorbing material (such as boron steel or cadmium) are used to limit the intensity of neutrons and control the chain reaction. When the rods are withdrawn, the multiplication factor becomes greater than one; the reactor operates. Control is achieved by moving the rods in and out.

In so-called *thermal reactors*, a moderator, such as graphite, is used to slow down most of the secondary neutrons so that they will fission ^{235}U atoms and not be captured by ^{238}U atoms. Some capture does occur, however, so that plutonium is made within the piles, and has, eventually, to be separated.

$$^{238}_{92}U + ^{1}_{0}n \longrightarrow ^{239}_{92}U \longrightarrow ^{239}_{94}Pu + 2_{-1}^{0}e$$

The plutonium is used for defence purposes, in making bombs, and can also be used in *fast reactors*.

The high intensity of neutrons within a reactor can be used for neutron diffraction experiments (p. 169), and for making many artificially radioactive isotopes by neutron bombardment of non-radioactive elements.

a British designs Britain pioneered the development of commercial nuclear reactors after World War II, and opened the first Magnox reactor at

Calder Hall in 1956. Rods of natural uranium containing both ^{235}U and ^{238}U atoms were used as the fuel, carbon dioxide under pressure as the coolant, and graphite as the moderator. This design developed into what is now known as the advanced gas-cooled reactor (AGR). It has a carbon dioxide coolant, a graphite moderator and a fuel of uranium oxide enriched with ^{235}U. The reactor operates at a higher temperature than the Magnox reactors and the fuel life is longer. Both Magnox and AGR reactors use moderators and are thermal reactors.

In the fast fission breeder reactor, being built at Dounreay, a core of plutonium, surrounded by ^{238}U, is used, and there are no moderators. The fast neutrons cause fission in the plutonium liberating energy; escaping neutrons are captured by the ^{238}U to make more plutonium. It can be arranged that more plutonium is made than is used up; hence the term 'breeder'. Molten sodium is used as the coolant.

b Pressurised water reactors (PWR) There have been many technical and economic difficulties in developing the British designs, and the major design, now being built more and more throughout the world, is an American reactor which uses pressurised water as the coolant, and water as the moderator. The fuel is enriched uranium oxide (Fig. 36).

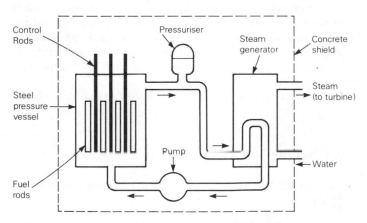

Fig. 36 Outline of a pressurised water reactor (PWR). The core is about 3 m in diameter and 4 m high.

10 Nuclear fusion

When a uranium atom is split into two or more lighter atoms there is a decrease in mass and a corresponding release of energy. The binding energy curve (p. 99) shows that there would be a similar result if two light nuclei could be combined together to give a heavier one. Combination, or fusion, of two nuclei is not, however, at all easy. They are both positively charged and repel each other. At high temperatures, however, nuclei may have sufficient energy to unite (fuse).

a Stellar energy Von Weizsacker and Bethe proposed, in 1938, that the vast amount of energy emitted by the sun and by stars might originate from a process involving nuclear fusion. The modern view is that there are two cycles, one producing the energy in very hot stars, and the other taking place in cooler stars, such as the Sun.

The overall change in both is thought to be the same, i.e. conversion of hydrogen into helium with the emission of γ-rays and two positrons,

$$4\,^1_1\text{H} \longrightarrow \,^4_2\text{He} + 2\,^0_1\text{e} + \gamma$$

b The hydrogen bomb The temperature attained when ^{235}U or plutonium atoms undergo fission in an explosive chain reaction is high enough to initiate fusion, or thermonuclear, reactions involving hydrogen isotopes. Such reactions have been brought about in hydrogen bombs. In effect, a ^{235}U or plutonium fission explosion acts as a detonator for a fusion reaction.

c Controlled thermonuclear fusions Efforts to control a useful thermonuclear fusion reaction have not yet been successful. If they ever are, a new source of energy will become available. For successful fusion reactions, which would liberate more energy than is needed to initiate them, it seems likely that temperatures of the order of 10^8 K will have to be achieved, and no constructional material will withstand such temperatures. The problem may not, however, be insoluble, and several hopeful lines are being investigated.

At the very high temperatures involved, all matter exists as a *plasma* (p. 33).

Questions on Chapter 10

1 Write full equations for the nuclear reactions summarised as follows:
(a) $^{23}_{11}\text{Na}\,(\alpha, \text{p})\,^{26}_{12}\text{Mg}$
(c) $^{26}\text{Mg}\,(\text{n}, \alpha)\,^{23}\text{Ne}$
(b) $^{12}\text{C}\,(\text{d}, \text{n})\,^{13}\text{N}$
(d) $^{26}_{12}\text{Mg}\,(\text{p}, \text{n})\,^{26}_{13}\text{Al}$.

2 What does X stand for in the following equations:
(a) $^{14}_7\text{N} + \,^4_2\text{He} \longrightarrow \,^{17}_8\text{O} + \text{X}$
(c) $2\,^2_1\text{D} \longrightarrow \,^3_2\text{He} + \text{X}$
(b) $^1_1\text{H} + \,^7_3\text{Li} \longrightarrow 2\text{X}$
(d) $^8_4\text{Be} + \,^4_2\text{He} \longrightarrow \text{X}$

3 What does Y stand for in the following:
(a) $^7_3\text{Li}\,(\text{p}, \text{n})\,\text{Y}$
(c) $^7_3\text{Li}\,(\text{Y}, \text{n})\,^7_4\text{Be}$
(b) $^{43}_{20}\text{Ca}\,(\alpha, \text{p})\,\text{Y}$
(d) $^{59}_{27}\text{Co}\,(\text{Y}, \alpha)\,^{56}_{25}\text{Mn}$?

4 What energy, in electron-volts, corresponds to radiation of wavelength (a) $1\,\mu\text{m}$ and (b) $200\,\text{nm}$? What wavelength, in nanometres, corresponds to an energy of $2.48\,\text{eV}$?

5 Explain what is meant by the following relationships (a) $1\,\text{cm}^{-1} = 11.97\,\text{J}\,\text{mol}^{-1} = 1.239\,58 \times 10^{-4}\,\text{eV}\,\text{atom}^{-1}$ (b) $1\,\text{eV} = 99.16\,\text{kJ}\,\text{mol}^{-1}$.

6 ^{55}Mn is one of the stablest nuclei. If its physical relative atomic mass on the ^{12}C scale is 54.956, determine its binding energy. Take the relative masses of the proton and neutron as 1.0076 and 1.009 respectively.

7 Calculate the binding energy of ^{12}C, taking the masses of the proton, the neutron and ^{12}C as 1.0078, 1.0089 and 12.0000 respectively.

8 Compare the repulsive force between two protons separated by 10^{-10} m, as in a hydrogen molecule, with that when the distance of separation is only 10^{-15} m, as in an atomic nucleus. Comment on the significance of your answer.

9 Deuterium nuclei of atomic mass 2.014 can be split up into protons of mass 1.0076 and neutrons by γ-radiation of energy 2.26 MeV. Calculate the mass of the neutron, from this data, in atomic mass units.

10 How much energy, in MeV, would be liberated if an electron were completely converted into energy?

11 The proton bombardment of the isotope ^{11}B yields three alpha particles for each proton and each ^{11}B atom. Each alpha particle has energy of 2.85 MeV. If the atomic mass of the alpha particle is 4.0034 and that of the proton is 1.0081, calculate the atomic mass of ^{11}B.

12 The relative atomic mass of helium is 4.0039 and that of deuterium is 2.0147. If it were possible to make 1 mol of helium by fusing 2 mol of deuterium together, how many joules would be released?

13 ^{13}C can be converted into nitrogen by capture of a proton. If the relative atomic masses are 13.0075, 14.0075 and 1.0081, respectively, what is the heat of the reaction in $kJ\,mol^{-1}$?

14 If a ^{235}U atom fissions according to the equation

$$^{235}_{92}U + {}^{1}_{0}n \longrightarrow {}^{148}_{57}La + {}^{85}_{35}Br + 3{}^{1}_{0}n$$

and if the relative isotopic masses concerned are U = 235.124, n = 1.009, La = 147.961 and Br = 84.938, calculate the energy given out from one fissioned atom in MeV.

15 (a) Describe the composition of the atomic nucleus and discuss the consequences of nuclear instability (b) Explain why the emission of an alpha particle from ^{214}Po followed by the emission of two successive beta particles and another alpha particle produces lead (c) Describe and account for the nature of the atomic spectrum of hydrogen (d) Explain why the third row (Na − Ar) of the periodic table contains only eight elements. (JMB)

16 (a) Explain the terms mass deficit and alpha emission. For the following atoms of naturally occurring non-radioactive isotopes ^{1}H, ^{4}He, ^{7}Li, ^{9}Be, ^{11}B, ^{12}C, ^{14}N, ^{16}O, ^{19}F, ^{20}Ne, ^{23}Na, ^{24}Mg and ^{27}Al plot the number of protons against the number of neutrons in each nucleus. Using your graph show that ^{14}C and ^{24}Na are radioactive isotopes decaying by beta emission. Write equations for the decay processes of these two isotopes (b) The mass of the nucleus of an atom of ^{127}I is $2.106\,61 \times 10^{-25}$ kg. The masses of a single proton and single neutron are $1.672\,52 \times 10^{-27}$ kg and $1.674\,82 \times 10^{-27}$ kg respectively. Calculate the mass deficit per nucleon for an iodine nucleus. (AEB, 1979)

Introduction to Physical Chemistry

17 How does the density of nuclear material vary with the relative atomic mass of the nucleus concerned?

18* Taking the nuclear diameter as $1.4 \times 10^{-15} A_r^{0.33}$ m calculate the density of the nucleus of (a) Li and (b) Au. Calculate the approximate ratio of the volume of the nucleus to the volume of the atom in each case.

11
Ionic or electrovalent bonding

1 Introduction

An ionic or electrovalent bond is formed by a complete transfer of one or more electrons from one atom to another. The element that loses electrons is said to be electropositive, for it forms a positive ion. The element gaining electrons is electronegative; it forms a negative ion. Ionic compounds can, therefore, be represented as follows, showing only electrons in the outermost orbits.

$$Na\cdot \; + \; {}^{x}_{x}\!\overset{x\,x}{\underset{x\,x}{Cl}}\,{}_{x} \; \longrightarrow \; Na^{+} \left[{}^{x}_{\bullet}\overset{x\,x}{\underset{x\,x}{Cl}}\,{}_{x} \right]^{-}$$

$$Ca^{2+} \left[{}^{x}_{\bullet}\overset{x\,x}{\underset{x\,x}{Br}}\,{}_{x} \right]^{-}_{2} \qquad K^{+}{}_{2} \left[{}^{\bullet}\overset{x\,x}{\underset{x\,x}{S}}\,{}_{x} \right]^{2-}$$

The · and × signs show electrons originally belonging to the metal and the non-metal respectively.

The formulae for ionic compounds, as above, show only the relative numbers of each ion in the compound. There are no individual molecules, for all the ions are held together in an ionic crystal by strong electrostatic forces. The arrangement of the ions within the crystal, i.e. the crystal structure, is mainly determined by the charges and sizes of the ions concerned (p. 109).

2 General characteristics of ionic compounds

These are summarised as follows.
a Ionic compounds are invariably electrolytes for, in the presence of an ionising solvent such as water, the forces between the ions are so greatly reduced that the ions 'fall apart'. The free ions, in the resulting solution, are able to move under the influence of an electrical field as in electrolysis.
b Ionic compounds are often hard solids because the inter-ionic forces within an ionic crystal are usually strong.
c Ionic compounds generally have high melting points because a lot of thermal energy is required to break down the inter-ionic forces and form a liquid. Once an ionic compound has been melted, the melt can undergo electrolysis because it contains free ions. A high melting point is associated, too, with a high boiling point for ionic compounds.
d Ionic compounds are commonly soluble in water, or other ionising solvents, and insoluble in benzene or other organic solvents.

3 Ionic radii

The distance between the nuclei of two adjacent ions in an ionic crystal can be measured by X-ray analysis, and, if the ions are regarded as spheres, the distance will be the sum of two ionic radii.

Before any single ionic radius can be obtained from internuclear distances it is necessary to decide the value of one radius. Various rather arbitrary methods have been used but when one value is decided the others are easily obtainable from measurements on crystals. Typical figures, for simple ions, are shown in Fig. 37 and the following points arise.

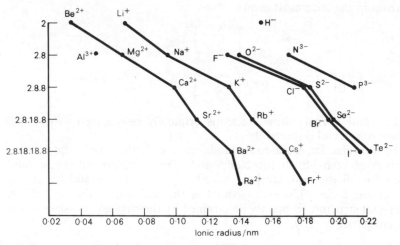

Fig. 37 Ionic radii of common ions. The ions on each horizontal line have the same arrangement of electrons.

a The ions of elements in any one group of the periodic table increase in size as the relative atomic mass increases.

b For ions with the same arrangements of electrons the size decreases as the atomic number increases. This is due to the increasing nuclear attraction for the electrons as the nuclear charge increases. Compare, for example, the ions O^{2-}, F^-, Na^+, Mg^{2+} and Al^{3+} (all with a 2.8 structure), and S^{2-}, Cl^-, K^+, Ca^{2+} and Sc^{3+} (all with a 2.8.8 structure).

c When two positively charged ions are formed by the same element it is the one with higher charge that is smaller. This is because the more highly charged ion has fewer electrons so that they are more tightly held. Compare, for example, Sn^{2+} and Sn^{4+}, Pb^{2+} and Pb^{4+}, Fe^{2+} and Fe^{3+}.

d The lowering of ionic size due to increase in ionic charge for positive ions is also shown by comparing the sizes of Na^+, Mg^{2+} and Al^{3+}. For negative ions, increase in charge, i.e. addition of more and more electrons, leads to increase in size, e.g. F^-, O^{2-} and N^{3-}.

e There is a small lowering in ionic size in passing along the d-block periods, or in the lanthanides.

f A cation radius is smaller than the covalent radius for the same element; an anion radius is larger. Anions are, in the main, larger than cations.

g Different methods of assessment give rather different values for ionic radii, and any values quoted tend, also, to refer to only one type of crystal structure. Numerical values must, therefore, be treated with some caution.

4 Ionic crystals

The electrostatic forces holding the ions together in a crystal are non-directional and the arrangement of ions within the crystal is mainly controlled by the sizes and charges of the ions.

a **Sodium chloride structure** A pair of Na^+ and Cl^- ions (Fig. 38) will have a strong electrical field and will attract a second ion pair as shown in Fig. 39. Four ion-pairs will arrange themselves as in Fig. 40, and a larger number will take up the arrangement shown in Fig. 41, which gives the crystal structure of sodium chloride.

Each Na^+ ion is surrounded by six Cl^- ions, and each Cl^- ion by six Na^+ ions.

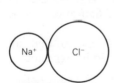

Fig. 38 Ion pair of sodium chloride.

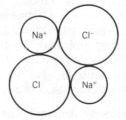

Fig. 39 Two ion pairs of sodium chloride.

Fig. 40 Four ion pairs of sodium chloride. The comparative sizes of the ions are not shown in this figure, but they are the same as in Fig. 39.

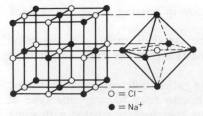

Fig. 41 Crystal structure of sodium chloride, showing (right) the octahedral arrangement of six sodium ions round one chloride ion.

b Coordination number This is the number of ions, atoms or molecules immediately surrounding a particular ion, atom or molecule in a crystal structure. In sodium chloride, it is 6.

In a binary ionic compound, A^+B^-, the coordination numbers of A^+ and B^- must be equal so that the crystal as a whole is electrically neutral. In $A^{2+}B_2^-$ compounds, the coordination number of A^{2+} will be twice that of B^-.

c The radius ratio The value of the coordination number depends on the relative sizes of the ions concerned, i.e. on the radius ratio, r^+/r^-. Cations are generally smaller than anions (Fig. 37) and it is a matter of how many anions can pack round one cation. The anions will repel each other and take up positions as far apart as possible.

For a very small cation and a very large anion, i.e. for a very small radius ratio, it may only be possible to pack two anions round one cation; the $A^+X_2^-$ grouping would be linear with A having a coordination number of 2. As the radius ratio increases it can be shown, by simple geometrical calculations, that the coordination number would rise to 3 (triangular), then to 4 (tetrahedral), 6 (octahedral) and 8 (cubical). The limits are triangular (0.1555–0.225), tetrahedral (0.225–0.414), octahedral (0.414–0.732) and cubical (0.732–1.000). These theoretical figures generally, but not invariably, apply in practice.

The radius ratio for Na^+Cl^- is 0.095/0.181 or 0.525 so that an octahedral structure with coordination number of 6 would be expected.

d Caesium chloride structure The radius ratio for CsCl is 0.169/0.181 or 0.93 and the crystal structure is cubical with a coordination number of 8 (Fig. 42).

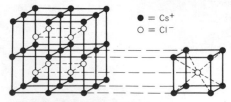

Fig. 42 Crystal structure of caesium chloride showing (right) the cubic arrangement of eight caesium ions round one chloride ion.

e Structures of CA$_2$ ionic crystals When the charge on the cation is twice that on the anion the crystal can be represented as CA$_2$. The two commonest crystal structures of this type are the fluorite, CaF$_2$, and the rutile, TiO$_2$, structures.

In the fluorite structure of CaF$_2$, each Ca^{2+} ion is surrounded by eight F$^-$ ions at the corners of a cube, and each F$^-$ ion is surrounded by four Ca^{2+} ions arranged tetrahedrally. The coordination numbers are 8 and 4.

In the rutile structure of TiO$_2$, each Ti^{4+} ion is surrounded by six O^{2-} ions arranged octahedrally, and each O^{2-} ion is surrounded by three Ti^{4+} ions arranged triangularly. The coordination numbers are 6 and 3.

A third closely related structure is the antifluorite structure which is the normal fluorite structure with the anions and cations interchanged.

5 Ionisation energy and ionisation enthalpy

Energy is required to remove an electron from a free, gaseous atom against the attraction of the nucleus, and, for a single atom, this is called the first atomic ionisation energy. For a mole of atoms it is called the first molar ionisation energy, or, as a simplification, the first ionisation energy. That is

$$X(g) \longrightarrow X^+(g) + e^- \qquad \Delta U_m = \text{first ionisation energy}$$

It is important to realise that this refers to the formation of a free, gaseous ion from a free, gaseous atom, and not to the formation of an ion from an element in its normal state (except for the noble gases), and that it is a ΔU, and not a ΔH, value (p. 245).

Successive ionisation energies refer to the loss of a second, third, . . . nth electron. Thus the second ionisation energy of X is the first ionisation energy of X$^+$, and the sum of the first n ionisation energies gives the energy change for the formation of the X^{n+} ion,

$$X \xrightarrow{-e^-} X^+ \xrightarrow{-e^-} X^{2+} \qquad X \xrightarrow{-ne^-} X^{n+}$$

Ionisation energies are quoted in kJ mol^{-1}, or, as ionisation potentials in eV mol^{-1} (1 eV mol^{-1} = 96.486 kJ mol^{-1}). Some typical values for first and second ionisation energies, in kJ mol^{-1} are listed below.

H							He
1311							2372
							5250

Li	Be	B	C	N	O	F	Ne
520	899	800	1086	1402	1314	1681	2081
7298	1757	2427	2353	2856	3388	3375	3952

Na	Mg	Al	Si	P	S	Cl	Ar
495.8	738	578	786	1012	1000	1251	1521
4563	1451	1817	1577	1903	2258	2296	2665

Ionisation energy values (ΔU_m at 0 K) can be converted into ionisation enthalpy values (ΔH_m at 298 K) by adding approximately $6\,\mathrm{kJ\,mol^{-1}}$. There is, however, a relatively small percentage difference between the ionisation energy and the ionisation enthalpy. The two terms and their numerical values are, sometimes, used interchangeably.

6 Measurement of ionisation energy

Ionisation energies can be measured by spectroscopic methods (p. 180), by using a mass spectrometer (p. 92), or by measuring the current passing through a discharge tube containing a monatomic gas or vapour as the applied voltage is gradually increased. At certain voltages there are marked rises in the current passing and these correspond to the points at which atoms of the gas or vapour lose one, two, three or more electrons.

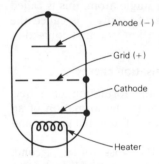

Fig. 43 A thyratron.

A thyratron (Fig. 43), which is similar to a radio valve filled with a gas at low pressure, can be used to measure the first ionisation energy of a noble gas. The thyratron filled, for example, with xenon, has a negative potential applied to the anode. Electrons emitted by heating the cathode, which is coated with metallic oxides, are attracted by, and accelerated towards, the positively charged grid. The electrons are repelled by the negatively charged anode, so that little or no current passes between cathode and anode.

As the positive potential on the grid is raised, there comes a time when the electrons from the cathode have sufficient energy to ionise some of the xenon atoms by impact. The Xe^+ ions formed

$$Xe(g) + e^-(\text{fast}) \longrightarrow Xe^+(g) + 2e^-(\text{slow})$$

are attracted by the negatively charged anode so that a current begins to flow between cathode and anode. This current (I) increases linearly as the positive potential on the grid (V) increases. The potential at which the current first passes, found from a plot of I against V, is a measure, in volts, of the first ionisation energy of xenon. To convert it into $kJ\,mol^{-1}$ it is necessary to multiply by 96.486.

7 Values of ionisation energies

The way in which ionisation energies vary with atomic number (Fig. 44) raises the following points:

a The noble gases have the highest first ionisation energies and are least likely to form X^+ ions. This is because of the stability of their electronic structures.

b The alkali metals (Group 1), with one electron more than a noble gas, have the lowest first ionisation energies; they form X^+ ions most readily.

c There is a decrease in the value of the ionisation energy in passing down Group 1A, so that caesium, the biggest atom, forms an ion more easily than lithium, the smallest. The same trend is noticeable in other groups, all the more so if the energy required to form X^{n+} ions is considered. This is to be expected, for the outermost electrons are further away from the positively charged nucleus in the bigger atoms.

In Group 1A, caesium ionises most readily; in Group 2A, barium ionises most readily. In Group 3B, boron is not big enough to form B^{3+} ions, and in Group 4B neither C^{4+} nor Si^{4+} exist.

d There is a general rise in ionisation energy in passing from left to right across a period, though Be, Mg, N and P have irregularly high values.

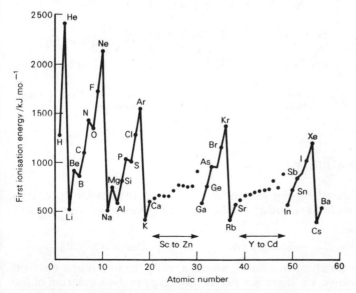

Fig. 44 Variation of first ionisation energy with atomic number.

These arise because extra energy is needed to remove an electron from a full s orbital (Be and Mg) or from a half full p-orbital (N and P).

e More energy is required to form ions with higher charge. This is because each succeeding electron has to be withdrawn against the attractive force of an increasingly strongly charged positive ion. Cations are therefore limited to a charge of four units and such highly charged ions are only formed by large atoms such as tin and lead.

The figures below show how it is increasingly difficult to remove successive electrons from atoms, and also indicate the existence of energy levels within atoms. For any atom there is a steady rise in the successive ionisation energy values. For electrons with the same principal quantum number the rise is not marked, but it becomes very much greater when an electron of lower principal quantum number is involved.

n	1		2								3		
H	1.3												
He	5.2	2.4											
Li	12	7.3	0.5										
Be	21	15	1.8	0.9									
B	33	25	3.7	2.4	0.8								
C	47	38	6.2	4.6	2.4	1.1							
N	64	53	9.5	7.5	4.6	2.9	1.4						
O	84	71	13	11	7.5	5.3	3.4	1.3					
F	106	92	18	15	11	8.4	6.0	3.4	1.7				
Ne	131		23	20	15	12	9.4	6.1	4.0	2.0			
Na	159	141	29	26	20	17	13	10	7.0	4.6	0.5		
Mg		170	36	32	26	22	18	14	11	8	1.5	0.7	
Al			42	39	32	28	23	18	15	12	2.7	1.8	0.6

Approximate values for successive ionisation energies/$(kJ\,mol^{-1}) \times 10^{-3}$ (n = principal quantum number). The right-hand figure gives the first ionisation energy.

8 Electron affinity

Anions are formed by the gain of electrons and the associated energy change per mole is known as the electron affinity of the element concerned, e.g.

$$\text{atom (A)} + e^- \longrightarrow \text{anion (A}^-) \qquad \Delta U_m = \text{electron affinity}$$

The value for the electron affinity, which is really the first molar electron affinity, is negative, i.e. there is a release of energy in the formation of the anion.

Once an atom has gained one electron the negative charge opposes the addition of a second so that the second electron affinity, i.e.

$$\text{anion (A}^-\text{)} + e^- \longrightarrow \text{anion (A}^{2-}\text{)} \qquad \Delta U_m = \text{second electron affinity of A}$$

has a positive value.

As for ionisation energy, values are quoted in $kJ\,mol^{-1}$ or in $eV\,mol^{-1}$; some typical values in $kJ\,mol^{-1}$ are given below:

H -72	O -142	O$^-$ 844	S -200	S$^-$ 532
F -333	Cl -348	Br -328	I -295	

The figures show that it is easier to form an A^- anion than an A^{2-} one, and, as a broad generalisation, it is the smallest atom in any one group that most readily forms the anion. The formation of an A^- ion is exothermic; that of an A^{2-} ion endothermic, e.g.

$$Cl(g) + e^- \longrightarrow Cl^-(g) \qquad \Delta U_m = -348\,kJ\,mol^{-1}$$
$$O(g) + 2e^- \longrightarrow O^{2-}(g) \qquad \Delta U_m = +702\,kJ\,mol^{-1}$$

To convert ΔU_m values into ΔH_m values at 298 K, it is necessary to add approximately $-6\,kJ\,mol^{-1}$.

9 Energy changes in forming ionic crystals

The formation of an ionic compound from its elements in their normal states involves the standard molar enthalpy of formation (p. 252), e.g.

$$Na(s) + \tfrac{1}{2}Cl_2(g) \longrightarrow NaCl(s) \qquad \Delta H_{m,f}^{\ominus}(298\,K) = -411\,kJ\,mol^{-1}$$

and the more negative its value the more likely is the reaction to be feasible (p. 247). The factors which lead to a highly negative enthalpy of formation are best understood by breaking the reaction down into stages, as follows.

a Atomisation of the elements The formation of free, gaseous atoms of sodium from solid sodium, and of free, gaseous atoms of chlorine (Cl) from gaseous chlorine (Cl_2), both involve the expenditure of energy, i.e.

$$Na(s) \longrightarrow Na(g) \qquad \Delta H_a^{\ominus}(298\,K) = 108\,kJ\,mol^{-1}$$
$$\tfrac{1}{2}Cl_2(g) \longrightarrow Cl(g) \qquad \Delta H_a^{\ominus}(298\,K) = 121\,kJ\,mol^{-1}$$

The numerical values quoted are the enthalpies of atomisation per mole of atom (p. 254).

b Formation of positive ions (cations) Formation of a free, gaseous Na^+ ion from a free, gaseous Na atom involves the withdrawal of an electron

Introduction to Physical Chemistry

against the nuclear attraction, i.e.

$$Na(g) \longrightarrow Na^+(g) + e^- \qquad \Delta H(298\,K) = 502\,kJ\,mol^{-1}$$

The numerical value quoted is known as the first ionisation enthalpy (p. 112) of sodium.

c Formation of negative ions (anions) Energy is given out when a free gaseous atom takes up an extra electron, e.g.

$$Cl(g) + e^- \longrightarrow Cl^-(g) \qquad \Delta H(298\,K) = -354\,kJ\,mol^{-1}$$

The value quoted is known as the electron affinity (p. 114) of chlorine.

d Formation of a crystal from free ions The ions in an ionic crystal are strongly bound together by electrostatic attractive forces between the ions, and there is a considerable release of energy when such a crystal is formed, i.e.

$$Na^+(g) + Cl^-(g) \longrightarrow NaCl(s) \qquad \Delta H(298\,K) = -788\,kJ\,mol^{-1}$$

The value quoted is known as the lattice enthalpy (p. 117).

e The Born-Haber cycle The relationship between the standard enthalpy of formation of NaCl(s) and the enthalpy changes in the five stages in the formation is best shown in an enthalphy cycle known as a Born-Haber cycle (Fig. 45). The only term in the cycle that cannot be measured experimentally is the lattice enthalpy, and the cycle enables its value to be found.

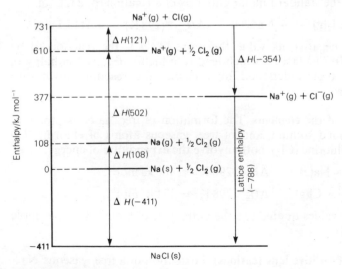

Fig. 45 The Born–Haber cycle used to find the lattice enthalpy of sodium chloride.

In this example, it will be seen that it is the high negative value of the lattice enthalpy which makes the largest contribution to the negative value for the enthalpy of formation, and this is so for other ionic compounds. Low enthalpies of atomisation, low ionisation enthalpies and a high negative electron affinity also favour a high negative value for the enthalpy of formation and, therefore, make the formation of an ionic compound more feasible.

f The stoichiometry of ionic compounds In the preceding example, a Born-Haber cycle has been used to find the value of a lattice enthalpy, all other values being measured experimentally. The cycle can also be used to get a value for the enthalpy of formation of a hypothetical compound. The lattice enthalpy for $NaCl_2$, for example, can be calculated (p. 118) or estimated from the known value for $MgCl_2$. By fitting this into a Born-Haber cycle, together with the other necessary values, the enthalpy of formation of $NaCl_2$ can be obtained. The figures, summarised below,

$$Na(s) \longrightarrow Na(g) \qquad \Delta H_a(298\,K) = 108\,kJ\,mol^{-1}$$

$$Na(g) \longrightarrow Na^{2+}(g) + 2e^- \qquad \Delta H(298\,K) = 5069\,kJ\,mol^{-1}$$

$$Cl_2(g) \longrightarrow 2Cl(g) \qquad \Delta H_a(298\,K) = 242\,kJ\,mol^{-1}$$

$$2Cl(g) + 2e^- \longrightarrow 2Cl^-(g) \qquad \Delta H(298\,K) = -708\,kJ\,mol^{-1}$$

$$Na^{2+}(g) + 2Cl^-(g) \longrightarrow NaCl_2(s) \qquad \Delta H(298\,K) = -2525\,kJ\,mol^{-1}$$

$$Na(s) + Cl_2(g) \longrightarrow NaCl_2(s) \qquad \Delta H_f(298\,K) = 2186\,kJ\,mol^{-1}$$

give a value of $2186\,kJ\,mol^{-1}$ for the enthalpy of formation of $NaCl_2$. The corresponding value for $NaCl_3$ is $5919\,kJ\,mol^{-1}$. NaCl is clearly more stable than $NaCl_2$ or $NaCl_3$.

Similarly, the enthalpies of formation of the hypothetical MgCl and $MgCl_3$ can be shown to be -130 and $3909\,kJ\,mol^{-1}$ respectively. $MgCl_2$, with a value of $-642\,kJ\,mol^{-1}$, is clearly the most stable.

10 Lattice enthalpy and lattice energy

These are the enthalpy and energy changes for the formation of 1 mol of an ionic solid from its component ions in the free gaseous state, e.g.

$$Na^+(g) + Cl^-(g) \longrightarrow NaCl(s) \qquad \Delta H(298\,K) = -788\,kJ\,mol^{-1}$$
$$\Delta U(298\,K) = -782\,kJ\,mol^{-1}$$

The values, on this basis, are always negative because energy has to be put in to break up the strong bonds between ions in a crystal. Care must be taken, however, for positive values are sometimes quoted; in this case it is the reverse process that is being considered, e.g.

$$NaCl(s) \longrightarrow Na^+(g) + Cl^-(g) \qquad \Delta H(298\,K) = 788\,kJ\,mol^{-1}$$

Lattice enthalpy values can be obtained from Born-Haber cycles (p. 116), or they can be calculated. Some typical values from Born-Haber cycles are as follows, in $kJ\,mol^{-1}$.

NaF	−919	NaCl	−788	$MgCl_2$	−2525	MgO	−3902
NaCl	−788	KCl	−718	$CaCl_2$	−2253	CaO	−3510
NaBr	−748	RbCl	−689	$SrCl_2$	−2152	SrO	−3331
NaI	−704	CsCl	−660	$BaCl_2$	−2052	BaO	−3167

They show, as would be expected, that the energy required to break up a crystal increases as the charge on the ions do, and as the distance between the ions gets smaller. Lattice enthalpy values play an important part in a consideration of the solubility of ionic solids (p. 265).

a Calculated lattice energies Lattice energy values can be calculated, at least approximately, by regarding the ions concerned as charged spheres and by considering the attractive and repulsive forces between them in a crystal structure. Certain assumptions have to be made and the basic calculation can be done in a number of ways depending on the degree of refinement required, so that there are many different quoted values. Some typical ones, in $kJ\,mol^{-1}$, are given below together with the experimental value in parentheses.

NaCl	−776 (−782)	AgI	−807 (−885)
CsI	−592 (−597)	CuBr	−882 (−956)

The agreement is very close for alkali metal halides, but less so in other cases. The discrepancy is attributed to the polarisation of the ions, i.e. the effect of the electrical field of one ion on that of its neighbours. This introduces some measure of covalent bonding so that the crystal is not fully ionic. Indeed, the best way of defining an ionic compound may well be to state that it is one in which the measured and calculated lattice energies are equal.

b Polarisation The polarising power of an ion and the extent to which it might be polarised by another ion (its polarisability) both depend on the size, charge and electronic structure of the ion.

Cations are smaller and more compact than anions and they have the lowest polarisabilities and the highest polarising powers. The smaller they are and the higher their charge the greater the polarising power. That is why the lithium, beryllium and aluminium compounds are so much more covalent than other similar compounds in the same groups. Cations with noble gas structures also have smaller polarisabilities than those with d-electrons in their outer orbits. The d-electrons screen the nuclear charge less effectively and are more diffusely spread than s- or p-electrons. That is why, for example, copper, silver and gold compounds are more covalent than those of the alkali metals, or those of zinc, cadmium and mercury than those of the alkaline earths.

Anions have high polarisabilities but low polarising power. The polaris-ability increases as the size of the ion does for the electrons become more diffusely spread and further from the nucleus. The polarisability also increases as the ionic charge does, due to mutual repulsion of the electrons. Thus fluorides are more ionic than other halides, and oxides are more ionic than sulphides or selenides.

The size of the discrepancy between calculated and measured lattice energy, as in **a**, is in agreement with these general ideas.

c Fajans' rules The effect of polarisation was summarised in Fajans' rules (1924). These state, in a modernised form, that an ion is most easily formed if the electronic arrangement of the ion is stable, if the charge on the ion is small, and if the atom from which the ion is formed is small for an anion and large for a cation. Covalent bonding occurs when the conditions for ionisation are not favourable.

11 The magnetic moments of ions

Much information has been obtained about the arrangement of electrons in ions from magnetic measurements.

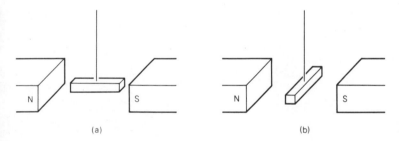

(a)   (b)

Fig. 46 (a) Paramagnetic substance. **(b)** Diamagnetic substance.

Substances may be classified as *paramagnetic* or *diamagnetic*. Paramagnetic substances are drawn into a strong magnetic field, i.e. they take up a position parallel to a magnetic field (Fig. 46(a)). Diamagnetic substances tend to be drawn out of a magnetic field, i.e. they set themselves at right angles to the field (Fig. 46(b)). Substances normally regarded as magnetic, e.g. iron, steel, cobalt, nickel and magnetic alloys, are paramagnetic, but the degree of magnetism possessed by such substances is much greater than that of any others. They are said to be *ferromagnetic*.

An ion containing an unpaired electron is paramagnetic, whilst one in which all the electrons are paired is diamagnetic. A spinning electron is equivalent to an electric current in a circular conductor and, as such, behaves as a magnet. The moment of such a magnet is measured in magnetons (see Appendix II, p. 474) and the magnetic moment of a paramagnetic substance containing n unpaired electrons can be shown to be $\sqrt{n(n+2)}$ magnetons. This enables the number of unpaired electrons in any ion to be determined by measurement of its magnetic moment.

(Questions on this chapter will be found on p. 154.)

12
The covalent bond

1 Introduction

Ionic bonds are formed by a complete transfer of electrons from one atom to another. In covalent bonding, electrons are shared, in pairs, between two bonded atoms.

In a single covalent bond between two atoms, one electron from each is held in common by both. The shared pair of electrons constitutes a single covalent bond, which can be represented in a simple way as in the following typical examples.

$$H \overset{x}{.} H \qquad \overset{..}{\underset{..}{:}} Cl \overset{x\,x}{\underset{x\,x}{\times}} Cl \overset{x}{\underset{x}{\times}} \qquad H \overset{x\,x}{.} Cl \overset{x}{\underset{x\,x}{\times}}$$

$$H{-}H \qquad Cl{-}Cl \qquad H{-}Cl$$

Double and triple covalent bonds involving two or three pairs of shared electrons are also common, e.g.

$$\overset{x\;\;x\;\;\circ}{\underset{x\;\;\circ\;\;\circ}{O\overset{\circ}{\underset{x}{:}}O}} \qquad \overset{x}{\underset{x}{x}}N\overset{\circ}{\underset{\circ}{:}}N\overset{\circ}{\underset{\circ}{\circ}} \qquad \overset{H.\;\;x\;\;\circ H}{\underset{H^x\;\;\circ\;\;\circ H}{C\overset{\circ}{\underset{x}{:}}C}} \qquad H\overset{x}{\underset{x}{:}}C\overset{\circ}{\underset{\circ}{:}}C\overset{x}{\underset{\circ}{:}}H$$

$$O{=}O \qquad N{\equiv}N \qquad \text{Ethene} \qquad \text{Ethyne}$$

The shared pair of electrons in a covalent bond may also be formed by one of the two bonded atoms providing both electrons. If so, the bond is sometimes called a dative bond but, as it is similar to a covalent bond once formed, the two are not always distinguished. The atom providing the two electrons to make up the dative bond is known as the donor. It must, of course, have an 'unused' pair of electrons available and such a pair is known as a lone pair. The atom sharing the pair of electrons from the donor is known as the acceptor.

When it is not necessary to distinguish between a dative and a covalent bond the — symbol is used for both. Two other symbolisms are, however, in use and have certain points in their favour. The first shows a dative bond between two atoms A and B as A → B, A being the donor and B the acceptor. This indicates, in a convenient way, the origin of the electrons making up the bond. The second shows an A → B bond as $A^{\oplus} \to B^{\ominus}$ and this indicates the electrical charges which develop as a result of the donor partly transferring two electrons to the acceptor.

Typical examples are as follows.

(i) Aluminium chloride   (ii) Isocyanomethane   (iii) Ammonium ion

$$\begin{array}{c} Cl \\ \\ Cl \end{array}\!\!\!\diagdown\!\!\!\diagup\!\!\! Al \!\!\!\diagdown\!\!\!\diagup\!\!\!\begin{array}{c} Cl \\ \\ Cl \end{array}\!\!\!\diagdown\!\!\!\diagup\!\!\! Al \!\!\!\diagdown\!\!\!\diagup\!\!\!\begin{array}{c} Cl \\ \\ Cl \end{array}$$

$$H-\underset{\underset{H}{|}}{\overset{\overset{H}{|}}{C}}-N\overset{\rightleftharpoons}{\equiv} C$$

$$\left\{ H \overset{\bullet}{\underset{\times}{\times}} \overset{\times}{\underset{\times}{N}} \overset{\times}{\underset{\bullet}{\times}} H \right\}^{+}$$

(iv) Chloric(vii) acid   (v) Nitromethane   (vi) Oxonium ion

$$H-O-\overset{\overset{\displaystyle O}{\uparrow}}{\underset{\underset{\displaystyle O}{\downarrow}}{Cl}}\!\!\rightarrow\! O$$

$$H-\underset{\underset{H}{|}}{\overset{\overset{H}{|}}{C}}-N\!\!\diagup\!\!\!\!\begin{array}{c} O \\ \\ \\ O \end{array}$$

$$\left\{ \begin{array}{c} H \\ \overset{\times\,\bullet}{\underset{\times}{O}}\, H \\ H \quad ^{\bullet\bullet} \end{array} \right\}^{+}$$

2 Electron pair repulsion

The shape of many simple molecules and ions of non-transitional elements can be predicted using the idea that electron pairs distributed round a central atom will repel each other and take up positions as far apart as possible. The electron pairs concerned may be bonded or shared pairs constituting covalent bonds, or unbonded lone pairs. As the latter are closer to the central atom they cause greater repulsion than shared pairs and the repulsions are in order

lone pair/lone pair > lone pair/shared pair > shared pair/shared pair

When multiple bonds occur they act as a single shared pair of electrons.

a Two electron pairs Beryllium chloride, $BeCl_2$, is linear because the two bonded pairs round the beryllium atom repel each other as fully as possible. The situation is summarised in Fig. 47, and other examples include $HgCl_2$, $O{=}C{=}O$, $S{=}C{=}S$, $H{-}C{\equiv}N$ and the $-C{\equiv}$ bond as in ethyne.

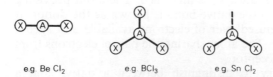

e.g. Be Cl_2 e.g. BCl_3 e.g. Sn Cl_2

Fig. 47 Shapes of molecules with two or three electron pairs.

b Three electron pairs Boron has three bonded electron pairs in all its BX_3 compounds. As these pairs repel each other equally the molecules, e.g. BCl_3 and $B(CH_3)_3$ are planar with bond angles of 120° (Fig. 47). Other examples of similar groupings include SO_3, $COCl_2$, NO_3^-, CO_3^{2-} and the $>C{=}$ bond as in ethene.

In tin(II) chloride, $SnCl_2$, in the vapour state, the tin atom is surrounded by two bonded pairs and one lone pair. The molecule is V-shaped (Fig. 47) with a bond angle smaller than $120°$ because of the increased repulsion caused by the lone pair. Other similar examples are SO_2, NOCl and O_3.

c Four electron pairs, all shared Four electron pairs will be distributed round a central atom tetrahedrally (Fig. 48) and this accounts for the tetrahedral shape of a methane, CH_4, molecule with a bond angle of $109° 28'$. Other similar species include $CHCl_3$, SO_2Cl_2, $POCl_3$, $Pb(C_2H_5)_4$, SO_4^{2-}, CrO_4^{2-}, PO_4^{3-}, ClO_4^-, MnO_4^- and NH_4^+.

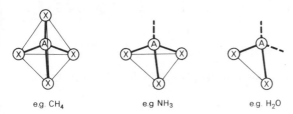

e.g. CH_4 e.g NH_3 e.g H_2O

Fig. 48 Shapes of molecules with four electron pairs.

d Four electron pairs, including one lone pair With three bonded pairs and one lone pair the resulting AX_3 molecule will be trigonal pyramidal in shape (Fig. 48). The increased repulsion caused by the lone pair will lead to a bond angle smaller than the $109° 28'$ found in methane. Ammonia, NH_3, for example, has a bond angle of $107°$.

The bond angle in phosphine, PH_3, is still lower ($93° 20'$). This is because phosphorus has a lower electronegativity than nitrogen so that the shared pairs are closer to the nitrogen atom in NH_3 than they are to the phosphorus atom in PH_3. This causes an increased repulsion between the shared pairs in NH_3 compared with PH_3. The bond angles in AsH_3 ($91° 51'$) and SbH_3 ($91° 20'$) are lower still.

Other examples of similar species include the halides of phosphorus, e.g. PCl_3, SO_3^{2-} and ClO_3^-.

e Four electron pairs, including two lone pairs With two lone pairs the resulting AX_2 molecule will be V-shaped (Fig. 48). The increased repulsive effect of one lone pair lowers the bond angle from $109° 28'$ in methane to $107°$ in ammonia. In the H_2O molecule the effect of two lone pairs is to lower the angle to $104° 31'$. The bond angles for H_2S ($93°$), H_2Se ($91°$) and H_2Te ($89° 30'$) are lower still because of the decrease in electronegativity in passing from sulphur to tellurium.

Other species with shapes like H_2O include Cl_2O and ClO_2^-.

f Five, six and seven electron pairs. Summary The possible arrangements of higher numbers of electron pairs which are actually found in some

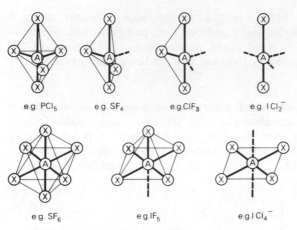

e.g. PCl₅   e.g. SF₄   e.g. ClF₃   e.g. I Cl₂⁻

e.g. SF₆   e.g. IF₅   e.g. I Cl₄⁻

Fig. 49 Shapes of molecules with five or six electron pairs.

molecules are shown in Fig. 49 and details are included in the following summary.

no. of shared pairs	no. of lone pairs	shape of molecule	example
2	0	linear	$BeCl_2$
3	0	triangular	BCl_3
2	1	V-shaped	$SnCl_2$
4	0	tetrahedral	CH_4
3	1	trigonal pyramidal	NH_3
2	2	V-shaped	H_2O
5	0	trigonal bipyramid	PCl_5
4	1	irregular tetrahedron	SF_4
3	2	T-shaped	ClF_3
2	3	linear	ICl_2^-
6	0	octahedral	SF_6
5	1	square pyramid	IF_5
4	2	square planar	ICl_4^-
7	0	pentagonal bipyramid	IF_7

The shapes of many of these molecules can also be considered from the point of view of hybridisation (p. 134).

3 Bond lengths and covalent radii

The length of a covalent bond is the distance between the nuclei of the two bonded atoms. It can be measured by X-ray analysis of crystals, by the diffraction of X-rays, electrons or neutrons by gases or vapours, or by spectroscopic methods. Values are usually expressed in nanometres.

Half the bond distance between the like atoms joined by a single covalent bond is known as the single bond covalent radius of the atom concerned. Values for double- and triple-bond covalent radii are obtained in the same way.

For many bonds, too, the sum of the covalent radii of two different atoms, A and B, gives a value equal to the actual bond distance in AB, so that once some individual covalent radii have been fixed it is possible to obtain values for other atoms even if they do not form diatomic molecules, A_2 or B_2.

Some compromise is necessary to obtain covalent radii which add to give good values for all bond distances, but the following values do this reasonably well.

H   0.037

Li 0.123	Be 0.089	B 0.080	C 0.077	N 0.074	O 0.074	F 0.072
Na 0.157	Mg 0.136	Al 0.125	Si 0.117	P 0.110	S 0.104	Cl 0.099
K 0.203	Ca 0.174		Ge 0.122	As 0.121	Se 0.117	Br 0.114
Rb 0.216	Sr 0.191		Sn 0.140	Sb 0.141	Te 0.137	I 0.133
Cs 0.235	Ba 0.198		Pb 0.154	Bi 0.152		

Single bond covalent radii/nm

C=	C≡	Si=	N=	N≡	O=	S=
0.067	0.060	0.107	0.060	0.055	0.060	0.094

Multiple bond covalent radii/nm

The single bond radii decrease in passing along a horizontal series. Each succeeding atom contains one more electron but the increasing positive charge on the nucleus causes the atom to contract in size. In passing down a group the atoms contain more and more electrons in outer and outer shells; there is, then, an increase in size.

When there is a marked discrepancy between the measured and calculated bond distances in a molecule it is usually an indication of polarity in the bond (p. 132) or of delocalisation (p. 138).

4 Atomic or covalent crystals

These are crystals in which atoms are held together by covalent, or predominantly covalent, bonds. Three main, simple structures are involved.

a The diamond structure Each C atom is surrounded by four others arranged tetrahedrally (Fig. 50). The atoms are linked by sp^3 hybrid bonds (p. 137), and the interlocking network of atoms accounts for the extreme hardness of diamond, and the high enthalpy of atomisation (p. 254). The structures are sometimes referred to as giant lattice structures because the whole can be regarded as a giant molecule. Germanium, silicon and grey-tin have the same structure.

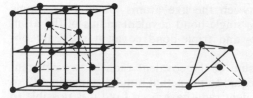

Fig. 50 The crystal structure of diamond showing (right) the tetrahedral arrangement of four carbon atoms round a central carbon atom.

b The zinc blende structure Zinc blende is one form of zinc sulphide. The crystal structure (Fig. 51) is related to that of diamond, the only difference being that adjacent atoms are different. Thus each Zn atom is surrounded by four S atoms arranged tetrahedrally, and each S atom by four Zn atoms similarly arranged.

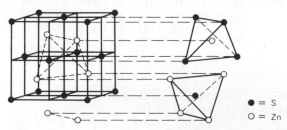

● = S
○ = Zn

Fig. 51 The crystal structure of zinc blende showing (upper right) the tetrahedral arrangement of four sulphur atoms about a zinc atom and (lower right) the tetrahedral arrangement of four zinc atoms about a sulphur atom. Compare the crystal structure of diamond (Fig. 50).

c The wurtzite structure Wurtzite is another form of zinc sulphide, and the structure (Fig. 52) is closely related to the zinc blende structure above. Each atom is surrounded by four nearest neighbours, arranged tetrahedrally, as in zinc blende. It is only when the second nearest neighbours are considered that the wurtzite structure differs from that of zinc blende.

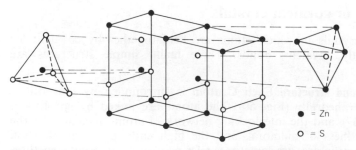

● = Zn
○ = S

Fig. 52 The crystal structure of wurtzite showing (right) the tetrahedral arrangement of four zinc atoms round a sulphur atom, and (left) the tetrahedral arrangment of four sulphur atoms around a zinc atom.

126

5 Average bond enthalpy

For a diatomic molecule, A—B, the bond enthalpy is defined as the enthalpy change for the process

$$A—B(g) \longrightarrow A(g) + B(g) \qquad \Delta H = \text{bond enthalpy}$$

For a polyatomic molecule, e.g. AB_3, the average bond enthalpy is defined as one third of the enthalpy change in the process

$$\begin{array}{c} B \\ | \\ B—A—B(g) \end{array} \longrightarrow A(g) + 3B(g)$$

Further details of the terms used, and methods of measurement are given on page 255.

Some typical values for average bond enthalpies for both single and multiple bonds are summarised below.

Average single bond enthalpies (kJ mol^{-1})

		H	C	N	O	F	Si	P	S	Cl	Br	I
	H	436	413	391	464	566	323	322	344	431	366	299
Group 4	C	413	346	305	358	485	301	264	272	339	284	218
	Si	323	301		368	582	226	213	226	391	310	234
Group 5	N	391	305	163	201	272		209		193		
	P	322	264	209	351	450	213	209	230	319	264	184
Group 6	O	464	358	201	146	190	368	351		205		201
	S	344	272			326	226	230	226	255	213	
Group 7	F	566	485	272	190	158	582	490	326	255	239	
	Cl	431	339	193	205	255	391	319	255	242	218	209
	Br	366	284			239	310	264	213	218	193	180
	I	299	218		201		234	184		209	180	151

Average multiple bond enthalpies (kJ mol^{-1})

C=C	N=N	O=O	C=O	C≡C	N≡N
611	410	497	749	835	945

			C≡O	C≡N	P≡P
			1071	890	488

The quoted values must be used with care, for they are average values and there can be a wide variation for the energy of what might appear to be the same bond in different compounds.

6 Molecular orbitals

It is clear that ions are held together by electrostatic attraction in an ionic bond, but it is less easy to understand why a shared pair of electrons should hold two atoms together. The best understanding comes from a consideration of molecular orbitals. These are orbitals within a molecule, very much like the atomic orbitals within an atom (p. 73). All, or nearly all, of the electrons associated with the component atoms of a molecule are supposed to enter the various molecular orbitals, which fill up according to certain rules just as atomic orbitals do (p. 66). The Pauli principle (p. 63) applies to molecular orbitals, as well as to atomic orbitals, and no molecular orbital can contain two precisely similar electrons. This means that any particular molecular orbital can only contain two electrons, and that these two must have different spins.

The nomenclature s, p and d used for atomic orbitals is replaced by that of σ, π and δ for molecular orbitals, σ and π being the most important. Electrons in some of these molecular orbitals contribute towards the binding together of the atoms in a molecule; such orbitals are known as *bonding orbitals*. Others cause repulsion between atoms and are known as anti-bonding orbitals.

The nomenclature most commonly used indicates (a) the nature of the molecular orbital, i.e. σ or π, occupied by electrons in a molecule, (b) the atomic orbital from which the electrons originated, and (c) whether the molecular orbital is bonding or anti-bonding. A $\sigma 1s$ orbital, for example, is a σ orbital made up by interaction of 1s atomic orbitals; it is a bonding orbital. A $\sigma^* 1s$ orbital is the corresponding anti-bonding orbital.

7 σ bonds

In a hydrogen molecule, H_2, the two 1s electrons, one from each of the two hydrogen atoms concerned, are present in a $\sigma 1s$ molecular orbital. This is a bonding orbital and constitutes the single covalent bond in the hydrogen molecule. Since the orbital can only hold two electrons if they have different spins, it is clear that a molecule will only be formed from two hydrogen atoms containing electrons with opposed spins. This is an important point. A covalent bond is not simply a shared pair; *it is a shared pair of electrons with opposed spins*. This means that only single, unpaired electrons, in atoms, can form covalent bonds.

An atomic orbital can be represented as a charge' cloud of varying density or, more conveniently, by mapping out the boundary surface within which the electron might be said to exist (p. 74). The same procedure can be adopted for molecular orbitals, as is illustrated for the hydrogen molecule in Fig. 53(a).

It is the accumulation of negative charge between the two positively charged atomic nuclei which is responsible for holding them together.

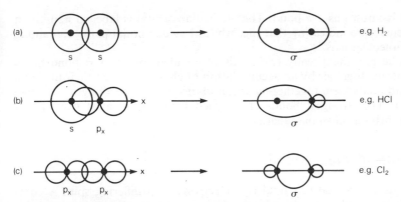

Fig. 53 The formation of σ bonds by overlap of **(a)** two s orbitals, **(b)** one s and one p orbital, or **(c)** two p orbitals end-on. The charge build-up is between the two nuclei.

Such an accumulation of charge occurs when two s orbitals containing electrons with opposed spins overlap. If electrons with the same spin are involved, one of them must enter a higher energy, antibonding orbital (p. 130).

The bonding molecular orbital formed from any pair of s electrons with different spins is plum-shaped (Fig. 53). It is symmetrical about the line joining the two nuclei and has no nodal plane, i.e. a plane in which the probability of finding an electron is zero (p. 74). The bond which it forms is called a σ bond, and the electrons which occupy a σ orbital are referred to as σ electrons.

Similar σ bonds can be formed by overlap of s and p atomic orbitals as in an HCl molecule (Fig. 53(b)), or, by end-on overlap of two p_x orbitals as in a Cl_2 molecule (Fig. 53(c)).

8 π bonds

Two p atomic orbitals can overlap end-on to form σ bonds but they can also overlap broadside-on, as shown for two $2p_z$ orbitals in Fig. 54. The accumulation of negative charge alongside the molecular axis constitutes

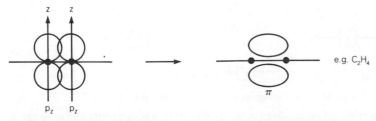

Fig. 54 The formation of a π bond by overlap of two p orbitals broadside-on. The charge build-up is alongside the two nuclei.

what is known as a π bond. The molecular orbital is labelled π2p, the π nomenclature indicating that the orbital has one nodal plane containing the molecular axis.

There are, then, really two kinds of covalent bond. A σ bond holds two atoms together by an accumulation of charge between their nuclei; a π bond holds them together by accumulation of charge alongside the two nuclei. The strongest bonds are formed from the greatest overlap between orbitals of similar energy.

9 Anti-bonding orbitals

Any atomic orbital can hold two electrons with differing spin. A combination of two atomic orbitals must, therefore, give two molecular orbitals so that all the four available electrons can be accommodated.

Two 1s atomic orbitals can form a σ1s molecular orbital; it is a bonding orbital with a build-up of charge between the nuclei. The two 1s atomic orbitals can, however, also be combined to give a molecular orbital which, because of its high energy level and because there is no charge build-up between the nuclei, is an anti-bonding orbital; it is labelled σ*1s (Fig. 55).

In the H_2 molecule, two electrons with opposed spins occupy the σ1s molecular orbital; the σ*1s anti-bonding orbital is not occupied (Fig. 56). In helium, with four available electrons, two electrons would have to occupy the σ1s orbital and two the σ*1s orbital if the He_2 molecule was to exist. A filled anti-bonding orbital more than counteracts a filled

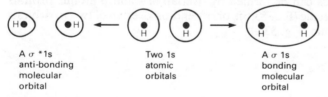

A σ *1s
anti-bonding
molecular
orbital

Two 1s
atomic
orbitals

A σ 1s
bonding
molecular
orbital

Fig. 55 The formation of two molecular orbitals, of different type, from two s atomic orbitals.

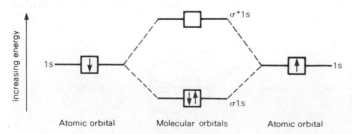

Fig. 56 The molecular orbital diagram for H_2. The σ*1s orbital, with the higher energy, is an antibonding orbital. The σ1s orbital is a bonding orbital. The two available electrons occupy the σ1s orbital.

bonding one, so that He_2 does not exist. That is why the noble gases are monatomic.

Similar molecular orbital diagrams can be drawn for molecules with more electrons.

10 Bond polarity

In a covalent bond between two unlike atoms the shared pair will not be shared equally, for the atom with the greatest attraction for electrons will 'pull' them in its direction. The resulting displacement of charge will give the bond some ionic character or cause it to be polar. The effect can readily be demonstrated by bringing an electrically charged piece of plastic (simply rub it on the sleeve) close to a jet of water; the water, with polar H_2O molecules, is attracted quite strongly.

Polarity in a bond, taking HCl as an example, can be shown by writing the bond as either $H^{\delta+}$—$Cl^{\delta-}$ or as H→Cl. The intermediate nature of the bond can also be illustrated in the molecular orbital shape shown in Fig. 57.

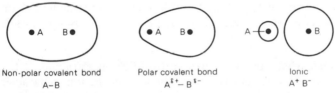

Non-polar covalent bond
A–B

Polar covalent bond
$A^{\delta+}$– $B^{\delta-}$

Ionic
$A^+ B^-$

Fig. 57 Representation of molecular orbital shapes for non-polar covalent bonds (as in Cl_2), polar bonds (as in HCl) and ionic bonds as in NaCl.

In an older method, the molecule was regarded as a resonance hybrid (p. 140) between H—Cl and H^+Cl^- and it was written as

$$H—Cl \longleftrightarrow H^+Cl^-$$

a Electronegativity values The degree of polarity in any covalent bond will depend on the relative electronegativities of the bonded atoms. The electronegativity of an atom is a measure of the extent to which it attracts electrons in a covalent bond. Various attempts have been made to allot numerical values to the electronegativities of atoms. All the methods are arbitrary, though they give results which agree reasonably well. Pauling's values for some common elements are summarised below.

H 2.1

Li 1.0	Be 1.5	B 2.0	C 2.5	N 3.0	O 3.5	F 4.0
Na 0.9	Mg 1.2	Al 1.5	Si 1.8	P 2.1	S 2.5	Cl 3.0
K 0.8	Ca 1.0		Ge 1.8	As 2.0	Se 2.4	Br 2.8
Rb 0.8	Sr 1.0					I 2.5
Cs 0.7						

There is a general fall in value in passing down a periodic table group, and a rise in value in passing along a period. The halogens (particularly fluorine) have high, and the alkali metals (particularly caesium) have low, values.

Pauling's electronegativity values, n_A and n_X, for two atoms, A and X, are based on the difference between the square root of $\{B(A—A) \times B(X—X)\}$ and the actual measured value of $B(A—X)$, where $B(A—A)$ represents the bond energy of the A—A bond, and so on. The geometric mean of $B(A—A)$ and $B(X—X)$ is an attempt to estimate what the bond energy of the A—X bond would be if it were *purely* covalent. The actual measured bond energy of A—X reflects its actual partial ionic character. The difference between the two was called the ionic resonance energy, Δ, by Pauling, and he then chose electronegativity values based on the relationship

$$96(n_A - n_X)^2 = \Delta$$

when the value of Δ is expressed in $kJ\,mol^{-1}$.

Mulliken arrived at a similar electronegativity scale by taking the electronegativity of an atom as the arithmetic mean of its ionisation energy and its electron affinity. Allred and Rochow base their figures on the ratio of the effective nuclear charge to r^2, where r is the atomic radius.

b Percentage ionic character The greater the difference, $(n_X - n_A)$, between the electronegativities of two atoms the greater the ionic character of the bond A—X. An $(n_X - n_A)$ value of 1.7 leads to 50 per cent ionic character; a value of 2.3 to 73 per cent. Estimates for typical bonds are

H—C	H—N	H—O	H—F	H—Cl	H—Br	H—I
4%	19%	39%	60%	19%	11%	4%

11 Effects of bond polarity

A covalent bond with some ionic character differs from a truly covalent bond in three main ways.

a Bond length The sum of the single bond covalent radii of two atoms A and B only gives the bond length, A—B, when the bond is purely covalent. A covalent bond with some ionic character will have a slightly different length, and there are various empirical formulae relating the difference in length to the electronegativities of the atoms concerned.

b Chemical reactivity A covalent bond with some ionic character has electrical charges associated with it. Such charges render the bond more liable to attack by other charged atoms or groups and, therefore, affect chemical reactivity and the detailed mechanism of chemical reactions.

c Dipole moments A magnet has a magnetic moment equal to ml, where m is the pole strength of the magnet and l the distance between the poles. In a similar way, two equal and opposite, but separated, electrical charges constitute an electrical dipole moment measured by the charge multiplied by the distance between the two charges.

Dipole moments are expressed in coulomb metre (C m) units, though an older unit, the Debye (D), is still used; it is equal to $3.335\,640 \times 10^{-30}$ C m. As the charge on an electron is approximately 1.6×10^{-19} C and molecular distances are of the order 10^{-10} m it follows that dipole moments are generally somewhat smaller than 1.6×10^{-29} C m or less than about 5 D. The direction of a dipole is conventionally regarded as being directed from the negative to the positive end.

Any bond which has any degree of polarity will have a corresponding dipole moment, though it does not follow that compounds containing such bonds will have dipole moments, for the polarity of the molecule as a whole is the vector sum of the individual bond moments. The C—Cl bond, for instance, has definite polarity, C$\rightarrow$Cl or $C^{\oplus}$—$Cl^{\ominus}$, and a definite dipole moment. Tetrachloromethane, however, has no dipole moment for the resultant of the four C—Cl moments is zero. By comparison, mono-, di- and tri-chloromethane all have dipole moments.

The methods of dipole moment measurement are outside the scope of this book, but such measurements provide a mass of data which can be used in a variety of ways to solve chemical problems. The ionic character of a bond, for instance, can be estimated from the values of its dipole moment and its bond length. The dipole moment for HCl, for example, is 3.436×10^{-30} C m. If the bond was fully ionic the expected dipole moment would be equal to the charge on the electron multiplied by the bond distance, i.e. $(1.602 \times 10^{-19}) \times (1.29 \times 10^{-10})$ or 20.67×10^{-30}. The actual ionic character of the H—Cl bond may, therefore, be taken as $343.6/20.67$ per cent, i.e. approximately 17 per cent. This result is in good agreement with that obtained from the electronegativities of hydrogen and chlorine (p. 131).

(Questions on this chapter will be found on p. 154.)

13
Hybridisation and delocalisation

1 Hybridisation

s, p and d atomic orbitals, of comparable energy, can be combined, by superimposition within an atom, to form what are known as hybrid orbitals. These have different shapes from s, p and d orbitals and they form stronger bonds by giving greater overlap in the bonding process.

Hybridisation, or mixing, is simply a matter of taking the wave functions of s, p and d orbitals and combining them together (by adding or subtracting) to give new wave functions representing new hybrid orbitals.

a sp (linear) hybridisation The arrangement of electrons in the beryllium atom, with both 1s and 2s orbits completely full, *i*,

		1s	2s	2p
i	Be	↓↑	↓↑	
ii	Be*	↓↑	↓	↓

would suggest that beryllium might form no compounds as it has no unpaired electron, whereas it is well known to be divalent. Prior to chemical combination, one of the two 2s electrons is promoted into a 2p orbital to give an excited beryllium atom, denoted by an asterisk, *ii*. If the available 2s and 2p electrons were used separately for bond formation, two different types of bond would be formed, whereas the two bonds in BeX_2 compounds are alike. The two available 2s and 2p electrons are, in fact, combined to form two collinear sp hybrid orbitals, with a shape as shown in Fig. 58. The large lobes of the two sp orbitals protrude along the axis further than the corresponding s and p orbitals so that the sp hybrids can form stronger bonds. In $BeCl_2$, for example, the two sp hybrids overlap with two $3p_x$ orbitals from two chlorine atoms to form two σ bonds (Fig. 59). The energy required to form the excited atom is recovered in the formation of stronger bonds.

sp hybridisation is also common in carbon compounds. The structures of the carbon atom in its ground and excited states are shown below:

	1s	2s	$2p_x$	$2p_y$	$2p_z$
C (ground state)	↓↑	↓↑	↓	↓	
C* (excited)	↓↑	↓	↓	↓	↓

One of the 2p electrons, in the excited state, can form sp hybrid orbitals with the 2s electron; two 2p atomic orbitals remain. In ethyne, C_2H_2, for example, each carbon atom has two collinear, sp hybrid orbitals. These

Fig. 58 Two collinear sp orbitals.

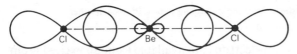

Fig. 59 The formation of two σ bonds in $BeCl_2$ by overlap of $3p_x$ orbitals from two chlorine atoms with two hybrid sp orbitals from a beryllium atom.

overlap to form a σ bond between the two C atoms, and each of the two sp hybrids also overlap with a 1s atomic orbital of a hydrogen atom to form σ bonds between C and H. It is because sp hybrids are involved that C_2H_2 is a collinear molecule (Fig. 60). Each of the two C atoms has two remaining 2p atomic orbitals which interact to form two π bonds between the C atoms, these two bonds being in planes at right-angles to each other. The C≡C bond consists, therefore, of a σ bond and two π bonds (Fig. 61). It is the π bonds which prevent free rotation about the C≡C bond.

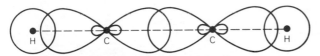

Fig. 60 σ bonds in ethyne formed by overlap of two sp hybrid orbitals (between the two carbon atoms) and by overlap of an sp hybrid and an s orbital (between the carbon and hydrogen atoms).

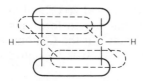

Fig. 61 The bonding in an ethyne molecule. Two σ bonds between carbon and hydrogen atoms. One σ bond and two π bonds (in planes at right angles to each other) between carbon atoms.

b sp² (trigonal) hybridisation Combination of one s and two p atomic orbitals gives rise to three sp² hybrid orbitals which are coplanar and directed at angles of 120° to each other (Fig. 62).

Fig. 62 Three sp² orbitals in the same plane but at angles of 120° to each other.

In boron, one s and two p electrons are available from an excited atom

	1s	2s	2p$_x$	2p$_y$	2p$_z$
B	⇅	⇅	↓		
B*	⇅	↓	↓	↓	

and the resulting sp² hybridisation means that the BX_3 compounds of boron are planar with bond angles of 120°.

In ethene, C_2H_4, each of the two C atoms forms bonds through sp² orbitals. Two of the orbitals of each atom form σ bonds with the 1s orbitals of hydrogen atoms. The remaining sp² orbital of each C atom forms a σ bond between the C atoms. The two C and four H atoms are all in the same plane and the bond angles are 120°.

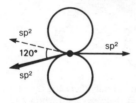

Fig. 63 The 2p orbital remaining after the formation of three coplanar sp² orbitals.

At right-angles to this plane there remains the unchanged 2p orbital of each C atom (Fig. 63), and these two 2p orbitals interact to form a π bond between the two C atoms. The double bond between the C atoms consists, therefore, of a σ bond and a π bond (Fig. 64). It is the latter which prevents free rotation.

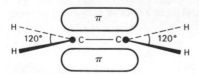

Fig. 64 The bonding in an ethene molecule. Four σ bonds between carbon and hydrogen atoms. One σ bond and one π bond between carbon atoms.

c sp³ (tetrahedral) hybridisation The four hybrid orbitals formed by hybridisation of one s and three p orbitals from, for example, an excited C atom are known as sp^3 orbitals; they are directed towards the corners of a tetrahedron (Fig. 65). The bond angles in the resulting tetrahedral molecules, e.g. CH_4, are always close to the expected theoretical angle of 109° 28'.

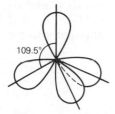

Fig. 65

d hybridisation involving d orbitals In larger atoms, d orbitals can be combined with s and p orbitals to form other hybrids as, for example, in PCl_5 and SF_6.

Phosphorus pentachloride exists as PCl_5 molecules, in the vapour state. sp^3d hybrid orbitals are involved, formed by the excited P atom; the shape of the molecule is trigonal bipyramidal (Fig. 49). SF_6 is octahedral (Fig. 49) and involves six sp^3d^2 hybrid orbitals formed by the excited S atom.

	1s	2s	2p	3s	$3p_x$	$3p_y$	$3p_z$	3d
P	2	2	6	⇅	↓	↓	↓	
P*	2	2	6	↓	↓	↓	↓	↓
S	2	2	6	⇅	⇅	↓	↓	
S*	2	2	6	↓	↓	↓	↓	⇊

Similar hybrid orbitals are of importance in accounting for the shape of many complexes of d-block elements.

2 Delocalised bonds

In a σ bond, the concentration of charge is between the bonded atoms. The bond is said to be localised (between the atoms) and it is reasonably well represented, at least positionally, by the traditional A—X or A ⋮ X symbolism.

In a π bond between two atoms, the concentration of charge is alongside the atoms, though a *single* π bond is still localised between them. When more than two adjacent atoms have available p orbitals, however, the overlap can take place across *all the orbitals* concerned to form what is known as a delocalised bond. Such bonds can account much more effectively than localised bonds for the properties of many molecules.

The formation of delocalised bonds is very well marked in benzene, C_6H_6 (Fig. 66). Each of the six C atoms in the hexagonal ring forms three σ bonds (two to adjacent C atoms and one to an H atom) using sp^2 hybrid orbitals; because the sp^2 orbitals are planar, all the six C and six H atoms are in the same place, and the bond angles are 120°. Each C atom still has an available p orbital, with an axis at right-angles to the plane of the C atoms, and the six p orbitals overlap with each other to form delocalised bonds all round the ring.

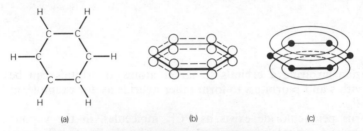

(a)   (b)   (c)

Fig. 66 Bonding in benzene. **(a)** The σ bonds. **(b)** The p orbitals on each carbon atom and (dotted lines) the way in which they can overlap. **(c)** The delocalised molecular orbital all round the ring.

a Bond lengths The bonds between C atoms in benzene are, therefore, neither C—C nor C=C bonds as used to be suggested in earlier methods of formulating the molecule. The actual bonds consist of a σ bond with some part of the delocalised π bonding. Bond length measurements bear this out. The carbon–carbon bonds are all of equal length (0.1397 nm), shorter than the C—C bond in ethane (0.154 nm) and longer than the C=C bond in ethene (0.134 nm).

b Delocalisation energy The enthalpy of formation of benzene from its free atoms, as measured experimentally, is $-5505\,\text{kJ mol}^{-1}$, whereas that calculated (using bond enthalpy values) for the older formulation, containing separate single and double bonds (Fig. 67), is $-5349\,\text{kJ mol}^{-1}$. The actual molecule in benzene, with delocalised bonds, is stabler by $156\,\text{kJ mol}^{-1}$; this is known as the delocalisation (or resonance) energy.

A similar figure can be obtained, too, from enthalpies of hydrogenation (p. 257). The expected value for benzene, if it had the structure in Fig. 67, would be $-361.5\,\text{kJ mol}^{-1}$, i.e. three times the enthalpy of hydrogenation of cyclohexene (with one C=C bond in the ring). The actual measured enthalpy of hydrogenation of benzene is, however, only -208.4. The delocalisation energy from these figures is $153.1\,\text{kJ mol}^{-1}$.

The relationship between the various figures is summarised in Fig. 68.

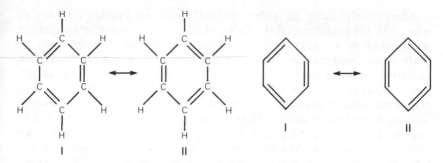

Fig. 67 Two of the commonest canonical forms for benzene; they are sometimes called Kekulé structures.

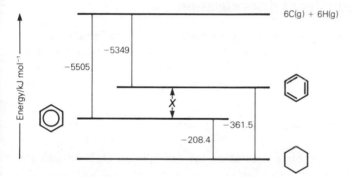

Fig. 68 Showing the relationship between enthalpies of formation and hydrogenation, and the delocalisation energy (X) for benzene.

c Representation of delocalised bonds The $—$, $=$ and $≡$ symbolism, for representing covalent bonds, has to be modified to indicate the existence of delocalised bonds. The general method adopted is to add a dotted line. so that the benzene molecule might be written as in Fig. 69(a). To facilitate the writing of this very common structure it has been modified into the structure shown in Fig. 69(b), but the dotted line symbolism is used in many other examples.

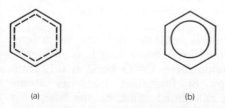

(a)   (b)

Fig. 69 Modern way of showing the existence of delocalised bonds. As **(b)** is easier to write it is generally used.

Alternatively, using an older method based on Pauling's theory of *resonance*, the actual structure of a molecule, which cannot be adequately represented in a single structure using —, = and ≡ symbolism, is written as a combination of two or more structures which can be written in traditional ways, but which do not, in fact, exist. Benzene, for example, is written as in Fig. 67. The actual structure in benzene is referred to as a resonance hybrid between the structures I and II; these nonexistent structures are known as canonical forms or structures. The ↔ is used to indicate the fact that the actual structure is somewhere between the two canonical structures. Alternatively, the term *mesomeric structures* has been used, the concept being known as *mesomerism* ('between the parts').

3 Other examples of delocalisation

Delocalisation can occur when adjacent atoms have available p orbitals all in the same plane so that they can overlap. Some typical situations where this can occur are

$$-C=C-C=C- \qquad -C=C-\overset{..}{\underset{..}{X}}$$

$$\langle\bigcirc\rangle-C=C- \qquad \langle\bigcirc\rangle-\overset{..}{\underset{..}{X}}$$

where X has an available p orbital as, for example, in O, N or halogen atoms.

The greater the extent of delocalisation the greater the delocalisation energy. The value for benzene, with one ring, is $153\,kJ\,mol^{-1}$; for naphthalene, with two rings, it is $297\,kJ\,mol^{-1}$; and for anthracene, with three rings, it is $435\,kJ\,mol^{-1}$.

a Graphite The C atoms in the layers of the layer lattice structure (p. 153) are linked by σ bonds, using sp^2 hydrid orbitals, as in benzene. Each C atom has a p orbital and all these orbitals in any one layer can overlap to give extensively delocalised bonding, above and below the layer, throughout the structure.

This extensive delocalisation accounts for the electrical conductivity of graphite, and repulsion between the delocalised bonds in each layer is partially responsible for the large distance between the layers.

b Carbon dioxide The CO_2 molecule is collinear and was, for a long time, written as O=C=O. The measured bond length is, however, $0.115\,nm$ whereas the calculated value for a C=O bond is $0.122\,nm$. Moreover, the measured enthalpy of formation from its atoms $(-1602\,kJ\,mol^{-1})$ is more than the calculated value, on the basis of a O=C=O structure $(-1447.6\,kJ\,mol^{-1})$, i.e. the delocalisation energy is $154.4\,kJ\,mol^{-1}$.

The molecule is best represented as containing σ bonds between C and O atoms, made from sp hybrid orbitals of the C atom (as in ethyne) overlapping with p orbitals of the O atoms. All three atoms have remaining p orbitals, which can provide two π bond systems in perpendicular planes. Delocalisation occurs within the π-bond systems. This can be summarised as

$$O\!\equiv\!\!\equiv\!C\!\equiv\!\!\equiv\!O$$

Alternatively, the molecule can be written as a resonance hybrid between the following canonical structures,

$$O\!=\!C\!=\!O \quad\longleftrightarrow\quad O\!\leftarrow\!C\!\equiv\!O \quad\longleftrightarrow\quad O\!\equiv\!C\!\rightarrow\!O$$

c The nitrate and carbonate ions All the bonds in the planar nitrate ion are of equal length (0.121 nm), between that calculated for the N—O bond (0.136 nm) and the N=O bond (0.115 nm); the bond angles are 120°. The N atom forms σ bonds with the three O atoms, using sp² hybrid orbitals, and the remaining p-orbitals on the N and O atoms give delocalised π bonding; the delocalisation energy is 188 kJ mol⁻¹. The ion can be represented as shown (Fig. 70), and the carbonate ion has a very similar structure (Fig. 71).

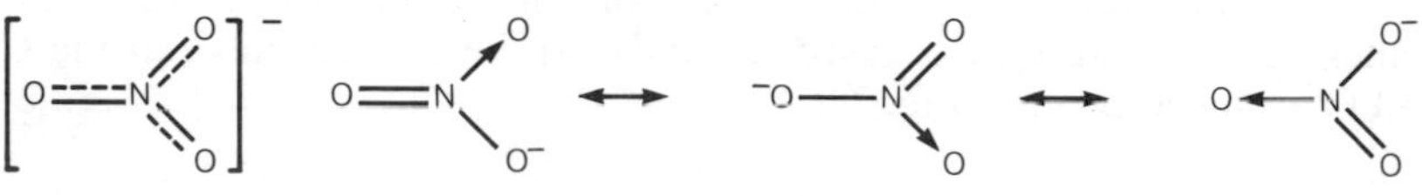

Fig. 70 The structure of the nitrate ion.

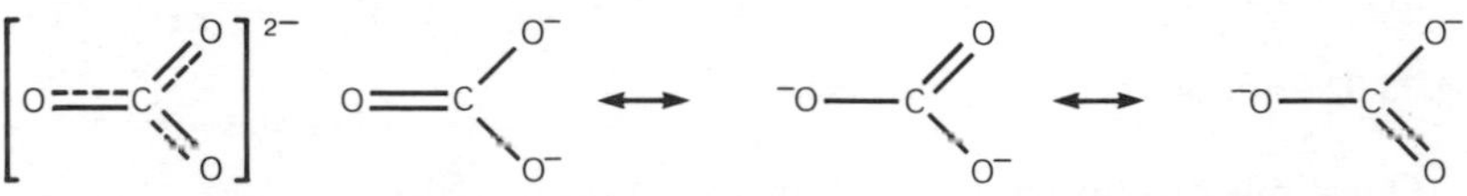

Fig. 71 The structure of the carbonate ion.

d Chloroethene and chlorobenzene In chloroethene, the C atoms form σ bonds, using sp² hybrid orbitals, and the p orbitals on the C and Cl atoms overlap to give a delocalised π bond, which is spread out over the C—C—Cl chain length. The bonding in the molecule consists, then, of localised σ bonds together with delocalised π-bonds (Fig. 72).

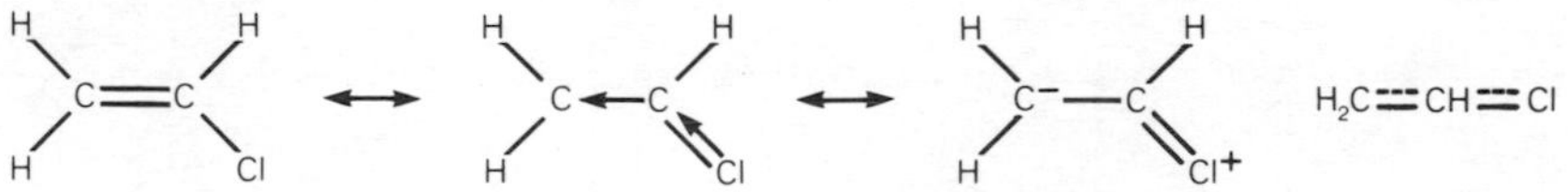

Fig. 72 The structure of chloroethene.

The effect of the delocalised bonding is to make the carbon–carbon bond rather less of a double bond than it is in ethene (where delocalisation is not possible as the H atoms have no available p orbitals), and to give some double bond character to the carbon–chlorine bond as compared with the bond in C_2H_5Cl (where delocalisation is not possible as the carbon atoms have no available p orbitals). The relevant bond lengths bear this out. C—Cl is 0.178 nm in chloroethane; C⚌Cl is 0.169 nm in chloroethene. C⚌C is 0.134 nm in ethene; C⚌C is 0.138 nm in chloroethene.

A similar situation exists in chlorobenzene, with delocalisation all over the molecule caused by overlap of the six carbon p orbitals with the p orbital of the Cl atom. The C⚌Cl bond length is 0.169 nm, as in chloroethene.

In both chlorobenzene and chloroethene, the Cl atom is less easily replaced than in chloroethane because of the partial double-bond nature of the C⚌Cl bonds caused by delocalisation.

The C⚌N and C⚌O bonds in phenylamine and phenol are, similarly, different from those in methylamine and ethanol.

e The ethanoate ion, $CH_3.COO^-$ The carbon–oxygen bonds are both of equal length (0.127 nm), shorter than the C—O bond in methanol (0.143 nm) and longer than the C⚌O bond in ethanal (0.122 nm). This comes about because delocalised π bonds can be formed between the C and O atoms in the ion (Fig. 73).

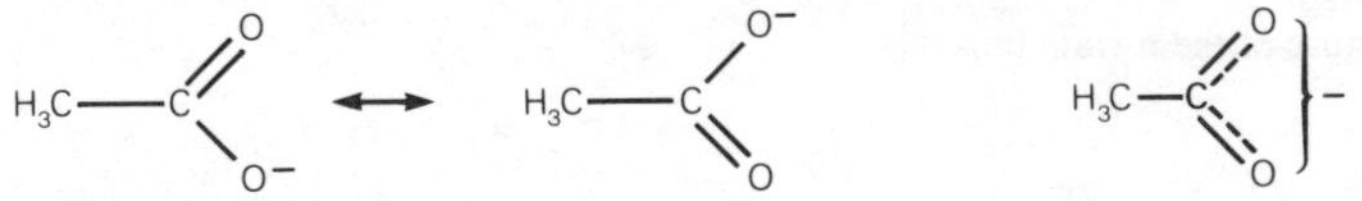

Fig. 73 The structure of the ethanoate ion.

(Questions on this chapter will be found on p. 154.)

14
Metallic, hydrogen and van der Waals' bonding

The nature of the bonding in the majority of chemicals can be described in terms of ionic or covalent bonds. The situation in metals requires, however, the postulation of a special type of metallic bond. There are, also, a number of hydrogen compounds in which hydrogen bonding plays a large part, and the bonding between molecules in molecular crystals, and in liquids, involves van der Waals' forces.

The metallic bond

Metals have very distinctive properties and to account for them, particularly for electrical conductivity, the idea of a special metallic bond is necessary. The general picture of the state of affairs in the crystal of a metal is that cations of the metal are packed in a regular array within the crystal and that the cations are surrounded by electrons which are relatively mobile. The cause of the mobility of the electrons is interpreted in terms of the band model of electronic energy levels in metallic crystals.

1 The band model

The arrangement of electrons in a sodium atom is

1s 2s 2p 3s
2   2   6   1

each electron being in a particular energy level. If two sodium atoms come close together, they will interact with each other so that the energy levels in each atom will be slightly affected. In a crystalline array of close-packed atoms, this interaction will cause the single, discrete energy levels of the single atoms to be replaced by bands of closely related energy levels. For two atoms there would be two levels within a band; for n atoms there are n possible levels. The bands must be regarded as belonging to the crystal as a whole and not to any individual atom.

In a sodium crystal n levels within the 1s band can be postulated. As each of the n atoms has two 1s electrons, all the n levels will be full, and this will be so, too, for all the levels within the 2s and 2p bands. For the n possible levels in the 3s band, however, there are only n electrons, whereas the band could hold $2n$ electrons. All the levels in the 3s band are not, therefore, full. As the energy difference between the levels in the

3s band are very small, these 3s electrons can move about within the 3s band very easily. In a crystal under ordinary conditions, equal numbers of 3s electrons will move in all directions but, under an applied potential difference, the electrons will move in one direction and a current will flow. It is this movement of electrons, within what is sometimes known as a conduction band, which accounts for electrical conductivity.

For a metal with an even number of valency electrons, e.g. Mg, 2.8.2, it might be expected that all the levels in the 3s band would be full because each atom contributes two 3s electrons. Magnesium is, however, still a good conductor because some of the levels in the 3p band overlap those in the 3s band and provide opportunities for movement of electrons.

Electrical conductivity is associated, then, with either partially filled bands or with the overlapping of an unfilled band with a full one.

It might be expected that increase in temperature would cause more electrons to be 'promoted' into conducting bands with a consequent increase in conductivity or decrease in resistance. In general, however, the conductivity of a metal decreases with increase in temperature, i.e. the resistance increases. This is because increasing temperature produces increased thermal vibration within a metal crystal. This upsets the regularity within the crystal and interferes with the ease of movement of electrons through the crystal. It is rather like comparing movement through the ranks of a battalion of soldiers on parade with that through a London crowd. Similarly, the introduction of an impurity may upset the regular array and so cause increased resistance. The resistance of copper, for example, is greatly increased by even traces of impurities.

2 Semiconductors

Some pure and impure metals have rather low conductivities which increase with temperature. Such conductivity comes about in three different ways.

a Intrinsic semiconductors In magnesium there is an overlap between the 3s and 3p bands. In substances like germanium or grey tin there is no actual overlap of bands but the energy difference between the highest filled band and the next empty one is very small. It might, therefore, be possible for an electron to gain enough energy to pass from the full to the empty band; conductivity would result. The chance of an electron in the full band being able to pass into the empty band would increase as the temperature increased, i.e. as the energy of the electrons increased. With these substances, called intrinsic semiconductors, the conductivity increases with increase in temperature.

b n-type semiconductors The addition of an impurity to a metal will provide additional energy levels and, if these levels are correctly related to the bands within the pure metal, conductivity may result.

If the impurity contains a full energy level which is just below that of an empty band in the pure metal, the electrons from the impurity might have enough energy to pass into the empty conducting band in the pure metal. This happens when arsenic or antimony are added to germanium. The passage of electrons from an energy level in the arsenic or antimony into one in the germanium causes the germanium to become negatively charged. It is, therefore, known as n-type germanium.

c p-type semiconductors If the impurity contains an empty energy level just above that of a full band in the pure metal, the electrons from the full band in the pure metal might be able to pass into the empty level of the impurity. This happens when gallium or indium are added to germanium. Passage of electrons from the germanium to the gallium or indium causes the germanium to become positively charged. It is called p-type germanium.

Semiconductors of n- and p-type have conductivities which, like that of germanium, increase with temperature. They are of great importance in making transistors.

The various types of conductivity are summarised in Fig. 74.

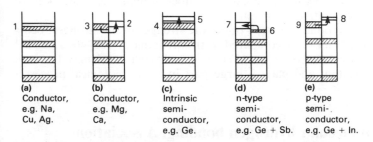

(a)	(b)	(c)	(d)	(e)
Conductor,	Conductor,	Intrinsic	n-type	p-type
e.g. Na,	e.g. Mg,	semi-	semi-	semi-.
Cu, Ag.	Ca,	conductor,	conductor,	conductor,
		e.g. Ge.	e.g. Ge + Sb.	e.g. Ge + In.

Fig. 74 Electrical conductivity. Full bands are shaded. **(a)** Conductivity due to partially filled band 1. **(b)** Conductivity due to overlap of empty band 2 with full band 3. **(c)** Conductivity due to narrow gap between full band 4 and empty band 5. **(d)** Conductivity due to full band 6 in impurity just below empty band 7 in pure metal. **(e)** Conductivity due to empty band 8 in impurity just above full band 9 in pure metal.

Hydrogen bonding

3 Introduction

A hydrogen atom normally forms one bond, but there are certain compounds in which it can form two.

Hydrogen fluoride, for example, is known, from relative molecular mass measurements, to be associated, i.e. to exist as $(HF)_n$, and the acid salt, potassium hydrogendifluoride, KHF_2, must be derived from the acid

H_2F_2 and contain the hydrogendifluoride, HF_2^- ion. The ion is envisaged as containing two negatively charged fluoride ions linked together by a positively charged hydrogen ion (proton). The proton is thought to be able to exert a sufficiently strong electrostatic attraction to do this because of its small size. The resulting bonding is called hydrogen bonding or, sometimes, proton bonding.

To distinguish a hydrogen bond, it is best to write it as a dotted line so that the hydrogendifluoride ion becomes $(F \cdots H—F)^-$. In general, when a hydrogen bond links two atoms, A and B, the structure is represented as A—H$\cdots$B or as a resonance hybrid between A—H$\cdots$B and A$\cdots$H—B.

That the electrostatic mechanism for the formation of a hydrogen bond is probably correct is shown by the fact that a hydrogen bond A—H$\cdots$B is formed most easily if A and B have high electronegativities. Thus the tendency of an A—H bond to form a hydrogen bond with another atom, B, increases greatly from C—H through N—H and O—H to F—H, and it decreases in passing from O—H to S—H or from F—H to Cl—H. This shows that the bond A—H has the greatest tendency to form hydrogen bonds when the ionic character of the bond is greatest, i.e. when the bond has the greatest polar character, $A^{\delta-}—H^{\delta+}$.

Fluorine, with the highest electronegativity, forms the strongest hydrogen bonds, and by far the greatest number of hydrogen bonds known are those which unite two oxygen atoms.

The hydrogen bond is a weak bond, the strength of the strongest being about $20–40 \, kJ \, mol^{-1}$ as compared with strengths of $120–400 \, kJ \, mol^{-1}$ for normal covalent bonds (p. 127). The hydrogen bond may be weak but it is important and plays a large part in many biological processes (p. 150).

4 Inter-molecular hydrogen bonding. Association

Examples of compounds containing hydrogen bonds between two or more molecules, with some of the evidence which points to the existence of such bonds, are given below.

a Hydrides of fluorine, oxygen and nitrogen Association like that in hydrogen fluoride is found in water (Fig. 75) and in ammonia and shows itself in the high relative permittivity (dielectric constant) of these three hydrides and also in their abnormal melting and boiling points

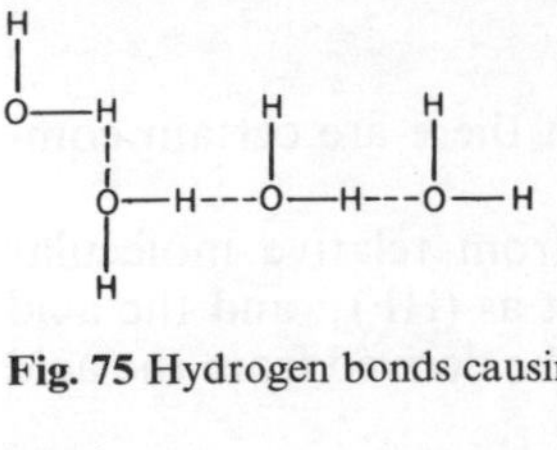

Fig. 75 Hydrogen bonds causing association in water.

compared with other hydrides in the same groups of the periodic table (Fig. 76).

Methane has normal values for its melting and boiling points. It is not associated, as carbon is not sufficiently electronegative to be linked by hydrogen bonds.

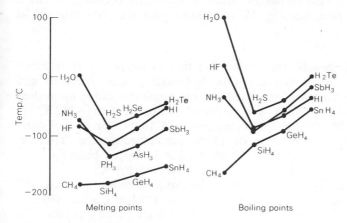

Fig. 76 Melting and boiling points of hydrides of Group IV, V, VI and VII elements.

b Ice The crystal structure of ice shows a tetrahedral arrangement of water molecules similar to that found in the wurtzite structure (p. 126). Each oxygen atom is surrounded, tetrahedrally, by four others. Hydrogen bonds link pairs of oxygen atoms together as shown in Fig. 77.

The distance between adjacent oxygen atoms is 0.276 nm and this suggests that the hydrogen atom linking the two oxygen atoms together is not midway between them, for the O—H distance in water vapour is 0.096 nm and not half 0.276 nm. Distance measurements on other compounds, too, indicate that the hydrogen atom in a hydrogen bonded pair of atoms is not equidistant from the two atoms.

Fig. 77 The crystal structure of ice. The central oxygen atom, A, is surrounded tetrahedrally by the oxygen atoms, 1, 2, 3 and 4. All other oxygen atoms are arranged similarly. The hydrogen atoms are shown as small circles, and the dotted lines indicate the hydrogen bonds.

147

The arrangement of the water molecules in ice is a very open structure and this explains the low density of ice. When ice melts, the structure breaks down and the molecules pack more closely together so that water has a higher density but this breaking down process is not complete until a temperature of 4 °C is reached.

Hydrogen sulphide shows no signs of hydrogen bonding; it has a normal boiling point (see Fig. 76) and a close-packed crystal structure in the solid state.

c Alcohols and phenols Association like that in water occurs in many other compounds containing —OH (hydroxyl) groups. Alcohols and phenols, for example, are associated like water and this shows up in their high boiling points; 64.5 °C for methanol compared with −162 °C for methane or −24.2 °C for chloromethane, and 182 °C for phenol compared with 80 °C for benzene or 132 °C for chlorobenzene. The corresponding sulphur compounds, with weaker hydrogen bonding, have lower boiling points; CH_3SH (5.8 °C) and C_6H_5SH (169 °C).

In compounds with two or more hydroxyl groups, there are still greater opportunities for hydrogen bonding. Ethane–1,2–diol (197 °C) and propane–1,2,3–triol (290 °C) both have high boiling points. The increase in viscosity as the number of —OH groups increases is also due to the increased hydrogen bonding.

d Copper(II) sulphate-5-water Gentle heating of copper(II) sulphate pentahydrate will convert it into the monohydrate: it is, in fact, efflorescent in hot, dry climates. A much higher temperature is required, however, to remove the final molecule of water of crystallisation from the monohydrate. This suggests that one of the molecules of water of crystallisation is different from the other four, and leads to the writing of the formula as $Cu(H_2O)_4SO_4.H_2O$. The crystal structure is shown in Fig. 78.

$Cu(NH_3)_4SO_4.H_2O$ exists and has a structure like that of $Cu(H_2O)_4SO_4.H_2O$ but $CuSO_4.5NH_3$ does not exist. This is probably

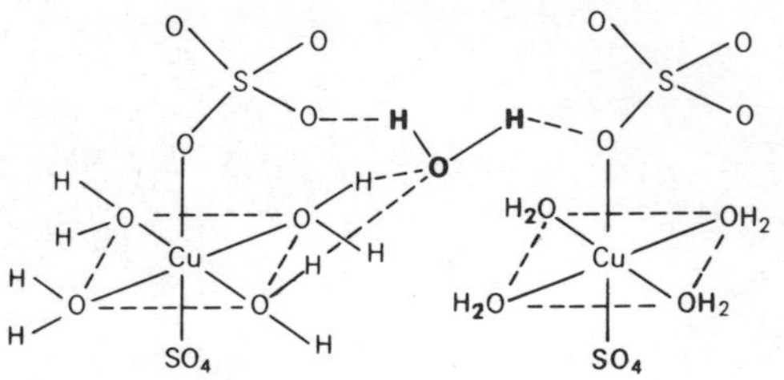

Fig. 78 The arrangement in $CuSO_4.5H_2O$. Each Cu^{2+} ion is surrounded octahedrally by four H_2O molecules and two SO_4^{2-} ions. One H_2O molecule (shown in bold type) is linked between two such octahedral groups by hydrogen bonding.

due to the fact that NH_3 molecules will not form hydrogen bonds so readily as H_2O molecules.

e Carboxylic acids Some carboxylic acids associate into dimers, i.e. two molecules link together, both in the vapour state and in certain solvents. Partition coefficient measurements (p. 200) of the distribution of ethanoic acid between water and benzene show, for instance, that the acid is present as a dimer in the organic solvent. This dimer is written as:

$$H_3C - C \underset{\displaystyle O \cdots H - O}{\overset{\displaystyle O - H \cdots O}{\Big\langle}} \Big\rangle C - CH_3$$

and the presence of an eight-membered ring is confirmed by electron diffraction studies. Relative density measurements also show the presence of double molecules of ethanoic acid in the vapour state.

In aqueous solution, the molecules of a carboxylic acid link up with water molecules rather than form dimers.

f Amines The association and basic strength of amines are both explained in terms of hydrogen bonding.

Primary and secondary amines are associated to some extent, though not greatly because nitrogen does not form hydrogen bonds very readily. Tertiary amines are not associated at all for they have no hydrogen atom capable of forming hydrogen bonds. Thus trimethylamine, which is not associated, has a lower boiling point $(4\,^\circ C)$ than dimethylamine $(7\,^\circ C)$ even though it has a higher relative molecular mass.

In aqueous solution, amines react with water molecules as shown,

$$CH_3NH_2 + H_2O \rightleftharpoons CH_3\overset{\displaystyle H}{\underset{\displaystyle H}{N}} - H \cdots O - H \rightleftharpoons (CH_3NH_3)^+ + OH^-$$

The resulting solution will contain some CH_3NH_3OH molecules together with some $CH_3NH_3^+$ and OH^- ions. In a solution of a quaternary base, e.g. tetramethylammonium hydroxide, $[(CH_3)_4N]OH$, however, there are no hydrogen atoms which could form hydrogen bonds. As a result the solution contains only $(CH_3)_4N^+$ and OH^- ions. This explains why quarternary bases are very much stronger than primary, secondary or tertiary amines. It was, in fact, the marked basic strength of quaternary bases which first led Moore and Winmill (1912) to suggest the possible existence of hydrogen bonds.

5 Intra-molecular hydrogen bonding

Association occurs when hydrogen bonding takes place between two or more molecules. Similar bonding may, however, also take place within a single molecule, as in 2-nitrophenol; it is known as intra-molecular bonding.

2-nitrophenol boils at 214 °C as compared with 290 °C for 3- and 279 °C for 4-nitrophenol. 2-nitrophenol is, moreover, volatile in steam, and less soluble in water than the other two isomers. All these facts can be accounted for on the assumption that 2-nitrophenol contains an internal hydrogen bond represented as:

This intra-molecular hydrogen bonding prevents inter-molecular bonding between two or more molecules. But in the 3- and 4-isomers, intra-molecular bonding is not possible because of the size of the ring which would have to be formed. Inter-molecular bonding therefore takes place, causing association, which accounts for the higher boiling points of the 3- and 4-isomers.

The low solubility of 2-nitrophenol in water may be explained in two ways. The formation of an internal hydrogen bond 'suppresses' the hydroxylic character of the compound, or it prevents hydrogen bonding between 2-nitrophenol and water.

The effect of hydrogen bonding in 2-nitrophenol is also shown spectroscopically. A normal —OH group is found to give rise to a particular line in the infrared absorption spectrum (p. 184) of the substance concerned. The spectra of 3- and 4-nitrophenol show this line, but that of 2-nitrophenol does not. There is not the same difference between the three methyl ethers of the nitrophenols because hydrogen bonding cannot take place in these ethers.

Other compounds in which intramolecular hydrogen bonding plays the same part as in 2-nitrophenol include 2-hydroxybenzaldehyde, 2-chlorophenol and 2-hydroxybenzoic acid.

6 Hydrogen bonding in living organisms

Both inter- and intramolecular hydrogen bonding occur in living organisms.

Proteins, for example, consist of chains of amino acids, the linear sequence of the acids being known as the *primary structure*. The *secondary structure* refers to the detailed configuration of the chain, and one of the commonest configurations is a spiral form known as the α-helix form. The amino acid units are arranged in a helix, like a stretched coil spring. The arrangement is stabilised by hydrogen bonding between the N—H group of each amino acid and the fourth C=O group following it along the chain (Fig. 79). The hydrogen bond is relatively weak so that the helical arrangement can readily be converted into an extended chain; this probably accounts for the fact that many protein materials can stretch so freely.

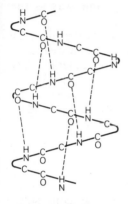

Fig. 79 Diagrammatic representation of the hydrogen bonding in the α-helix structure of a protein molecule.

The *tertiary structure* of a protein involves the way in which helical or extended molecules are arranged in relation to each other. A number of helices might twist together, like the wires in an electric cable, or extended molecules might pack together side by side. In both cases, intermolecular hydrogen bonding may hold the structure together.

In nucleic acids, two intertwining helices (a double helix) are held together by hydrogen bonding.

Van der Waals' forces

Ions are held together by electrostatic attraction in ionic crystals (p. 109), and atoms are held together by covalent bonds in atomic or covalent crystals (p. 125). The nature of the forces holding molecules together in a molecular crystal are, however, of a different nature, as are the forces holding molecules together in a liquid. These forces are known as van der Waals' forces after the man who first took them into account in modifying the gas equation (p. 29).

7 Nature of the forces

When the molecules held together in a molecular crystal are polar in nature, e.g. hydrogen chloride and ammonia, there will be a dipole–dipole attraction between the molecules when they are correctly orientated. Such attractive forces will also be increased by the fact that one dipole will induce another in a neighbouring molecule.

With non-polar substances, such as the halogens and the noble gases, there are no initial dipoles; the molecules or atoms are electrically symmetrical. The forces in these cases are caused by slight and temporary displacements between nuclei and electrons, giving rise to temporary dipoles, which lead to attractive forces in the same way as the more permanent dipoles in polar molecules.

Van der Waals' forces in solids are ten to twenty times weaker than the forces involved in ionic or covalent bonding, and they are still weaker in liquids. The weakness of the bonding is shown by the generally low melting and boiling points of substances held together by van der Waals' forces. The forces are still weaker in gases, though it is their existence in gases that is partially responsible for the deviation of gases from the ideal gas laws.

8 Examples of van der Waals' bonding

a **The noble gases** These gases are monatomic in the gaseous state, existing as single, discrete atoms. The relative lack of interaction between the atoms is shown by the fact that the noble gases approach ideal behaviour more closely than most other gases. In the solid state, the atoms are arranged in cubic close-packed structures, each atom being surrounded by twelve equidistant neighbours, i.e. the coordination number is 12.

The very weak van der Waals' forces holding the crystals together are reflected in the low melting points of the noble gases. Helium, indeed, cannot be solidified without application of an external pressure of 26 atmospheres; it then solidifies at $-272.1\,°C$.

As the atoms get bigger the melting point rises, i.e.

	He	Ne	Ar	Kr	Xe	Rn
m.p./°C		-249	-189	-157	-112	-71
b.p./°C	-269	-245	-186	-152	-109	-62

This is a general phenomenon. Unless hydrogen bonding interferes (p. 147), it is generally true that the melting and boiling points of like substances increase as the size of the atom or molecule concerned increases. It can be seen in the noble gases and also the halogens, the members of most organic homologous series, and in many other groups. It is attributed to the increased possibility of electrical asymmetry within larger atoms and molecules and also with their increased tendency to be polarised (p. 118).

b The halogens The crystal structures of the halogens are not all exactly alike, but they all consist of diatomic molecules. Iodine is typical, with a close-packed structure (Fig. 80) very much like that of the noble gases. There is a rise in melting and boiling point in passing from the gaseous fluorine to the solid iodine, as there is in passing from helium to radon in the noble gases.

ICl and Br_2 have about the same relative molecular masses, but the former has the higher boiling point as its molecule is polar.

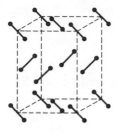

Fig. 80 The crystal structure of iodine. I_2 molecules are held together by van der Waals' forces.

c Graphite. Layer lattice structures Crystals of graphite (Fig. 81) contain hexagonally arranged carbon atoms and, within the layers, the carbon atoms are linked by covalent bonds like those in benzene. The bond distance in graphite is 0.142 nm. Adjacent layers are held together, 0.335 nm apart, by van der Waals' forces. The weakness of the forces allows the layers to slide over each other, which is why graphite is soft and acts as a solid lubricant. The distance between the layers is also such that a number of graphitic 'compounds' can be formed in which other atoms or molecules, e.g. K, O, F and H_2SO_4, are positioned between the layers (p. 162).

Similar layer lattice structures with layers held together by van der Waals' forces are found in boron nitride, many di- and trihalides, e.g. $CdCl_2$, sulphides such as MoS_2, and some hydroxides.

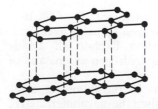

Fig. 81 The layer lattice structure of graphite. Adjacent layers of C atoms are held together by van der Waals' forces.

d van der Waals' radii Half the distance between the nuclei of adjacent atoms in the crystal of a noble gas measures what is known as the van der Waals' radius of the atom concerned. It gives the best value for the size of an atom as it measures how close two atoms will approach when not attracted by any very strong bond.

For a crystal, such as that of iodine, containing diatomic, covalent molecules the closest distance between the nuclei of atoms in adjacent molecules will be equal to twice the van der Waals' radius for iodine.

Some typical values for van der Waals' radii are given below, in nm:

H 0.12   He 0.12

N 0.15	O 0.140	F 0.135	Ne 0.160
P 0.19	S 0.185	Cl 0.180	Ar 0.192
As 0.20	Se 0.200	Br 0.195	Kr 0.197
Sb 0.22	Te 0.220	I 0.215	Xe 0.217

The values are significantly greater than the corresponding covalent radii.

Questions on chapters 11–14

1 Cite briefly two pieces of evidence which help to indicate the real existence in matter of discrete entities described as atoms and molecules.

Explain by reference to the theory of atomic structure what is meant by the terms (i) atomic number (ii) isotope (iii) neutron.

Discuss how far the properties of the following compounds may be inferred from the modes of linkage of their component atoms (a) magnesium oxide (b) tetrachloromethane (c) ammonium chloride (d) copper(II) sulphate-5-water. (JMBS)

2 In what ways are a sodium ion and a neon atom alike? How do they differ?

3 Discuss the arrangement of valency electrons in the oxides of (i) nitrogen (ii) sulphur (iii) carbon.

4 What is the arrangement of electrons in the following ions: Ag^+, Ca^{2+}, Fe^{2+}, Hg^{2+}, Al^{3+}, Fe^{3+}, Bi^{3+}? Which of these ions would you expect to be most readily formed from its atoms?

5 Give the arrangement of electrons in the following ions: F^-, Cl^-, Br^-, I^-, O^{2-} and S^{2-}. Which of these ions is the strongest reducing agent?

6 Hydrogen can form both an anion and a cation. Explain why this is so, and give examples of compounds containing hydrogen anions and of those containing hydrogen cations.

7 Show the arrangement of valency electrons in the following compounds: ethane, ammonium bromide, potassium hexacyanoferrate(II), trichloromethane and nitric acid.

8 The number of water molecules attached to a hydrated cation in aqueous solution increases as the size of the cation and its charge increase. Illustrate this statement.

9 Suggest a reasonable arrangement of bonds in the following ions: CO_3^{2-}, SiO_4^{4-}, NO_3^-, PO_4^{3-}, SO_4^{2-}, SO_3^{2-}, $S_2O_3^{2-}$, ClO_3^-, ClO_4^-.

10 Show the probable arrangement of bonds in the following: BF_3, NH_3, SiH_4, ClF, $Ag(NH_3)_2^+$, HCN, $CH_3.NC$, $C_6H_5NO_2$.

11 What is the arrangement of valency electrons in the following: H_2S, PH_3, BCl_3, NH_4^+, the nitronium ion NO_2^+, the peroxide ion O_2^{2-}, the carbide ion?

12 What is the probable arrangement of electrons in the following compounds of iodine: HI, NaI, ICl_3, IF_5, IF_7, I_3^-, CH_3I and NI_3?

13 Discuss briefly in electronic terms (a) the formation of ions of the type MO^{n-} where n is 1, 2 or 3 (b) the existence of a stable compound BCl_3NH_3 (c) the valency types shown in ammonium carbonate.

14 What experiments would you carry out in order to try to decide whether a pure white solid was an ionic or covalent compound?

15 Examine the ease of formation and stability of the Cu^+ and Zn^{2+} ions from the point of view of Fajans' rules. Why do you think it is that the Cu^+ ion is found so rarely?

16 State Fajans' rules and illustrate their application.

17 Fajans' rules suggest that electrovalency is favoured by small ionic charge and by large cation radius. If Z is the charge on a cation, and r its radius, examine the statement that the oxide of a metal is basic if $\sqrt{Z/r}$ is less than 2.2, and acidic if $\sqrt{Z/r}$ is greater than 3.2.

18 Show how the electronic theory of valency provides an explanation of the regular variation of the valency of the typical elements in passing from group to group of the periodic classification of the elements.

19 The first, second and third ionisation energies of sodium, magnesium and aluminium are tabulated below in $kJ\,mol^{-1}$.

Na	494	4561	6920
Mg	732	1443	7680
Al	577	1812	2745

Comment on any significant points.

20 Explain and illustrate the meaning of the terms electropositive and electronegative.

21* Calculate the expected volume occupied by a water molecule from (a) the bond angle and bond distances in a water molecule (b) the fact that the density of water at $4\,°C$ is $1\,g\,cm^{-3}$. Comment on the results.

22 Suggest reasons for the following (a) magnesium oxide has a much higher melting point than sodium fluoride (b) anhydrous aluminium chloride is soluble in benzene whereas the hydrated form is insoluble (c) liquefied hydrogen chloride is nonconducting, whereas a solution of hydrogen chloride in water conducts electricity.

23 Discuss the arrangement of bonds in H_2CO_3, HNO_3, H_2SiO_3, H_3PO_4, H_2SO_4, H_2SO_3, $HClO_3$ and $HClO_4$.

24 Give varied examples to show that it is an over-simplification to say that atoms combine to attain an octet of electrons in their outermost orbit.

25 By considering the orbitals available for bond formation predict the geometrical arrangement in the following: NH_4^+, CH_4, NH_3, H_2S and BCl_3.

26 What factors affect the shape of simple molecules? Give illustrative examples.

27 How can measurement of dipole moments help in deciding the shape of a molecule?

28 'The tendency of electrons of like spin to avoid each other does more than any other single factor to determine the shapes and properties of molecules.' Comment.

29 The C—H bond lengths in ethyne, ethene and methane are 0.1057, 0.1079 and 0.1094 nm respectively. Suggest a reason for these changes.

30 Comment on the statement that all compounds of carbon, nitrogen, oxygen and fluorine, containing only σ-bonds, can be regarded as tetrahedral molecules if lone pairs be considered as bonding electrons.

31 If the first ionisation energy of hydrogen is taken as being 1 unit, the values for the elements from helium to argon, inclusive, are 1.81, 0.40, 0.69, 0.61, 0.83, 1.07, 1.00, 1.37, 1.59, 0.38, 0.56, 0.44, 0.60, 0.80, 0.76, 0.96 and 1.15. Plot these figures against atomic number, and comment on any points of significance.

32 What type of bond hybridisation would you expect in simple compounds of (a) boron (b) beryllium? What effect would the hybridisation have on the geometry of the molecules concerned?

33 Explain, with illustrative examples, what is meant by the term delocalisation.

34 Describe and explain some examples of the effect that delocalisation can have on the properties of an organic molecule.

35 The fact that *trans*-butenedioic (fumaric) acid has a higher melting point than *cis*-butenedioic (maleic) acid, with which it is isomeric, can be accounted for in terms of hydrogen bond formation. How might this be done?

36 *m*- and *p*-compounds differ markedly from the corresponding *o*-compounds in being (a) less volatile (b) more miscible with water, and (c) less miscible with benzene. Illustrate and explain this statement.

37 How would you expect intermolecular hydrogen bonding to affect (i) the relative molecular mass (ii) the boiling point (iii) the vapour pressure (iv) the surface tension of a compound? Use these criteria to discuss the possibility of hydrogen bonding in water.

38 How do you think hydrogen bonds might be formed in (a) chlorophenol (b) the enol form of ethyl 3-oxobutanoate (c) 2-hydroxybenzylaldehyde (d) 2,2,2-trichloroethane-1,1-diol or chloral hydrate?

39 Illustrate the meaning of hybridisation by considering the bonding in the following substances: $BeCl_2$, BF_3, CF_4, PF_5, SF_6 and IF_7.

40 Suggest reasons why the nitrate and carbonate ions are planar whilst the chlorate ion is pyramidal.

41 Compare the nature of the bonding in (a) benzene and borazole, $B_3N_3H_6$ (b) graphite and boron nitride (c) sulphur dioxide, nitrogen dioxide and carbon dioxide (d) sulphur dichloride oxide, sulphur dichloride dioxide and phosphorus trichloride oxide.

42 The measured dipole moments of the hydrogen halides are given below.

	HF	HCl	HBr	HI
Dipole moment/$C\,m \times 10^{30}$	6.371	3.436	2.602	1.268

Comment on these figures. Taking the charge on an electron as $1.602 \times 10^{-19}\,C$ and the bond distance in hydrogen chloride as $1.29 \times 10^{-10}\,m$, estimate the ionic character of the bond in hydrogen chloride.

43 Examine the statement that bonds are to be regarded as ionic when the electronegativity difference between the two bonded elements is greater than 1.7.

44 Give an account of the various bonds formed between ions, atoms and molecules in the solid state. (CC)

45 To what extent is it possible to predict or at least explain the structures of simple molecules? To what extent can the bonding also be explained? Illustrate your answer with as many different examples as possible. (OJSE)

46 The bulk properties of matter are a reflection of the interactions between the atomic constituents of that matter. Discuss this statement by referring to some of the following pairs: diamond and graphite, sugar and salt, oil and vinegar, rubber and clay, ice and glass. (OJSE)

47 Discuss the nature of the bonding in metals and show how it may be used to account for the main physical properties of metals. What is a semiconductor and how may semiconductance be explained? Give one example of a semiconductor and state for what it might be used. (AEB, 1976).

48 In the early days of the electronic theory of valency, chemists postulated the 'octet' theory, i.e. that in a chemical bond atoms gained, lost or shared electrons until they had eight of them in their outer shells. (a) Discuss in modern terms why the octet theory was seemingly so important. (b) Discuss cases where the octet theory is useless, irrelevant or misleading. (LS)

49 In elementary work, chemical bonding can be regarded as resulting either from the transfer of electrons ('electrovalency') or from the sharing of pairs of electrons ('covalency'). Write a systematic account of cases where this simple picture is inadequate or inappropriate.

50 Give a molecular description of (a) gases (b) liquids (c) solids. In the case of solids, discuss, giving specific examples, the different types of bonding by which the constituent species may be held together and the influence this has upon the physical properties of the solids. (WS)

51 Show how Hess's law may be used (i.e. the Born-Haber cycle) in calculations of lattice energies of alkali halides. State clearly what quantities are used in the calculations. Using the following enthalpies of formation (in $kJ\,mol^{-1}$), of crystal lattices, discuss the factors governing the magnitude of lattice energies: LiF, -612; LiI, -271; NaF, -569; NaI, -288; CaO, -621; CaF$_2$, -1203. (WS)

52 Discuss the nature of the interaction between: (i) the hydrogen atoms in the hydrogen molecule (ii) the argon atoms in liquid argon (iii) the ions in crystalline LiI (iv) solute and solvent in aqueous $HClO_4$. (CC)

53* Sketch the Born-Haber cycles for the formation of (a) MgO (b) CaO (c) $MgCl_2$ (d) $CaCl_2$ from their elements. Include the numerical values used as in Fig. 45 (p. 116).

54* Select some values of lattice energies to show how the values depend on the charge and size of the ions concerned.

55* Record the dipole moments of
(a) CH_3Cl, C_6H_5Cl and $C_6H_5.CH_2Cl$
(b) H_2O, CH_3OH, C_2H_5OH and C_6H_5OH
(c) $C_6H_5CH_3$ and $C_6H_5.C_2H_5$
(d) 1,1-dichloroethane and 1,2-dichloroethane.
Comment on the values.

56* Calculate the values of the difference between the dipole moments of CH_3-X and C_6H_5-X when X is NO_2, Cl, OH, I, CHO and CN. Comment on the figures.

57* Use ionic radii values to predict the coordination number* in the crystals of MgO, TlCl, CaS, AgI and FeO.

58* Explain the difference between ionic radius, metallic radius, van der Waals' radius and covalent radius. Illustrate your answer with selected numerical values.

15
Crystal structure

1 Crystal systems

The crystals of a given substance have plane surfaces, known as faces, and the angle between the faces is constant, however irregularly the crystal may have grown. Different crystals of the same substance may not look alike, because different faces can grow at different rates, but the corresponding faces always intersect at the same angle. Moreover, cleavage or splitting of a crystal occurs along definite planes so that the same interfacial angles are found even when a crystal is split.

The number of different crystal shapes is large, and some substances can crystallise in two or more different forms. Sulphur, for example, forms crystals of α- and β-sulphur, under different conditions, and sodium chloride, which usually forms cubic crystals, gives octahedral crystals in the presence of carbamide (urea). Crystalline shape may, too, be affected by temperature. Ammonium chloride, for example, gives one type of crystal below its transition temperature, and another above. This occurrence of the same chemical substance in different crystalline forms is known as polymorphism, or allotropy if only elements are concerned (p. 171).

The wide variety of crystals may be classified into seven crystal systems, according to the set of axes which must be used to characterise the crystal faces. A crystal which is a perfect cube, for example, can be characterised by three axes, at right angles to each other, and with $a = b = c$. On this basis, the seven crystal systems may be summarised as follows, where α is the angle between b and c, β that between a and c, and γ that between a and b.

system	axes	angles	maximum no. of planes of symmetry	example
cubic	$a = b = c$	$\alpha = \beta = \gamma = 90°$	9	sodium chloride
tetragonal	$a = b \neq c$	$\alpha = \beta = \gamma = 90°$	5	white tin
orthorhombic	$a \neq b \neq c$	$\alpha = \beta = \gamma = 90°$	3	rhombic sulphur
monoclinic	$a \neq b \neq c$	$\alpha = \gamma = 90°; \beta \neq 90°$	1	monoclinic sulphur
rhombohedral or trigonal	$a = b = c$	$\alpha = \beta = \gamma \neq 90°$	3	calcite
hexagonal	$a = b \neq c$	$\alpha = \beta = 90°; \gamma = 120°$	7	quartz
triclinic	$a \neq b \neq c$	$\alpha \neq \beta \neq \gamma \neq 90°$	0	copper(II) sulphate

Each system includes a number of forms or classes. The cubic system, for example, includes the cube and the regular octahedron. Crystals in each system have a maximum number of planes of symmetry, and other symmetrical features based on axes and centres of symmetry. The detailed symmetry of the forms or classes within one system vary, and there are, altogether, 32 crystal forms or classes.

2 Space lattices and unit cells

The external shape of a crystal, and its symmetry, reflect an orderly array of particles within the crystal. These particles may be ions, atoms or molecules, and the arrangement of these units within the crystal is represented by a space lattice or a unit cell.

The structural units are held together by electrostatic forces in ionic crystals (p. 109), by van der Waals' forces in molecular crystals (p. 153), by covalent bonds in atomic or covalent crystals (p. 125), by metallic bonds in metallic crystals, or by hydrogen bonds (p. 145).

A space lattice is a regular pattern of points, the points representing the positions of the structural units which, when in position, make up the crystal structure. The space lattice extends in all directions, but it is only necessary to consider a specimen portion of it, which is representative of the whole. Such a portion is known as a unit cell. It is defined as the smallest portion of a space lattice, which, by moving a distance equal to its own dimensions in various directions, can generate the complete space lattice.

Three cubic unit cells are shown in Fig. 82. Other similar cells occur corresponding with the other six crystal classes. There are, in fact, seven groups of unit cell with 14 different types; three cubic, two tetragonal, two monoclinic, four orthorhombic, and one each of rhombohedral, hexagonal and triclinic. The 14 types are known as *Bravais lattices*; taken together they can account for all crystal structures. Those lattices with points only at corners are called *primitive*; those with points at the centre are *body-centred*; those with points in the faces are *face-centred* (Fig. 82).

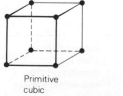

Primitive
cubic

Body-centred
cubic

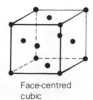

Face-centred
cubic

Fig. 82 Cubic unit cells.

It is important to realise that unit cells can pack together in different ways, and grow at different rates in different directions, so that the superficial shape of a crystal does not always immediately reflect the symmetry of the unit cell. Fig. 83 illustrates the way in which cubic unit cells can give different crystal faces.

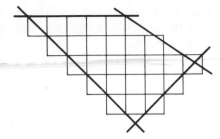

Fig. 83 Showing the development of different crystal faces from a cubic unit cell.

It is also important to realise that the structural units within a unit cell are shared with its neighbours. At first sight it might appear that the primitive cubic unit cell contains eight atoms (or ions or molecules), one at each corner. But each corner is shared by eight unit cells so that each one contains only the equivalent of one atom. For the body-centred cubic cell, it is two atoms per unit cell.

3 Isomorphism

Substances forming crystals in which geometrically similar units are arranged in similar ways are said to be isomorphous. In the simplest cases, as in the three groups below, they form crystals with the same external shape, form overgrowths on crystals of each other, and will form mixed crystals (see below).

$K_2SO_4.Al_2(SO_4)_3.24H_2O$ $CaCO_3$ $ZnSO_4.7H_2O$
'potash alum'

$FeCO_3$ $MgSO_4.7H_2O$

$(NH_4)_2SO_4.Fe_2(SO_4)_3.24H_2O$
'iron alum'

$MgCO_3$ $NiSO_4.7H_2O$

When two substances crystallise with identical space lattices it is to be expected that the points in the lattice might be occupied by structural units from the two substances so long as the geometrical differences between the units is not too great, and so long as they are similar in type.

This was recognised in the law of isomorphism (Mitscherlich, 1819) which states that substances with similar chemical compositions are isomorphous, but it is only partially true. CsCl and RbCl are superficially similar, but they are not isomorphous as they have different crystal structures because of the difference in size of the Cs^+ and Rb^+ ions. $CaCO_3$ and $NaNO_3$, or $BaSO_4$ and $KMnO_4$, are not, chemically, alike, but they are isomorphous because they are made up of ions of similar size and shape.

Two substances with similar ions can form *mixed crystals* or *solid solutions* containing some ions from each substance. The Cl^- and Br^- ions are almost the same size, so that KCl and KBr are miscible in all proportions and form a continuous series of mixed crystals containing anything from 0 to 100 per cent of Cl^- or Br^-. Cl^- and I^- differ in size by about 21 per cent; they are only partially miscible.

4 Other types of mixed crystals

The random replacement of one ion by another in the mixed crystals formed from potassium chloride and potassium bromide gives rise to what are sometimes referred to as *substitutional* solid solutions.

Other crystal structures can be formed in which atoms or ions get 'trapped' in spaces between the structural units of a regular crystal lattice. This gives rise to *interstitial* solid solutions; the products are also known as *inclusion compounds*. It is also possible to obtain crystals in which some of the sites for structural units are vacant.

a Metallic hydrides Many transitional metals can absorb large amounts of hydrogen, and the absorbed gas can generally be liberated by pumping at a high enough temperature. During the absorption, the crystal lattice of the metal is expanded but not greatly distorted and it is thought that the small hydrogen atoms are situated between the metallic atoms in interstitial compounds. The amount of hydrogen absorbed varies with the conditions but hydrides with 'formulae' such as $TiH_{1.73}$, $TaH_{0.76}$, $CeH_{2.8}$, $LaH_{2.8}$ and $VH_{0.6}$ have been reported.

b Iron(II) sulphide and iron(II) oxide Chemical analysis of different crystalline specimens of iron(II) sulphide show that they vary in composition from FeS to $FeS_{1.14}$. Such 'formulae' suggest that the crystals contain excess sulphur and that they may be interstitial in nature, but density measurements show quite clearly that the crystals are really deficient in iron so that the formulae variation of FeS to $Fe_{0.88}S$ is perhaps preferable. To maintain an electrical balance within the crystal, some of the Fe^{2+} ions are converted into Fe^{3+} ions.

Similarly, crystalline iron(II) oxide shows a composition varying from $FeO_{1.055}$ to $FeO_{1.19}$, again because of a deficiency of iron.

c Graphite compounds Graphite, with its layer-lattice structure (p. 153), can form a number of 'compounds' in which other atoms or molecules take up a position between the layers of carbon atoms in the graphite structure.

Graphite absorbs liquid potassium, for example, to form 'alloys' with detectable compositions represented by KC_8, KC_{16}, KC_{24} and KC_{40}. Graphite also reacts with strong oxidising agents, such as nitric acid or potassium chlorate, to form graphitic oxides with compositions varying from $C_{2.9}O$ to $C_{3.5}O$. Similar 'compounds' are formed between graphite and fluorine, e.g. $(CF)_n$, and between graphite and acids such as sulphuric acid, e.g. $C_{24}HSO_4.2H_2SO_4$.

d Clathrates Benzene-1,4-diol (quinol), $C_6H_4(OH)_2$, crystallises from an aqueous solution, in the presence of argon at 40 atmospheres, with a structure in which an atom of argon is 'trapped' inside a 'cage' of hydrogen bonded quinol molecules. The argon can be liberated by melting or dissolving the crystalline product. The 'compound', known as a clathrate, has a formula $[C_6H_4(OH)_2]_3Ar$.

Xenon, krypton, hydrogen sulphide, hydrogen chloride, hydrogen cyanide, sulphur dioxide and carbon dioxide, all with molecules of the correct size to fit inside the benzene-1,4-diol 'cage', form similar clathrates.

Benzene molecules can also be 'trapped' by shaking with an ammoniacal solution of nickel(II) cyanide. The resulting clathrate compound has a formula $Ni(CN)_2 \cdot NH_3 \cdot C_6H_6$.

e Molecular sieves Some zeolites (naturally occurring or synthetic aluminosilicates) lose water on heating to give structures containing many holes, the size of which can be varied by altering the composition of the solid. The products can be used as molecular sieves for they can absorb gases and liquids whose molecules will fit into the holes whilst not absorbing substances with larger molecules.

5 Diffraction of light

When a beam of light is passed through a diffraction grating, which consists of a large number of very fine opaque lines, parallel to each other and of equal width, drawn on a piece of glass, a series of spectra can be observed on either side of the original path of light. If monochromatic light is used, the spectra are replaced by a series of bright images on a dark background.

A small portion of a grating is shown in Fig. 84. If monochromatic light of wavelength λ is incident perpendicularly on the upper face of the grating all the clear spaces act as secondary sources of light and emit rays in all directions. The rays in any one direction from each of the spaces will interfere with each other. In simple terms, where a crest of one wave coincides with a crest of another the resulting displacement will

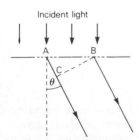

Incident light

Fig. 84 A small portion of a diffraction grating.

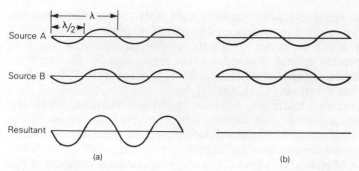

·Fig. 85 (a) Reinforcement and (b) neutralisation of one source by another.

be increased, i.e. the light will be brighter. Similarly, when a crest of one wave coincides with a trough of another there will be no resultant displacement, i.e. there will be no visible light. The conditions leading to maximum or zero displacements are summarised in Fig. 85.

The ray from A will reinforce the ray from B, to give a maximum displacement, only if the distance AC is equal to $n\lambda$, where n has any integral value. As AC is equal to $AB\sin\theta$, and AB is equal to the spacing of the grating, d, it follows that, for reinforcement, $d\sin\theta$ must equal $n\lambda$.

If the grating has 550 lines to the mm, then AB (d) is equal to approximately 1.81×10^{-3} mm. If the wavelength of light used is 5.8×10^{-4} mm then $\sin\theta = 0.3197n$, the values of θ given by $n = 0, \pm 1, \pm 2$ and ± 3 being $0°, \pm 18° 39', \pm 39° 45'$ and $\pm 73° 33'$ respectively.

The light emerging from such a grating, if viewed through a movable telescope, will show up as a series of bright and dark lines as in Fig. 86 where bright lines are indicated by B and dark ones by D.

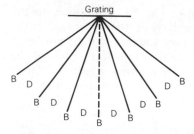

Fig. 86

6 Diffraction by X-rays

In 1912, von Laue suggested that X-rays should be diffracted in the same way as light waves, if they were electromagnetic waves of short wavelength, so long as a diffraction grating with a small enough spacing could be found. A crystal with a regular array of lattice planes was considered to be capable of acting as such a grating, and Friedrich and Knipping found this was so. They passed a beam of X-rays through a thin section of a zinc blende crystal, and on photographing the emergent rays they found that the plate showed a bright central spot surrounded symmetrically by other bright spots caused by diffraction of the X-rays.

This original diffraction of X-rays was a diffraction of transmitted rays, but it is rather easier, both theoretically and experimentally, to treat the diffraction of reflected X-rays.

The atoms or ions in a crystal are arranged in a series of planes, and the lines KL and MN in Fig. 87 may be taken as representing two such parallel planes. When a beam of X-rays is incident at a glancing angle of θ some reflection takes place at each plane and diffraction is caused by the interference of the reflected rays. The path difference between the reflected rays from KL and those from MN is equal to AYB. Taking the distance between the two planes as d, it follows that AY is equal to $d\sin\theta$. For the reflected rays to reinforce, the necessary condition is that AYB (the path difference) must be equal to an integral number of wavelengths. If, therefore, the wavelength of the X-rays is λ, the condition for reinforcement is that

$$n\lambda = 2d\sin\theta$$

This is the basic relationship of X-ray analysis of crystals; it is generally referred to as the *Bragg equation*.

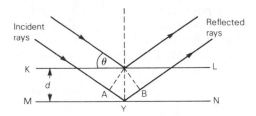

Fig. 87

Experimental measurement of n, λ and θ enables values of d to be obtained. It is, therefore, possible to determine the positions of atoms, ions or molecules within a crystal structure. It is also possible to measure the bond distances and bond angles in molecules or compound ions in crystals. Values of bond lengths can be measured to an accuracy of ± 0.001 nm, and complex molecules such as nucleic acids and proteins can be studied. X-ray crystallography has developed extremely rapidly since about 1955, as computers have become available to unravel the often very complicated experimental evidence.

X-ray diffraction methods are, in the main, limited to solids, and they are unable to detect hydrogen atoms. Neutron or electron diffraction (p. 169) can overcome such problems.

7 Experimental methods

A beam of X-rays is obtained by bombarding a metal with electrons. X-rays with different wavelengths come from different metals (p. 50); those corresponding to the K_α lines in the X-ray spectra of copper and molybdenum, with wavelengths of 0.154 and 0.071 nm respectively, are commonly used.

The beam is directed onto a single crystal, or onto a powdered crystal contained in a thin-walled glass capillary, deposited on a metal or moulded into a wire. Reflected X-rays are recorded on a photographic plate surrounding the crystal, or by an ionisation chamber rotating round the crystal.

a Powder methods Some of the tiny crystals within a powder will be orientated so that a set of reflecting planes (X) are at an angle (θ) to the incident beam such that $n\lambda$ is equal to $2d\sin\theta$. They will give a diffracted beam at an angle of 2θ to the incident beam (Fig. 88). Similar planes in the power will also be at an angle of θ to the incident beam but they will be at different angles to its line of approach, i.e. they will be rotated about the line of approach. The sets of planes (X) will, then, give diffracted radiation lying on the surface of a cone with an angle of 4θ. This conical envelope of radiation will cut the surrounding photographic film in two slightly curved lines.

Other sets of planes in the powder will give other cones with different angles, so that typical, simple X-ray photographs are as shown in Fig. 88. The angle for each cone can be obtained from the position of the lines on the photograph. It is, then, a matter of deciding which planes within the crystal correspond with each cone. This is known as indexing

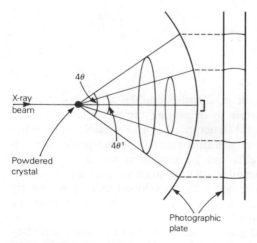

Fig. 88 Diffraction of X-rays by a powdered crystal showing how lines are formed on the photographic plate.

the reflection, for each set of planes is labelled by what are known as *Miller indices*. Once this is done, λ and θ being known, the distances, d, between the various planes can be calculated.

The indexing of powder photographs from crystals with simple symmetry, e.g. metallic crystals and simple ionic crystals, is relatively easy. It can be helped by taking the intensity of the diffracted beams into account.

The scattering of X-rays is caused by the electrons round an atom or ion and its intensity depends on the electron density. It is, therefore, greater for heavier atoms or ions, and very low for light atoms such as hydrogen. It is for this reason that X-ray analysis does not show up the positions of H atoms. The intensity of a diffracted beam will depend, then, on the nature of the atoms or ions in the planes concerned. It is related to what is known as a structure factor.

In a sodium chloride crystal, for example, some of the planes will contain only Na^+ ions, some will contain only Cl^- ions and some will contain both ions. The intensity of reflection from the Cl^- ions is greater than that from the Na^+ ions, and this assists in indexing. The diffraction by potassium chloride is similar in kind, but different in detail, from that of sodium chloride because the K^+ and Cl^- ions have almost the same electron densities and scattering power.

b Single crystal methods In these methods, a small, single crystal (about $0.1 \times 0.1 \times 0.1$ mm) is mounted on a goniometer head, which enables angles to be measured accurately. A monochromatic beam of X-rays impinges on the crystal and the diffracted beams are recorded on a surrounding photographic film (Fig. 89).

If the crystal is mounted with, say, its c-axis along the camera axis, and it is rotated or oscillated, all the planes which are parallel to the c-axis

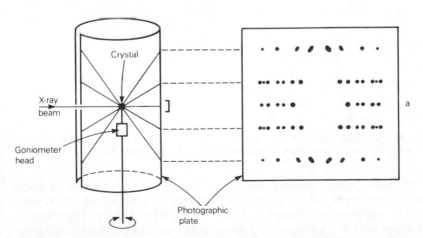

Fig. 89 Diffraction of X-rays by a single crystal showing the layer lines of spots produced.

will cause diffraction when $n\lambda$ is equal to $d\sin\theta$. Different d and θ values will be involved for different sets of planes and a *layer line* of spots (**a** in Fig. 89) will occur in a plane containing that of the incident beam. Other layer lines of spots, above and below, will be formed by sets of planes which are tilted to the c-axis of the crystal or the axis of the camera. The crystal can be studied when mounted along other axes.

As before, it is a matter of trying to decide which planes cause which spots. In a complicated structure, considerable ingenious trial and error and computer analysis may be involved to match the actual photograph with that to be expected from likely, trial structures.

The analysis is helped by taking into account the intensities of the various spots. It may, also, be possible to introduce a heavy atom into a compound without changing its crystal structure. Comparison of the two isomorphous (p. 161) crystals can then be used. In effect, the planes containing the heavier atom are being 'labelled'.

In a Weissenberg camera, more detailed photographs are taken by introducing shields so that only one layer line of spots is photographed, and, as the crystal is rotated, the photographic film is moved along the camera axis. In this way the spots in one layer line are spread over the whole film and provide more detailed information. A four-circle diffractometer provides a still more sophisticated method.

c The structure of molecules If each unit in a crystal is a molecule or a compound ion, rather than a simple atom or ion, the whole molecule cannot lie at one point and there will be many more planes within the crystal (Fig. 90).

The distance between some of the planes depends on the bond distances and angles within the molecule concerned so that these quantities can be measured if the detailed crystal structure can be elucidated.

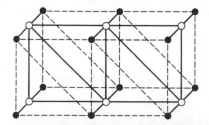

Fig. 90 Illustrating the greater number of planes in a crystal composed of a collinear, triatomic molecule (dashed lines) compared with a single atom (solid lines).

d Electron density contour maps The nature of the diffraction by a crystal is determined by the varying electron density within the crystal, for the scattering of X-rays depends on electron density. The electron densities within a particular crystal or molecule can be determined from the diffraction pattern obtained and by calculations involving structure

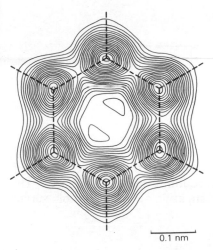

Fig. 91 Electron density contour map for benzene, C_6H_6.

factors (p. 167); a mathematical technique known as Fourier analysis has to be used. The results are conveniently summarised on electron density contour maps, or Fourier maps, which show contour lines of equal electron density (Fig. 91).

8 Neutron diffraction

A beam of neutrons with wavelength 0.1 nm can be obtained from an atomic pile and is diffracted in much the same way as X-rays of similar wavelengths. The scattering of the neutrons, however, is caused by atomic nuclei and not by electrons, so that hydrogen atoms can be detected.

As neutrons have a magnetic moment, neutron diffraction can also provide information about the magnetic characteristics of ions or molecules with unpaired electrons, which are, themselves, magnetic.

9 Electron diffraction

Beams of electrons with wavelengths between 0.5 and 0.0001 nm can be obtained by accelerating the electrons between low or high potential differences. An electron beam with wavelength about 0.01 nm is diffracted on passing through a gas or vapour at low pressure. The diffraction is similar to that of X-rays by powdered crystals, and cones of diffracted electrons are produced. They are detected on a photographic plate as concentric rings (Fig. 92). For simple molecules in the gas or vapour involved, it is possible to measure bond lengths to an accuracy of ± 0.0003 nm, and hydrogen atoms can be detected.

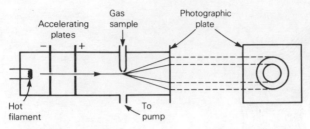

Fig. 92 General arrangement in an electron diffraction apparatus.

A beam of low energy electrons (high wavelength) is not very penetrating, but it can be diffracted by reflection from solid surfaces. This provides a good method of investigating a number of surface phenomena.

10 Polymorphism

Substances which can crystallise in more than one form are said to be polymorphic. If the number of crystalline forms is two, they are dimorphic; if three, trimorphic; and so on. The term polymorphic applies to crystalline forms of both elements and compounds. The term allotropy is also used to describe different forms of the same *element*, in the same state, whether they be crystalline forms or not. Polymorphism, then, includes all cases of allotropy caused by variation in crystalline form, but some examples of allotropy, e.g. ozone and oxygen, are caused by different arrangements of atoms in molecules and do not involve crystal structure at all.

Some polymorphic substances change from one structure to another at a definite temperature (for a particular pressure) known as the *transition temperature*, e.g.,

$$HgI_2(red) \overset{126\,°C}{\rightleftharpoons} HgI_2(yellow)$$

$$ZnO(white) \overset{258\,°C}{\rightleftharpoons} ZnO(yellow)$$

$$\underset{\text{(NaCl structure)}}{NH_4Cl(s)} \overset{184.3\,°C}{\rightleftharpoons} \underset{\text{(CsCl structure)}}{NH_4Cl(s)}$$

They are said to be *enantiotropic*, from the Greek meaning opposite change. One structure is stable below the transition temperature; and one above that temperature. The rate of change may, however, be very slow. Yellow mercury(II) iodide, for example, reverts to the stabler red form only very slowly unless it is disturbed by touching.

The change from one enantiomorph to another may result in an evolution or absorption of heat or in some change of physical property, e.g., colour, density, solubility or viscosity. Investigation of such properties at different temperatures enables the transition point to be measured.

Other polymorphs do not have any specific transition temperature; they are said to be *monotropic*, from the Greek meaning one change. One of the structures is more stable than the others at all temperatures, and the less stable forms revert to the more stable one at all temperatures, but the change may be very slow. Graphite, for example, is the stable form of carbon, but the change from diamond to graphite is, perhaps fortunately, extremely slow.

The distinction between enantiotropy and monotropy is made clearer by a consideration of vapour-pressure-temperature curves as given for sulphur and phosphorus on pp. 93–4.

11 Allotropy

When an element exists in two or more forms, in the same state, it is said to exhibit allotropy. When the forms have different crystal structures the elements are, really, polymorphic, and the allotropy they exhibit can be either enantiotropy or monotropy, e.g.,

  enantiotropic elements: sulphur and tin
  monotropic elements: phosphorus and carbon

Allotropy may also be caused by differences in the arrangement of atoms in molecules, as with oxygen and ozone (trioxygen).

a Enantiotropy of sulphur The transition temperature between the two crystalline forms of sulphur, rhombic (α) and monoclinic (β) is 95.5 °C,

$$\text{rhombic}(\alpha)\ S \overset{95.5\,°C}{\rightleftharpoons} \text{monoclinic}(\beta)\ S \overset{120\,°C}{\rightleftharpoons} \text{liquid } S$$

There are, also, some liquid allotropes (p. 174).

Slow heating of rhombic sulphur changes it to monoclinic sulphur at the transition point, and further heating converts the monoclinic sulphur into liquid sulphur at the melting point of monoclinic sulphur (120 °C). On slowly cooling liquid sulphur, the reverse changes occur (Fig. 93).

If rhombic sulphur is heated rapidly, however, the dotted line (ODA) will be followed, as the change from rhombic to monoclinic is slow. Rhombic sulphur will be found to melt at 114 °C. Similarly, rapid cooling of liquid sulphur may cause the curve ADO to be followed, with rhombic sulphur being formed direct from the liquid form.

It will be seen that the transition temperature is below the melting points of *both* allotropes. This is always so for substances exhibiting enantiotropy.

BO, OA and AE represent the vapour pressure curves (p. 215) for rhombic, monoclinic and liquid sulphur respectively. OC represents the effect of pressure on the transition point between rhombic and monoclinic sulphur. AC and CF represent the effect of pressure on the melting points of monoclinic and rhombic sulphur respectively. The slopes of the

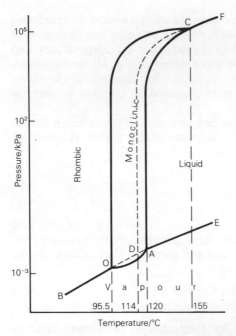

Fig. 93 Vapour pressure–temperature curves for rhombic, monoclinic and liquid sulphur, typical of an enantiotropic system. Compare Fig. 94.

lines OC, AC and CF, at high pressures, show that rhombic sulphur has a higher density than monoclinic, and that monoclinic and rhombic sulphur both have higher densities than liquid sulphur. (Compare Fig. 117, p. 217).

Within the areas FCAE, EAOB, FCOB and COA only one phase can exist, as shown in the diagram. Monoclinic sulphur can only exist by itself, for instance, within the area OAC. Point O is the triple point (p. 216) for rhombic and monoclinic sulphur with sulphur vapour; point A, the triple point for monoclinic sulphur with liquid sulphur and sulphur vapour; point C, the triple point for monoclinic, rhombic and liquid sulphur.

The dotted lines in the diagram represent metastable systems which, in many cases, can be obtained quite readily because the attainment of equilibrium in the system is very slow.

α-sulphur is the stabler allotrope at room temperature (p. 249),

$$\alpha\text{-S} \rightleftharpoons \beta\text{-S} \qquad \Delta H^{\ominus}(298\,\text{K}) = 0.3\,\text{kJ}\,\text{mol}^{-1}$$
$$\Delta G^{\ominus}(298\,\text{K}) = 0.1\,\text{kJ}\,\text{mol}^{-1}$$

As the temperature is increased, the free energies of the two allotropes decrease, but at different rates. That of β-sulphur decreases more rapidly because it has a higher entropy than α-sulphur. At the transition point,

95.5 °C, the free energies become equal so that ΔG is 0. That is why the two allotropes can coexist in equilibrium. Above 95.5 °C the free energy of α-sulphur becomes greater than that of β-sulphur, so that the latter becomes the stabler form.

b Enantiotropy of tin Tin has three solid allotropes, with two transition points, as summarised below:

$$\text{grey tin} \underset{}{\overset{13\,°C}{\rightleftharpoons}} \text{white tin} \underset{}{\overset{161\,°C}{\rightleftharpoons}} \text{rhombic tin} \underset{}{\overset{232\,°C}{\rightleftharpoons}} \text{liquid tin}$$

The main point of interest is that the density of grey tin is smaller than that of white tin. When, therefore, white tin changes into grey tin the increase in volume causes the metal to expand and crumble in places. The effect is known as tin plague. Grey tin is formed only in the severest persistent winter conditions as the change from white tin is slow, so that white tin can exist below the transition point in a metastable condition.

c Monotropy of phosphorus The allotropes of phosphorus are red and white phosphorus, with the former being the stable one,

$$P_4(\text{white}) \rightleftharpoons P_4(\text{red}) \qquad \Delta H^{\ominus}(298\,\text{K}) = -73.64\,\text{kJ mol}^{-1}$$

White phosphorus is slowly changing into red phosphorus at any temperature, so there is no transition point. At room temperature the change is very slow so that white phosphorus can be kept in a metastable state for long periods. The change to the red form takes place in a matter of days at 250 °C, and it can also be speeded up by adding a catalyst such as iodine. Red phosphorus cannot change directly into white phosphorus.

The vapour pressure curves are shown in Fig. 94. If red phosphorus is heated it will melt at T_1; white phosphorus will melt at T_2. If liquid phosphorus is cooled rapidly, white phosphorus will be formed. Thus, to convert red phosphorus into white it is necessary to melt or vaporise the red form, and to cool the resulting liquid or vapour quickly.

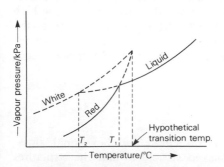

Fig. 94 Vapour pressure–temperature curves for white, red and liquid phosphorus, typical of a monotropic system. Compare Fig. 93.

The transition point between red and white phosphorus has no real meaning for it lies at a temperature *above* the melting points of the two allotropes, i.e. at a point when they have both ceased to be solid allotropes at all. This is typical of all monotropic systems, which also have vapour pressure-temperature curves like those of phosphorus.

d Monotropy of carbon The allotropes of carbon are graphite and diamond, with graphite being just the stable form at room temperature,

$$C(\text{diamond}) \rightleftharpoons C(\text{graphite}) \quad \Delta H^{\ominus}(298\,\text{K}) = -1.9\,\text{kJ mol}^{-1}$$
$$\Delta G^{\ominus}(298\,\text{K}) = -2.9\,\text{kJ mol}^{-1}$$

The change from diamond to graphite at room temperature is, however, infinitesimally slow and has never been observed.

Graphite can be converted into diamond, as a commercial process, at about 2000 °C and 70 000 atmospheres pressure in the presence of metal catalysts. These catalysts contaminate the diamonds so that they can be distinguished from natural ones.

e Allotropy dependent on different molecular species The allotropy of oxygen is not dependent on crystal structure; it depends, instead, on the existence of two different molecules, O_2 for dioxygen and O_3 for trioxygen (ozone). Dioxygen is the stabler allotrope,

$$3O_2(g) \rightleftharpoons 2O_3(g) \quad \Delta H^{\ominus}(298\,\text{K}) = 284.5\,\text{kJ mol}^{-1}$$
$$\Delta G^{\ominus}(298\,\text{K}) = 326\,\text{kJ mol}^{-1}$$

and it has to be highly energised, in an ozoniser, to convert it into trioxygen. Once formed, trioxygen, as the unstable allotrope, reverts at all temperatures to oxygen. The reversible changes between di- and trioxygen are quite rapid. The type of allotropy is sometimes called *dynamic allotropy*.

Sulphur also exhibits dynamic allotropy in the liquid and gas phases. Both α- and β-sulphur have crystals made up of S_8 puckered-ring molecules. On melting, these rings break up into S_8 chains, with unpaired electrons at each end. The chains can link together into longer chains and it is the formation of these chains that accounts for the high viscosity of liquid sulphur around 180 °C. At higher temperatures the chains break up, and the viscosity is lower. Plastic sulphur also contains long chains of S atoms. In the vapour phase, S_8, S_2 and S species occur at different temperatures.

Questions on Chapter 15

1 Explain the relationship between (a) cubic and hexagonal close-packed structure (b) cubic close-packed structures and the structures found in sodium chloride, zinc blende and fluorite (c) hexagonal close-packed structures and the wurtzite structure.

2 By considering structures made up of spheres of unit radius calculate the relative sizes of the spheres which would just fit into (a) a triangular site (b) a tetrahedral site (c) an octahedral site.

3 What is meant by saying that a tetrahedral site is smaller than an octahedral site?

4 What percentage of the available space is occupied in a cubic close-packed structure?

5 Draw diagrams to show that a tetrahedron results (a) when half the corners of a cube are connected (b) when half the faces of an octahedron are extended.

6 Describe what you would do to make a single crystal of potash alum such as would be suitable for displaying its regular external shape.

7 How would you attempt to establish, in a school laboratory, whether two substances were isomorphous?

8 Apply Le Chatelier's principle in a discussion of the effect of change of pressure on the changes (a) water to ice (b) rhombic to monoclinic sulphur (c) grey to white tin.

9 What are ortho- and para-hydrogen? Are they to be regarded as allotropes of hydrogen?

10 How would you demonstrate, experimentally, that sodium and potassium chlorides can form mixed crystals?

11 Give an account of interstitial compounds.

12 Write an account on one of the following topics: (i) mica (ii) naturally occurring silicates (iii) the piezoelectric effect (iv) polymorphism (v) types of symmetry.

13 Discuss (a) any one example of allotropy (b) any one example of polymorphism.

14 How is the fact that sulphur can appear to have two different melting points explained?

15 Write notes on (a) the law of constant composition (b) clathrates (c) silica (d) the allotropy of tin.

16 Compare and contrast the allotropy of dioxygen and trioxygen with that of diamond and graphite.

17 Describe the principles of a method which is used to determine the arrangement of atoms in a crystal of a metal. Explain what is meant by face-centred cubic, body-centred cubic and hexagonal close packing. Suggest briefly how measurements of the interionic distances in a crystal of sodium chloride could be used to find the Avogadro constant, L.

18 Describe, in outline, how X-rays can be used to discover crystal structures.

19 Sketch the electron density maps you would expect to find for (a) a hydrogen molecule (b) sodium chloride (c) dimethylbenzene.

16
Spectroscopy

1 Energy changes in atoms and molecules

The total energy of a substance, called its internal energy (U) (p. 245), is the sum of four different types of energy: translational, rotational, vibrational and electronic. Absorption or emission of energy may cause changes in some or all of these types of energy, and spectroscopy provides an excellent method of measuring some of the changes involved.

a Translational energy is concerned with the overall movement of atoms or molecules in three dimensions (as in a swarm of bees). It is only significant in gases and, to a lesser extent, in liquids, as the translational energy in solids is minute.

b Rotational energy involves the spinning of molecules about their centre of gravity (Fig. 95) or of parts of molecules.

c Vibrational energy is associated with vibrations within molecules, e.g. the stretching or bending of bonds (Fig. 95).

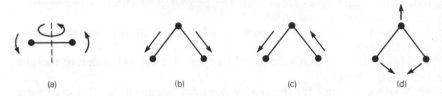

(a)   (b)   (c)   (d)

Fig. 95 Rotation and vibration within molecules. **(a)** shows rotation about two different axes in a diatomic molecule. **(b)**, **(c)** and **(d)** show vibrations in a triangular molecule. **(b)** is referred to as symmetric stretching, **(c)** as asymmetric stretching, and **(d)** as bending.

d Electronic energy involves changes in the distribution of electrons as in the splitting of chemical bonds or the promotion of electrons into higher energy levels.

e Total energy Chemical energy is the total of all four types of energy, with electronic energy making the largest contribution. For CO_2, for example, at $25\,°C$, the translational, rotational and vibrational energies are approximately 3.7, 2.5 and $26.2\,kJ\,mol^{-1}$ respectively, zero values being allotted to a CO_2 molecule which is not moving or rotating or vibrating. The electronic energy is approximately $-425.8\,kJ\,mol^{-1}$ on the basis of zero values for C(s) and O_2(g).

Chemicals also contain *nuclear energy* but this remains essentially constant in the course of normal chemical reactions (p. 97).

f Quantisation of energy Each of the four types of energy is quantised, i.e., there is a range of permitted energy levels (p. 60); energy changes of each type can only take place in discrete amounts. It is rather like the changes in energy that are possible in falling down, or climbing up, a flight of stairs with many steps of varying height.

The difference between the energy levels (the quantum) for electronic energy changes is about $450 \, \text{kJ mol}^{-1}$; for vibrational energy changes, it is generally between 5 and $40 \, \text{kJ mol}^{-1}$; for rotational changes, about $0.02 \, \text{kJ mol}^{-1}$; and for translational changes, it is so small that the changes can be regarded as not quantised.

2 Energy change and radiation

Any quantised energy change within an atom or molecule is associated with the emission or absorption of radiation (visible, ultraviolet or infrared light) of a particular frequency or wavelength or wave-number. According to the quantum theory (p. 59)

$$\text{energy change} = h\nu = hc/\lambda = hc\sigma$$

where h is the Planck constant, c is the velocity of light, and ν, λ and σ are the frequency, wavelength and wave-number, respectively, of the radiation involved.

Frequency is expressed in vibrations per second and is generally quoted in hertz $(1 \, \text{Hz} = 1 \, \text{s}^{-1})$ or MHz. Wavelengths are quoted in metres, centimetres or nanometres. Wave-number (the reciprocal of wavelength) is generally quoted in cm^{-1}.

The lower the wavelength, or the higher the frequency or wave-number, the higher the energy of the radiation. X-rays or γ-rays (with very short wavelengths) can damage human tissue; sun-light can cause sun-burn; radio waves (with very long wavelengths) are harmless to the skin. Some typical average values are summarised below.

	infrared	visible		ultraviolet
		red	violet	
wavelength/nm	1000	750	420	250
frequency/10^{14} Hz	2.99	4.00	7.12	11.96
wave-number/cm^{-1}	10 000	13 333	23 809	40 000
energy/kJ mol^{-1}	119.6	159.5	284.7	478.4

The energy may also be expressed in electron volts $(1 \, \text{eV mol}^{-1} = 96.486 \, \text{kJ mol}^{-1})$. The relationships are

$$\text{wavelength/nm} = \frac{10^7}{\text{wave-number/cm}^{-1}} = \frac{2.99 \times 10^3}{\text{frequency/}10^{14} \, \text{Hz}}$$

$$= \frac{119.6 \times 10^3}{\text{energy/kJ mol}^{-1}}$$

3 Types of spectra

Both atoms and molecules can be made to absorb or emit energy, and they do this in ways characteristic of the atom or molecule concerned. *Atomic* or *molecular spectra* of various types result.

When the energy change* within an atom or molecule is from a higher to a lower energy level (falling down the stairs), energy is emitted and an *emission spectrum* results. To obtain such a spectrum a substance must, first, be energised or activated (moved upstairs) and this can be done by heating, or by electrical excitation using an electric arc, spark or discharge. Absorption of energy by atoms or molecules causes energy changes from lower to higher levels; radiation is absorbed and an *absorption spectrum* results. Black lines occur in the spectrum corresponding to the wavelengths absorbed.

The type of spectrum (Fig. 96) depends on the magnitude of the energy change involved. High energies bring about electronic energy changes; the spectra, known as *electronic spectra*, are within the visible, ultraviolet or X-ray regions. Lower energy changes bring about vibrational changes in molecules, producing *vibrational spectra* in the near infrared (near to the visible region). Still lower energy changes cause rotational changes in molecules, with *rotational spectra* in the far infrared or microwave regions.

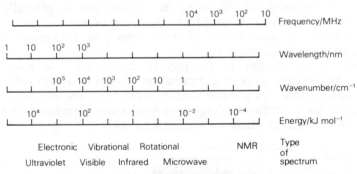

Fig. 96 Types of spectrum associated with different radiation. See, also, Fig. 17 on page 59.

In order to use numbers of reasonable size, different units are used for different types of spectra. In X-ray, ultraviolet and visible spectra the radiation is generally quoted as a wavelength in nm or as a wave-number in cm^{-1}. In infrared spectra, wave-numbers in cm^{-1} are generally used. In microwave or nuclear magnetic resonance spectroscopy (p. 189) frequencies are expressed in MHz.

---

*Some apparently possible energy changes within atoms and molecules are forbidden. Those that are *allowed* are summarised in *selection rules*, based on symmetry.

When only a few discrete energy changes take place, the emission or absorption spectrum consists of a series of lines; it is known as a *line spectrum*. Energy changes which are large enough to bring about electronic changes can, however, also bring about vibrational and rotational changes so that some spectra are *electronic-vibrational-rotational spectra*. They consist of bands and are known as *band spectra*. Similarly, energy changes capable of bringing about vibrational changes can also cause rotational changes so that *vibrational-rotational spectra* result. *Pure rotational spectra* involve only small energy changes, which cannot, at the same time, bring about vibrational or electronic changes.

Each electronic level has associated vibrational levels, and each vibrational level has associated rotational levels (Fig. 97). If too many changes take place at the same time, the resulting spectrum is too complex to analyse and interpret. A study of the various individual types of spectra can, however, give valuable information about the available energy levels in atoms and molecules, and can be used to measure molecular dimensions in many cases. As each atom or molecule gives characteristic spectra, spectroscopy can, also, be used very effectively in analysis.

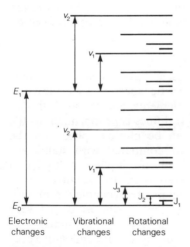

Electronic changes   Vibrational changes   Rotational changes

Fig. 97 Diagrammatic representation of possible quantised energy changes within a molecule. The diagram is not to scale. The ratio of $E_{EL}:E_{VIB}:E_{ROT}$ is about $10\,000:500:1$.

4 Atomic spectra

Atoms give characteristic line emission spectra in the ultraviolet, visible and infrared regions, and it has already been explained how the interpretation of the lines leads to an understanding of the various electronic energy levels within an atom (p. 60). When larger energy changes are involved, X-ray spectra are produced (p. 50).

a Measurement of ionisation energy When sufficient energy is absorbed by an atom, electrons can be removed completely and ions formed. The energy required to remove one electron from an atom in its ground state (lowest energy state) is known as the first ionisation energy (p. 111).

The value, for hydrogen, can be found by using the formula (p. 59) relating the wavelengths of the lines in the various spectral series, i.e.

$$\sigma = 1/\lambda = R_\infty(1/n^2 - 1/m^2)$$

The Lyman series ($n = 1$) involves changes from the ground state, so that the value of the first ionisation energy for hydrogen is obtained by putting $n = 1$ and $m = \infty$. This gives values of $109\,737\,cm^{-1}$ for the wave-number and $1314\,kJ\,mol^{-1}$ for the ionisation energy.

The values can also be obtained by plotting the wave-number differences between successive lines in the Lyman spectrum against the wave-number of either the higher or lower line involved. By extrapolation, the wave-number when the difference is zero can be found.

The line spectra of the alkali metals are made up of series of lines similar to those in the hydrogen spectrum. At least four series known as the sharp, principal, diffuse and fundamental series are known. The lines in these series can be related in formulae similar to that for the hydrogen spectrum, but with n and m not having whole number values. Ionisation energies can be found as for hydrogen.

b Use in analysis The atoms of each element have different electronic energy levels and give different characteristic spectra. Line emission spectra can, therefore, be used to identify elements, and the analysis can be quantitative as the intensity of the spectral lines is proportional to the concentration of atoms. The spectrum can be photographed and the intensity of the lines can be measured or compared with standards. Alternatively, a particular line can be directed onto a photomultiplier so that the intensity of radiation can be measured as an electric current.

Highly automated, commercial spectrometers are available for analytical purposes. They enable a variety of elements to be estimated very rapidly, even if the elements are only present in very small amounts.

Still greater sensitivity can be obtained using atomic emission spectroscopy. The sample being tested for an element, X, is vapourised by injection into a burning mixture (propane and air, for example) in a nebulizer. Radiation from X, obtained from a vapour discharge or hollow-cathode lamp, is passed through the flame region and the absorption by atoms of X in the flame is recorded automatically.

c Flame tests Flame tests involve a very simple spectral analysis in the visible region. Lithium compounds give a red colour, potassium compounds a purple colour and sodium compounds a yellow colour. The chlorides give the strongest colours because they are most volatile at the temperature of the Bunsen burner flame. That is why flame tests are done by dipping the unknown substance in concentrated hydrochloric acid.

The flame colours are caused by strong spectral lines in the visible region. That for lithium is due to a transition from a 2p to a 2s level; for sodium it is a 3p to 3s change; for potassium 4p to 4s.

5 Infrared molecular spectra

Infrared absorption spectra are obtained on infrared spectrophotometers, which are, nowadays, standard equipment in advanced laboratories. An infrared source is provided by a tungsten filament lamp in the near infrared, a hot rod of metallic oxides (a Nernst glower) in the mid infrared or a high pressure mercury-arc lamp in the far infrared.

The radiation is split into two beams (Fig. 98) so that one can be passed through the sample under test whilst the other is used as a reference beam. Gases are contained in rock-salt cells; liquids are studied as films on rock-salt plates or in solution. Solids can be studied in solution, as mulls in liquids such as Nujol, or by compression into discs with potassium bromide.

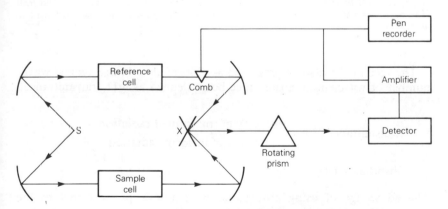

Fig. 98 Simplified diagram of the arrangement in a double-beam infrared spectrophotometer. S is the source, and X a pair of reciprocating or chopping mirrors.

Each beam is directed alternately onto a rotating rock-salt prism or a system of diffraction gratings, by a pair of reciprocating or 'chopping' mirrors, and then onto a detector. This enables the frequencies present in each beam, and those absorbed, to be measured.

Automatic recording is achieved by having a comb, which can move in or out of the reference beam and which is connected both to the detector and to a pen recorder. When each of the two beams falling on the detector have equal intensity there has been no absorption by the sample under test and the comb is moved completely out of the reference beam; 100 per cent transmittance is recorded on the trace. When the sample absorbs a particular frequency the comb is moved into the reference beam until the two beams are restored to equal intensity. The amount the comb moves measures the percentage transmittance. The result is a recorded plot of percentage transmittance against wave-number, as shown in Fig. 99.

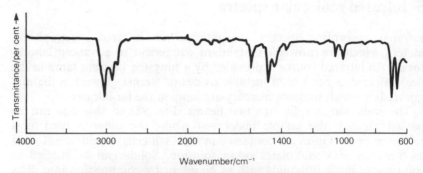

Fig. 99 The infrared absorption spectrum for liquid toluene. The main peaks are caused by aromatic C—H stretching (3020), aliphatic C—H stretching (2900), C≡C stretching (1600), aliphatic C—H bending (1500), and aromatic C—H out-of-plane bending (725 and 690). The particular patterns around 1800 and 700 are typical of a monosubstituted benzene derivative.

The spectra can also be presented as plots of absorbance against wave-number or wavelength, absorbance being defined as $\lg(1/\text{transmittance})$, i.e.,

$$\text{transmittance } (T) = \frac{\text{intensity of transmitted radiation}}{\text{intensity of incident radiation}}$$

$$\text{absorbance } (A) \quad = \lg 1/T$$

The advantage of using absorbance is that it is proportional to the concentration of the absorbing solution (p. 186).

a Pure rotation spectra These can only be obtained from molecules with a permanent dipole moment and by using infrared (or microwaves) below $300\,\text{cm}^{-1}$ as a source. The dipole moment is necessary so that the rotating molecule can give rise to an electric field which can interact with the incident radiation. A low energy source is needed to prevent any vibrational or electronic changes.

The quantised energy levels of a rotating molecule depend on its moment of inertia and this can be calculated from the wave-numbers of the spectral lines. Bond lengths and angles can be calculated from the moment of inertia.

These bond lengths can be found very accurately for diatomic molecules. The bond lengths in a linear, triatomic molecule, A—B—C, can also be obtained, but as two lengths are involved, it is necessary to compare the spectra of the A—B—C molecule with that of an isotopically substituted molecule, e.g., A*—B—C, where A* is an isotope of A. For more complicated molecules it may be possible to measure bond lengths and bond angles if the molecule has a dipole moment and if it is symmetric.

b Vibration-rotation spectra Absorption of radiation of wave-number $650{-}4000 \, cm^{-1}$ causes both vibrational and rotational changes within molecules. In the simplest case of a diatomic molecule it must be heteronuclear so that vibration causes a change in bond length and, hence, a change in dipole moment.

At low resolution, only lines due to vibrational changes may be apparent, but at higher resolution each vibrational line splits into a band, with closely spaced lines on each side of the vibrational line. Each vibrational energy change has associated rotational energy changes so that a band system appears as illustrated in Figs. 100 and 101. A single molecule may produce many band systems, which may overlap; the result is known as a band spectrum.

In simple cases, bond lengths can be found, as in **a**, from measurements on the rotational lines, and bond dissociation energies can be calculated from the vibrational lines.

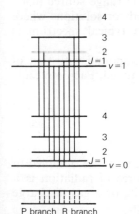

Fig. 100 Diagrammatic representation (not to scale) of the formation of a band system associated with a change from a lower vibrational state ($v = 0$) to a higher one ($v = 1$). The rotational lines appear on each side of the vibrational line in what are known as the R and P branches. In the R branch the rotational changes in J are $0 \to 1$, $1 \to 2$, $2 \to 3$ and $3 \to 4$; in the P branch the changes are $1 \to 0$, $2 \to 1$, $3 \to 2$ and $4 \to 3$.

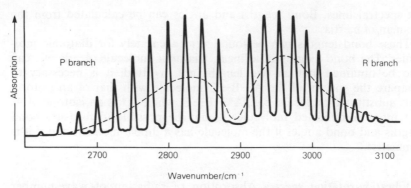

Fig. 101 The fundamental band system for gaseous hydrogen chloride. The fundamental vibration at $2890\,cm^{-1}$ is forbidden (p. 178). The dotted line shows the spectrum at low resolution. The rotational lines in the P and R branches appear at higher resolution. At still higher resolution these lines split into two for the $H^{35}Cl$ and $H^{37}Cl$ molecules.

c Analysis by infrared spectroscopy In a complex molecule there are many bonds and many possible vibrational and rotational energy changes. A detailed interpretation of the band spectrum may not be possible but it can still be very useful in chemical analysis, particularly of organic substances.

Many of the bonds in such molecules will absorb specific frequencies so that the presence of such bonds can be established. A free O—H bond in alcohols and phenols, for example, absorbs around $3700\,cm^{-1}$. If the O—H bond is hydrogen bonded, the bond is weakened and absorbs at a lower wave-number around $3400\,cm^{-1}$.

Such infrared spectra are examined using a wide range source (650–$4000\,cm^{-1}$), and they are best carried out on liquids or solutions where rotational absorption is not possible (p. 179). A typical spectrum is shown in Fig. 99.

Comparison of the spectrum of an unknown compound with that of known compounds enables the unknown to be identified. Each molecule has its own 'fingerprint'.

6 Ultraviolet and visible spectra

Absorption spectra in the ultraviolet and visible regions are obtained in much the same way as infrared spectra. The source of radiation is a heated metallic filament, as in an electric light bulb, or a gas discharge lamp. Samples are generally investigated in solution.

Absorption between 13 333 and $24\,000\,cm^{-1}$ (417 and 750 nm) is involved in the visible region, and up to $50\,000\,cm^{-1}$ (200 nm) in the far ultraviolet. The energies involved (160–598 kJ mol^{-1}) can bring about electronic, vibrational and rotational changes, but, in the liquid state, the vibrational and rotational fine structures are largely obliterated.

The energies towards the top end of the range are bigger than many bond energies. That is why visible and ultraviolet light can cause dissociation in photochemical reactions (p. 330). For simple molecules, the ultraviolet absorption spectra can be used to measure bond energies.

When any visible light frequencies are absorbed from incident white light by a substance, it will be coloured. Substances absorbing blue-green light, for example, look red: those absorbing red look blue-green, and so on.

Typical examples of visible spectra are given in Fig. 102 for solutions of methyl orange at different pH values.

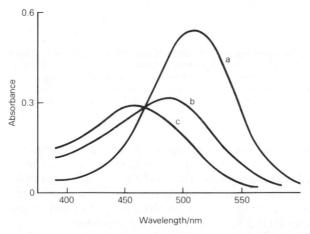

Fig. 102 The visible spectra of solutions of methyl orange. **a** pH 1, **b** pH 3.8 and **c** pH 13. At pH 1 the main absorption is of blue-green light so that the solution looks red. At pH 13 the main absorption is of blue-violet light so that the solution looks yellow. At pH 3.8 (the mid-point) the main absorption is of green-blue light so that the solution looks orange.

a Chromophores Molecules which give ultraviolet or visible spectra generally contain particular bonds, or groups of atoms, known as chromophores. Examination of the spectra can establish the existence of such bonds or groups. Unsaturated bonds, e.g., C=C, C≡C, C=O and N=N are particularly common chromophores; it is the electrons in π-orbitals that are involved in ultraviolet and visible absorption. C=O bonds, for example, absorb wavelengths around 180 and 290 nm; C=C bonds absorb around 180 nm.

When two or more such bonds are conjugated (p. 140), the absorption is at a higher wavelength. Benzene, for example, absorbs at 184, 203 and 255 nm (Fig. 103). If the conjugation is extensive, the absorption may be in the visible region. Carotene, for example, has eleven conjugated C=C bonds and absorbs around 450 nm. It is the main yellow colour in carrots, butter and egg-yolk. Lycopene, with a similar structure, provides the red colour in tomatoes, rose hips and many berries.

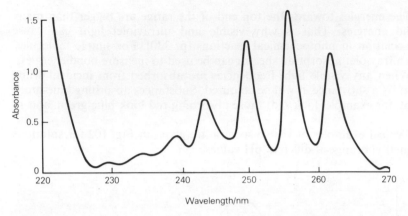

Fig. 103 Ultraviolet absorption spectrum for a solution of benzene in cyclohexane. There are strong absorptions at 184 and 203 nm and a band, as shown, around 255 nm, which is characteristic of aromatic hydrocarbons. If a substituent group in a benzene ring causes increased conjugation the wavelengths absorbed are lengthened so that absorption may take place in the visible region. That is why many substituted benzene derivatives are coloured.

b Absorption by transition metals Transition metal complexes give ultraviolet or visible spectra because electronic transitions in the d-orbitals (from t_{2g} to e_g) absorb in these regions. The $Ti(H_2O)^{3+}$ ion, for example, absorbs around 500 nm (20 000 cm^{-1}), i.e. green light; that is why the complex looks purple. The energy change involved is 239.1 kJ mol^{-1} and this measures the energy level difference between the d-orbitals.

Similarly, the $Cu(H_2O)_4^{2+}$ ion looks blue because it absorbs yellow light; $Cu(NH_3)_4^{2+}$ is violet because it absorbs yellow-green light; Cu^{2+} and $Cu(CN)_4^{2-}$ are both colourless as they do not absorb in the visible region.

c Analysis Because different molecules or transition metals have their own particular ultraviolet or visible spectrum, a study of the spectrum can be used in analysis. This can be done quantitatively because the *Beer–Lambert law* states that the extent of the absorption by a substance, for a given wavelength, depends on the thickness of the absorbing layer and the concentration of the substance in the layer. Thus

$$\text{absorbance} = \lg\left[\frac{\text{intensity of incident light}}{\text{intensity of transmitted light}}\right] = \varepsilon cl$$

where ε is a constant for a substance, c is its concentration and l the thickness of the absorbing layer. If c is measured in mol m^{-3} and l in metres, ε will be expressed in m^2 mol^{-1}; it is called the *molar extinction coefficient*.

Once the value of ε for any substance has been measured the absorbance of any particular frequency by that substance measures the concentration of the solution involved. Changes of concentration with time can also be measured in kinetic studies (p. 306). These methods can be applied to any substance giving an ultra-violet or visible spectrum.

d Colorimetry A colorimeter is a simple piece of equipment used to measure the concentration of coloured substances in solutions. It measures the extent of the absorption of a particular wavelength of visible light by the substance.

Light is passed through a filter to provide a suitable wavelength which is then passed through a solution of the substance under test. The intensity of the transmitted light is measured on a photosensitive detector, such as a selenium or cadmium sulphide cell connected to a meter (Fig. 104). If necessary, the apparatus can be calibrated against known solutions.

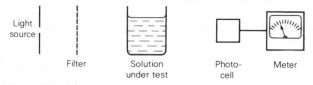

Light source   Filter   Solution under test   Photo-cell   Meter

Fig. 104 The essential features of a colorimeter.

Colorimeters can be operated very easily and give reasonably accurate results. They are widely used in routine analysis. The concentrations of coloured ions are readily determined, and most colourless ions can be converted into coloured ones by complexing.

Typical uses of colorimetric methods include the estimation of traces of ammonia in water by adding Nessler's solution (p. 397) to give a colour; measurement of rates of reaction such as those between MnO_4^- and $C_2O_4^{2-}$ ions or between iodine and propanone; and the measurement of the value of x in the formula of the coloured complex ion $Ni(EDTA)_x^{2-}$ by finding out what mixtures of Ni^{2+} and EDTA solutions will give the maximum colour intensity.

7 Raman spectra

When radiation of wavelength, λ, is passed through a gas, liquid or solid which does not absorb it, a small proportion is scattered if the particles of the gas, liquid or solid are comparable in size to λ. This is known as *Tyndall scattering*; it is the cause of the scattered light seen when sunlight passes through a dusty atmosphere or a colloidal solution (p. 457).

When similar radiation is scattered by molecules or atoms with the necessary vibrational or rotational energies, the scattered light shows some lines of higher and some of lower frequency than that of the incident radiation. This arises because the incident radiation can lose energy in exciting the molecules or atoms, or, alternatively, gain energy from molecules and atoms already excited.

The resulting spectrum is known as a Raman spectrum. The lines of lower frequency are called Stokes lines; those of higher frequency are anti-Stokes lines. Measurements on these lines can, in some cases, be used to find bond lengths. Raman spectra are useful, as molecules which do not give normal vibrational or rotational spectra may give Raman spectra.

The experimental method is different because the lines are very weak, but the use of lasers for the incident radiation promises to make the technique easier and of wider application.

8 Fluorescence and phosphorescence

Radiation absorbed by an atom or molecule energises or activates or excites the atom or molecule.

In fluorescence the absorbed radiation is re-emitted at a higher wavelength. Fluorescein, for example, absorbs blue and ultraviolet light but emits yellow-green light. Some of the absorbed energy is transferred into vibrational, rotational and translational energy amongst neighbouring solvent molecules; the remainder is re-emitted at a lower energy (higher wavelength). Fluorescence ceases as soon as the incident light source is removed.

Phosphorescence is a delayed fluorescence and it continues after the removal of the incident light. Molecules or atoms which phosphoresce do so because they can exist in metastable states from which the absorbed energy is re-emitted only slowly.

9 Lasers

Both fluorescence and phosphorescence are spontaneous processes. Lasers (light amplification by stimulated emission of radiation) depend on a stimulated process. A ruby crystal (aluminium oxide plus a little Cr^{3+}) provides a simple example. The Cr^{3+} ions are excited by light absorption, some of the energy is transferred to the aluminium oxide and the remainder is re-emitted as dark red light giving the normal colour of ruby.

In a laser, this emitted radiation is used to stimulate further emission. This is achieved by making a ruby crystal in the form of a rod with flat, parallel ends, one of which is completely silvered and the other partially silvered. The rod is excited by a powerful optical flash. The emitted radiation is reflected backwards and forwards along the rod stimulating more and more emission so that a very high intensity of radiation builds up. When it is high enough, it passes out through the partially silvered end, as a laser beam.

Laser beams can also be obtained by initial electrical stimulation, and a variety of materials (e.g. glasses containing neodymium, and mixtures of helium and neon) can be used.

The beams provide a source of high intensity, monochromatic radiation. It can be used in telecommunications, and in cutting very hard materials. In chemistry, laser beams are finding increasing use as, for example, in flash photolysis (p. 309), in Raman spectroscopy (p. 187), in vapourising and ionising material in mass spectroscopy (p. 88), in atomising samples in atomic and molecular absorption spectroscopy (p. 180), and in converting material into plasma (p. 33).

10 Nuclear magnetic resonance (NMR) spectroscopy

NMR spectroscopy is particularly useful for giving information about the hydrogen atoms in an organic molecule. The hydrogen atom nucleus, i.e. the proton, contains an odd number of nucleons* (one) and can spin in two opposite directions. The spinning nucleus (proton) has a magnetic moment so that it can be orientated in an external magnetic field either in the same direction as the field or in the opposite direction. Protons aligned in the same direction as the external field have lower energy than those opposed to the field, the energy difference being proportional to the field strength.

Protons in organic molecules can absorb radiation when its energy is of the right value to raise them from their lower to their higher energy level, i.e., to invert their spins. The frequency of radiation necessary to do this depends on the strength of the external magnetic field and on the nature of the hydrogen atom concerned. For a field strength of about 1 tesla, 1 T (10^4 gauss) the frequencies absorbed lie in the range of radio waves between 10 and 100 MHz.

For any fixed magnetic field the precise frequency of radiation absorbed depends on the nature of the hydrogen atom concerned. The hydrogen atoms in —OH, —CH_3, —CHO and other groups are all surrounded by slightly different arrangements of electron density so that they are shielded differently from any external magnetic field. The nature of the hydrogen atoms in organic molecules can, therefore, be determined either by measuring what frequencies of radiation the molecule will absorb when in a fixed magnetic field, or what magnetic field strengths will cause the absorption of a fixed frequency of radiation. It is, in practice, easier to use the latter technique and only relatively small

---

* The NMR spectra associated with the ^1H nucleus are easily the most important but other nuclei with odd mass numbers, e.g. ^{19}F and ^{31}P, will also give rise to NMR spectra. Nuclei with even mass numbers and even atomic numbers, e.g., ^{12}C and ^{16}O give no NMR spectra.

Introduction to Physical Chemistry

changes in magnetic strength are needed. The number of any particular type of hydrogen atom in a molecule is indicated by the amount of radiation of a particular frequency that is absorbed.

a Experimental An NMR spectrometer is provided with an oscillator producing radio waves of a fixed frequency (typically 40, 60 or 100 MHz), a powerful electromagnet whose field can be changed slightly by altering the current, and a detector to record the level of radiation absorbed as the magnetic field is changed (Fig. 105). The poles of the electromagnet are very close together to give a homogeneous field and the sample under test (either in liquid form or in solution) is spun between the poles.

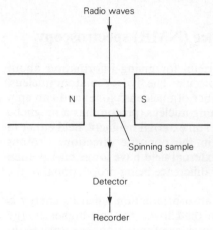

Fig. 105 Essential features of an NMR spectrometer.

Tetramethylsilane (TMS), $(CH_3)_4Si$, is commonly used as a standard. It contains only one type of hydrogen atom which absorbs a fixed frequency of radiation at one particular magnetic field strength (H) which is taken as the standard. Other protons in other organic molecules will absorb the same frequency at different field strengths (H_1) and the relative change, i.e. $(H - H_1)/H$ is a constant for any particular proton. This constant is generally multiplied by 10^6, i.e., expressed as parts per million (p.p.m.); it is then known as the chemical shift, δ, for the proton concerned, the value for TMS being zero. Alternatively TMS is given an arbitrary value of 10 p.p.m. and chemical shifts are expressed in tau (τ) units, with $\tau = 10 - \delta$.

Values for the chemical shifts of hydrogen atoms in typical organic groups are given opposite and typical NMR spectra are shown in Figs. 106 and 107. The spectra can be plotted either at low or high resolution. A modern NMR spectrometer can also record automatically the areas under each peak thus indicating the number of protons concerned.

Chemical shifts (δ) for protons in p.p.m. relative to TMS = 0

type of proton	δ	type of proton	δ
$R-CH_3$	0.9	$R-O-CH_3$	3.8
$\begin{array}{c} R \\ \diagdown \\ CH_2 \\ \diagup \\ R \end{array}$	1.3	$R-O-H$	about 5
$\begin{array}{c} R \\ \diagdown \\ R-CH \\ \diagup \\ R \end{array}$	2.0	$\begin{array}{c} R-C-H \\ \\ O \end{array}$	9.7
$R-CH_2-I$	3.2	$R-COOH$	about 11

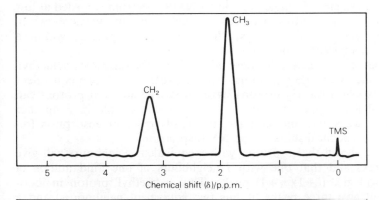

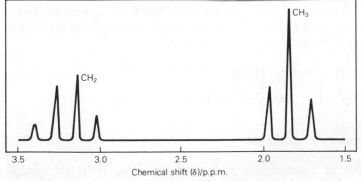

Fig. 106 Typical NMR spectra for iodoethane. The upper diagram shows the peaks for CH_3 and CH_2 groups at low resolution. At higher resolution (lower diagram) the CH_3 splits into three peaks (it has two equivalent H neighbours) and the CH_2 group splits into four peaks (it has three equivalent H neighbours).

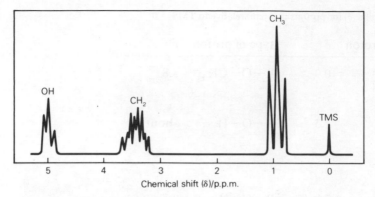

Fig. 107 High resolution NMR spectrum for ethanol. The OH and CH_3 groups each give three peaks as they have two equivalent H neighbours. The CH_2 group gives eight peaks as it has three CH_3 hydrogen neighbours and one OH hydrogen neighbour.

b Spin-spin coupling A single peak in an NMR spectrum recorded at low resolution may split into more peaks at higher resolutions as shown in Figs. 106 and 107. This arises when protons are present within a molecule on adjacent carbon atoms.

In iodoethane, for example, there are what may be regarded as 'methyl' (H_3) and 'methylene' (H_2) protons. The H_3 protons have two equivalent neighbours whilst the H_2 protons have three. The adjacent protons can have parallel or opposed spins so that they can couple in a number of different ways and thus give rise to slightly different absorption frequencies, i.e. chemical shifts, in the NMR spectrum.

It can be shown that a proton with n equivalent neighbours will give $(n+1)$ peaks and that one with l neighbours of one kind and m of another will give $(l+1)(m+1)$ peaks. Thus a 'methyl' proton in iodoethane will give three peaks (it has two equivalent neighbours) and a 'methylene' proton will give four peaks (it has three equivalent neighbours). Furthermore, if n is 2 the intensity of the peaks will be in the ratio $1:2:1$; if n is 3 the ratio is $1:3:3:1$ (Fig. 106).

$$
\begin{array}{cccc}
& H\ \ H & & H\ \ H \\
& |\ \ \ | & & |\ \ \ | \\
H- & C-C & -I \qquad H- & C-C-O-H \\
& |\ \ \ | & & |\ \ \ | \\
& H\ \ H & & H\ \ H \\
& \text{iodoethane} & & \text{ethanol}
\end{array}
$$

In ethanol, at a sufficiently high resolution (Fig. 107), the 'methyl' protons give three peaks ($n = 2$); the 'hydroxyl' proton gives three peaks ($n = 2$); and the 'methylene' proton gives eight peaks (it has three 'methyl' protons and one 'hydroxyl' proton as neighbours).

High resolution NMR spectroscopy can, therefore, provide invaluable information about the types of hydrogen atoms in an organic molecule, about the number of each type, and about the neighbouring hydrogen atoms.

11 Electron spin resonance (ESR) spectroscopy

Electrons possess spin and they can be made to resonate in a magnetic field in much the same way as atomic nuclei in NMR spectroscopy. For a field around 0.35 tesla the frequencies concerned are about 10 GHz, i.e. wavelengths around 3 cm.

Unpaired electrons in different species give different ESR spectra just as hydrogen atoms in different molecules give different NMR spectra. It is, therefore, possible to detect and identify many species containing unpaired electrons. The method can be applied to odd molecules, e.g. NO_2 and ClO_2, free radicals and transition-metal ions.

Questions on Chapter 16

1 Give an account of the ways in which the interactions of electromagnetic radiation (e.g. X-rays, ultraviolet, infrared, visible light, etc.) with matter have been of use in chemistry. (LS)

2 Write an essay on the chemical aspects of colour. (OJSE)

3 Chemists make extensive use of 'physical methods' to elucidate molecular structures. Briefly review the application of *three* of the following techniques for this purpose (a) dipole moment measurement (b) infrared spectroscopy (c) mass spectroscopy (d) X-ray analysis. (No details about instrumentation or measurement techniques are required. Instead, emphasis should be given to the kind of information that can be obtained and to the way in which this helps towards the elucidation of molecular structures.) (LS)

4 How can electromagnetic radiation be absorbed or emitted by (a) atoms (b) molecules? How, in general, do the spectra of atoms differ from those of diatomic molecules?

5 Calculate the wave-numbers of the first six lines in the Lyman spectrum of hydrogen. Plot the differences between the wave-numbers of successive lines against the wave-number with the higher value of the pair. What is the wave-number when the difference is zero? What is its significance?

6 Describe the uses of spectroscopy in analysis.

7 Convert the following: (a) $96.486 \, kJ \, mol^{-1}$ into $J \, molecule^{-1}$ (b) $1 \, cm^{-1}$ into eV (c) $1 \, m^{-1}$ into cm^{-1} (d) $10^{14} \, Hz$ into $kJ \, mol^{-1}$.

8 The limits of the spectral lines formed by transitions from the ground state occur at 229.9 nm for lithium and 285.6 nm for potassium. Calculate the molar ionisation energies of the two elements. If the ionisation energy of sodium is $496 \, kJ \, mol^{-1}$, calculate the wavelength of the convergence limit for sodium.

9 Write short notes on (a) flame tests (b) 'fingerprinting' (c) the Beer–Lambert law.

10 Give the names and structures of any six highly coloured organic compounds.

11 Sketch the visible absorption spectra between 400 and 700 nm for (a) phenolphthalein and (b) litmus, in solutions of pH 1 and pH 13.

12 Describe how you would use a colorimeter to study (a) the rate of the reaction between a solution of potassium manganate(VII) and sodium ethanedioate (b) the value of x in $Ni(NH_3)_x^{2+}$.

13 How is nuclear magnetic resonance spectroscopy carried out and why is it important?

14* Record the wavelengths of some of the lines in the Balmer series of the hydrogen spectrum, and use the values to calculate the value of the Rydberg constant.

17
Solutions of solids and gases in liquids

1 Measurement of solubility of solids

The solubility of a solid in a liquid is usually defined as *100 times the maximum mass of the solid (solute) which will dissolve in 1 unit mass of the liquid (solvent) at the temperature concerned, and in the presence of excess, undissolved solid.* Typical units used are $g(100 \, g \, solvent)^{-1}$.

a Saturated solution The presence of excess, undissolved solid in contact with a solution ensures that the solution is saturated. There is a dynamic equilibrium between the undissolved and the dissolved solid (p. 286). The solvent has dissolved as much of the solute as it can, at the particular temperature.

Solutions containing more dissolved solid than a saturated solution can be made; they are called *supersaturated* solutions. The conditions have to be peculiar. On heating sodium thiosulphate, $Na_2S_2O_3.5H_2O$, for example, a solution of the salt in its own water of crystallisation can be obtained. The solution is, theoretically, unstable, and shaking, or the addition of dust particles or another crystal, causes rapid crystallisation. With care, however, the supersaturated solution can be kept; it is said to be metastable.

b Measurement of amount of dissolved solid Measurement of the solubility of a solid which is reasonably soluble requires, first, a known mass of a saturated solution of the solid. This is then analysed to find the mass of solute it contains. The mass of solvent is found by subtraction, so that the solubility can be calculated in terms of grams of solute per 100 g of solvent.

A saturated solution is made by stirring the chosen solvent with excess of the solute in a container immersed in a thermostat at the temperature required. After undissolved solid has had time to settle, a portion of the supernatant saturated solution is withdrawn and placed in a container of known mass. The total mass is then measured to give the mass of the saturated solution. In the withdrawal of this sample of saturated solution it is important not to include any solid. This can be done, with care, using a pipette, the tip being covered with a plug of glass wool if necessary. The pipette used must be at the same temperature as the saturated solution.

Analysis of the known mass of saturated solution is best done volumetrically if suitable reagents are available.

The measurement of the solubility of sparingly soluble substances requires special methods mentioned on pp. 395–6.

c Solubility curves The variation of solubility with temperature is summarised graphically in what is known as a solubility curve (Fig. 108). As the solution of a solid in a liquid is usually endothermic the solubility generally rises with increase in temperature, according to Le Chatelier's principle (p. 296). Those solids whose solubility in water decreases with temperature, e.g., anhydrous sodium sulphate (Fig. 108), evolve heat on dissolving.

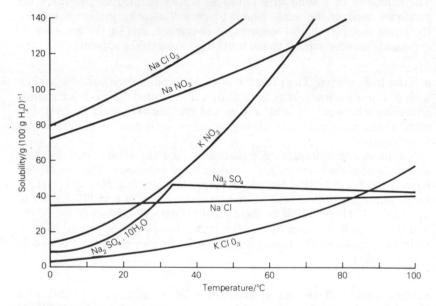

Fig. 108 Typical solubility curves.

Solubility curves may show marked transition points due either to a polymorphic change (p. 170) or to a change in hydration. The transition point at 34 °C (Fig. 108) for sodium sulphate, for example, is due to a change in hydration. Two separate solubility curves for different forms of the same solid are really being plotted on one diagram.

The solubility curves shown are limited to a temperature range of 0–100 °C. A more complete representation over a wider temperature range is given on page 210.

2 Fractional crystallisation

This process makes use of the different solubilities of two or more solids in the same solvent to separate the solids. With large solubility differences, and for solids with steep solubility curves, it is relatively easy. Otherwise, it can be very lengthy.

a Separation of potassium nitrate and potassium chlorate(V) Fig. 108
shows the solubilities of these two salts in gram per 100 gram of water as

	KNO$_3$	KClO$_3$
20 °C	31	8
50 °C	86	18

If a mixture of 20 g of potassium nitrate and 18 g of potassium
chlorate(V) is dissolved in water at 50 °C, the solution will be just
saturated with potassium chlorate(V) but not with potassium nitrate,
neglecting any effect the solubility of one salt may have on the other. On
cooling to 20 °C, 10 g of potassium chlorate(V) will crystallise out, but
there will be no crystals of potassium nitrate for the solution is still not
saturated with that salt. Thus pure potassium chlorate(V) can be obtained in one crystallisation

b Separation of solids with similar solubilities The conditions chosen in
the preceding example are extremely favourable. The main difficulties of
fractional crystallisation arise when it is necessary to deal with very small
quantities of material and/or when the two components to be separated
are so chemically similar that they have very similar solubilities in all
solvents. Fractional crystallisation, under such conditions, involves a lot
of careful but rather tedious work. Mme Curie, for example, took about
four years to separate radium bromide from barium bromide in her
isolation of radium.

The general principle adopted is outlined in Fig. 109. The original
mixture of A and B is dissolved in a hot solvent in which A is slightly less
soluble than B. Quantities are taken such that, on cooling, about half the

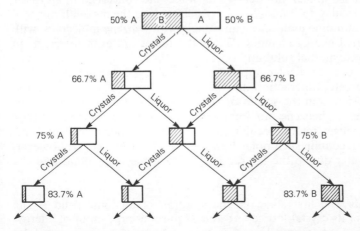

Fig. 109 Diagrammatic representation of fractional crystallisation of a mixture of A and B.
A is slightly less soluble than B, and the relative amounts of B and A are shown by the
shaded and unshaded areas.

197

dissolved solid will crystallise out. The crystals resulting will be enriched in the less soluble A, whilst the mother liquor will be enriched in the more soluble B.

The enriched crystals are redissolved and the solution is recrystallised, yielding a crop of crystals still richer in A, and a mother liquor (C) still richer in B. The original mother liquor gives, on evaporation and recrystallisation, another B-rich liquid and an A-rich solid (D). C and D are mixed together as shown on Fig. 109. By repeated treatment, it is possible to obtain pure A and pure B, but the crystallisation process may have to be repeated very many times.

3 Soluble or insoluble?

A complete understanding of the reasons why some substances are soluble and others insoluble is not yet possible, but there are some relatively simple broad trends. It is, for example, well known as a generalisation that 'like dissolves like' or 'birds of a feather flock together', i.e., ionic (polar) solids tend to dissolve in polar (ionising) solvents whilst non-polar (covalent) solids tend to dissolve in non-polar (covalent) liquids.

The overriding necessity for a solid to dissolve in a liquid is that the free energy change for the process (p. 270) must be negative. In simple terms, the interaction between the solute and solvent must be greater than, or at least similar to, the interactions in the pure solute and solvent for solubility to be likely.

a Covalent solids Covalent compounds tend to dissolve in non-polar solvents because similar van der Waals' forces are concerned in all cases. Benzene molecules, for example, can penetrate into the molecular crystal of iodine, and iodine molecules can mingle with benzene molecules, with relatively small energy changes. When the forces are even lower, as in gases, complete mutual solubility occurs.

Solubility of a covalent compound in a polar solvent is inhibited by the much greater interaction in the pure solvent than between the covalent molecules in the solute or the covalent and polar molecules in solution. Neither benzene nor iodine, for instance, can effectively break down the stronger intermolecular forces in water. Many covalent compounds which contain such polar bonds as O—H or N—H are, however, soluble in water due to the interaction between the polar bond and the water.

b Ionic solids The first stage in the solution of an ionic solid is the breakdown of its crystal lattice and this requires a large input of energy. This input may, however, be offset by a release of energy caused by solvation (hydration if the solvent is water) of the ions. Whether or not an ionic solid will dissolve is, broadly, a matter of the balance between these two energy terms. If the lattice energy is much higher than the hydration energy the process will be endothermic and, therefore, not likely to take place (Fig. 138, p. 265). When the hydration energy is greater than the lattice energy, solution is highly likely. A more detailed analysis, however, requires a consideration of entropy, as well as enthalpy changes (p. 266).

The break down of an ionic crystal lattice is favoured by solvents with high relative permittivity (dielectric constant). The attractive force between two ions in a crystal is given by

$$F = q_1 q_2 / 4\pi\varepsilon r^2$$

where q_1 and q_2 are the charges on the ions, r is the interionic distance, and ε is the permittivity of the medium separating the ions. In the solid state, the ions are separated by a vacuum with a permittivity of $8.85 \times 10^{-12} \, Fm^{-1}$. When water, with a permittivity 80 times higher than that of a vacuum, is used as a solvent the interionic forces are considerably lowered so that the crystal begins to 'fall to pieces' and gives free ions.

The solvation of ions is favoured by polar solvents, for the solvation involves the attraction of the solvent molecules by the charged ions. Water, for instance, hydrates ions by forming a sheath around both positive and negative ions.

The ideal solvent for an ionic solid would, then, have both high permittivity and high polarity; the latter, which can be measured by dipole moments, is the more important. Four good ionising solvents are shown below (on the left); three typical non-polar solvents, which will not dissolve ionic solids, are shown on the right.

	H_2O	HF(l)	HCN(l)	HCOOH	C_6H_6	CCl_4	C_6H_{14}
relative permittivity	80.1	84	114	58	2.3	2.2	1.9
dipole moment $(10^{-30} \, Cm)$	6.3	6.7	10	5	0	0	0

Solvents such as ethanol and propanone (acetone) lie in between.

4 Distribution of a solid between two immiscible solvents

If a solid or liquid, X, is added to a mixture of two immiscible liquids, in both of which it is soluble, it will distribute itself between the two liquids according to the distribution or partition law,

$$\frac{\text{concentration of X in solvent A}}{\text{concentration of X in solvent B}} = K(\text{a constant})$$

K is an equilibrium constant generally known as the *partition coefficient*. It is only a constant for given substances, and it varies with temperature.

The solute, moreover, must be in the same molecular state in both solvents.

It is important to realise that it is the ratio of the *concentrations* which matters and not the ratio of the total masses of solute. Clearly, too, the same units of concentration must be used for each solvent.

5 Effect of association on partition law

The partition law only holds if the same molecular species is considered in each phase. If a solute associates (or dissociates) in one or both of the solvents concerned, the partition law must be modified, though it still holds so long as identical molecular species are considered.

Consider a solute, AB, which is associated, to a degree α, into double molecules in solvent 2 but is not associated in solvent 1. If the total concentration of solute, as measured by analysis, in solvent 2 is c_2, the concentration of unassociated molecules will be $c_2(1-\alpha)$, and that of associated molecules will be $\frac{1}{2}c_2\alpha$,

$$2\text{AB} \rightleftharpoons (\text{AB})_2$$
$$c_2(1-\alpha) \qquad \frac{1}{2}c_2\alpha$$

Application of the equilibrium law (p. 287) shows that

$$\frac{[(\text{AB})_2]}{[\text{AB}]^2} = K = \frac{c_2\alpha}{2[c_2(1-\alpha)]^2}$$

The concentration of unassociated molecules in solvent 2, $c_2(1-\alpha)$, is, therefore, equal to $\sqrt{c_2\alpha/2K}$.

If the concentration of unassociated molecules in solvent 1 is c_1, then

$$c_1/\sqrt{c_2\alpha/2K} = K' \quad \text{or} \quad c_1/\sqrt{c_2\alpha} = K''$$

When α is 1, i.e. when there is complete association into double molecules in solvent 2,

$$c_1/\sqrt{c_2} = K$$

6 Solvent extraction

Organic compounds are generally much more soluble in such organic solvents as ethoxyethane (ether) or benzene than in water, and these organic solvents are also immiscible with water. An organic compound can often be extracted from an aqueous solution or suspension by adding ethoxyethane or benzene, shaking, separating the two layers in a separating funnel, and finally distilling off the ethoxyethane or benzene to leave the purified compound required.

In such a solvent extraction it is advantageous to use a given volume of organic solvent in small lots rather than in one whole.

Suppose that $100\,cm^3$ of benzene is available for extracting a solute, X, dissolved in $100\,cm^3$ of water, and that the partition coefficient of X between benzene and water is 5, i.e.

$$\frac{\text{concentration of X in benzene}}{\text{concentration of X in water}} = 5$$

If the whole of the benzene is added to the $100\,cm^3$ of solution, X will distribute itself between the benzene and the water. If W_B gram pass into the benzene layer, and W_W gram remain in the water, then

$$\frac{W_B/100}{W_W/100} = 5 \quad \text{or} \quad \frac{W_B}{W_W} = 5 \quad \text{or} \quad \frac{W_B}{W_B + W_W} = \frac{5}{6}$$

This means that the $100\,cm^3$ of benzene has extracted $\frac{5}{6}$ of the total amount of X originally present in the water.

Using only $50\,cm^3$ of benzene:

$$\frac{W_B/50}{W_W/100} = 5 \quad \text{or} \quad \frac{W_B}{W_W} = \frac{5}{2} \quad \text{or} \quad \frac{W_B}{W_B + W_W} = \frac{5}{7}$$

This means that $50\,cm^3$ of benzene will extract $\frac{5}{7}$ of the original amount of X. $\frac{2}{7}$ of X will, therefore, remain in the water, but addition of a second $50\,cm^3$ portion of benzene will extract $\frac{5}{7}$ of the X which is still present, i.e. $\frac{5}{7}$ of $\frac{2}{7}$, or $10/49$ of the original amount of X.

The first $50\,cm^3$ of benzene, therefore, extracts $\frac{5}{7}$ and the second $50\,cm^3$ portion extracts a further $10/49$ of the original amount of X. The total amount of X extracted by two $50\,cm^3$ portions of benzene will be $45/49$, and this is more than the $\frac{5}{6}$ extracted by one $100\,cm^3$ portion of benzene.

A still greater proportion could be extracted by using four $25\,cm^3$ portions of benzene.

7 Chromatography

Chromatography, first used by Tswett in 1906 to separate the pigments in leaves (hence the name), and extensively developed in the last forty years, depends on the different partitioning of different solutes between the same pair of liquids, and on differences in absorption. It provides a technique which has greatly facilitated chemical research, particularly in its ability to separate small amounts of material. Many different ways of applying the technique are used.

a Column chromatography If a solution containing various solutes is passed through a glass tube packed with a powdered material it is commonly found that the solutes concentrate at different levels of the column giving what is known as a *chromatogram*.

By pushing the whole column out of the glass tube and cutting it into separate bands each solute can be extracted separately. Alternatively, the various solutes can be washed through the column, in turn, by using more solvent or a different solvent. This is known as *elution* and the solvents as *eluants*.

If coloured solutes are involved the separation can be achieved visually. For colourless substances, ultraviolet illumination may be necessary, or the column may have to be treated with some chemical to distinguish one solute from another.

Many different packing materials, e.g. aluminium oxide, calcium carbonate, magnesium oxide, ion-exchange materials, and many proprietary mixtures, can be used. Typical solvents, which may be mixed together, include water, propanone, ethanol, benzene and hexane. A satisfactory combination of packing material and solvent can be found for most problems.

The separation achieved depends on the fact that the packing material in the column may itself be wet with one liquid, e.g., water, which is held stationary, and the various solutes partition differently between this stationary liquid and the moving solvent. When this is the major effect, it is known as *partition column chromatography*. If the packed column is dry, any separation depends on the fact that different solutes may be adsorbed differently on the column; if this is the major effect it is known as *adsorption column chromatography*. In practice both partitioning and adsorption may take place together.

b Paper chromatography Column chromatography requires quite large amounts of material and the same type of separation can be achieved with much smaller amounts by using specially made grades of absorbent paper instead of the packed column.

A very small spot of solution is placed near the bottom of a strip of paper which is then dipped into a suitable solvent. As the solvent front rises up the paper by capillary action the various solutes present are carried forward to differing extents. When the solvent front reaches almost to the top of the paper, the paper is removed and dried. If the solutes are coloured their positions on the paper can be seen. For colourless solutes, the paper can be treated with some chemical to convert the solutes into coloured compounds. The paper can then be cut up and each solute extracted separately.

In other applications, the solvent front can be arranged to pass down the paper from a trough, or radially outwards from the centre of a circle of paper.

A mixture of adsorption and partitioning is generally involved. The cellulose of the paper acts as an adsorbent but it is also wet with water which allows for partitioning if a non-aqueous solvent is used. For any particular set of conditions, each individual substance has a fixed R_f value given by

$$R_f = \frac{\text{distance moved by substance}}{\text{distance moved by solvent front}}$$

The identity of any unknown substance can be established by comparison with known materials under the same conditions.

Two substances with equal R_f values cannot be separated by one-way chromatography but a two-way process may be possible. The original mixture is spotted onto one corner of a square piece of paper and a one-way separation is carried out using one solvent. The paper is then dried and turned through 90° before carrying out a second separation with a different solvent.

c Thin layer chromatography Paper chromatography is limited to separations that can be achieved on a cellulose base but the same principle can be applied by using a thin layer of material deposited evenly on a glass plate. The plate is then used in the same way as a sheet of paper. Many materials can be used and there is the normal wide choice of solvent, and even smaller amounts of material can be dealt with than in paper chromatography.

d Gas chromatography Gas chromatography is like column chromatography except that a gas is used instead of a liquid solvent. It is used for separating gases or volatile liquids.

A column is packed with a material which may simply act as an adsorbent or which may be wetted with an involatile oil so as to facilitate partitioning. A steady flow of a carrier gas, such as hydrogen, nitrogen or carbon dioxide, is then passed through the column, heated if necessary. The mixture of gases or vapours to be analysed is fed into the stream of carrier gas, via a microsyringe inserted through a rubber cap, before it enters the column (Fig. 110).

Some constituents of the mixture are soon carried right through the column, whereas others pass through only very slowly. Each constituent will, in fact, pass through the column at a different rate and will emerge

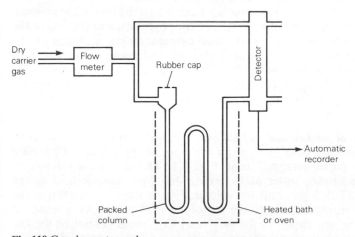

Fig. 110 Gas chromatography.

at the end of the column at different times. A detector at the end shows the emergence of the various constituents of the original mixture, which is therefore analysed.

The commonest type of detector is a thermal conductivity gauge. The gas emerging from the end of the column is passed round a heated filament. The heat loss from this filament depends on the thermal conductivity of the gas mixture surrounding it, and its electrical resistance, which is measured, changes as the gas mixture changes. The presence of different gases can be detected in this way.

If the carrier gas is hydrogen or if some hydrogen is added to the carrier gas at the end of the column a flame ionisation detector can also be used. The emergent gas is burnt at a metal jet which acts as a negative electrode. The flame impinges on a metal plate above it which acts as a positive electrode; a potential difference is maintained between the two electrodes. The current passing between the electrodes depends on the concentration of ions within the flame and this varies with the composition of the gas being burnt.

Both methods of detection can be connected to automatic recording devices.

8 Applications of chromatography

a Estimation of carotene in grass The pigments in a weighed sample of dry grass are extracted by warming with petroleum ether. The solution obtained is green, the yellow colour of the carotene being masked by the green of the chlorophyll.

A vertical glass tube with a pad of cotton wool and a clip at the bottom is packed with a 50:50 mixture of aluminium oxide and sodium sulphate(VI) by pouring in a slurry of the mixed solids with petroleum ether. The column is drained but not allowed to dry.

The cooled solution of pigments is poured on to the top of the column, whereupon a green band of adsorbed chlorophyll forms at the top of the column with a yellow band of adsorbed carotene below it. When all the pigment solution has been poured through the column, the carotene layer is eluted by passing a 2 per cent solution of propanone in petroleum ether through. The carotene solution coming from the bottom of the column is collected, and the amount of carotene it contains is estimated colorimetrically.

b Separation of nickel and cobalt ions A small drop of an aqueous solution containing both Ni^{2+} and Co^{2+} ions is applied to a 1 cm wide strip of filter paper arranged as in Fig. 111. The solvent, a mixture of propanone (acetone), water and concentrated hydrochloric acid in the proportions 87:5:8 per cent by volume, rises up the paper. When the solvent front reaches nearly to the top of the strip, the paper is removed and dried. The separation of the Ni^{2+} and Co^{2+} ions cannot be seen at this stage, but it becomes visible when the paper strip is sprayed with, or

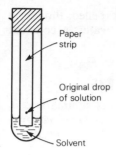

Fig. 111

dipped into, a solution of rubeanic acid. A yellow-brown spot, showing the position of the Co^{2+} ions, is seen well above a blue-violet spot, showing the position of the Ni^{2+} ions.

9 Solubility of gases in liquids

Care must be taken in using any tables of gas solubility figures for a number of different methods are used. Modern tables tend to express the solubility of a gas, at any particular temperature and pressure, as the *mass* of gas dissolved in unit *mass* of the solvent in a saturated solution. Alternatively, the number of moles of gas dissolved in 100 grams of solvent may be used.

What is known as the *absorption coefficient* of a gas is also used. It is the volume of a gas in cm^3 (or m^3) which will just saturate $1\,cm^3(1\,m^3)$ of solvent, at a particular temperature and pressure, the gas volume being expressed at s.t.p. When the gas volume is measured at the temperature and pressure at which the solubility measurement is being made (and not at s.t.p.) the value is known as the *coefficient of solubility*.

Some gases, e.g. NH_3, HCl and SO_2 are very soluble in water and some, e.g. CO_2, H_2S and Cl_2 are reasonably soluble; all these gases react with, or ionise in, water. Other gases, e.g. O_2, N_2 and CO have very low solubilities; so low that they are often regarded as insoluble. At 0 °C and 1 kPa pressure, the absorption coefficients of NH_3, CO_2 and O_2 are 1300, 1.713 and 0.0489 respectively; the ratio of the solubilities is approximately $27\,000:35:1$.

10 Effect of temperature on the solubility of a gas

When gases dissolve there is generally an evolution of heat, and it would therefore be expected, on the basis of Le Chatelier's principle (p. 296), that the solubility of a gas would decrease with increase in temperature. For most gases, this is found to be so, though hydrogen and the inert gases behave differently over some ranges of temperature.

On heating a solution of a gas, the gas is usually expelled, though in some cases, e.g. a solution of hydrogen chloride in water, a constant boiling mixture (p. 224) may be formed.

11 Effect of pressure on the solubility of a gas

The variation in the solubility of a gas with pressure is summarised in Henry's law (1803) which states that *the mass of a gas dissolved by a given volume of liquid, at a constant temperature, is proportional to the pressure of the gas*, i.e.

mass dissolved $\propto$ pressure

If x g of gas is dissolved by V cm^3 of liquid at a pressure p, then $2x$ g will be dissolved at $2p$. As the volume of $2x$ g of gas measured at $2p$ is equal to the volume of x g of gas measured at p, it follows that *the volume of a gas dissolved, measured at the pressure used, is independent of the pressure*. This is an alternative statement of Henry's law.

The law can be expressed in a third, and much more general, form. If a unit volume of liquid is considered, the mass of gas dissolved in it will be equal to the concentration of the solution. Moreover, the pressure of a gas is only a convenient way of expressing its concentration. The proportionality between the mass dissolved and the pressure can, therefore, be rewritten as:

$$\text{concentration of gas in solution} \propto \text{concentration of gas above the solution}$$

This can be widened into the general statement that *the concentrations of any single molecular species in two phases in equilibrium are in a constant ratio to each other at a fixed temperature*. This applies not only to the solution of a gas in a liquid but, also, to the solution of solids in liquids (the concentration of the solid in the solid phase being constant), and to the distribution of a solid between two immiscible liquids (p. 199).

12 Solution of two or more gases in the same liquid

Henry's law applies separately to a mixture of two gases dissolved in the same liquid, so long as the partial pressure of each gas is used in considering the gas.

The absorption coefficient of oxygen, at 15 °C and 100 kPa is approximately 0.05, and that of nitrogen is 0.025; the oxygen is twice as soluble as the nitrogen. In air, at 100 kPa, the approximate partial pressures of oxygen and nitrogen are 20 and 80 kPa respectively, and the corresponding values for the absorption coefficients will be 0.05×20, i.e., 1 for oxygen, and 0.025×80, i.e., 2 for nitrogen. The ratio by volume of oxygen to nitrogen in dissolved air will therefore be 1:2, i.e., there will be 33.3 per cent of oxygen and 66.6 per cent of nitrogen, and a gas of this composition will be obtained by boiling a solution of dissolved air.

If a diver were to breathe air under high pressure, large quantities of nitrogen would dissolve in his body fluids and cause problems. Deep sea divers therefore breathe a mixture of approximately 5 per cent oxygen with 95 per cent helium, which has a very low absorption coefficient.

Questions on Chapter 17

1 What is the mole fraction of ethanol in a mixture of 10 g of it with 10 g of water? If the mole fraction of ethanol in a mixture of it with water is 0.5, what is the percentage of ethanol by mass?

2 A solution of ethanoic acid ($M_r = 60.1$) containing 80.8 g dm^{-3} of acid has a density of 1.0097 g cm^{-3}. Calculate the mole fraction of ethanoic acid in the solution. Describe, with some experimental detail, how you would plot the solubility curve of potassium chloride.

3 Describe, as precisely as possible, what happens when sodium chloride dissolves in water. What energy changes take place?

4 When common salt is added to water it dissolves. When it is heated, it melts. What have these two processes in common and how do they differ? In what ways do the dissolving and melting of (a) sucrose (b) oxygen differ from common salt?

5 Make a list of actual chemical processes in which fractional crystallisation is used.

6 Potassium salts often have 'steeper' solubility curves than the corresponding sodium salts. To what extent is this true, and what bearing does it have on the fact that potassium salts are commonly used in the laboratory in preference to the corresponding sodium salt, even though the sodium salt may be cheaper?

7 Describe, with experimental detail, how you would obtain a sample of pure benzenecarboxylic (benzoic) acid from an impure sample by recrystallisation.

8 What is meant by solubility and partition coefficient? Explain the principles of (a) recrystallisation   (b) extraction with ethoxyethane.

9 Benzenecarboxylic (benzoic) acid is distributed between water and benzene according to the following figures. What conclusions can you draw?

| conc. in water | 0.0150 | 0.0195 | 0.0302 |
| conc. in benzene | 0.2420 | 0.4120 | 0.9700 |

10 The solubility of iodine in water is 0.7 per cent of that of iodine in carbon disulphide. An aqueous solution of iodine containing 0.1 g of iodine per 100 cm^3 is shaken with carbon disulphide. To what value does the concentration of the aqueous solution sink   (i) when 1 dm^3 of it is shaken with 50 cm^3 of carbon disulphide   (ii) when 1 dm^3 of it is shaken successively with five separate quantities of carbon disulphide of 10 cm^3 each? (CS)

Introduction to Physical Chemistry

11 The partition coefficient of a solute between water and ethoxyethane is D, the solute being more soluble in ethoxyethane. If the solute is extracted from an aqueous solution using an equal volume of ethoxyethane in two successive portions prove that the fraction of the solute extracted is $D(D + 4)/(D + 2)^2$.

12 X is a weak tribasic acid ($M_r = 210$) which is soluble in benzene and in water. After 2.800 g of X had been shaken with 100 cm^3 of benzene and 50 cm^3 of water until completely dissolved, it was found that 25 cm^3 of the aqueous layer required 14.50 cm^3 of M sodium hydroxide solution for neutralisation. Calculate the ratio of the concentrations of X which are contained in equal volumes of benzene and water. (JMB)

13 Describe the experimental method of measuring the solubility, at room temperature, of either (i) nitrogen (ii) hydrogen chloride or (iii) sulphur dioxide.

14 Henry's law can be expressed in the form $p = C \times K$ where p is the gas pressure in kPa, C is the mole fraction of gas in the solution and K is a constant. The value of K for carbon dioxide in water at 25 °C is 1.667×10^5 kPa. Calculate the solubility of carbon dioxide in water at 101.325 kPa pressure, expressing the answer in mol dm^{-3}.

15 Discuss the effect of temperature, pressure, the nature of the gas, and the presence of other gases on the solubility of a gas in a liquid. Describe how you would determine the solubility of ammonia in water at the temperature of the laboratory, being given a supply of aqueous ammonia.

16 By dissolving air in water and boiling off the dissolved gas a mixture enriched in oxygen is obtained. This mixture can, theoretically, be still further enriched in oxygen by redissolving and subsequent boiling off. How many times must this dissolving-boiling off process be repeated before the oxygen content exceeds 90 per cent?

17 Taking air as containing 78 per cent nitrogen, 21 per cent oxygen and 1 per cent argon by volume calculate the percentage composition of the gas boiled off from a saturated solution of air in water. The absorption coefficients of nitrogen, oxygen and argon are 0.0239, 0.0489 and 0.0530 respectively.

18 Discuss the factors which determine whether or not a solid will dissolve in a liquid. Comment on the following. (a) Sucrose, $C_{12}H_{22}O_{11}$, will dissolve in water; the solution is not an electrolyte (b) Barium sulphate, an ionic compound, is hardly soluble in water (c) Hydrogen chloride is soluble in toluene, and also gives an aqueous solution which is a good electrolyte (d) High polymers, though covalently bonded, are often not soluble in organic liquids. (LS)

18
Solidification of solutions. Eutectics

When a single substance, in the liquid state, is cooled, it solidifies (freezes) into a single solid, and there are very few complications, but a solution contains at least two components and there are various possibilities on cooling a solution until it wholly or partially freezes.

1 Solidification of aqueous solutions of solids

On cooling a hot solution of sodium nitrate in water, crystals of sodium nitrate may separate out above 0 °C, if the solution is concentrated enough ever to become saturated, but what will happen if the solution be cooled below 0 °C?

If a *dilute* solution of sodium nitrate is cooled below 0 °C, pure ice will form when the freezing point of the solution is reached. This freezing point will be less than 0 °C, because the added solute lowers the freezing point of the water (p. 233). As ice is formed, the solution becomes more concentrated so that its freezing point becomes lower still, but further cooling will deposit more ice. Eventually, at −17.5 °C, the solution remaining will become saturated. Any further cooling will deposit a mixture of ice and solid sodium nitrate, the temperature remaining constant at −17.5 °C until the whole system solidifies.

If a *concentrated* solution of sodium nitrate is cooled, crystals of pure sodium nitrate will be deposited until, at a temperature of −17.5 °C, a mixture of ice and sodium nitrate will again crystallise out at a constant temperature.

Such results, together with other useful information, can be summarised in a temperature–composition diagram (Fig. 112). The line BE represents the solubility curve of sodium nitrate in water (compare Fig. 108, p. 196); the line AE shows the way in which the freezing point of water is lowered as more and more sodium nitrate is added to it.

Point E is known as the *eutectic point*; it is the lowest temperature which can be reached before the whole system solidifies. It is also the only point at which ice, solid sodium nitrate and saturated solution of sodium nitrate in water are in equilibrium. At the eutectic point the crystals deposited from the solution have the same composition as the solution. This composition is 38.62 per cent by mass of sodium nitrate; and a mixture of this composition is known as the *eutectic mixture* or the cryohydrate.

The eutectic mixture may, by chance, have a composition corresponding to a simple chemical formula, but microscopic examination shows the existence of two kinds of crystal in the mixture. The composition of the mixture, and its melting point, also change with change of pressure.

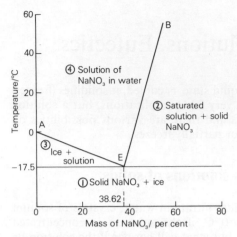

Fig. 112 Temperature–composition diagram for the sodium nitrate–water system.

The temperature–composition diagram for the sodium nitrate–water system is divided into different areas. In area 1, below the eutectic point, only ice and solid sodium nitrate can coexist; everything is solid. In area 2, solid sodium nitrate is in equilibrium with a saturated solution. In area 3, ice is in equilibrium with a solution. In area 4, a solution exists, with no solid present.

2 Solidification of mixtures of two liquids without compound formation

A mixture of two liquids very often forms a eutectic mixture on freezing, in the same way as an aqueous solution of a solid, and this is particularly important in a study of alloy systems.

Pure zinc will melt on heating, and resolidify, at its freezing point of 419 °C, on cooling; pure, molten cadmium will resolidify at its freezing point of 321 °C. Addition of a little cadmium to molten zinc will lower the freezing point, but, on cooling, pure zinc will still separate out. The freezing point will continue to be lowered as more and more cadmium is added, or as more and more zinc separates out until, eventually, a solid mixture of zinc and cadmium will separate. Similarly, addition of zinc to molten cadmium lowers the freezing point, and when sufficient zinc has been added, a solid mixture of zinc and cadmium again forms on cooling.

The temperature–composition diagram for a zinc–cadmium mixture is shown in Fig. 113. Point A shows the freezing point of pure zinc, and point B that of pure cadmium. The line AE shows the way in which the freezing point of zinc is lowered by adding cadmium; line BE shows the lowering of the freezing point of cadmium on adding zinc.

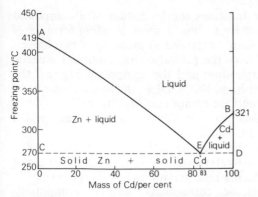

Fig. 113 Temperature–composition diagram for a zinc–cadmium mixture.

Point E is the eutectic point (270 °C) at which an alloy of zinc and cadmium first separates out, and the eutectic mixture formed has a composition of 17 per cent zinc and 83 per cent cadmium.

At all points above AEB the system is entirely liquid; in the area ACE, solid zinc is present with molten mixture; in BDE, solid cadmium is present with molten mixture; below CED the system is entirely solid.

Other mixtures which form eutectics in the same way, include tin and lead, antimony and lead, gold and thallium, potassium and silver chlorides, bromomethane and benzene, and camphor and naphthalene.

3 Solidification of mixtures of two liquids with compound formation

If two liquids form a solid compound on cooling, the temperature–composition diagram is of the type shown for a magnesium–zinc mixture in Fig. 114.

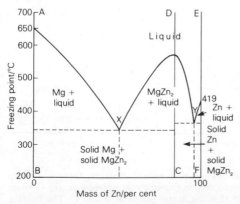

Fig. 114 Temperature–composition diagram for a magnesium–zinc mixture.

The maximum in the curve indicates the formation of a compound with a composition represented by C, and a corresponding formula of $MgZn_2$. The diagram is most simply regarded as made up of two halves. The left-hand half, ABCD, shows the formation of a normal eutectic mixture from a mixture of magnesium and the compound $MgZn_2$; the eutectic point is at X. The right-hand half, CDEF, shows a eutectic at Y, formed from a mixture of zinc and the compound $MgZn_2$.

The diagram can be divided up into various areas within which different phase relationships hold, as indicated.

The rounded maximum in the diagram indicates that the $MgZn_2$ compound is not very stable; it tends to dissociate into magnesium and zinc. If a stabler compound is formed the maximum is a sharper peak.

The formation of intermetallic compounds, and of compounds between pairs of organic substances, e.g. phenol and aniline, is not uncommon, though the plotting of a temperature–composition diagram is often the only practical method of detecting the formation of such a compound.

More than one compound can be formed and this will be shown by more than one maximum in the temperature–composition diagram.

4 Cooling curves

The idealised cooling curve for a pure substance is shown in Fig. 115(a); in reality, the dotted line may well be followed because of supercooling. A similar cooling curve is found with any eutectic mixture; although it is a mixture, it has a sharp freezing and melting point.

For any mixture, other than a eutectic mixture, cooling curves as in Fig. 115(b) are found. The liquid mixture at point A (temperature, T_1)

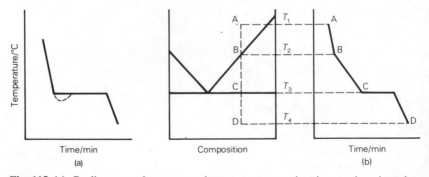

Fig. 115 (a) Cooling curve for a pure substance or a eutectic mixture; there is a sharp freezing point, but there may be some supercooling as shown by the dotted line. **(b)** Cooling curve for a mixture other than a eutectic showing arrest points at B and C. The freezing takes place over a range of temperature.

begins to freeze at point B (T_2). As the deposition of solid changes the composition of the mixture there is a fall in temperature until point C (T_3) is reached. Here the eutectic mixture is formed and the temperature remains constant until freezing is completed. Further cooling of the solid occurs until point D (T_4) is reached. The mixture, then, freezes over a range of temperature. Points B and C are known as *arrest points*. By measuring the temperatures corresponding to these arrest points for a number of mixtures of different compositions, but equal total mass, it is possible to construct an equilibrium diagram.

This can also be done by measuring, for similar mixtures, both the temperature at which melting first starts on warming the solid $(T_3$ at point C) and the temperature at which the melting is completed (point B at temperature T_2).

Questions on Chapter 18

1 The following data apply to a mixture of gold and tellurium:

Te/%	0	10	30	40	42	50	56.4	60	70	82.5	90	100
m.p./°C	1063	940	710	480	447	458	464	460	448	416	425	453

Plot the phase diagram for the system. Sketch the cooling curves which would be obtained from melts containing (a) 50 per cent (b) 82.5 per cent of tellurium.

2 The freezing point and the melting point of a pure substance are equal, but this is not generally so for most mixtures of two pure substances. Why is this? For what mixtures are the freezing point and melting point equal?

3 The boiling point of a liquid can be defined as the temperature at which the vapour pressure of the liquid becomes equal to the external pressure on the liquid. What would be the comparable definition for the melting point of a solid?

4 A mixture of salt and sand is scattered on frozen roads in winter months. What does it do?

5 The data below represent the temperatures at which solid begins to separate from fused alloys of zinc and antimony of the compositions shown:

Sb/%	0	2	5	10	20	30	40	50	60	70	75	80	90	100
T/°C	443	420	428	456	502	535	553	565	566	530	507	510	561	632

Plot the figures on squared paper. What can you infer from the graph? What would happen if an alloy with 65 per cent antimony were gradually heated from 400 °C to 600 °C? (OS)

6 The freezing points of mixtures of bismuth and cadmium are as follows:

Cd/%	90	80	70	60	50	40	30	20	10	0
$T/°C$	300	280	255	225	195	145	175	205	235	270

Draw and interpret the freezing point curve. Describe and explain what will be observed on cooling each of the following from 400 °C to 100 °C (a) pure Cd   (b) a mixture of 60 per cent Bi and 40 per cent Cd   (c) a mixture of 75 per cent Bi and 25 per cent Cd. (JMB)

7 How would you find out, experimentally, whether two organic compounds, A and B, formed a compound AB_2?

8 Sodium chloride and water form a eutectic mixture which melts at −22 °C and contains 23.6 per cent by mass of NaCl. Draw the temperature–composition diagram for NaCl and H_2O.

9 Sketch the temperature–composition diagram for a mixture of tin and lead. Find out what you can about (a) plumber's solder and (b) tinsmith's solder.

19
The vapour pressure of liquids and mixtures of liquids

The escape of molecules from a liquid or a liquid system, which causes a build up of vapour and a resulting vapour pressure (p. 31), involves the breaking down of weak forces within liquids. Vapour pressure measurements can, therefore, throw some light on the nature of these forces. The vapour pressures of pure liquids and of pairs of liquids are considered in this chapter. Chapter 20 considers the vapour pressure of solutions containing non-volatile (p. 231) solids in liquids.

1 The vapour pressure of water and ice

Water can exist as a liquid, as a solid (ice), or as a vapour (it is called steam when it is above 100 °C). The various equilibria that can exist between the three phases is conveniently summarised in a *phase diagram* in which temperature is plotted against pressure. The diagram shown (Fig. 116) is not drawn to scale so that the various areas can be shown more clearly.

Line OA shows how the vapour pressure of water changes with temperature. It summarises the conditions of temperature and pressure under which liquid water and water vapour are in equilibrium. In passing along the line wxy the temperature is being raised at a constant pressure. At point w only liquid water can exist; at point x it boils; at y only vapour can exist. A similar series of changes takes place, at constant

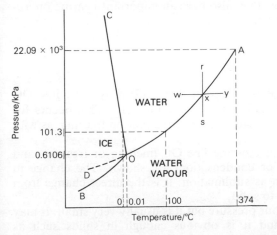

Fig. 116 Phase diagram for the water system (not to scale).

temperature, in passing down the line rxs. The upper limit of line OA, at A, is the *critical point* (p. 28), beyond which liquid and vapour phases cannot be distinguished.

Line OB shows how the vapour pressure of ice varies with temperature; it summarises the conditions under which solid ice and water vapour can be in equilibrium. The lower limit will be at 0 K.

Line OC shows the effect of pressure on the melting point of ice or the freezing point of water. It is the equilibrium between ice and water that is concerned here. So far as is known there is no upper limit to the line but complications arise at very high pressure because of the existence of different polymorphic forms of ice.

At O, which is called a *triple point*, solid, liquid and vapour can all be in equilibrium. This can only happen at one particular pressure (610.6 Pa) and one particular temperature (273.16 K). The triple point temperature is not the same as the ordinary melting point of ice (273.15 K) which is the temperature at which ice and water are in equilibrium under an applied pressure of 101 325 Pa.

Liquid water can be cooled below its freezing point without solidifying; this is known as *supercooling*. When water is supercooled, the line AO is extended along OD, but the system existing at any point on OD is *metastable*; it is really unstable, but the change to a stabler condition might be very, very slow.

The slope of the line OC shows that the melting point of ice decreases as the pressure increases. This is rather unusual, and is only shown by bismuth and antimony amongst other common substances but it has important consequences. The decrease of melting point with increase of pressure is related, by Le Chatelier's principle (p. 296), to an increase in volume, i.e., decrease in density, on solidification. If pressure is applied to ice it will melt, because the decrease in volume on melting will tend to relieve the applied pressure. Such considerations show why pipes burst and why ice floats on water; they also have an important bearing on ice-skating and the flow of glaciers.

2 Sublimation

A more general form of phase diagram (Fig. 117) shows the line OC sloping in the opposite direction to that in Fig. 116. This occurs for substances which increase in volume on melting (not on freezing, as with water).

The changes occurring in crossing line OC (melting or freezing) and OA (evaporation, boiling or condensation) are shown. The change in crossing line OB is known as sublimation; it is the direct change from solid to vapour, or vapour to solid.

Every solid exerts a vapour pressure but it is usually very small. It may not be small, however, and it is obvious enough in solids such as naphthalene which evaporate quite markedly, and in the recognisable

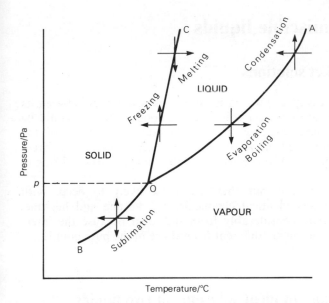

Fig. 117 The general form (not to scale) of phase diagram for a substance which increases in volume on melting (compare Fig. 116). p is the triple point pressure, and the various changes of state from one phase to another are summarised.

smell of many solids. Like the vapour pressure of liquids, that of solids increases with temperature. If the vapour pressure of a solid reaches the external pressure before it melts, then the solid will vaporise before it melts, i.e. it will sublime.

a Iodine The triple point of iodine is 114 °C and 12 kPa pressure. If solid iodine is heated quickly in a container, e.g. a test tube, from which the vapour cannot escape easily, the vapour pressure of the iodine might build up to 12 kPa. If so, liquid iodine will form as soon as the temperature reaches 114 °C. But, if solid iodine is heated slowly in an open vessel from which the vapour can escape, it is unlikely that the vapour pressure will reach 12 kPa, which is why the iodine will sublime. Similarly, iodine vapour will only sublime to the solid on cooling if it is so diluted with air that its vapour pressure is below 12 kPa.

b Carbon dioxide The triple point is −56.4 °C and 517.5 kPa pressure. This means that liquid carbon dioxide cannot exist at ordinary atmospheric pressure (101.3 kPa), and solid carbon dioxide sublimes when left exposed to a warm atmosphere. Cylinders of carbon dioxide generally contain the liquid, but a pressure of 6783 kPa is required at 25 °C.

c Hoar-frost The triple point of water is 273.16 K and 610.6 Pa. A fall in temperature when the water vapour pressure in the atmosphere is below 610.6 Pa leads to sublimation. Hoar-frost is formed at ground level; snow in the upper atmosphere.

Completely miscible liquids

3 Ideal or perfect solutions

An ideal or perfect gas (p. 28) would have no cohesive forces between its molecules. In an ideal or perfect solution, the cohesive forces would be just the same as those existing in the separate components of the solution. A solution made from A and B would only be ideal if the forces existing in the solution of A and B were just the same as those existing in pure A and pure B.

Ideal solutions are rare, but they are most likely to occur with mixtures of two almost identical chemicals, e.g. hexane and heptane. Most solutions deviate considerably from the ideal because the interactions within the solution are different from those in the pure liquids.

4 Vapour pressure of ideal solutions of two liquids

The vapour in equilibrium with a mixture of two liquids is a mixture of two vapours, and the total vapour pressure is the sum of two partial vapour pressures. All three pressures vary with temperature, and with the composition of the solution. The change with composition, for ideal solutions at a fixed temperature, is described by Raoult's law (1886) which states that *the partial vapour pressure of A in a solution, at a given temperature, is equal to the vapour pressure of pure A, at the same temperature, multiplied by the mole fraction (p. 471) of A in the solution.*

In an ideal solution, components A and B will have just the same tendency to pass into the vapour phase as they have in pure A and pure B, because the internal forces within the pure liquids and the solution are alike. There will, however, be relatively fewer particles of A in a solution containing both A and B than in pure A, so that the partial vapour pressure of A above the solution might be expected, ideally, to be proportional to the mole fraction (p. 471) of A in the solution. Similarly the partial vapour pressure of B above the solution would be proportional to the mole fraction of B. The total vapour pressure above the solution would be equal to the sum of the partial vapour pressures of A and B.

This is illustrated in Fig. 118. The vapour pressure of pure B is 50, but it is only 25 when the mole fraction of B, in a solution with A, is 0.5. Similarly the vapour pressure of pure A is 60, but only 30 at a mole fraction of 0.5. The total vapour pressure of a mixture of A and B at a mole fraction of 0.5 will, therefore, be 25 plus 30, i.e. 55. Numerical results of this type are given only by ideal solutions, e.g. hexane and heptane or bromoethane and iodoethane.

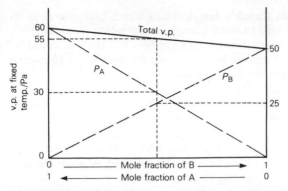

Fig. 118 The vapour pressure–composition diagram for an ideal solution.

5 Vapour pressure of non-ideal solutions of two liquids

The straight line relationship between vapour pressure and mole fraction (composition) for an ideal solution of two liquids becomes curved when the liquids deviate from ideal behaviour. Four different examples are shown in Fig. 119.

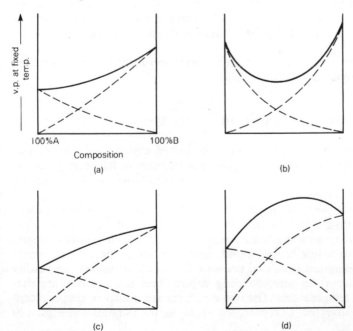

Fig. 119 Vapour pressure–composition diagrams for non-ideal solutions of A and B. **(a)** and **(b)** are for liquids showing a negative deviation from Raoult's law. **(c)** and **(d)** show a positive deviation. The dotted lines show the partial pressures of A and B.

219

a Negative deviation from Raoult's law Liquids which give curves as in (a) or (b) on Fig. 119 are said to have a negative deviation from Raoult's law. The total vapour pressure above the liquids is *less* than it would be if the liquids were ideal. There is less tendency for molecules to escape from the solution than from the pure liquids.

This must indicate stronger attractive forces between the molecules in solution than between those in the pure liquids, which may be due to association of one or both of the components in solution, or to some degree of compound formation between the components of the solution, commonly caused by hydrogen bonding. It is related to a contraction in volume and an evolution of heat in making the solution, whereas there is no volume or heat change in making an ideal solution.

Water and nitric acid provide an example of a pair of liquids showing negative deviation from Raoult's law. Other pairs are given on p. 224.

b Positive deviation from Raoult's law Liquids giving vapour pressure–composition curves as in (c) or (d) in Fig. 119, are said to have a positive deviation from Raoult's law. The total vapour pressure is *greater* than would be expected for ideal liquids, the molecules in the solution having a greater tendency to escape from the solution than from the pure liquid. This must be due to weaker attractive forces between the molecules in solution than between those in the pure liquids. It is commonly caused by a breaking down of the hydrogen bonding in the pure liquids, and it is associated with an increase in volume and an absorption of heat on mixing.

Ethanol and water show a positive deviation from Raoult's law, and other pairs are mentioned on p. 225.

6 Boiling point–composition diagrams

The way in which the total vapour pressure of a mixture of two liquids varies with composition is of importance because of its bearing on the possibility of separating the components of a mixture of two liquids by fractional distillation.

For this purpose, it is most convenient to use boiling point–composition diagrams at a fixed pressure, instead of vapour pressure–composition diagrams at a fixed temperature. As a high vapour pressure corresponds to a low boiling point, and vice versa, the boiling point–composition diagram at a fixed pressure for a pair of liquids can readily be obtained from the corresponding vapour pressure–composition diagram at fixed temperatures. The two diagrams are similar in shape, except that they are inverted. Compare, for example, the diagrams in Figs. 120 and 121.

The shape of the boiling point–composition diagram depends on the nature and the degree of deviation from Raoult's law of the two liquids concerned. There are three important types of diagram:

i No maximum or minimum This type corresponds with the vapour pressure–composition diagrams in Figs. 118, 119(a) and 119(c). Any deviation from Raoult's law is relatively small.

ii A maximum boiling point Corresponding with vapour pressure–composition diagrams as in Fig. 119(b), and a large negative deviation from Raoult's law.

iii A minimum boiling point Corresponding with vapour pressure–composition diagrams as in Fig. 119(d), and a large positive deviation from Raoult's law.

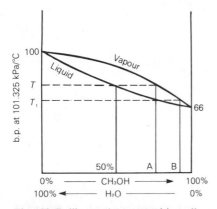

Fig. 120 Boiling point–composition diagram at a fixed pressure for methanol–water mixtures. Compare Fig. 121.

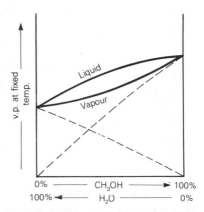

Fig. 121 Vapour pressure–composition diagram at a fixed temperature for methanol–water mixtures.

7 Boiling point–composition diagram with no maximum or minimum

A diagram of this type is given by methanol–water mixtures (Fig. 120). The liquid line shows the way in which the boiling point of a methanol–water mixture varies with composition, at a fixed pressure.

For a liquid mixture of any one composition, the vapour with which it is in equilibrium will be richer in the more volatile component, i.e. in methanol. The liquid line has, therefore, an associated vapour line. The vapour pressure–composition diagram corresponding with this boiling point–composition diagram is shown in Fig. 121.

When a mixture of methanol and water containing 50 per cent of each is boiled, it will boil at temperature T. The vapour coming from it will have a composition represented by A, and on condensing, this vapour will form a liquid with the same composition. If this liquid is boiled again, it will now boil at temperature T_1, giving a vapour of composition B, and this vapour will condense into a liquid whose composition is also

B. By repeating this boiling–condensing–boiling process, pure methanol could be obtained, but the method would be tedious, and the same result can be obtained in one operation by fractional distillation using a fractionating column (Fig. 122).

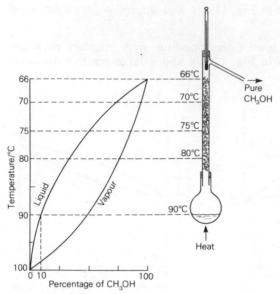

Fig. 122 Idealised and simplified representation of the fractional distillation of a mixture of methanol (10 per cent) and water (90 per cent) using a fractionating column.

8 Fractional distillation

Mixtures with a boiling point–composition diagram as in Fig. 120 can be separated into their component parts by fractional distillation, and this is most commonly done by using a fractionating column.

A simple and effective column for laboratory use consists of a long glass tube packed with short lengths of glass tubing, glass beads or specially made porcelain rings. The aim is to obtain a large surface area, and there are many patent designs of column. Industrially, a fractionating tower is used. Such a tower is divided into a number of compartments by means of trays set one above the other (Fig. 123). These trays contain central holes, covered by bubble caps, to allow vapour to pass up the tower, and overflow pipes to allow liquid to drop down.

At each point in a column or at each plate in a tower, an equilibrium between liquid and vapour is set up and this is facilitated by (a) an upward flow of vapour and downward flow of liquid, (b) a large surface area, (c) slow distillation. It is also preferable to maintain the various levels of the column or tower at a steady temperature so that external lagging or an electrical heating jacket is often used.

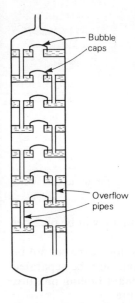

Fig. 123 A fractionating tower.

The state of affairs existing in an idealised and simplified distillation of a mixture of methanol and water, containing 10 per cent by mass of methanol, is shown in Fig. 122. The figure shows five liquid–vapour equilibria which are set up at different temperatures in the fractionating column. In reality, the liquid–vapour equilibria change continuously in passing up the column. The purpose of the fractionating column is to facilitate the setting up of these equilibria.

Mixtures of varied compositions can be drawn off from different points on the column or tower as is done, for instance, in the fractional distillation of crude oil in a refinery.

9 Boiling point–composition diagram with a maximum

The vapour pressure–composition diagram for nitric acid–water mixtures shows a minimum (as in Fig. 119(b)) and the corresponding boiling point–composition diagram, with a maximum, is shown in Fig. 124.

On distilling a mixture of nitric acid and water containing less than 68.2 per cent nitric acid, the distillate will consist of pure water and the mixture in the flask will become more and more concentrated until it contains 68.2 per cent nitric acid. At this stage, the liquid mixture will boil at a constant temperature because the liquid and the vapour in equilibrium with it have the same composition, i.e. 68.2 per cent nitric acid.

Mixtures containing more than 68.2 per cent nitric acid will give a distillate of pure nitric acid until the residue in the flask reaches the 68.2

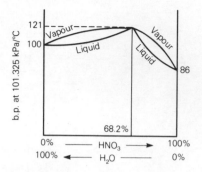

Fig. 124 Boiling point–composition diagram at a fixed pressure for nitric acid–water mixtures.

per cent nitric acid composition. Thereafter, the distillate will be 68.2 per cent nitric acid, as before.

A mixture with this type of boiling point–composition curve cannot be completely separated by fractional distillation. It can only be separated into one component and what is known as the constant boiling mixture, maximum boiling point mixture, or azeotropic mixture.

Maximum boiling point mixtures are also obtained from mixtures of water with hydrofluoric, hydrochloric, hydrobromic, hydriodic, sulphuric and methanoic acids.

10 Boiling point–composition diagram with a minimum

Ethanol and water give a vapour pressure–composition diagram with a maximum, as in Fig. 119(d). The corresponding boiling point–composition diagram, with a minimum, is shown in Fig. 125. It is not possible to get a complete separation of ethanol and water by fractional distillation. A mixture containing more than 95.6 per cent ethanol can be separated into pure ethanol and a minimum boiling point mixture, with a composition of 96.5 per cent ethanol. A mixture containing less than 96.5 per

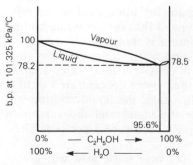

Fig. 125 Boiling point–composition diagram at a fixed pressure for ethanol–water mixtures.

cent ethanol can be separated into pure water and the same boiling point mixture.

Water with propanol or butanol or pyridine, and ethanol with trichloromethane or methyl benzene, also give minimum boiling point mixtures.

11 Separation of azeotropic mixtures

An azeotropic mixture may have either a maximum or a minimum boiling point but, at any one pressure, it has a fixed composition. It is unusual for this composition to correspond with that of any simple chemical formula for the mixture, and there is definitely no compound formation because the composition of the mixture does depend on pressure. Moreover, the mixture can be separated into its component parts fairly easily. Such separation can be brought about by the following methods.

a By distillation with a third component The azeotropic mixture of ethanol and water contains 95.6 per cent of alcohol at normal atmospheric pressure. If benzene is added, distillation yields, first a ternary azeotropic mixture of ethanol, water and benzene, then a binary azeotropic mixture of ethanol and benzene, and finally absolute ethanol.

b By chemical methods Quicklime may be used to remove the water from an azeotropic mixture of ethanol and water. Concentrated sulphuric acid will remove aromatic or unsaturated hydrocarbons from mixtures with saturated hydrocarbons in the refining of petrols and oils.

c Adsorption Charcoal or silica gel may adsorb one of the components.

d Solvent extraction One component can be extracted by a solvent.

Immiscible liquids

12 Vapour pressure of immiscible liquids

Some liquids will not mix at all and simply separate out into two distinct layers according to their densities. Examples are provided by mercury and water or by paraffin and water. Water floats on top of mercury: paraffin floats on top of water.

Each liquid behaves almost independently of the other and the vapour pressure above an agitated mixture of two immiscible liquids, at any temperature, is equal to the sum of the vapour pressures of the individual liquids at the same temperature. Agitation of the mixture is necessary to enable each liquid to establish its vapour phase. The vapour pressure above the mixture will be independent of the amount of each liquid present, so long as there is enough to give a saturated vapour.

As the temperature is increased, the vapour pressure of each liquid rises, so that the total vapour pressure above the mixture also rises. When this total vapour pressure becomes equal to the external pressure, the mixture will boil, and this boiling point will be lower than the boiling points of either of the two separate liquids.

A mixture of phenylamine (aniline) (b.p. 184 °C) and water (b.p. 100 °C), for example, will boil at 98 °C, under a pressure of 101.325 kPa. At 98 °C, the vapour pressure of phenylamine is 7.065 kPa and that of water is 94.260 kPa. The combined vapour pressure is, therefore, 101.325 kPa, and the mixture will continue to boil at 98 °C so long as both phenylamine and water are present in it. The vapour coming from the boiling mixture will contain both phenylamine vapour and water vapour, in volumes proportional to their partial vapour pressures.

The relative masses of phenylamine and water in the vapour may be calculated as follows, P standing for phenylamine and W for water.

$$\frac{\text{vol. of P vapour}}{\text{vol. of W vapour}} = \frac{\text{no. of molecules of P}}{\text{no. of molecules of W}} = \frac{\text{vapour pressure of P}}{\text{vapour pressure of W}}$$

$$\frac{\text{mass of P}}{\text{mass of W}} = \frac{\text{relative density of P} \times \text{vol. of P vapour}}{\text{relative density of W} \times \text{vol. of W vapour}}$$

$$\therefore \frac{\text{mass of P}}{\text{mass of W}} = \frac{\text{relative density of P} \times \text{vapour pressure of P}}{\text{relative density of W} \times \text{vapour pressure of W}}$$

$$\therefore \frac{\text{mass of P}}{\text{mass of W}} = \frac{46.5 \times 7.065}{9 \times 94.260} = \frac{1}{2.5} \text{ (approx)}$$

13 Steam distillation

Steam distillation depends on the properties of immiscible liquids and is a useful process for separating a liquid or a solid from a mixture. It is most successful when the liquid or the solid to be separated (a) is immiscible with or insoluble in water (b) has a high relative molecular mass (c) exerts a high vapour pressure at about 100 °C. Any impurities present must be non-volatile under the conditions used.

Steam distillation is particularly useful for the purification of substances that decompose at temperatures near their normal boiling points. For, in steam distillation, a substance is distilled at a temperature considerably below its normal boiling point.

The method is to pass steam through the mixture in an apparatus as shown in Fig. 126. The distillate, collected in the receiver, consists of

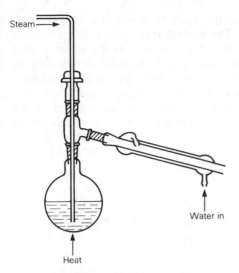

Fig. 126 Steam distillation.

water and the solid or liquid required. The solid or liquid can be isolated by filtration, by extraction with a solvent (p. 200), or by using a separating funnel and drying with a drying agent.

If any mixture of a substance, X, with water and other non-volatile impurities is subjected to steam distillation, the argument in section 12, shows that the distillate collected will contain X and water in the proportion

$$\frac{\text{mass of X}}{\text{mass of water}} = \frac{\text{relative density of X} \times \text{vapour pressure of X}}{\text{relative density of water} \times \text{vapour pressure of water}}$$

It is desirable to get as much X as possible in the distillate, and that is why a high relative density (or relative molecular mass) and a high vapour pressure for X is helpful. The use of superheated steam, i.e. steam at high pressure, also helps by increasing the vapour pressure of X in relation to that of water, and the low relative density of water is an important factor.

Partially miscible liquids

14 Critical solution temperature

Phenol and water are completely miscible, forming one solution, above 66 °C; but two immiscible solutions may form below that temperature, depending on the composition of the mixture. One of the solutions will be a solution of phenol in water; the other, a solution of water in phenol. They are called *conjugate solutions*.

Introduction to Physical Chemistry

The effect of composition and temperature is shown in a temperature–composition diagram (Fig. 127). The temperature above which phenol and water are always completely miscible is known as the upper critical solution temperature. At any point above the curve there will only be one layer, i.e. one solution. Below the curve, two layers will always form and the curve will give the compositions of the two conjugate solutions making up the two layers. A mixture of 50 per cent phenol and 50 per cent water, for example, at ·50 °C, will form two layers whose compositions are given by A and B. The line YZ is known as a tie-line. The ratio YX/XZ is equal to the ratio of the mass of the phenol layer (of composition B) to that of the mass of the aqueous layer (of composition A).

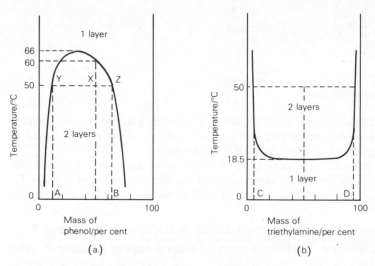

Fig. 127 Temperature–composition diagrams. **(a)** Phenol and water. **(b)** Triethylamine and water.

The complete miscibility of phenol and water with increasing temperature comes about because their mutual solubilities increase as the temperature does. The curve in Fig. 127(a) can be regarded as made up of two halves, one being the solubility curve of water in phenol and the other the solubility curve of phenol in water.

With triethylamine and water the mutual solubilities decrease as the temperature is increased. This leads to a temperature–composition diagram with a lower critical solution (or consolute) temperature of 18.5 °C (Fig. 127(b)). A 50 : 50 mixture will be completely miscible at 10 °C but will separate into two layers, with compositions C and D, at 50 °C.

Mixtures of nicotine and water are very unusual as they have both an upper (208 °C) and a lower (61 °C) critical solution temperature.

Conjugate solutions have the same total vapour pressure and the same vapour composition; that is why they can coexist together.

Questions on Chapter 19

1 Write short notes on (a) sublimation (b) triple point (c) supercooling (d) hoar-frost.

2 Compare and contrast the pressure–temperature diagrams for ice and iodine.

3 Explain, fully, what happens when (a) a little iodine (b) a little ammonium chloride is heated in a test tube.

4 How is the triple point of water related to the Kelvin scale of temperature?

5 Explain, carefully, what happens when (a) a liquid freezes (b) a vapour condenses (c) a solid melts (d) a solid sublimes.

6 Water is, in many ways, an unusual liquid. Explain some of the ways in which it is not a typical liquid.

7 What are the main characteristics of (i) an ideal solution of one liquid in another (ii) an ideal gas?

8 If water has a vapour pressure of x Pa at a certain temperature, what will be the vapour pressures above (a) a mixture of 18 g of water and 0.1 mol of sucrose (b) a mixture of 36 g of water and 5.85 g of sodium chloride (c) a mixture of 18 g of water and 0.1 mol of a liquid with a vapour pressure of y Pa at the same temperature? Assume all solutions are ideal.

9 A liquid, A, with a vapour pressure of 16 kPa is mixed with an equimolecular portion of another liquid, B. The vapour pressure of the mixture of A and B is found to be 101.325 kPa. What is the vapour pressure of pure B?

10 The vapour pressures of methanol and ethanol at 20 °C are 12.530 and 5.866 kPa respectively. If 20 g of methanol give an ideal solution when mixed with 100 g of ethanol calculate (a) the partial pressure exerted by each component of the mixture (b) the total vapour pressure above the mixture (c) the composition of the vapour.

11 What will be the composition of the vapour over an ideal solution of A in B, if the mole fraction of A is 0.25 and the vapour pressures of pure A and pure B are 8 and 13.332 kPa respectively?

12 At 101.325 kPa pressure, the constant boiling mixture from hydrochloric acid and water contains 20.22 per cent of hydrogen chloride by mass and has a density of $1.0962 \, \mathrm{g \, cm^{-3}}$. What volume of this constant boiling mixture would be required to make $1 \, \mathrm{dm^3}$ of a 1 M solution of hydrochloric acid?

13 How do an azeotropic mixture and a eutectic mixture (i) resemble (ii) differ from each other?

14 How would you obtain, experimentally, the data which would enable you to draw Fig. 120?

15 Explain two different methods for obtaining absolute ethanol.

16 The constant boiling mixture of hydrochloric acid and water has a composition approximating to the formula $HCl.8H_2O$. How would you try to convince a sceptic that the mixture was not a hydrate of hydrochloric acid?

17 Draw typical equilibrium diagrams showing the variation of the percentage composition of the vapour and liquid phases with boiling temperatures (at constant pressure) for mixtures of two completely miscible liquids A and B, which do not form a constant boiling mixture and of which A has the higher boiling point. Use the diagram to describe the changes which occur in the composition of the residual liquid and of the distillate when a 50 per cent mixture of A and B is distilled. Describe one useful practical application of constant boiling mixture formation. (W)

18 Explain with the aid of diagrams how two liquids may be separated by distillation. What is the advantage of a fractionating column? In what circumstances is it not possible to separate a mixture of two liquids by fractional distillation? (OS)

19 Write short notes on (a) azeotropic mixtures (b) consolute temperatures (c) theoretical plate (d) drip point.

20 Give examples of the use of steam distillation.

21 Discuss (i) the concept of ideality as applied to gases and solution (ii) the relationship which is common to both ideal systems. Explain, with the aid of a composition–boiling point diagram, why mixtures of ethanol and water cannot be separated to pure ethanol and pure water by fractional distillation. Explain why it is very difficult to liquefy helium. (AEB, 1978)

22 How is crude oil refined?

23* Plot a graph of the vapour pressure of water against temperature. At what temperature will the water boil if the external pressure is (a) $10\,kPa$ (b) $100\,kPa$ (c) $120\,kPa$?

24* Sketch boiling point–composition diagrams for mixtures of (a) HBr and H_2O (b) CCl_4 and C_2H_5OH (c) CH_3COOH and C_6H_6.

25* A compound X is insoluble in water. A mixture of it with water steam distils at $96\,°C$ and the distillate contains 68 per cent by mass of water. Calculate the relative molecular mass of X.

26* Water and nitrobenzene are immiscible. A mixture of the two distils at $98\,°C$. What percentage by mass of nitrobenzene will there be in the distillate?

20
Vapour pressure, boiling point and freezing point of solutions with non-volatile solutes

There are certain properties which are common to all solutions with non-volatile solutes and which vary with the composition of the solution in the same sort of way. The most important of these are the vapour pressure, the boiling point, the freezing point and the osmotic pressure. Each of them can be measured for solutions of different concentration and the results obtained depend, essentially, on the total number of solute particles present in a fixed amount of solvent. For this reason the properties are sometimes known as *colligative* properties.

The laws governing these colligative properties are generally known as the *laws of dilute solutions*. The laws only hold at all accurately for dilute solutions because it is only at low concentrations that the solutions approximate to ideal solutions (p. 218). The vapour pressure, boiling point and freezing point of dilute solutions are considered in this chapter. Osmotic pressure is considered in Chapter 21.

1 Raoult's law

The way in which the vapour pressure of an ideal solution (p. 218) changes with composition of the solution is summarised in Raoult's law (p. 218), which can be stated in three ways.

First, *the partial vapour pressure of A in a solution, at a given temperature, is equal to the vapour pressure of pure A, at the same temperature, multiplied by the mole fraction of A in the solution.*

For solutions made up of volatile solvents and non-volatile solutes, the total vapour pressure of the solution will be equal to the partial vapour pressure of the solvent, for the solute will not contribute to the vapour pressure. Thus

$$p_{\text{soln}} = p_{\text{solv}} \times \text{mole fraction of solvent}$$

$$\text{or} \quad 1 - \frac{p_{\text{soln}}}{p_{\text{solv}}} = 1 - \text{mole fraction of solvent}$$

$$\text{or} \quad \frac{p_{\text{solv}} - p_{\text{soln}}}{p_{\text{solv}}} = 1 - \text{mole fraction of solvent}$$

$$= \text{mole fraction of solute}$$

This is the second form in which Raoult's law can be expressed. In words, *the relative lowering of the vapour pressure of a solution containing a non-volatile solute is equal to the mole fraction of the solute in the solution.*

If n is the number of molecules, or moles, of solute in a solvent, and N the number of molecules, or moles, of solvent, Raoult's law is expressed as

$$\frac{p_{\text{solv}} - p_{\text{soln}}}{p_{\text{solv}}} = \frac{n}{n + N}$$

For dilute solutions, n will be small compared with N, so that $(n + N)$ will be approximately equal to N. Moreover, for any given solvent, p_{solv} will be constant, so that

$$p_{\text{solv}} - p_{\text{soln}} \propto n/N$$

and, for a given quantity of solvent, N will also be constant so that

$$p_{\text{solv}} - p_{\text{soln}} \propto n$$

This means that *equimolecular quantities of any non-volatile solute dissolved in the same quantity of the same solvent will produce the same lowering of vapour pressure*, and this is the third statement of Raoult's law.

This only applies to dilute solutions containing non-volatile solutes. Moreover, it does not apply for solutes which dissociate or associate in solution. If a solute dissociates, more particles will be present in solution than if it did not dissociate, and the lowering of vapour pressure will be proportionately greater. Association of a solute gives fewer particles in solution with a proportionately smaller lowering of vapour pressure.

If a solute with molecules AB caused a lowering of vapour pressure of x without dissociating or associating it would cause a lowering of $2x$ if it completely ionised into A^+ and B^- or a lowering of $x/2$ if it completely associated into $(AB)_2$ particles. The lowering of vapour pressure is really proportional to the number of dissolved particles in a solvent, and these particles can be molecules, ions or associated molecules, or a mixture of all three.

2 The boiling point and freezing point of solutions with non-volatile solutes

The vapour pressure–temperature diagram for a pure solvent, and a solution of a non-volatile solute in the solvent, are shown in Fig. 128 (compare Fig. 116, p. 215).

When the vapour pressure of the pure solvent reaches the external pressure, E, (which will, generally, be atmospheric pressure), the solvent will boil, at T_b. The freezing point of the pure solvent will be T_f.

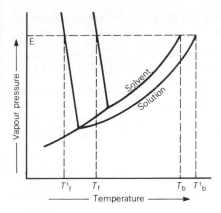

Fig. 128

The solution will have a lower vapour pressure than the pure solvent, and this will lead to an elevation of boiling point (to T_b') and a depression of freezing point (to T_f'). Addition of more solute will give a more concentrated solution with a still lower vapour pressure and correspondingly higher boiling point and lower freezing point.

It can be shown, experimentally and theoretically, that the elevation of the boiling point and the depression of the freezing point are both proportional to the lowering of the vapour pressure, i.e.

$$\begin{matrix} \text{lowering of} \\ \text{vapour pressure} \end{matrix} \quad \propto \quad \begin{matrix} \text{elevation of} \\ \text{boiling point} \end{matrix} \quad \propto \quad \begin{matrix} \text{depression of} \\ \text{freezing point} \end{matrix}$$

This is only true for dilute solutions, which deposit pure, solid solvent on freezing, and which contain non-volatile solutes, which neither dissociate nor associate and do not react with the solvent.

3 Determination of relative molecular masses from boiling point measurements

Because equimolecular quantities of any non-volatile solute dissolved in a given quantity of solvent produce the same lowering of vapour pressure and because lowering of vapour pressure and elevation of boiling point are proportional, it follows that equimolecular quantities of any non-volatile solute dissolved in a given quantity of solvent produce the same elevation of boiling point.

The elevation of boiling point produced by dissolving 1 mol of any non-volatile solute, which does not dissociate or associate, in 1 kg of a particular solvent is known as the *boiling point constant* or the *ebullioscopic constant* or the *molar elevation constant* for that solvent. Some experimental values for common solvents, obtained by using solutes of known relative molecular masses, are

water	$0.52\,\mathrm{K\,mol^{-1}\,kg}$	benzene	$2.53\,\mathrm{K\,mol^{-1}\,kg}$
propanone	$1.7\ \ \mathrm{K\,mol^{-1}\,kg}$	ethanol(95.6%)	$1.15\,\mathrm{K\,mol^{-1}\,kg}$

The constants are, sometimes, quoted for $100\,\mathrm{g}$ or $1\,\mathrm{dm^3}$ of solvent, and care must be taken in using quoted values.

A solution of $1\,\mathrm{mol}$ ($M_r\,\mathrm{g}$) of a solute in $1\,\mathrm{kg}$ of water will give a boiling point elevation of $0.52\,\mathrm{K}$. It follows that the elevation ($T\,\mathrm{K}$) caused by $x\,\mathrm{g}$ of solute in $W\,\mathrm{kg}$ of water will be given by

$$T = 0.52x/WM_r$$

If T, W and x are known, M_r can be calculated.

4 Measurement of boiling point elevation. Cottrell's method

Measurement of boiling point or boiling point elevation is simple, in principle, but complicated in detail because of the precautions which have to be taken to ensure steady and uniform heating, to avoid superheating and to avoid the loss of any solvent. Cottrell's method provides the simplest, reasonably accurate method (Fig. 129).

A small funnel is placed in the boiling liquid or solution in order to direct a stream of vapour and boiling liquid over a thermometer bulb or a thermistor. The glass container has a platinum wire sealed through it

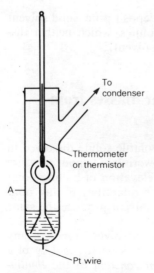

Fig. 129 Cottrell's apparatus.

at the base, or a small piece of porous pot is added. It is heated very gently by a small flame from below.

A known mass of solvent (W kg) is placed in A and its boiling point is measured. Pellets of solute, of known mass, are then added through the condenser and the boiling points of the resulting solutions are measured to find the elevation (T K). Results can be averaged by plotting x against T and taking the slope, x/T, for

$$M_r = K_b x / WT$$

where K_b is the boiling point constant of the solvent used in $K\,mol^{-1}\,kg$.

5 Determination of relative molecular mass by freezing point measurement

As lowering of freezing point and elevation of boiling point are proportional it follows that the argument given in section 3 can be applied to the lowering of freezing point in just the same way as to the elevation of boiling point.

The freezing point lowering caused by 1 mol of a solute in 1 kg of solvent is called the *freezing point constant*, the *cryoscopic constant* or the *molar depression constant* for the solvent. Some experimental values are

water	$1.86\,K\,mol^{-1}\,kg$	benzene	$5.12\,K\,mol^{-1}\,kg$
ethanoic acid	$3.9\;\;K\,mol^{-1}\,kg$	camphor	$40\;\;\;K\,mol^{-1}\,kg$

As for boiling constants, figures may be quoted in terms of $100\,g$ or $1\,dm^3$ of solvent instead of $1\,kg$, and care must be taken.

Freezing point measurements can be used in just the same way as boiling point measurements to determine relative molecular masses. Using a solvent with a cryoscopic constant K_f, 1 mol of any solute dissolved in 1 kg of the solvent will give a freezing point depression of K_f. $x\,g$ of solute, of relative molecular mass M_r, dissolved in W kg of solvent will give a depression, T K, given by

$$T = K_f x / W M_r$$

M_r values can be found from quoted values of K_f and measured values of T, x and W.

6 The effect of dissociation or association

The colligative properties of solutions, i.e. the lowering of vapour pressure, the elevation of boiling point, the depression of freezing point and the osmotic pressure, all depend on the total number of solute particles present in a solution.

A solute which dissociates in solution by splitting up, either completely or partially, into ions, will provide *more* particles than would otherwise be present and it will cause an increased effect. One which associates, by the linking together, either completely or partially, of solute molecules, will provide *fewer* particles and a decreased effect.

A relative molecular mass calculated from a measured effect which, on account of ionisation is too high, will be too low. A relative molecular mass calculated from a lowered effect, due to association, will be too high.

If a molecule, AB, ionises completely into A^+ and B^- ions, one molecule will produce two ions.

$$AB \longrightarrow A^+ + B^-$$

before ionisation	1 particle	
after complete ionisation	0 particles	2 particles

A solution made from AB molecules will, therefore, produce twice the effect of an equal amount of solute which did not dissociate. A molecule, e.g. AB_2, ionising completely into A^{2+} and two B^- ions will provide three ions, giving three times the effect, and so on.

When the ionisation is not complete but only partial, it is measured by the *degree of ionisation*. This can be expressed as a percentage or as a fraction. For 1 mol of a solute AB, giving A^+ and B^- ions, the situation, if the degree of ionisation is α, will be as follows.

$$AB \longrightarrow A^+ + B^-$$

original amount before ionisation	1 mol	0	0
amount after ionisation to a degree, α	$(1 - \alpha)$ mol	α mol	α mol

The number of particles present is proportional to the number of moles, so that ionisation causes a proportional change in the number of particles from 1 to $(1 + \alpha)$. The ratio of the expected effect, i.e. that resulting if there was no ionisation, to the actual effect caused by ionisation will be

$$\frac{\text{expected effect if no ionisation}}{\text{actual effect caused by ionisation}} = \frac{1}{1 + \alpha}$$

Such a relationship applies to any of the four colligative effects. If any one of these effects is measured, and a relative molecular mass calculated from the measurements, then

$$\frac{\text{real relative molecular mass}}{\text{apparent relative molecular mass}} = \frac{1 + \alpha}{1}$$

Alternatively, the degree of ionisation of a solute whose real relative molecular mass is known can be obtained by measuring its apparent relative molecular mass.

Remember, however, that the relationship given applies only to AB compounds. Different expressions can be derived for AB_2, AB_3 etc.

Similar arguments involving the degree of association can be applied to those substances that associate in solution.

Questions on Chapter 20

1 The lowering of vapour pressure of a solution of 108.2 g of a substance X in 1 kg of water at 20 °C is 24.79 Pa. The vapour pressure of water at 20 °C is 2.338 kPa. Calculate the relative molecular mass of X.

2 Explain why a liquid at a given temperature has a definite vapour pressure. How would you attempt to measure the vapour pressure of benzene at 60 °C? (OS)

3 How would you determine the vapour pressure of water at various temperatures between 20 °C and 100 °C? Explain what use can be made of measurements of vapour pressures of pure liquids, solutions, and mixtures. (OS)

4 What is Raoult's law? An acidic substance was made up into two solutions, A and B, A containing 15.0 g of the substance in 100 g of water, and B containing 6.9 g of the substance in 100 g of benzene. These solutions, A and B, had vapour pressures of 99.13 and 99.02 kPa respectively at the boiling points of the pure solvents at 101.325 kPa pressure. Calculate the apparent relative molecular mass of the substance in each case. Suggest a reason or reasons why the results differ in the two solvents, and outline briefly an experiment to confirm your suggestions. (CS)

5 Two solutions containing respectively 4.45 g of anthracene (M_r 178) and 6.42 g of sulphur in 1 kg of carbon disulphide have the same vapour pressure at the same temperature. What is (a) the molecular formula of the dissolved sulphur (b) the relative lowering of the vapour pressure?

6 The boiling point of ethanol is 78 °C and its molar elevation constant is $1.15 \, \text{K mol}^{-1} \, \text{kg}$. A solution of 0.56 g of camphor in 16 g of ethanol had a boiling point of 78.278 °C. Calculate the relative molecular mass of camphor.

7 You are given a clinical thermometer graduated at intervals of 0.2 from 94 °F to 110 °F. Devise a simple small-scale apparatus, using the thermometer and the ordinary materials of a laboratory, for determination of the relative molecular mass of a substance soluble in ethoxyethane. Ethoxyethane boils at 95 °F and has a molecular elevation of $2.11 \, \text{K mol}^{-1} \, \text{kg}$.

8 A solution of 0.06 g of a non-electrolyte AB, of high boiling point, in 100 g of water had a freezing point 0.006 K below that of pure water. A 0.1 g sample of AB, vaporised at atmospheric pressure, occupied a volume of $36.52 \, \text{cm}^3$ at 200 °C and $45.74 \, \text{cm}^3$ at 240 °C. What can you deduce from these data? (CS)

9 If a volatile solute is added to a solvent what effects may be noticed in the total vapour pressures of the mixture? What difference would there be if the solute were non-volatile? Explain the importance of these observations either on the separation of two volatile liquids by distillation or on the determination of relative molecular mass in solution. (OCS)

10 Solution of 0.142 g naphthalene in 20.25 g benzene caused a lowering of freezing point of 0.284 K. The molar depression constant of benzene is 5.12 K mol^{-1} kg. Calculate the relative molecular mass of naphthalene.

11 The melting point of camphor is 177.5 °C, whilst that of a mixture of 1 g naphthalene (M_r 128) and 10 g camphor is 147 °C. What is the cryoscopic constant for camphor? The melting point of a mixture of 1 g of N-phenylethanamide (acetanilide) and 10 g of camphor is 148.5 °C. What is the relative molecular mass of N-phenylethanamide?

12 The freezing point of a solution of 20 g of mercury(II) chloride in 1 kg of water was -0.138 °C. Would you expect this solution to conduct an electric current? Give reasons for your answer.

13 What advantages and disadvantages has ethane-1,2-diol over methanol as an antifreeze for use in car radiators? A car radiator contains 12 dm^3 of water. Addition of 5 kg of ethane-1,2-diol lowers the freezing point by 12 K. What mass of (a) propane-1,2,3-triol or glycerol (b) methanol would be required to give the same protection against freezing?

14* The boiling point constant of a liquid is given by $RT^2/10^3 L$, where R is the gas constant, T the boiling point of the liquid in K, and L the specific latent heat of vaporisation in J g^{-1}. Examine the validity of this for any five liquids. What would you expect the similar relationship for freezing point constants to be?

15* The freezing point constant for benzene is 5.12 K mol^{-1} kg. Calculate its value in K mol^{-1} dm^3.

21
Osmotic pressure

1 Diffusion

Gases diffuse quite rapidly (p. 6) but liquids, solutions and even solids will also diffuse, if more slowly. Two miscible liquids, for instance, diffuse throughout the whole mixture until it is homogeneous. A solution will also diffuse into another solution, or into pure solvent, so long as the pairs of substances used are miscible.

It is possible, with care, to pour a saturated aqueous solution of copper(II) sulphate through a tube into a vessel containing water so that the blue saturated solution, with a higher density than water, forms a well defined lower layer. On standing, the two layers slowly merge until a homogeneous mixture of uniform concentration is obtained. Water diffuses into the copper(II) sulphate solution, and copper(II) sulphate diffuses into the water layer.

Similar diffusion will take place if two solutions of unequal concentration are in contact. Solvent will pass from the dilute to the concentrated solution, and solute will pass from the concentrated to the dilute, until equality of concentration is achieved. This is illustrated in Fig. 130.

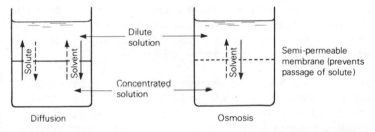

Fig. 130 Diffusion and osmosis. The bold arrows indicate quicker rates of diffusion than the dotted arrows. In diffusion, there is an overall passage of solvent into the concentrated solution, and of solute into the dilute solution. In osmosis, there is an overall passage of solvent into the concentrated solution. Passage of solute is prevented by the semi-permeable membrane.

2 Osmosis

If the diffusion process is limited to the movement of *solvent* molecules, the phenomenon is known as osmosis. The passage of any solute molecules can be prevented by what are known as *semi-permeable membranes*. Quite how these function is not well understood but many substances can act in this way, e.g. parchment paper, cellophane, copper(II) hexacyanoferrate(II), $Cu_2Fe(CN)_6$, and many plant and animal membranes.

Osmosis is the name given to the diffusion ` of solvent through a semi-permeable membrane from a solvent into a solution, or from a dilute to a more concentrated solution. The passage of solvent is not simply one-way, but the rate of diffusion from low to high concentration is greater than in the opposite direction (Fig. 130).

3 Examples of osmosis

Osmosis was first observed by the Abbé Nollet in 1784. He found that a pig's bladder filled with wine and immersed in water swelled and eventually burst. The bladder was acting as a semi-permeable membrane. Other examples are described below:

a The hard, outer shell of an egg can be removed by dissolving it in hydrochloric acid. If two eggs are treated in this way, one placed in water will swell, whilst the other placed in strong brine will shrink. The membrane beneath the outer shell acts as a semi-permeable membrane. Slices of beetroot or carrot, segments of orange, or dried fruits behave similarly.

b Crystals of copper(II) sulphate will form weird 'growths' when placed in a solution of potassium hexacyanoferrate(II). This is because the crystals become coated with a semi-permeable layer of copper(II) hexacyanoferrate(II). Water then passes through this layer, stretching it until it bursts. As soon as it does burst, more copper(II) hexacyanoferrate(II) layer is formed, and so on.

c Weird 'growths', as in **b** are also obtained if crystals of many salts, e.g. iron(II) sulphate, nickel(II) chloride, cobalt(II) nitrate and iron(III) chloride, are placed in a solution of water glass (sodium silicate). The layers of metallic silicates formed by double decomposition reactions are semi-permeable. The growths resulting form what is commonly called a chemical garden.

d If a beaker containing a concentrated solution of sugar in water and one containing pure water are placed under a bell jar, water will pass, through the vapour phase, from the pure water to the concentrated solution. Air, in this instance, acts as the semi-permeable membrane.

4 Osmotic pressure

A thistle funnel can be sealed by cellophane so that it can hold an aqueous solution. If such a funnel is immersed in water (Fig. 131) the

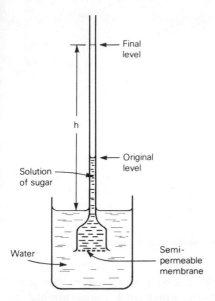

Fig. 131 Osmotic pressure. Osmosis takes place from the water into the sugar solution until the hydrostatic pressure due to the column of sugar solution becomes equal to the osmotic pressure of the solution, when osmosis ceases.

level of liquid in the thistle funnel will rise as water passes through the semi-permeable membrane into the solution. As the liquid level rises, it builds up a pressure and, when this reaches a certain value, osmosis stops. The value of this pressure is known as the osmotic pressure of the solution in the thistle funnel.

The term is misleading if it is not understood: a solution, by itself, cannot exert any pressure due to osmosis. *The osmotic pressure of a solution is the minimum pressure required to prevent osmosis when the solution is separated from pure solvent by a semi-permeable membrane.*

5 Measurement of osmotic pressure

Variations of the apparatus shown in Fig. 131 can be used to measure osmotic pressure, the pressure being recorded as the height of a column of solution of measured density. The method is most useful for small osmotic pressure values. Modern versions of the apparatus are known as osmometers.

a Berkeley and Hartley's method In this method, the pressure necessary to prevent osmosis is measured. The apparatus is shown diagrammatically in Fig. 132, but the complex jointing systems are omitted. A porous

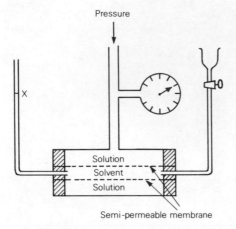

Fig. 132 Berkeley and Hartley's apparatus for measuring osmotic pressure.

tube containing copper(II) hexacyanoferrate(II) was used as the semi-permeable membrane, and the pressure which had to be applied to maintain a steady liquid level at X was measured on a pressure gauge.

b Comparison of osmotic pressures The osmotic pressures of two solutions can be compared using a method introduced by de Vries. If a plant or animal cell is placed in a solution it will shrink (undergo plasmolysis) or swell, unless the osmotic pressure of the external solution is equal to that of the solution within the cell. The shrinking or swelling can be observed under a microscope. The epidermal cells of leaves, or red blood corpuscles, can be used.

The osmotic pressures of two solutions can be compared by measuring how much each has to be diluted to prevent shrinking or swelling of the cells placed in them. If the osmotic pressure of one is known, that of the other can be found.

6 Effect of concentration and temperature on osmotic pressure

For dilute solutions, with solutes that neither dissociate nor associate, the osmotic pressure is proportional to the concentration (at a fixed temperature), and to the absolute temperature (for a fixed concentration). Van't Hoff (1877) pointed out the similarity between these relationships and Boyle's law and Charles's law, and it eventually became clear that the same laws govern both gas and osmotic pressure. Thus

for n mol of a gas with volume, V,
pressure, p, and temperature, T (p. 6) $pV = nRT$

for n mol of a solute in a solution
of volume V, at a temperature, T, $\pi V = nRT$
and osmotic pressure, π

The gas constant, R, applies in both cases. If pressure is expressed in Pa, volume in m^3, and temperature in K, its value is $8.31441\,J\,K^{-1}\,mol^{-1}$.

The osmotic pressure of a dilute solution is, in fact, equal to the pressure which the solute would exert if it existed as a gas, at the same temperature, occupying the same volume as that of the solution. 1 mol of any gas occupying $22.4\,dm^3$ at $0\,°C$ has a pressure of $101.325\,kPa$; 1 mol of any solute in a solution of volume $22.4\,dm^3$ at $0\,°C$ has an osmotic pressure of $101.325\,kPa$.

It does not matter what the solute is so long as it is not associated or dissociated. Like vapour pressure, freezing point depression and boiling point elevation, the osmotic pressure of a solution is a colligative property depending simply on the total number of solute particles in a fixed amount of solution. Ionisation produces an increased number of particles and an increase in osmotic pressure; association produces fewer particles and lower osmotic pressure. The arguments given on page 235 apply.

7 Determination of relative molecular mass by osmotic pressure measurement

If the osmotic pressure of a solution containing x g of solute in $V\,m^3$ of solution is measured at a temperature, T, the number of moles, n, of solute can be obtained from the expression

$\pi V = nRT$

where π is the osmotic pressure in Pa and R is $8.31441\,J\,K^{-1}\,mol^{-1}$. It follows that M_r equals x/n without the units.

There are easier methods of measuring the relative molecular mass of most substances but this method is useful when substances of very high relative molecular mass, e.g. proteins or polymers, are involved. A 1 per cent aqueous solution of a protein of relative molecular mass 50 000, for example, would have an osmotic pressure of about 500 Pa. This can be measured as a height of a column of water of about 5 cm. The freezing point depression of the same solution would be only 0.000 37 K, which is too small for accurate measurement.

8 Reverse osmosis

A semi-permeable membrane allows the passage of a solvent into a solution and, as a result, a pressure can be built up. If a solution is separated from a solvent by a semi-permeable membrane and a pressure, higher than the osmotic pressure, is applied to the solution, solvent will pass through the membrane from the solution. This phenomenon is known as reverse osmosis.

It provides a potentially useful method of purifying water, and some commercial plants are in operation for treating brackish water. It is, however, difficult to manufacture suitable membranes and to get good flow rates. The problems are particularly difficult in the case of sea water, which contains quite a high concentration of dissolved solids.

Questions on Chapter 21

1 A 3.42 per cent solution of sucrose (M_r 342) was found to have the same osmotic pressure as a 5.96 per cent solution of raffinose. Calculate the relative molecular mass of raffinose.

2 A saturated aqueous solution of a non-electrolyte of relative molecular mass 180 exerted an osmotic pressure of 52.82 kPa at 22 °C. What is the solubility of the substance in $g\,dm^{-3}$ of water at this temperature?

3 Compare and contrast the phenomenon of osmosis with the behaviour of gases.

4 Explain what is meant by the osmotic pressure of a solution. How would you demonstrate its existence? How can it be measured? A solution of potassium chloride containing $5\,g\,dm^{-3}$ was found to have an osmotic pressure of 304 kPa at 18 °C. What conclusions may be drawn from this observation?

5 (a) Why is osmotic pressure measurement sometimes advantageous for measuring very high relative molecular masses? (b) The osmotic pressure, at 25 °C, of a solution containing 1.346 g of a protein in $100\,cm^3$ of solution is 971.8 Pa. Calculate the relative molecular mass of the protein.

6 A solution of 0.608 g of haemoglobin in $100\,cm^3$ of water gave an osmotic pressure of 202.6 Pa at 0 °C. Assuming the solution to be ideal, calculate the relative molecular mass of haemoglobin.

7 The osmotic pressure of a solution of a sample of synthetic polyisobutylene in benzene at 25 °C was 20.66 Pa. If the solution had a concentration of $2\,g\,dm^{-3}$ what is the relative molecular mass of the polyisobutylene?

8 Describe the part played by osmosis in biological processes.

9 What do the terms isotonic and isosmotic mean?

10 Discuss the principles underlying methods for determining relative molecular masses in solution. What methods would you consider most suitable for measuring the relative molecular masses of sucrose, polystyrene and benzoic acid in solution? Give reasons for your choice of solvent in each case. (OJSE)

11 Discuss the principles underlying the uses of ion exchange and osmosis in the purification of water. What are the main difficulties to be overcome in any industrial application of the methods?

12 Distinguish between the meaning of the terms diffusion, osmosis, dialysis, reverse osmosis and sedimentation.

22
Enthalpy (heat) changes. Thermochemistry.

1 Internal energy, and enthalpy, changes. ΔU and ΔH

In most chemical and physical changes there is an associated energy change, and it is generally possible to measure the value of the *change* even though it may not be possible to determine the absolute energy levels before and after the change.

The energy change may show up as heat evolved or absorbed; as electrical energy (as, for example, in a battery); as light energy, in a chemiluminescent reaction; or, when there is an increase or decrease in volume (ΔV) as work done in expansion or contraction against or by an external pressure, p. 249.

In general,

$$\begin{array}{c}\text{change in}\\\text{energy}\end{array} = \begin{array}{c}\text{heat absorbed}\\\text{or evolved } (q)\end{array} - \begin{array}{c}\text{work done in}\\\text{expansion or}\\\text{contraction } (p\Delta V)\end{array} + \begin{array}{c}\text{other work}\\\text{done } (w)\end{array}$$

Heat evolved, work done in expansion (when ΔV is positive, i.e. when there is an increase in volume), and other work done by a reaction are all regarded as negative; they all cause a *lowering* of energy. Heat absorbed, work done in contraction (when ΔV is negative, i.e. when there is a decrease in volume), and work done on a reaction are positive; they cause an *increase* in energy.

a Internal energy change, ΔU For a reaction carried out at constant volume (in a sealed container), or for reactions in which there is no change in volume, and if no other work is done, the change in energy is called the internal energy change, ΔU, and

$$\Delta U = q_v \text{ (at constant volume)}$$

When heat is absorbed (q_v positive), there is an increase in internal energy (ΔU, positive): when heat is evolved (q_v negative) there is a decrease in internal energy (ΔU, negative).

b Enthalpy change, ΔH For a reaction carried out at constant pressure (usually at atmospheric pressure in an open container), and if no other work is involved, the heat evolved or absorbed, q_p, is called the enthalpy change, ΔH, i.e.

$$\Delta U = q_p - p\Delta V \qquad q_p = \Delta H \qquad \Delta H = \Delta U + p\Delta V$$

When ΔV is zero, ΔU and ΔH are equal (p. 279).

2 Standard molar enthalpy change, $\Delta H_m^{\ominus}$

It is enthalpy changes that are most widely used in chemistry for chemical reactions are, generally, carried out in open vessels.

As the enthalpy change depends on the conditions, and on the amount of material, it is commonly expressed as a *standard molar enthalpy* change. The standard conditions chosen are a pressure of 101.325 kPa and a temperature of 298 K. If solutions are involved, concentrations of unit activity (approximately $1 \, mol \, dm^{-3}$) are specified. Standard molar enthalpy changes are then symbolised by $\Delta H_m^{\ominus}(298 \, K)$, but the simpler symbol, ΔH, is sometimes used when there is no ambiguity. Subscripts may be used to indicate the nature of the change; $\Delta H_{m,c}^{\ominus}(298 \, K)$, for example, refers to a molar standard enthalpy change of combustion.

The units used for standard molar enthalpy changes are $kJ \, mol^{-1}$, but it is important to represent what the mol^{-1} refers to as clearly as possible. This is best done by relating $\Delta H_m^{\ominus}(298 \, K)$ to an equation.

A reaction evolving heat to the surroundings is *exothermic*, e.g.

$$N_2(g) + 3H_2(g) \longrightarrow 2NH_3(g) \qquad \Delta H_m^{\ominus}(298 \, K) = -92.4 \, kJ \, mol^{-1}$$

This is to be interpreted as meaning that 92.4 kJ of heat are evolved when 1 mol of N_2 reacts with 1 mol of ($3H_2$), i.e. 3 mol of H_2, to form 1 mol of ($2NH_3$), i.e. 2 mol of NH_3 (p. 297). The negative value for the enthalpy change means that the system loses heat and finishes with a lower enthalpy. ΔH values for exothermic reactions are negative.

The absorption of heat in an *endothermic* reaction leads to an increase in enthalpy and a positive value for the enthalpy change, e.g.

$$N_2(g) + O_2(g) \longrightarrow 2NO(g) \qquad \Delta H_m^{\ominus}(298 \, K) = +180.5 \, kJ \, mol^{-1}$$

Such enthalpy changes are conveniently summarised in enthalpy diagrams as in Fig. 133. The enthalpy change is the sum of the enthalpies of the products minus the sum of the enthalpies of the reagents. Though it is always an enthalpy *change* it is commonly referred to, as a simplification, as an enthalpy of reaction.

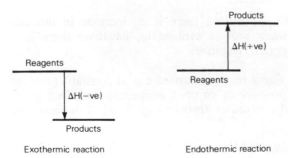

Fig. 133 Simple enthalpy diagrams for exothermic and endothermic reactions

In a simple way, an exothermic reaction can be regarded as a 'downhill' reaction; an endothermic one is 'uphill'. The sign of the enthalpy change for a reaction is, indeed, a pointer to its feasibility. The more negative the value of ΔH for the reaction (the more 'downhill') the more likely it is to be feasible. Reactions with high positive ΔH values ('uphill' reactions), are not likely to be feasible. Deductions made from ΔH values must, however, be treated with caution, for it is the free energy change (ΔG) in a reaction that gives an accurate indication of its feasibility (p. 270).

Enthalpy changes for some reactions can be measured directly by calorimetric methods (pp. 250–1), the precise method depending on the nature of the reaction. Alternatively, the enthalpy changes can be calculated from enthalpies of formation (p. 254) or bond enthalpies (p. 256).

3 Hess's law of constant heat summation

Hess's law (1840) states that *the enthalpy change in a chemical reaction is the same whether the reaction is brought about in one stage or through intermediate stages.*

Thus the enthalpy change in a reaction A to C is the same whether the reaction takes place in one stage as A → C or in two stages as A → B → C. This is a particular application of the law of conservation of energy. If Hess's law did not hold it would be possible to obtain energy without doing any work. If a change from A to B evolved 100 J when carried out in one way and only 90 J in another way it would be possible, by going from A to B by the first path and then back to A by the second, to obtain 10 J of heat from nowhere. This is contrary to all scientific experience.

The validity of Hess's law can be demonstrated experimentally by measuring the enthalpy change when a reaction is brought about in two or more different ways. The result of making a solution of ammonium chloride from 36.5 g of hydrogen chloride, 17 g of ammonia and water in two ways are summarised below, and as an enthalpy diagram or cycle in Fig. 134, the enthalpies of reaction being given in kilojoules. The figures illustrate the truth of Hess's law.

Method 1		Method 2	
i Mix the two gases	$\Delta H = -176.1$	i Dissolve NH_3 in water	$\Delta H = -35.2$
ii Dissolve the product in water	$\Delta H = +16.3$	ii Dissolve the HCl in water	$\Delta H = -72.4$
		iii Mix the two solutions	$\Delta H = -52.3$
Overall reaction	$\Delta H = -159.8$	Overall reaction	$\Delta H = -159.9$

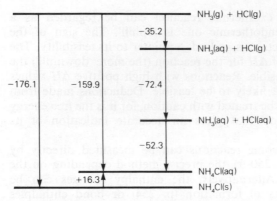

Fig. 134 Enthalpy diagram or cycle for formation of $NH_4Cl(aq)$ from $NH_3(g)$ and $HCl(g)$ by two methods.

4 Thermochemical calculations

Hess's law enables thermochemical equations to be added and subtracted, and this is the procedure to adopt in all simple thermochemical calculations. Given the enthalpies of reactions,

i $C(s) + O_2(g) = CO_2(g)$ $\qquad$ $\Delta H_m^{\ominus} = -393.5\,\text{kJ mol}^{-1}$
ii $2CO(g) + O_2(g) = 2CO_2(g)$ $\qquad$ $\Delta H_m^{\ominus} = -565.9\,\text{kJ mol}^{-1}$

it is very simple to find the enthalpy of the reaction

iii $2C(s) + O_2(g) = 2CO(g)$

This is done by taking $2 \times i$ and subtracting *ii*. The enthalpy of the reaction *iii* is, therefore, $2 \times (-393.5) - (-565.9)$, i.e. $-221.1\,\text{kJ mol}^{-1}$. And for half the quantities, so that the enthalpy of reaction can be expressed in terms of 1 mol of carbon monoxide,

$$C(s) + \tfrac{1}{2}O_2(g) = CO(g) \qquad \Delta H_m^{\ominus} = -\frac{221.1}{2} = -110.55\,\text{kJ mol}^{-1}$$

Not all thermochemical calculations are as easy as this but it is generally a matter of adding and/or subtracting the given equations, multiplied or divided throughout wherever necessary.

5 Factors affecting enthalpies of reactions

The enthalpy of a reaction depends on the conditions under which it is carried out as indicated below.

a State of reagents and products Latent heat is concerned in changes of state so that the state of the reagents and products affect the enthalpy of a reaction. The standard molar enthalpy of reaction between hydrogen gas and solid iodine to form hydrogen iodide gas, for example, is

$51.9\,kJ\,mol^{-1}$; using iodine vapour, the value is $-9.2\,kJ\,mol^{-1}$. The difference is due to the enthalpy change in vaporising iodine,

$$H_2(g) + I_2(s) \longrightarrow 2HI(g) \qquad \Delta H_m^{\ominus}(298\,K) = 51.9\,kJ\,mol^{-1}$$
$$H_2(g) + I_2(g) \longrightarrow 2HI(g) \qquad \Delta H_m^{\ominus}(298\,K) = -9.2\,kJ\,mol^{-1}$$
$$I_2(s) \longrightarrow I_2(g) \qquad \Delta H_m^{\ominus}(298\,K) = 61.1\,kJ\,mol^{-1}$$

b Allotropic modifications An enthalpy change is involved in the conversion of one allotrope into another, so that the particular allotrope used in a reaction affects the enthalpy of the reaction. For example,

			$\Delta H_m^{\ominus}(298\,K)$
S(rhombic)	$+ O_2 \longrightarrow$	$SO_2(g)$	$-297.0\,kJ\,mol^{-1}$
S(monoclinic)	$+ O_2 \longrightarrow$	$SO_2(g)$	$-297.3\,kJ\,mol^{-1}$
S(rhombic)	$\longrightarrow$	S(monoclinic)	$0.3\,kJ\,mol^{-1}$
C(graphite)	$\longrightarrow$	C(diamond)	$2.1\,kJ\,mol^{-1}$

6 ΔH and ΔU values

ΔH and ΔU have equal values for reactions in which there is no volume change (p. 279); otherwise they are different.

a Reactions in which there is an increase in volume For a reaction in which there is an increase in volume (ΔV) work will have to be done in expansion against the external pressure (p) if the reaction is carried out in an open vessel. The work will be equal to $p\Delta V$, which, for an increase in volume of n mol of gas, will be equal to nRT or $8.3nT$. For 1 mol of gas, at 25 °C, it will be 2.47 kJ.

If, for example, 1 mol of zinc reacts with an acid, at constant pressure in an open vessel, there is an evolution of heat (q_p) equal to $-152\,kJ$. (Heat evolved is, conventionally, taken as negative; heat absorbed as positive). Therefore,

$$q_p = \Delta H \text{ (at constant pressure)}$$
$$Zn(s) + 2H^+(aq) \rightarrow Zn^{2+}(aq) + H_2(g) \qquad \Delta H_m^{\ominus}(298\,K) = -152\,kJ$$

The system will also do 2.47 kJ of work in expansion against the external pressure. The total loss of energy by the system will, therefore, be $154.47\,kJ\,mol^{-1}$.

If the reaction is carried out at constant volume (in a closed vessel) no expansion work will be done so that more energy will be evolved as heat (q_v). The corresponding energy change is known as the *internal energy change*, ΔU, (p. 245). It follows that

$$q_v = q_p - 2.47 \qquad \Delta U_m = \Delta H_m - 2.47$$
$$Zn(s) + 2H^+(aq) \rightarrow Zn^{2+}(aq) + H_2(g) \qquad \Delta U_m^{\ominus}(298\,K) = -154.47\,kJ\,mol^{-1}$$

In general, for a reaction in which there is an increase in volume of n mol of gas,

$$\Delta H = \Delta U + p\Delta V \qquad \Delta H = \Delta U + 8.3nT$$

b Reactions in which there is a decrease in volume For a decrease in volume, ΔV will be negative so that, if the decrease is of n mol of gas,

$$\Delta H = \Delta U - 8.3nT$$

For example,

$$2CO(g) + O_2(g) \rightarrow 2CO_2(g) \qquad \Delta H_m^{\ominus}(298\,K) = -565.9\,kJ\,mol^{-1}$$
$$\Delta U_m^{\ominus}(298\,K) = -563.43\,kJ\,mol^{-1}$$

7 Molar enthalpy of combusion ($\Delta H_{m,c}$)

The molar enthalpy of combustion of a substance is the enthalpy change when 1 mol of it is completely burnt in oxygen.

Measurements are made in a bomb or a flame calorimeter.

a The bomb calorimeter (Fig. 135) A known mass of the substance under test is placed (in a thin glass tube if volatile) in a platinum crucible at the centre of a steel alloy, cylindrical bomb. The bomb is filled through a complicated valve system, with oxygen, at a pressure of about $2.5 \times 10^3\,kPa$ (25 atmospheres), to ensure rapid and complete combustion, and immersed in a calorimeter, usually containing water. The combustion of the substance is started by passing a current through a thin wire of iron or platinum in contact with the substance. The heat evolved on burning is determined by measuring the rise in temperature

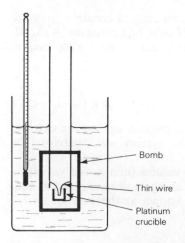

Bomb

Thin wire

Platinum
crucible

Fig. 135 Outline arrangement of a bomb calorimeter. The complicated engineering details necessary to maintain pressure inside the bomb are not shown.

of the liquid in the calorimeter. The amount of heat evolved can be found by measuring the amount of electrical energy that is required through a heating coil to give the same rise of temperature. Alternatively, the thermal capacity of the apparatus can be measured electrically, or by using a substance whose heat of combustion is known.

For a slow combustion, or when only a small amount of heat is liberated, it is more accurate to make readings adiabatically, the whole set-up being surrounded by a water jacket which is maintained at the same temperature as the inner part. This allows heat losses by radiation or convection to be neglected. The outer jacket can be maintained at the correct temperature by having one junction of a thermocouple in the liquid of the outer jacket and another in the liquid of the inner calorimeter. The thermocouples are arranged to heat or cool the outer jacket, if any temperature differences arise.

Measurements made in a bomb calorimeter, at constant volume, are ΔU values, but they can be converted into values at constant pressure, ΔH, using the equation on p. 250.

b The flame calorimeter In a flame calorimeter, the heat given out on burning a measured amount of gas at a jet (or liquid through a wick) is allowed to raise the temperature of a known mass of water. The heat evolved is calculated from the measured temperature rise. As the measurement is made at constant pressure the values obtained are ΔH values. The accuracy depends on the detailed type of calorimeter used.

8 Enthalpy of neutralisation (ΔH_n)

This is *the enthalpy change in the reaction between an acid and an alkali, the reaction being carried out in infinitely dilute solution (p. 262).*

Quoted values generally refer to quantities of acid and alkali yielding 1 mol of H^+ and 1 mol of OH^- ions, e.g.

$$HCl(aq) + NaOH(aq) \rightarrow NaCl(aq) + H_2O(l) \qquad \Delta H_{m,n}{}^{\ominus}(298\,K) = -57.1\,kJ\,mol^{-1}$$

$$HNO_3(aq) + \tfrac{1}{2}Ba(OH)_2 \rightarrow \tfrac{1}{2}Ba(NO_3)_2(aq) + H_2O(l) \qquad \Delta H_{m,n}{}^{\ominus}(298\,K) = -58.2\,kJ\,mol^{-1}$$

Enthalpies of neutralisation can be measured by mixing solutions of acids and alkalis in a calorimeter and measuring the rise in temperature.

Approximate values can be obtained by carrying out the reaction in a plastic cup or bottle. $100\,cm^3$ of M NaOH added to $100\,cm^3$ of M HCl might, for example, give a temperature rise of $6.5\,°C$. On the assumption that the solutions involved have the same specific heat capacity as water $(4.184\,J\,g^{-1}\,K^{-1})$ the heat evolved is $200 \times 6.5 \times 4.184\,J$, i.e. $5439\,J$. The enthalpy of neutralisation per mole of H^+ and OH^- will, therefore, be $-54.39\,kJ$. This is a low value due to heat losses.

Accurate values can only be obtained by paying very detailed attention to temperature measurement and heat losses. Insulated Dewar flasks can be used, and heat losses can be compensated for by measuring the electrical energy that has to be put into a mixture through a heating coil in order to produce the same rise in temperature as a neutralisation reaction.

For strong acids and strong alkalis (p. 341) the molar enthalpy of neutralisation is effectively constant at about $-57\,\text{kJ}\,\text{mol}^{-1}$. This is because strong acids and strong alkalis, and the salts they form, are all completely ionised in dilute solution (p. 343), so that the reaction is simply the formation of unionised water from H^+ and OH^- ions,

$$H^+(aq) + OH^-(aq) \rightarrow H_2O(l) \quad \Delta H_m^{\ominus}(298\,K) = -56.9\,\text{kJ}\,\text{mol}^{-1}$$

This constancy of the enthalpy of neutralisation of any strong acid and any strong alkali provides simple, but convincing, evidence that strong acids and alkalis are completely ionised.

With weak acids and/or alkalis, neutralisation produces an enthalpy change due to the formation of water from H^+ and OH^- ions but, in the course of the reaction, previously un-ionised acid and/or alkali has to be converted into ions and this involves enthalpy changes which may be either positive or negative. The enthalpy, of neutralisations involving either weak acids or alkalis may, therefore be greater or smaller than $-57\,\text{kJ}$, e.g.

$$HF(aq) + NaOH(aq) \rightarrow NaF(aq) + H_2O(l) \quad \Delta H_m^{\ominus}(298\,K) = -68.6\,\text{kJ}\,\text{mol}^{-1}$$

$$HCN(aq) + KOH(aq) \rightarrow KCN(aq) + H_2O(l) \quad \Delta H_m^{\ominus}(298\,K) = -11.7\,\text{kJ}\,\text{mol}^{-1}$$

9 Standard molar enthalpy of formation of compounds ($\Delta H_{m,f}^{\ominus}$)

The standard molar enthalpy of formation of a compound is the enthalpy change when 1 mol of the compound is formed from its elements under standard conditions.

The standard molar enthalpy of combustion of, say, carbon is also the standard molar enthalpy of formation of carbon dioxide, but, for those compounds which cannot be made directly from their elements, use has to be made of Hess's law to obtain their enthalpies of formation.

Methane, for example, cannot be made directly from carbon and hydrogen but its molar enthalpy of formation can be calculated from the molar enthalpies of combustion of methane, graphite and hydrogen, as follows:

$$\Delta H_{m,c}^{\ominus}(298\,K)$$

i $CH_4(g)$	$+ 2O_2(g)$	$\longrightarrow CO_2(g) + 2H_2O(l)$	$-890.4\,\text{kJ}\,\text{mol}^{-1}$
ii $C(\text{graphite}) + O_2(g)$		$\longrightarrow CO_2(g)$	$-393.5\,\text{kJ}\,\text{mol}^{-1}$
iii $2H_2(g)$	$+ O_2(g)$	$\longrightarrow 2H_2O(l)$	$-571.7\,\text{kJ}\,\text{mol}^{-1}$

Adding *ii* and *iii*, and subtracting *i*, gives the result

$$C(\text{graphite}) + 2H_2(g) \longrightarrow CH_4(g) \quad \Delta H_{m,f}^{\ominus}(298\ K) = -74.8\ \text{kJ mol}^{-1}$$

so that the standard molar enthalpy of formation of methane is $-74.8\ \text{kJ mol}^{-1}$.

The enthalpy of all elements in their standard states is conventionally taken as zero, so that the standard enthalpy of formation of a compound represents the enthalpy content of the compound on this arbitrary scale. The enthalpy content of methane is *lower* than that of its component elements; it is an *exothermic* compound. Ethene and ethyne, with positive standard enthalpies of formation have *higher* enthalpy content than their component elements; they are *endothermic* compounds. Such facts can be summarised in an enthalpy diagram (Fig. 136).

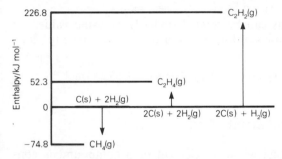

Fig. 136 Enthalpy diagram showing that methane is an exothermic compound whereas ethene and ethyne are endothermic.

a Stability of compounds The value of the standard molar enthalpy of formation of a compound gives a good indication of its stability *so far as decomposition into its elements is concerned*. Highly exothermic compounds, with high negative $\Delta H_{m,f}^{\ominus}$ values, are likely to be very stable and may require electrical methods to decompose them into their elements. Highly endothermic compounds, with high positive $\Delta H_{m,f}^{\ominus}$ values, will be unstable and tend to decompose, perhaps explosively, on very slight heating or even in the cold. For example,

stable compounds	unstable compounds
$Al_2O_3(-1675.7\ \text{kJ mol}^{-1})$	$Cl_2O_7(+75.73\ \text{kJ mol}^{-1})$
$NaCl(-411\ \text{kJ mol}^{-1})$	$NCl_3(+230\ \text{kJ mol}^{-1})$
$HF(-271.1\ \text{kJ mol}^{-1})$	$SnH_4(+162.8\ \text{kJ mol}^{-1})$

Still better indications of stability are given by standard free energies of formation (p. 272).

Beware, however, of using the word 'stable' loosely. A substance very stable so far as decomposition into its elements is concerned may be very reactive as a chemical reagent.

b Calculation of enthalpy of reaction from enthalpies of formation

Tabulated values of the standard enthalpies of formation of compounds can be used to calculate the enthalpy of a reaction under standard conditions, for

$$\left\{ \begin{array}{l} \text{standard} \\ \text{enthalpy} \\ \text{change of a} \\ \text{reaction} \end{array} \right\} = \left\{ \begin{array}{l} \text{sum of standard} \\ \text{enthalpies of} \\ \text{formation of} \\ \text{products} \end{array} \right\} - \left\{ \begin{array}{l} \text{sum of standard} \\ \text{enthalpies of} \\ \text{formation of} \\ \text{reagents} \end{array} \right\}$$

For example,

$$CO(g) \quad - \tfrac{1}{2}O_2(g) \rightarrow \quad CO_2(g) \qquad \Delta H_m^{\ominus}(298\,K) = -283.0\,kJ\,mol^{-1}$$
$$(-110.5) \qquad 0 \qquad\qquad (-393.5)$$

$$H_2O(g) + C(s) \rightarrow CO(g) + H_2(g) \qquad \Delta H_m^{\ominus}(298\,K) = +131.3\,kJ\,mol^{-1}$$
$$(-241.8) \qquad 0 \qquad (-110.5) \quad 0$$

Reactions involving ions can be treated similarly by using values for the standard molar enthalpies of formation of the ions involved (p. 262). For example,

$$Ag^+(aq) + Cl^-(aq) \rightarrow AgCl(s) \qquad \Delta H_m^{\ominus}(298\,K) = -65.2\,kJ\,mol^{-1}$$
$$(105.6) \qquad (-167.4) \qquad (-127)$$

10 Enthalpy of atomisation ($\Delta H_a^{\ominus}$)

This is the enthalpy change when an element or a compound is converted into free gaseous atoms under standard conditions.

Values for elements are generally quoted in kilojoule per mole of atoms of the element, e.g.

$$C(\text{graphite}) \longrightarrow C(g) \qquad \Delta H_{m,a}^{\ominus}(298\,K) = 715\,kJ\,mol^{-1}$$

$$Na(s) \qquad\qquad \longrightarrow Na(g) \qquad \Delta H_{m,a}^{\ominus}(298\,K) = 109\,kJ\,mol^{-1}$$

$$\tfrac{1}{2}H_2(g) \qquad\qquad \longrightarrow H(g) \qquad \Delta H_{m,a}^{\ominus}(298\,K) = 218\,kJ\,mol^{-1}$$

They represent the standard enthalpy of formation of 1 mol of the gaseous monatomic element in its standard state. The values are not easy to measure, and that for carbon is particularly difficult.

Values for the enthalpy of atomisation of compounds are quoted per mole of the compound. They can be obtained from the standard molar enthalpy of formation of the compounds and the molar enthalpies of atomisation of its component elements. For methane, for example, the standard enthalpy of formation gives

$$CH_4(g) \rightarrow C(\text{graphite}) + 2H_2(g) \qquad \Delta H_{m,f}^{\ominus}(298\,K) = 74.8\,kJ\,mol^{-1}$$

and the enthalpies of atomisation of graphite and hydrogen give

$$C(\text{graphite}) \rightarrow C(g) \qquad\qquad \Delta H_{m,a}^{\ominus}(298\,K) = 715\,kJ\,mol^{-1}$$

$$2H_2(g) \qquad\quad \rightarrow 4H(g) \qquad\qquad \Delta H_{m,a}^{\ominus}(298\,K) = 872\,kJ\,mol^{-1}$$

The molar enthalpy of atomisation of methane is, therefore, $1661.8\,\text{kJ mol}^{-1}$,

$$CH_4(g) \rightarrow C(g) + 4H(g) \qquad \Delta H_{m,a}{}^{\ominus}(298\,\text{K}) = 1661.8\,\text{kJ mol}^{-1}$$

$-1661.8\,\text{kJ mol}^{-1}$ represents the molar enthalpy of formation of methane *from its free atoms,*

$$C(g) + 4H(g) \rightarrow CH_4(g) \qquad \Delta H^{\ominus}(298\,\text{K}) = -1661.8\,\text{kJ mol}^{-1}$$

11 Bond enthalpy and bond energy

a Meaning of terms The splitting of a single covalent bond in diatomic molecules, e.g.

$$H_2(g) \longrightarrow 2H(g) \qquad \Delta H_m{}^{\ominus}(298\,\text{K}) = +436\,\text{kJ mol}^{-1}$$
$$HCl(g) \longrightarrow H(g) + Cl(g) \qquad \Delta H_m{}^{\ominus}(298\,\text{K}) = +431\,\text{kJ mol}^{-1}$$

involves standard molar enthalpy changes which are known as the bond enthalpies of the H—H and H—Cl bonds respectively. The corresponding $\Delta U_m{}^{\ominus}(0\,\text{K})$ values are known as the bond energies, and there is a small difference between the two values.

In polyatomic molecules, with more than one bond, the bond enthalpy (or the bond energy) can be interpreted in two ways. The C—H bond enthalpy in methane, for example, might be taken as one quarter of the molar enthalpy of atomisation of methane, i.e. one quarter of $1661.8\,\text{kJ mol}^{-1}$,

$$CH_4(g) \longrightarrow C(g) + 4H(g) \qquad \Delta H_m{}^{\ominus}(298\,\text{K}) = 1661.8\,\text{kJ mol}^{-1}$$

Such a bond enthalpy is really an *average bond enthalpy.* Its value, in methane, is $415.45\,\text{kJ mol}^{-1}$.

Alternatively, the C—H bond enthalpy in methane might be taken as the enthalpy of the reaction

$$CH_4(g) \longrightarrow CH_3(g) + H(g) \qquad \Delta H_m{}^{\ominus}(298\,\text{K}) = 431\,\text{kJ mol}^{-1}$$

This value is correctly called the *bond dissociation enthalpy,* and it is generally represented as D_{CH_3-H}. There are three other similar bond dissociation enthalpies in methane.

$$CH_3(g) \longrightarrow CH_2(g) + H(g) \qquad \Delta H_m(298\,\text{K}) = 364\,\text{kJ mol}^{-1}$$
$$CH_2(g) \longrightarrow CH(g) + H(g) \qquad \Delta H_m(298\,\text{K}) = 523\,\text{kJ mol}^{-1}$$
$$CH(g) \longrightarrow C(g) + H(g) \qquad \Delta H_m(298\,\text{K}) = 339\,\text{kJ mol}^{-1}$$

One quarter of the sum of these four bond dissociation enthalpies gives a value of $414.3\,\text{kJ mol}^{-1}$ for the average bond enthalpy in methane.

A value of $413\,\text{kJ mol}^{-1}$ is generally taken as the best average value to fit the C—H bond enthalpy in a wider range of compounds.

b Measurement of bond enthalpies The values for covalent bonds in diatomic molecules can, generally, be measured by spectroscopic methods (p. 183).

Average bond enthalpies are obtained from standard molar enthalpies of formation and molar enthalpies of atomisation, as with CH_4 in **a**. In more complicated molecules it is also necessary to assume the value of one bond enthalpy to find another. To obtain the bond enthalpy of the C—C bond in ethane, for example, it is necessary to make use of the C—H bond enthalpy value.

The standard molar enthalpy of formation for ethane is $-84.7 \, \text{kJ mol}^{-1}$. Together with the molar enthalpies of atomisation for carbon and hydrogen this gives a molar enthalpy of atomisation for ethane of $2822.7 \, \text{kJ mol}^{-1}$:

$$2C(s) + 3H_2(g) \rightarrow C_2H_6(g) \qquad \Delta H_m^{\ominus}(298 \, \text{K}) = -84.7 \, \text{kJ mol}^{-1}$$

$$2C(s) \rightarrow 2C(g) \qquad \Delta H_m^{\ominus}(298 \, \text{K}) = 1430 \, \text{kJ mol}^{-1}$$

$$3H_2(g) \rightarrow 6H(g) \qquad \Delta H_m^{\ominus}(298 \, \text{K}) = 1308 \, \text{kJ mol}^{-1}$$

$$C_2H_6(g) \rightarrow 2C(g) + 6H(g) \qquad \Delta H_m^{\ominus}(298 \, \text{K}) = 2822.7 \, \text{kJ mol}^{-1}$$

Breaking six mol of C—H bonds in 1 mol of C_2H_6 requires (6×413), i.e. 2478 kJ. The energy required to break the single C—C bond is, therefore $2822.7 - 2478$, i.e. $344.7 \, \text{kJ mol}^{-1}$.

The bond enthalpy of the C=C and C≡C bonds can be found, similarly, from data for C_2H_4 and C_2H_2.

c Bond enthalpy values Some typical values are given on page 127 for both single and multiple covalent bonds.

The values for bonds in diatomic molecules are bond dissociation enthalpies. They are closely related to enthalpies of atomisation, e.g.

$$\tfrac{1}{2}H_2(g) \longrightarrow H(g) \qquad \Delta H_m^{\ominus}(298 \, \text{K}) = 218 \, \text{kJ mol}^{-1}$$
(enthalpy of atomisation)

$$H_2(g) \longrightarrow 2H(g) \qquad \Delta H_m^{\ominus}(298 \, \text{K}) = 436 \, \text{kJ mol}^{-1}$$
(bond enthalpy)

For bonds occurring in polyatomic molecules, average bond enthalpies are given, and some compromise is necessary to choose values which fit well for different molecules.

12 Calculation of enthalpy changes in reactions from bond enthalpies

Any chemical reaction involves a rearrangement of atoms; some bonds have to be split and others re-formed. The overall enthalpy change comes about because the energy which has to be put in to split the bonds in the reagents to form free atoms is not, generally, equal to the energy

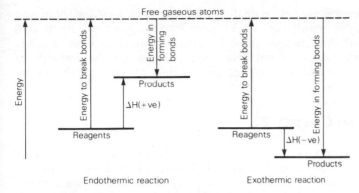

Fig. 137 Energy changes involved in breaking and forming bonds in endothermic and exothermic reactions.

given out when the free atoms recombine to form different bonds in the products. If less energy is required to break the bonds in the reagents than is liberated in forming the bonds in the products the reaction will be exothermic; otherwise it will be endothermic (Fig. 137).

Average bond enthalpy values can, therefore, be used to calculate the enthalpy change in a reaction so long as the nature of the bonds concerned is known and the *average* bond enthalpy values apply. In general

$$\begin{Bmatrix} \text{enthalpy} \\ \text{change in} \\ \text{reaction} \end{Bmatrix} = \begin{Bmatrix} \text{sum of bonds} \\ \text{enthalpies} \\ \text{of reagents} \end{Bmatrix} - \begin{Bmatrix} \text{sum of bond} \\ \text{enthalpies} \\ \text{of products} \end{Bmatrix}$$

For example, the reaction

$$\underbrace{\underset{945}{N_2(g)} + \underset{3(436)}{3H_2(g)}}_{2253} \longrightarrow \underbrace{\underset{6(391)}{2NH_3(g)}}_{2346}$$

involves the splitting of 1 mol of N≡N and 3 mol of H—H bonds, which requires 2253 kJ, and the formation of 6 mol of N—H bonds, which liberates 2346 kJ. The calculated enthalpy change for the reaction (twice the molar enthalpy of formation of ammonia) is $-93\,\text{kJ mol}^{-1}$ in agreement with the measured value of $-92.4\,\text{kJ mol}^{-1}$.

The reduction of an alkene to an alkane by hydrogen can be represented as

$$\underset{611}{\overset{\diagdown}{\underset{\diagup}{C}}=\overset{\diagup}{\underset{\diagdown}{C}}} + \underset{436}{H—H} \longrightarrow \underset{346}{-\overset{|}{\underset{|}{C}}-\overset{|}{\underset{|}{C}}-} + \underset{2(413)}{2C—H}$$

giving an enthalpy of reaction of $-125\,\text{kJ mol}^{-1}$, in good agreement with measured values (p. 138), known as the *enthalpy of hydrogenation*.

Introduction to Physical Chemistry

However, care is needed. In dealing with reactions involving solids and liquids, energy is required to convert them into the gaseous state (p. 249) and these enthalpy changes must be allowed for. Moreover, average bond enthalpy values do not give correct values for the enthalpy of reaction when the bonds concerned are not fully covalent (p. 132) or are delocalised (p. 138).

Questions on Chapter 22

1 Give an account of the experimental determination of the molar enthalpy of combustion of benzene. The heats evolved when 2 mol of benzene and 2 mol of ethyne are fully burnt are 6689 and 2595 kJ respectively. Calculate the enthalpy change of the reaction in which 1 mol of benzene is formed from 3 mol of ethyne. (OC)

2 Define molar enthalpy of combustion and molar enthalpy of formation. State Hess's law   (a) Calculate the enthalpy change when 1 mol of benzene is formed from 3 mol of ethyne   (b) Calculate the molar enthalpy of combustion of ethyne gas. You are given the following molar enthalpies of formation in $kJ\,mol^{-1}$; $C_6H_6(g)$, 82.84; $CO_2(g)$, -393.3; $C_2H_2(g)$, 226.8; $H_2O(g)$, -241.8. (OC)

3 Explain the terms: enthalpy of neutralisation, enthalpy of formation and enthalpy of combustion. Why is the enthalpy of neutralisation of a strong acid and a strong base constant? Given that the enthalpy of combustion of benzene is $-3278\,kJ\,mol^{-1}$ and that the enthalpies of formation of carbon dioxide and water are -393.4 and $-285.8\,kJ\,mol^{-1}$ respectively, calculate the enthalpy of formation of benzene. (OC)

4 The molar enthalpies of combustion of hydrogen, carbon and methane are -285.8, -393.5 and $-890.4\,kJ\,mol^{-1}$ respectively. Calculate the molar enthalpy of formation of methane.

5 If the molar enthalpy of combustion of hydrogen is $-285.8\,kJ\,mol^{-1}$ what mass of ice at 0 °C could be converted into steam at 100 °C by the combustion of 1000 dm^3 of hydrogen measured at s.t.p.? (Specific latent heat of fusion of ice = $330.5\,kJ\,kg^{-1}$. Specific latent heat of vapourisation of water = $2255\,kJ\,kg^{-1}$.)

6 State Hess's law. The molar enthalpy of combustion of ethanol is $-1430\,kJ\,mol^{-1}$ and the molar enthalpies of formation of carbon dioxide and water are -393.4 and $-285.8\,kJ\,mol^{-1}$ respectively. What is the molar enthalpy of formation of ethanol under the same conditions?

7 Do the following figures agree with Hess's law?

$H_2(g) + Cl_2(g) \longrightarrow 2HCl(g)$ $\qquad\qquad \Delta H_m^{\ominus} = -184.1\,kJ\,mol^{-1}$

$H_2(g) + I_2(s) \longrightarrow 2HI(g)$ $\qquad\qquad \Delta H_m^{\ominus} = +506\,kJ\,mol^{-1}$

$2HI(g) + Cl_2(g) \longrightarrow I_2(s) + 2HCl(g)$ $\qquad \Delta H_m^{\ominus} = -234.3\,kJ\,mol^{-1}$

8 Calculate the ΔU_m value for the change of 1 mol of $H_2O(l)$ at 100 °C into 1 mol of $H_2O(g)$ at 100 °C, given that the ΔH_m value at 100 °C is $40.7\,kJ\,mol^{-1}$.

9 The enthalpy changes in going from R to S, S to T, T to U and R to U are $+v$, $+w$, $-x$ and $-y$ respectively. Represent that information on an enthalpy diagram and say how v, w, x and y are related to each other.

10 An adiabatic calorimeter operating at constant pressure had a thermal capacity of $2\,kJ\,K^{-1}$. The temperature rose by 22.2 K when 0.6 g of ethane was burnt in the calorimeter. What is the molar enthalpy of combustion of ethane?

11 What temperature rise would be expected if $100\,cm^3$ of 0.5 M hydrochloric acid and $100\,cm^3$ of 0.5 M sodium hydroxide solution were mixed in a calorimeter of thermal capacity $150.6\,J\,K^{-1}$? Take all solutions as having a heat capacity of $4.2\,J\,K^{-1}\,cm^{-3}$, and neglect any heat losses.

12 The molar enthalpy of formation of AgCl(s) is $-127\,kJ\,mol^{-1}$ and the heat liberated when 1 mol of silver chloride is precipitated from silver nitrate and sodium chloride solutions is 65.48 kJ. Calculate the enthalpy change in the reaction

$$Ag(s) + \tfrac{1}{2}Cl_2(g) + H_2O \longrightarrow Ag^+(aq) + Cl^-(aq)$$

13 The thermal capacity of a bomb calorimeter is $2.259\,kJ\,K^{-1}$. When 1 g of benzenecarboxylic acid is completely burnt in the bomb, the rise in temperature of 1200 g of water in the calorimeter is 3.65 K. Calculate the approximate value of the molar enthalpy of combustion of the acid.

14 Given that

$$CO(g) + H_2(g) \longrightarrow H_2O(g) + C(s) \qquad \Delta H_m^{\ominus}(298\ K) = -131.3\,kJ\,mol^{-1}$$

calculate the enthalpy change at constant pressure and 25 °C when 1 g of carbon reacts completely with water vapour to form carbon monoxide and hydrogen. What would the value be at constant volume?

15 Calculate the enthalpy of the reaction between 2 mol of Al and 1 mol of Cr_2O_3 given that the molar enthalpies of formation of Al_2O_3 and Cr_2O_3 are -1590 and $-1130\,kJ\,mol^{-1}$ respectively. Comment on points of interest associated with the high molar enthalpy of formation of Al_2O_3.

16 State Hess's law and define the molar enthalpy of formation. The molar enthalpies of combustion of carbon and carbon monoxide are -393.3 and $-284.5\,kJ\,mol^{-1}$ respectively. Calculate the molar enthalpy of formation of carbon monoxide.

17 State and account for Hess's law of constant heat summation. Explain its value in the determination of the enthalpies of formation of organic compounds. The molar enthalpies of formation of water and carbon dioxide are -285.8 and $-405.4\,kJ\,mol^{-1}$ respectively, at 15 °C and constant pressure. The molar enthalpies of combustion of methane and ethane at 15 °C and constant pressure are -886.6 and $-1423\,kJ\,mol^{-1}$ respectively. Calculate the molar enthalpies of formation of methane and ethane at 15 °C   (a) at constant pressure   (b) at constant volume. (Molar gas constant, R, is $8.3\,J\,K^{-1}\,mol^{-1}$.) (W)

18 The molar enthalpy of formation of hydrogen selenide is $-811.7 \, kJ \, mol^{-1}$ if amorphous selenium is used, and $-1050 \, kJ \, mol^{-1}$ if metallic selenium is used. Calculate the enthalpy change on converting 1 mol of amorphous into metallic selenium.

19 Define (a) the molar enthalpy of combustion of a compound (b) the molar enthalpy of a reaction. Outline an experiment to measure (a).

$$Au(OH)_3 + 4HCl \rightarrow HAuCl_4 + 3H_2O \quad \Delta H_m^{\ominus}(298 \, K) = -96.23 \, kJ \, mol^{-1}$$

$$Au(OH)_3 + 4HBr \rightarrow HAuBr_4 + 3H_2O \quad \Delta H_m^{\ominus}(298 \, K) = -154 \, kJ \, mol^{-1}$$

On mixing 1 mol of tetrabromoauric(III) acid with 4 mol of hydrochloric acid 2.092 kJ are absorbed. Calculate the percentage of tetrabromoauric(III) acid which has been changed into tetrachloroauric(III) acid.

20 How can it be decided (a) which is the stabler of two allotropes (b) whether a molecule is likely to have a conjugated system of $C=C$ bonds?

21* What difference would you expect between the molar enthalpies of combustion of ethanol and methoxymethane in terms of bond enthalpy values?

22* Use standard enthalpies of formation to compare the stability so far as decomposition into the elements is concerned of (a) Al_2O_3 and Cl_2O_7 (b) NaCl and NCl_3 (c) CH_4 and SnH_4.

23* Compare the enthalpies of combustion of (a) graphite and diamond (b) red and white phosphorous. What information do the values give about the stability of a particular allotrope?

24* Draw an enthalpy diagram to show what values would require to be known in order to find the bond enthalpy of the H—Cl bond in HCl(g).

25* Write an account of the stability of compounds, explaining the significance of the term and giving some illustrative examples with associated numerical data.

26* Use standard enthalpies of formation to calculate the enthalpy changes in the following reactions:

(a) $C_2H_2(g) + H_2(g) \longrightarrow C_2H_4(g)$

(b) $H_2(g) + I_2(s) \longrightarrow 2HI(g)$

(c) $2C_2H_6(g) + 7O_2(g) \longrightarrow 4CO_2(g) + 6H_2O(l)$

(d) $Ag^+(aq) + Cl^-(aq) \longrightarrow AgCl(s)$

(e) $Zn^{2+}(aq) + Cu(s) \longrightarrow Zn(s) + Cu^{2+}(aq)$

27* Record the enthalpies of combustion of glucose and sucrose, and explain why the difference between the two values arises.

28* Use the necessary bond enthalpies to calculate the enthalpies of combustion of (a) $CH_4(g)$ (b) C(graphite) (c) $H_2S(g)$.

29* What is the main factor contributing to the difference in value for the enthalpies of formation of HF and HCl.

30* Trouton's rule states that the molar enthalpy of vaporisation (in J) for many compounds is approximately 88 times the boiling point in K. Examine the truth of this rule.

31* Estimate the amount of heat you would get by burning an average lump of sugar, and compare this with the amount you could get from equal masses of graphite, hydrogen and methanol.

32* The enthalpy change when 2 mol of hydrogen peroxide is converted into water and oxygen is $-192.4\,kJ$. Use this value, and any necessary bond enthalpy values, to calculate the bond enthalpy of the O—O bond in hydrogen peroxide.

23
Enthalpy changes in solutions of ions

1 Enthalpies of solution (ΔH_s)

There is an enthalpy change, known as the enthalpy of solution, when a solute dissolves in a solvent. The value depends on the concentration of the solution formed, for dilution of a solution also causes an enthalpy change known as the enthalpy of dilution.

The enthalpy of solution most commonly used is that at infinite dilution per mole of solute. It is called the *molar enthalpy of solution at infinite dilution* and is defined as *the enthalpy change when 1 mol of solute dissolves in such a large excess of solvent that addition of further solvent produces no further enthalpy change*. Thus, the molar standard enthalpy of solution for hydrogen chloride gas is $-75.1\,\text{kJ}\,\text{mol}^{-1}$, and that for sodium chloride is $+3.9\,\text{kJ}\,\text{mol}^{-1}$, i.e.

$$\text{HCl(g)} + \text{aq} \rightarrow \text{H}^+\text{(aq)} + \text{Cl}^-\text{(aq)} \qquad \Delta H_{m,s}^{\ominus}(298\,\text{K}) = -75.1\,\text{kJ}\,\text{mol}^{-1}$$

$$\text{NaCl(s)} + \text{aq} \rightarrow \text{Na}^+\text{(aq)} + \text{Cl}^-\text{(aq)} \qquad \Delta H_{m,s}^{\ominus}(298\,\text{K}) = +3.9\,\text{kJ}\,\text{mol}^{-1}$$

2 Standard molar enthalpies of formation of ions in solution

Combination of the standard molar enthalpies of formation and solution for HCl(g) gives a value for the standard molar enthalpy of formation of HCl(aq) from its component elements, i.e.

$$\tfrac{1}{2}\text{H}_2\text{(g)} + \tfrac{1}{2}\text{Cl}_2\text{(g)} \rightarrow \text{HCl(g)} \qquad \Delta H_{m,s}^{\ominus}(298\,\text{K}) = -92.3\,\text{kJ}\,\text{mol}^{-1}$$

$$\text{HCl(g)} + \text{aq} \rightarrow \text{H}^+\text{(aq)} + \text{Cl}^-\text{(aq)} \qquad \Delta H_{m,s}^{\ominus}(298\,\text{K}) = -75.1\,\text{kJ}\,\text{mol}^{-1}$$

$$\tfrac{1}{2}\text{H}_2\text{(g)} + \tfrac{1}{2}\text{Cl}_2\text{(g)} + \text{aq} \rightarrow \text{H}^+\text{(aq)} + \text{Cl}^-\text{(aq)} \qquad \Delta H_{m,f}^{\ominus}(298\,\text{K}) = -167.4\,\text{kJ}\,\text{mol}^{-1}$$

This standard enthalpy of formation is the sum of the standard enthalpies of formation of the $\text{H}^+\text{(aq)}$ and $\text{Cl}^-\text{(aq)}$ ions from their elements. It is not easy to measure or calculate the value for any single ion, but relative values can be obtained by taking the value for the standard enthalpy of formation of $\text{H}^+\text{(aq)}$ as zero. This gives a relative value for the $\text{Cl}^-\text{(aq)}$ ion of $-167.4\,\text{kJ}\,\text{mol}^{-1}$, i.e.

$$\tfrac{1}{2}\text{H}_2\text{(g)} + \text{aq} \longrightarrow \text{H}^+\text{(aq)} \quad \Delta H_{m,f}^{\ominus}(298\,\text{K}) = 0 \text{ (arbitrary choice)}$$

$$\tfrac{1}{2}\text{Cl}_2\text{(g)} + \text{aq} \longrightarrow \text{Cl}^-\text{(aq)} \quad \Delta H_{m,f}^{\ominus}(298\,\text{K}) = -167.4\,\text{kJ}\,\text{mol}^{-1}$$

For sodium chloride

$$\text{NaCl(s)} + \text{aq} \rightarrow \text{Na}^+\text{(aq)} + \text{Cl}^-\text{(aq)} \qquad \Delta H_{m,s}^{\ominus}(298\,\text{K}) = 3.9\,\text{kJ}\,\text{mol}^{-1}$$

$$\text{Na(s)} + \tfrac{1}{2}\text{Cl}_2\text{(g)} \rightarrow \text{NaCl(s)} \qquad \Delta H_{m,f}^{\ominus}(298\,\text{K}) = -411\,\text{kJ}\,\text{mol}^{-1}$$

$$\text{Na(s)} + \tfrac{1}{2}\text{Cl}_2\text{(g)} \rightarrow \text{Na}^+\text{(aq)} + \text{Cl}^-\text{(aq)} \qquad \Delta H_{m,f}^{\ominus}(298\,\text{K}) = -407.1\,\text{kJ}\,\text{mol}^{-1}$$

and if the relative enthalpy of formation of $Cl^-(aq)$ is $-167.4\,kJ\,mol^{-1}$ the value for $Na^+(aq)$ must be $-239.7\,kJ\,mol^{-1}$. Relative values for other ions can be obtained similarly.

(a) Use of relative enthalpies of formation of ions in calculating enthalpies of reactions The enthalpy of a reaction involving ions can be calculated in just the same way as that of a reaction involving compounds (p. 254). For example:

$$HCl(g) + aq \rightarrow H^+(aq) + Cl^-(aq) \qquad \Delta H_{m,s}{}^{\ominus}(298\,K) = -75.1\,kJ\,mol^{-1}$$
$$(-92.3) \qquad\quad 0 \qquad\quad (-167.4)$$

$$Ag^+(aq) + Cl^-(aq) \rightarrow AgCl(s) \qquad \Delta H_m{}^{\ominus}(298\,K) = -65.2\,kJ\,mol^{-1}$$
$$(105.6) \quad (-167.4) \quad (-127)$$

$$2Na(s) + 2H_2O(l) \rightarrow 2Na^+(aq) + 2OH^-(aq) + H_2(g) \qquad \Delta H_m{}^{\ominus}(298\,K) = -367.2\,kJ\,mol^{-1}$$
$$0 \qquad 2(-286) \qquad 2(-239.7) \quad 2(-229.9) \quad 0$$

(b) Absolute values of enthalpies of formation of ions These can only be obtained by using a calculated value for the enthalpy of hydration (see section 3) of $H^+(g)$, i.e.

$$H^+(g) + aq \longrightarrow H^+(aq) \qquad \Delta H_{m,hyd}{}^{\ominus}(298\,K) = -1075\,kJ\,mol^{-1}$$

It is then possible to calculate as follows:

$$\tfrac{1}{2}H_2(g) \xrightarrow{218} H(g) \xrightarrow{1317} H^+(g) \xrightarrow{-1075} H^+(aq)$$

i.e. $\tfrac{1}{2}H_2(g) \longrightarrow H^+(aq) \qquad \Delta H_f{}^{\ominus}(298\,K) = 460\,kJ\,mol^{-1}$

$$Na(s) \xrightarrow{108.4} Na(g) \xrightarrow{502} Na^+(g) \xrightarrow{-390} Na^+(aq)$$

i.e. $Na(s) \longrightarrow Na^+(aq) \qquad \Delta H_f{}^{\ominus}(298\,K) = 220.4\,kJ\,mol^{-1}$

Absolute values are obtained by adding 460 to the relative value for any ion.

3 Molar enthalpy of hydration of ions ($\Delta H_{m,hyd}$)

This is the enthalpy change when 1 mol of a gaseous ion is hydrated, i.e.

$$M^{n\pm}(g) + aq \rightarrow M^{n\pm}(aq) \qquad \Delta H_{m,hyd}{}^{\ominus}(298\,K) = \text{enthalpy of hydration}$$

It is relatively easy to obtain a value for the *sum* of the hydration enthalpies of two ions, but it is less easy to decide the value for any single ion very precisely. Quoted values are generally based on a calculated value for the enthalpy of hydration of $H^+(g)$, i.e.

$$H^+(g) + aq \rightarrow H^+(aq) \qquad \Delta H_{m,hyd}{}^{\ominus}(298\,K) = -1075\,kJ\,mol^{-1}\ \text{(calculated)}$$

The calculation cannot, however, be done very precisely and many figures other than -1075 are adopted. The problem is, sometimes, overcome by quoting *relative* values for enthalpies of hydration based on an arbitrary value of zero for $H^+(g)$. These relative values are obtained from the absolute values by adding $1075z\,kJ\,mol^{-1}$ where z is the charge on the ion (including its sign).

a The enthalpy of hydration of $Cl^-(g)$ This can be obtained from
  i   the bond enthalpy of $HCl(g)$,
 ii   the enthalpy of solution of $HCl(g)$,
iii the ionisation enthalpy of $H(g)$, and
iv the electron affinity of $Cl(g)$, as follows:

  i $H(g) + Cl(g) \rightarrow HCl(g)$ $\Delta H_{m,f}^{\ominus}(298\,K) = -431\,kJ\,mol^{-1}$

 ii $HCl(g) + aq \rightarrow H^+(aq) + Cl^-(aq)$ $\Delta H_{m,s}^{\ominus}(298\,K) = -75.1\,kJ\,mol^{-1}$

iii $H(g) \rightarrow H^+(g) + 1e^-$ $\Delta H_m^{\ominus}(298\,K) = 1317\,kJ\,mol^{-1}$

iv $Cl(g) + 1e^- \rightarrow Cl^-(g)$ $\Delta H_m^{\ominus}(298\,K) = -354\,kJ\,mol^{-1}$

From $(i + ii - iii - iv)$

$$H^+(g) + Cl^-(g) \rightarrow H^+(aq) + Cl^-(aq) \quad \Delta H_m^{\ominus}(298\,K) = -1469.1\,kJ\,mol^{-1}$$

which gives a value of $-394.1\,kJ\,mol^{-1}$ for the enthalpy of hydration of $Cl^-(g)$ if the value for $H^+(g)$ is taken as $-1075\,kJ\,mol^{-1}$, or a value of $-1469.1\,kJ\,mol^{-1}$ if the value for $H^+(g)$ is taken as zero.

b The enthalpy of hydration of $Na^+(g)$ The sum of the relative enthalpies of hydration of $Na^+(g)$ and $Cl^-(g)$ can be obtained from the enthalpy of solution and the lattice enthalpy of $NaCl(s)$, as follows:

$$NaCl(s) + aq \rightarrow Na^+(aq) + Cl^-(aq) \qquad \Delta H_{m,s}^{\ominus}(298\,K) = 3.9\,kJ\,mol^{-1}$$

$$Na^+(g) + Cl^-(g) \rightarrow NaCl(s) \qquad \Delta H_m^{\ominus}(298\,K) = -788\,kJ\,mol^{-1}$$

Therefore,

$$Na^+(g) + Cl^-(g) \rightarrow Na^+(aq) + Cl^-(aq) \qquad \Delta H_m^{\ominus}(298\,K) = -784.1\,kJ\,mol^{-1}$$

giving the value for the enthalpy of hydration of $Na^+(g)$ of $-390\,kJ\,mol^{-1}$ if the value or $Cl^-(g)$ is taken as $-394.1\,kJ\,mol^{-1}$ or a value of $685\,kJ\,mol^{-1}$ on the basis of $H^+(g)$ having a zero enthalpy of hydration.

c Typical values Some values for molar enthalpies of hydration of ions, based on $H^+(g)$ having a value of $-1075\,kJ\,mol^{-1}$ are given below, in $kJ\,mol^{-1}$.

H^+ -1075			
Li^+ -505	Be^{2+} -2462	Al^{3+} -4620	F^- -530
Na^+ -390	Mg^{2+} -1889		Cl^- -394
K^+ -305	Ca^{2+} -1545		Br^- -363
Rb^+ -275	Sr^{2+} -1414		I^- -322
Cs^+ -247	Ba^{2+} -1275		

It will be seen that the enthalpy of hydration gets more negative as the charge on a cation increases, or as the size of an ion decreases. The smaller the size and the higher the charge, the more strongly will an ion attract water molecules. The value for the very small H^+ ion is much higher than that for other monovalent ions, and the value for Al^{3+}, which is highly charged, is also very high.

4 Enthalpy of solution and solubility of ionic solids

The molar enthalpy of solution of an ionic solid, i.e.

$$AX(s) \rightarrow A^+(aq) + X^-(aq) \qquad \Delta H_{m,s}^{\ominus}(298 \text{ K}) = \text{enthalpy of solution}$$

is made up of two terms:

a The breakdown of the crystal lattice

i.e. $\quad AX(s) \rightarrow A^+(g) + X^-(g) \qquad \Delta H_m(298 \text{ K}) = -\text{lattice enthalpy}$

the ΔH_m value here being the lattice enthalpy (p. 117) with the sign reversed.

b The hydration of the gaseous ions (p. 263) taken together

i.e. $\quad A^+(g) + X^-(g) \longrightarrow A^+(aq) + X^-(aq)$
$\qquad\qquad\qquad\qquad \Delta H_m(298 \text{ K}) = \text{sum of two enthalpies of hydration}$

The relationship between the various terms is shown in Fig. 138. A high lattice enthalpy and low enthalpy of hydration makes the enthalpy of solution more highly positive, i.e. the process of solution is endothermic, so that solubility is low. Low lattice enthalpy and high enthalpy of hydration favour solubility.

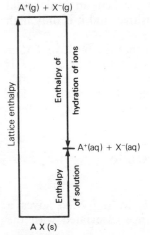

Fig. 138 A high lattice enthalpy and low enthalpy of hydration gives a high positive value for the enthalpy of solution and favours insolubility.

These generalisations must, however, be treated with caution. The factors that favour high lattice enthalpy (small ionic sizes and high ionic charges) also favour high enthalpies of hydration, so that the two tend to cancel each other out. The enthalpy of solution is the difference between two large numbers, and its value may be quite small. Moreover, the entropy change in the process of solution may be of more importance than the enthalpy change, and this is particularly true when the enthalpy change has a small value. Conclusions drawn about solubility from enthalpy values can, therefore, be very misleading. The free energy values must be considered.

5 Free energy of solution

An ionic solid will only dissolve in water if the free energy change is negative. Some typical values are given below:

	NaCl	NaF	AgCl	AgF
enthalpy of solution/kJ mol^{-1}	3.9	-3.5	65.5	-24
$T\Delta S$/kJ mol^{-1}	12.84	-3.9	9.9	-7.33
free energy of solution/kJ mol^{-1}	-8.94	0.4	55.5	-16.67
solubility/mol $(100 g\, H_2O)^{-1}$	0.615	0.099	1.3×10^{-6}	1.42

It will be seen that the ΔG values are made up from ΔH and $T\Delta S$ terms (p. 271),

$$\Delta G = \Delta H - T\Delta S$$

The make-up of the ΔH term has been discussed in the preceding section. As far as entropy is concerned, there is an increase in randomness when a crystal structure is broken down into free gaseous ions, but this is offset by the probable decrease in randomness when the gaseous ions are hydrated. The overall entropy change when an ionic solid dissolves can, therefore, be either negative or positive, and it is only high positive values that favour solubility.

Sodium chloride owes its solubility to the fact that the positive $T\Delta S$ is greater than the positive ΔH; silver fluoride is soluble because ΔH is negative and greater than the negative $T\Delta S$. Silver chloride is insoluble because the very high positive ΔH is only offset by a relatively low positive $T\Delta S$; sodium fluoride is insoluble because the negative $T\Delta S$ term is just greater than the negative ΔH term.

6 Enthalpy changes and standard electrode potential values

The standard electrode potential for a couple, $M^+(aq)/M(s)$, gives a measure, in electrical terms, of the energy change associated with the process

$$M(s) \longrightarrow M^+(aq) + 1e^- \qquad \Delta H = X\,kJ\,mol^{-1}$$

where X is the enthalpy of formation of $M^+(aq)$. The formation of the ion from its element involves three stages

$$M(s) \xrightarrow{\text{atomisation}} M(g) \xrightarrow{\text{ionisation}} M^+(g) \xrightarrow{\text{hydration}} M^+(aq)$$

and X is equal to (enthalpy of atomisation + ionisation enthalpy + enthalpy of hydration). The values involved in typical cases are summarised below.

a Monovalent ions

	Li	Na	H	Ag
$M(s) \longrightarrow M(g)$	161	108	218	286
$M(g) \longrightarrow M^+(g)$	526	502	1317	737
$M^+(g) \longrightarrow M^+(aq)$	-505	-390	-1075	-464
$M(s) \longrightarrow M^+(aq)$	182	220	460	559
elect. potential/V	-3.045	-2.714	0	$+0.80$

On this scale, the enthalpy of formation of $H^+(aq)$ is $460\,\text{kJ mol}^{-1}$ and elements having higher values have positive standard electrode potentials whilst elements with lower values have negative electrode potentials. The enthalpy of formation of $Li^+(g)$, $687\,\text{kJ mol}^{-1}$, is higher than that of $Na^+(g)$, $610\,\text{kJ mol}^{-1}$, but the high enthalpy of hydration of $Li^+(g)$ compared with $Na^+(g)$ means that the enthalpy of formation of $Li^+(aq)$ is lower than that for $Na^+(aq)$. Lithium has, therefore, a higher standard electrode potential than sodium and is higher in the electrochemical series. The data is summarised in Fig. 139.

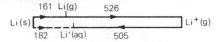

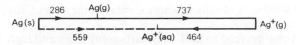

Fig. 139 Showing the contribution of enthalpy of atomisation, ionisation enthalpy and enthalpy of hydration to the enthalpy of formation of $M^+(aq)$ from M in its stable state.

b Zinc, copper and mercury These three metals all form $M^{2+}(aq)$ ions but there are big differences between their standard electrode potentials. The origin of the differences can be seen in the following summarised figures.

	Zn	Cu	Hg
$M(s) \longrightarrow M(g)$	126	339	61
$M(g) \longrightarrow M^{2+}(g)$	2652	2715	2828
$M^{2+}(g) \longrightarrow M^{2+}(aq)$	-2013	-2069	-1792
$M(s) \longrightarrow M^{2+}(aq)$	765	985	1097
elect. potential/V	-0.76	0.34	0.85

Because these ions are doubly charged it is those with enthalpies of formation greater than 920, i.e. (2×460), that have positive standard electrode potentials.

The difference between the values for zinc and copper is mainly caused by the lower enthalpy of atomisation of zinc, differences in ionisation enthalpy and enthalpy of hydration being small. For copper and mercury the differences in enthalpies of atomisation and hydration, rather than differences in ionisation enthalpy, are important.

c Aluminium The enthalpy of atomisation for aluminium is $324\,kJ\,mol^{-1}$, and the ionisation enthalpy to form $Al^{3+}(g)$ is $5158\,kJ\,mol^{-1}$. To form the ion from solid aluminium requires, then $5482\,kJ\,mol^{-1}$.

This is such a high value that the existence of the free Al^{3+} ion is doubtful. The heat of hydration is -4620 so that the enthalpy of formation of $Al^{3+}(aq)$ is $862\,kJ\,mol^{-1}$. On the basis of one electron this value is $287\,kJ\,mol^{-1}$, comparing with a value for zinc of 382.5 per electron. Aluminium is, therefore above zinc in the electrochemical series.

Questions on Chapter 23

1 Sketch the general shape of a graph plotting the enthalpy change against the concentration when sodium chloride is added to water.

2 Explain the difficulty in obtaining values for the enthalpies of formation of single aqueous ions from their elements. How is the problem resolved?

3 'The thermodynamic properties of an ion in a dilute solution are independent of other ions'. Comment on that statement.

4 Discuss the relation between the lattice enthalpy, the enthalpy of solution and the enthalpy of hydration for ionic solids.

5 Explain why it is that NaCl, NaF and AgF are soluble in water whereas AgCl is insoluble.

6 To what extent is the solubility of salts in water similar to their solubility in liquid ammonia?

7 'Small ions with high charges lead to both high lattice enthalpy and high hydration energies'. Comment on that statement. What effect does it have on the solubility of ionic solids?

8 Why can enthalpy values give very misleading results when applied to solubility problems?

9 How, in general, is the solubility of a substance affected by (a) its lattice enthalpy (b) temperature (c) hydration enthalpy (d) hydrogen bonding?

10 What general factors are likely to make a substance (a) soluble in water (b) soluble in benzene?

11 Discuss the general way in which the standard electrode potential of an element is related to the standard enthalpy of formation of its ion in aqueous solution.

12* Record the enthalpies of solution at infinite dilution of (a) LiCl and NaCl (b) NaCl and NaI. Some of the values are negative and some positive. What is the main reason for this?

13* Use Born–Haber cycles to obtain values for the enthalpies of hydration of (a) $Na^+(g) + Cl^-(g)$ (b) $Li^+(g) + Cl^-(g)$ (c) $Na^+(g) + I^-(g)$.

14* Discuss the enthalpy, free energy and entropy changes in the following processes:

(a) $NaCl(s) \longrightarrow Na^+(aq) + Cl^-(aq)$

(b) $HCl(g) \longrightarrow H^+(aq) + Cl^-(aq)$.

15* Compare and contrast the enthalpy, free energy and entropy changes in the conversion of 1 mol of (a) the halides of sodium (b) the halides of silver, into a solution in water.

16* Calculate the enthalpy changes in the following reactions:

(a) $Pb^{2+}(aq) + 2Cl^-(aq) \longrightarrow PbCl_2(s)$

(b) $H^+(aq) + OH^-(aq) \longrightarrow H_2O(l)$

(c) $HI(g) \longrightarrow HI(aq)$

(d) $Zn(s) + 2H^+(aq) \longrightarrow Zn^{2+}(aq) + H_2(g)$

17* Plot values of the hydration enthalpies (in $kJ\,mol^{-1}$) of any ten anions against z^2/r, and of any ten cations against $z^2/(r + 0.085)$, where z is the charge on the ion and r its radius in nm. What do you find?

24
Free energy and entropy

1 Introduction

In the preceding chapter (p. 245) the idea that a reaction was more likely to be feasible if it was exothermic (with negative ΔH) was introduced as a useful approximation. At one time, indeed, it was thought that the evolution of heat was the driving force behind a chemical change. Such ideas are now untenable, for it is well known that many endothermic reactions can take place spontaneously, and reversible reactions take place with an evolution of heat in one direction and an absorbtion in the other. Moreover, the heat evolved in an exothermic reaction is absorbed by the surroundings with the loss and gain of heat being equally spontaneous. Some changes take place with no heat change, e.g. the expansion of gas into a vacuum, or the mixing of two gases.

What then is the driving force behind a chemical change? Why do some changes go completely and others not go at all? The answers lie in a study of chemical thermodynamics and in an understanding of the ideas of enthalpy (H), free energy (G) and entropy (S), which are known as *state*, or *thermodynamic, functions*.

a Free energy change The heat that many reactions can evolve is very obvious for it causes a rise in temperature. But many reactions can, if carried out correctly, be made to do useful work. A reaction carried out in a cell, for example, can do electrical work. Burning petrol and air can drive a motor car.

It is the ability of a reaction to do useful work that provides the driving force, not the ability to evolve heat. If heat could always be completely converted into useful work the two criteria would be the same, but that is not so (p. 280). It is the *maximum* useful work that a reaction can do, if carried out correctly, that measures how feasible it will be.

The maximum work can only be done when the reaction is carried out *reversibly* (p. 280), and when this happens there is a corresponding decrease in free energy for the reaction, i.e.

$$\Delta G = \text{maximum useful (net) work done by a reaction}$$

As work done by a reaction is regarded as negative (p. 245), ΔG will be negative for reactions which can do useful work and which, therefore, are feasible. *The more negative the value of ΔG the more feasible the reaction and the higher its equilibrium constant.*

b Entropy Entropy is best considered initially (though this is an over-simplification) as a measure of disorderliness or randomness or dispersion; what Gibbs once called 'mixed-upness'. But it is a 'mixed-upness' of energy that is concerned; not spatial disorder.

As there is no thermal energy in a pure, crystalline substance at 0 K, there is no disordered energy so that the entropy is zero. Raising the temperature introduces random motion and the disorderliness increases, through the liquid state to that of a vapour, and the entropy increases accordingly.

Increase in entropy is, then, related to changes from solid to liquid, or from liquid to gas, or to absorption of heat. Other changes likely to lead to increased disorderliness and increase in entropy include the solution of a crystalline solid in a solvent, the mixing of two gases, the increase in the gaseous volume in a reaction, a reaction in which a few molecules are converted into a larger number, and reactions in which gases are formed from liquids or solids, or liquids are formed from solids. These are, however, only broad generalisations.

c $\Delta G = \Delta H - T\Delta S$ This is the important relationship (p. 282) between changes in enthalpy, free energy and entropy. Under standard conditions

$$\Delta G^{\ominus} = \Delta H^{\ominus} - T\Delta S^{\ominus}$$

ΔG can, therefore, be negative because of a negative value of ΔH or a positive value of ΔS. The value of ΔH gives some indication about the feasibility of a change because it provides, in many cases, the major contribution to the ΔG value. At 0 K, or for processes with no change in entropy ($\Delta S = 0$), ΔG and ΔH are equal. But at high temperatures and for changes with high increases in entropy, $T\Delta S$ may become just as significant as ΔH.

It is the interplay of ΔH and $T\Delta S$ values that accounts so elegantly for so many chemical facts.

2 Electrical measurement of free energy and entropy changes

When a chemical reaction can be made to take place in a cell, the electrical energy can be measured very easily. Moreover, the cell reaction can be made to take place reversibly (p. 280), at a fixed temperature and pressure, by balancing the e.m.f. of the cell against an external potential difference, using a potentiometer. Under these reversible conditions, the electrical work done is a maximum (p. 280) so that it is equal to $-\Delta G$.

The electrical energy is equal to the flow of charge (in coulombs) multiplied by the e.m.f. (in volts). For a flow of z mol of electrons it follows that

electrical energy $= -\Delta G = zFE$

where F is the Faraday constant, and the electrical energy (E) and free energy (G) are measured in $J\,mol^{-1}$.

Under standard conditions,

$$-\Delta G^{\ominus} = zFE^{\ominus}$$

so that $\Delta G^{\ominus}$ values can be obtained from $E^{\ominus}$ values. As $\Delta H^{\ominus}$ values can be measured by carrying out reactions in calorimeters it is possible to find $\Delta S^{\ominus}$ values.

a The reaction between zinc and M copper(II) sulphate solution The $E^{\ominus}$ value for the $Cu^{2+}(aq)/Cu$ couple is 0.337 V (p. 409) and z is 2. The $\Delta G^{\ominus}$ value is, therefore, $-65\,kJ\,mol^{-1}$. The corresponding values for the $Zn^{2+}(aq)/Zn$ couple are -0.763 V and $+147.2\,kJ\,mol^{-1}$. That is

$$Zn(s) + Cu^{2+}(aq) \rightarrow Zn^{2+}(aq) + Cu(s) \quad E^{\ominus} = 1.1\,V$$
$$\Delta G_m{}^{\ominus}(298\,K) = -212.2\,kJ\,mol^{-1}$$

The maximum heat available when the reaction is carried out in a calorimeter is $216.7\,kJ\,mol^{-1}$, so that $\Delta H_m{}^{\ominus}$ equals $-216.7\,kJ\,mol^{-1}$. It follows that $T\Delta S_m{}^{\ominus}$ equals $-4.5\,kJ\,mol^{-1}$ giving a value for $\Delta S_m{}^{\ominus}$ of $-15.1\,J\,K^{-1}\,mol^{-1}$, i.e. there is a decrease in entropy.

It is not possible in this reaction to convert all the available heat energy into useful work. When the reaction produces 212.2 kJ of electrical work it also produces 4.5 kJ of heat energy. This is transferred to the surroundings as heat causing an increase in the entropy of the surroundings (p. 282).

b The formation of water from hydrogen and oxygen By burning hydrogen in oxygen and by carrying out the reaction in a cell using hydrogen and oxygen electrodes, it is possible to measure the $\Delta H^{\ominus}$ and $\Delta G^{\ominus}$ values as follows:

$$H_2(g) + \tfrac{1}{2}O_2(g) \longrightarrow H_2O(l) \quad \Delta H_m{}^{\ominus}(298\,K) = -286\,kJ\,mol^{-1}$$
$$\Delta G_m{}^{\ominus}(298\,K) = -237\,kJ\,mol^{-1}$$

Once again, it is not possible to convert all the heat energy into useful work. By subtraction, $T\Delta S_m{}^{\ominus}$ equals $-49\,kJ\,mol^{-1}$ giving a value for $\Delta S_m{}^{\ominus}$ of $-164\,J\,K^{-1}\,mol^{-1}$.

3 Calculation of free energy and entropy changes

These calculations can be made, very easily, by using the values of standard free energies of formation and standard entropies obtainable from tables of thermodynamic data.

a Standard molar free energies of formation The standard molar free energy of formation of a compound $(\Delta G_m{}^{\ominus})$ is defined as the *free energy change when 1 mol of the compound is made from its elements in their* *standard states*. Conventionally, the elements in their standard states are taken as having zero free energy. Similar standard free energies of formation for hydrated ions can be obtained on the basis that the value for $H^+(aq)$ is zero.

These values can be used, in the same way as standard enthalpies of formation (p. 254), to obtain $\Delta G^\ominus$ values, for

$$\begin{Bmatrix} \text{standard free energy} \\ \text{change in a reaction} \end{Bmatrix} = \begin{Bmatrix} \text{sum of standard} \\ \text{free energies of} \\ \text{formation of} \\ \text{products} \end{Bmatrix} - \begin{Bmatrix} \text{sum of standard} \\ \text{free energies of} \\ \text{formation of} \\ \text{reagents} \end{Bmatrix}$$

For example,

	$\Delta G_m^\ominus(298\,K)$ $kJ\,mol^{-1}$	$\Delta H_m^\ominus(298\,K)$ $kJ\,mol^{-1}$
$H_2(g) + \tfrac{1}{2}O_2(g) \longrightarrow H_2O(l)$ 0　　　0　　　　　-237	-237	-286
$Zn(s) + Cu^{2+}(aq) \longrightarrow Zn^{2+}(aq) + Cu(s)$ 0　　65　　　　　-147.2　　0	-212.2	-216.7
$NaCl(s) \longrightarrow Na^+(aq) + Cl^-(aq)$ -384　　　　-261.8　　-131.2	-9.0	$+3.9$

The $\Delta G_m^\ominus$ values for all these changes are negative so that they are all feasible, but there are significant differences between the $\Delta G_m^\ominus$ and $\Delta H_m^\ominus$ values arising from the varying entropy changes involved. In the last example, in particular, this leads to a difference in sign between $\Delta G_m^\ominus$ and $\Delta H_m^\ominus$.

b Standard molar entropies Standard molar entropy values at 298 K are based on the convention that the entropy of an element in its stable state at 0 K is zero, or, for hydrated ions, that the standard entropy of $H^+(aq)$ is zero. The values are quoted in $J\,K^{-1}\,mol^{-1}$. They can be used to calculate the entropy change in any process for

$$\begin{Bmatrix} \text{entropy change} \\ \text{in a reaction} \end{Bmatrix} = \begin{Bmatrix} \text{sum of standard} \\ \text{entropies of} \\ \text{products} \end{Bmatrix} - \begin{Bmatrix} \text{sum of standard} \\ \text{entropies of} \\ \text{reagents} \end{Bmatrix}$$

Typical examples are:

$H_2(g) + \tfrac{1}{2}O_2(g) \longrightarrow H_2O(l)$　　　　　$\Delta S_m^\ominus(298\,K) = -163.6\,J$
131　　0.5(205)　　69.9

$Zn(s) + Cu^{2+}(aq) \longrightarrow Zn^{2+}(aq) + Cu(s)$　　$\Delta S_m^\ominus(298\,K) = -16.1\,J$
41.6　　-98.6　　　　-106.4　　33.3

$NaCl(s) \longrightarrow Na^+(aq) + Cl^-(aq)$　　　　　$\Delta S_m^\ominus(298\,K) = 43.1\,J$
72.4　　　　59　　　　56.5

Comparison of these $\Delta S_m^{\ominus}$ values with the $\Delta G_m^{\ominus}$ and $\Delta H_m^{\ominus}$ values in **a** shows that

$$T\Delta S_m^{\ominus} = \Delta H_m^{\ominus} - \Delta G_m^{\ominus}$$

4 Types of change

As both ΔH and ΔS can, theoretically, be zero, negative or positive there are nine theoretical types of change, but only five of them are feasible, i.e. can ever have a negative value for ΔG. These are summarised as follows:

ΔH	ΔS	ΔG	example of change
0	+	always −	the expansion of a gas into a vacuum or the mixing of two gases
−	0	always −	a smooth ball running down a smooth incline
−	+	always −	exothermic reactions with an increase in entropy are always feasible
−	−	− when $\Delta H > T\Delta S$	exothermic reactions with a decrease in entropy will be feasible at low temperatures but will reverse at high temperatures, e.g. $N_2(g) + 3H_2(g) \longrightarrow 2NH_3(g)$ $2SO_2(g) + O_2(g) \longrightarrow 2SO_3(g)$
+	+	− when $T\Delta S > \Delta H$	endothermic reactions will only take place if the temperature is high enough, e.g. $N_2(g) + O_2(g) \longrightarrow 2NO(g)$ $C(s) + H_2O(g) \longrightarrow CO(g) + H_2(g)$

5 Effect of temperature on the feasibility of a reaction

The relationship

$$\Delta G = \Delta H - T\Delta S$$

can be used to calculate the approximate temperature at which the ΔG value will become negative. The result is only approximate as it is assumed that the values of ΔH and ΔS do not change with temperature.

a Decomposition of carbonates The $\Delta H_m^{\ominus}$ and $\Delta S_m^{\ominus}$ values for the thermal decomposition of calcium carbonate are $177.8 \, \text{kJ mol}^{-1}$ and $160.4 \, \text{J K}^{-1} \, \text{mol}^{-1}$ respectively. The free energy change for the reaction will only become negative when $T\Delta S_m$ becomes greater than ΔH_m and the temperature at which this will happen is given by

$$T = \frac{\Delta H_m^{\ominus}(298\text{ K})}{\Delta S_m^{\ominus}(298\text{ K})} = \frac{177.8}{0.1604} = 1108\text{ K}$$

Values for other carbonates (in °C) are as follows:

Li_2CO_3	Na_2CO_3	$MgCO_3$	$CaCO_3$	$SrCO_3$	$BaCO_3$
1157	1863	404	835	1098	1370

b Ellingham diagrams Metallic oxides may be reduced by heating with carbon,

$$2MO(s) + 2C(s) \longrightarrow 2M(s) + 2CO(g) \qquad i$$

so long as a temperature at which the ΔG_m for the reaction is negative can be achieved. The overall reaction i is really a combination of two reactions,

$$2C(s) + O_2(g) \longrightarrow 2CO(g) \qquad\qquad ii$$

$$2M(s) + O_2(g) \longrightarrow 2MO(s) \qquad\qquad iii$$

and the ΔG_m for reaction i will be equal to that for reaction ii minus that for reaction iii.

In reaction ii there is an entropy increase (p. 271) because one mole of gas is converted into two; ΔS_m is, therefore, positive

	ΔH_m/kJ	ΔG_m/kJ	ΔS_m/J K^{-1}
$2C(s) + O_2(g) \longrightarrow 2CO(g)$	-221	-274.6	$+178.2$

As the value of $T\Delta S_m$ increases with rise in temperature the ΔG_m value gets more and more negative at higher temperatures (Fig. 140).

Conversely, the entropy change in reaction iii is negative because a solid is obtained from a gas. For magnesium oxide, for instance,

	ΔH_m/kJ	ΔG_m/kJ	ΔS_m/J K^{-1}
$2Mg(s) + O_2(g) \longrightarrow 2MgO(s)$	-1203.4	-1138.8	-226.7

As a result, the ΔG_m value becomes less negative as temperature rises (Fig. 140). The change of ΔG_m with temperature is not regular because entropy changes at the melting and, particularly, the boiling point of magnesium (650 °C and 1117 °C) cause abrupt changes in the ΔG_m value.

From Fig. 140, which is known as an Ellingham diagram, it will be seen that the ΔG_m value of reaction i will only become negative at a temperature of approximately 1900 °C. Magnesium oxide can, then, be reduced by carbon above that temperature. Below it, the ΔG_m value would be positive and the oxide would not be reduced by carbon. Indeed it would be the magnesium that would reduce the carbon monoxide.

Introduction to Physical Chemistry

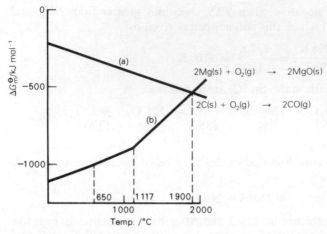

Fig. 140 Change of ΔG_m with temperature for (a) $2C(s) + O_2(g) \rightarrow 2CO(g)$ and (b) $2Mg(s) + O_2(g) \rightarrow 2MgO(s)$.

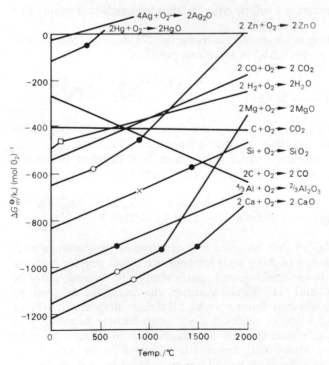

Fig. 141 Ellingham diagrams. ○ signifies m.p. of element, ● b.p. of element, □ b.p. of oxide, × transition temperature.

Data for a number of oxides is plotted on the Ellingham diagrams in Fig. 141 and it will be seen that discontinuities occur at the melting and, particularly, boiling points of the elements or oxides concerned. There may also be discontinuities at transition temperatures.

Ellingham diagrams provide an extremely useful method of summarising a lot of data. The ones given refer to the reduction of oxides; similar ones can also be drawn relating to the reduction of sulphides or chlorides.

6 Free energy change and equilibrium constants

As a feasible reaction is one with a high equilibrium constant, and as the feasibility depends on the $\Delta G_m{}^{\ominus}$ value it is not surprising that there is a relationship between free energy change and equilibrium constant, K (p. 287). This can be shown to be:

$$\Delta G_m{}^{\ominus}(298 \text{ K}) = -RT\ln K$$

where $\Delta G_m{}^{\ominus}$ is the standard free energy change at 298 K in kJ mol^{-1}, R is $8.314 \times 10^{-3} \text{ kJ K}^{-1} \text{ mol}^{-1}$ and T, at 25 °C, is 298. It follows that

$$\lg K = -0.175\Delta G_m{}^{\ominus}(298 \text{ K})$$

and the relationship between $\Delta G_m{}^{\ominus}$ and K and $\lg K$ is

$\Delta G_m{}^{\ominus}(298 \text{ K})$	-100	-57	-10	0	$+10$	$+57$	$+100$
K	3.4×10^{17}	10^{10}	57	1	0.018	10^{-10}	2.95×10^{-18}
$\lg K$	$+17.5$	10	1.75	0	-1.75	-10	-17.5

For practical purposes it is convenient to regard a 'reaction' with an equilibrium constant less than 10^{-10} ($\Delta G_m{}^{\ominus} > 57 \text{ kJ mol}^{-1}$) as one which will not take place, whilst one with an equilibrium constant greater than 10^{10} ($\Delta G_m{}^{\ominus} < -57 \text{ kJ mol}^{-1}$) will go completely. Reactions with intermediate values for the equilibrium constant can be regarded as reversible reactions.

The following examples, using standard molar free energy values, show the power of the method:

$$N_2(g) + 3H_2(g) \longrightarrow 2 NH_3(g) \qquad \Delta G_m{}^{\ominus}(298 \text{ K}) = -33.3 \text{ kJ mol}^{-1}$$
$$0 \qquad\quad 0 \qquad\qquad 2(-16.7)$$

$$\lg K_p = -0.175 \times (-33.3) = 5.828 \qquad\qquad \therefore\ K_p = 6.73 \times 10^5 \text{ atm}^{-2}$$

$$CaCO_3(s) \longrightarrow CaO(s) + CO_2(g) \qquad \Delta G_m{}^{\ominus}(298 \text{ K}) = +130.4 \text{ kJ mol}^{-1}$$
$$(-1129) \qquad\quad (-604.2) \quad (-394.4)$$

$$\lg K_p = -0.175 \times 130.4 = -22.81 \qquad\qquad \therefore\ K_p = 6.46 \times 10^{-22} \text{ atm}$$

The very low value for K_p shows that calcium carbonate is thermally stable at 25 °C.

$$Zn(s) + Cu^{2+}(aq) \longrightarrow Cu(s) + Zn^{2+}(aq) \qquad \Delta G_m^{\ominus}(298 \text{ K}) = -212.2 \text{ kJ}$$
$$0 \qquad (65) \qquad\qquad 0 \qquad (-147.2) \qquad\qquad\qquad\qquad \text{mol}^{-1}$$

$$\lg K_c = -0.175 \times (-212.2) = 37.13 \qquad\qquad \therefore \quad K_c = 1.35 \times 10^{37}$$

The calculation of K values at other temperatures is discussed on page 295.

7 The first law of thermodynamics

Thermodynamics, as its name implies, is concerned with a study of the heat and other energy changes, e.g. mechanical or electrical, which a system can undergo. Initially, thermodynamic considerations were applied to steam engines, and other heat engines, in an attempt to solve the problem of how to get the maximum useful work out of a given amount of heat. Its application to chemical problems was developed in the second half of the nineteenth century, mainly by J. Willard Gibbs.

The first law of thermodynamics is a generalisation from experience, supported, for instance, by the fact that no perpetual motion machine has ever been designed. It is best stated as: '*the total energy of an isolated system remains constant*'. An *isolated* system is one which can neither receive energy from, nor give energy to, its surroundings. A *closed* system is one in which energy can be transferred from or to the surroundings.

Energy changes are brought about by the evolution or absorption of heat, or by work being done on, or by, the system. Work can be done by a chemical reaction, for example, when it expands against the external pressure, when it produces electrical energy in a cell, or when it produces mechanical energy in an engine.

The first law can be summarised as:

$$\begin{array}{l}\text{increase in energy} \\ \text{of a system}\end{array} = \begin{array}{l}\text{heat absorbed by} \\ \text{the system } (q)\end{array} + \begin{array}{l}\text{total work done on} \\ \text{the system } (w)\end{array}$$

It is important to get the signs right (p. 245). Conventionally, heat absorbed is positive, whilst heat evolved is negative. Work done on the system is positive; work done by the system is negative. Using this convention,

$$\text{change in energy } (\Delta U) = q + w$$

If a system absorbs 50 J of heat and does external work equivalent to 100 J the change in energy will be -50 J, i.e. the energy will decrease. If a system evolves 50 J of heat, and work of 75 J is done on it the change in energy will be 25 J, i.e. the energy will increase.

Free energy The total work done on or by a system may be contraction/expansion work (equal to $p\Delta V$ if there is a change in volume of ΔV, p. 249) or net (useful) work, e.g. electrical work. It follows (p. 245) that

change in energy $(\Delta U) = q - p\Delta V + w_{net}$

$$\Delta U + p\Delta V = q + w_{net}$$

For a change at constant volume, $p\Delta V$ will be zero, so that

$$\Delta U = q_v + w_{net} \qquad \text{(at constant volume)}$$

and ΔU will equal q_v when there is no net work (p. 245).

For a change at constant pressure, $\Delta U + p\Delta V$ is equal to ΔH (p. 245) so that

$$\Delta H = q_p + w_{net} \qquad \text{(at constant pressure)}$$

When there is no net work, ΔH and q_p are equal (p. 245).

The maximum amount of net work, obtained when a change is carried out reversibly (p. 280), is equal to the change in free energy.

For reactions at constant pressure*, Gibbs free energy symbolised by G, is used, and

$$\Delta G = w_{net,max} \qquad \text{(at constant pressure)}$$

In a chemical process which does work, $w_{net,max}$ and ΔG will both be negative, and such a process will be feasible.

8 The second law of thermodynamics

The second law of thermodynamics is concerned with the conditions under which heat might be converted into work. It defines the conditions necessary to achieve *maximum* work.

The law originated from a consideration of the conversion of heat into work in heat engines, such as the steam engine. In particular, Carnot, in 1824, made a fundamental contribution by a theoretical consideration of an ideal heat engine in which all the changes could be carried out in a perfectly reversible cycle. The detailed argument is beyond the scope of this book, but Carnot drew the important conclusion that the most efficient cycle of changes in an ideal heat engine was the perfectly reversible cycle. This can be generalised into the conclusion that *the maximum amount of work is obtained from a perfectly reversible process.*

---

*For reactions at constant volume, Helmholtz free energy, symbolised by A (from the German for work, arbeit) is used. ΔA replaces ΔG.

The free energy change in a process is, then, the net work involved when that process is carried out reversibly.

a Reversible and irreversible processes From a thermodynamical point of view, an irreversible process is one in which a very slight alteration of conditions causes a spontaneous and complete change. Typical examples are the flow of water from a higher to a lower level, the flow of heat from a hotter to a colder body, the dissolving of a solute in a solvent, the diffusion of one gas into another or into a vacuum, and the reaction between ammonia and hydrogen chloride. These processes, once started, will continue until completed. They can be reversed, but only by doing work.

By comparison, a reversible change is one which can be made to take place in either direction by a slight change of conditions. The forces acting in both directions are nicely balanced; a slight increase in one direction causes movement in that direction, whilst a slight decrease causes movement in the other direction. A gas in a cylinder fitted with a piston can expand reversibly, for example, when the pressure inside the piston is very slightly greater than the pressure outside.

One of the simplest ways of carrying out a reaction reversibly (if it is possible) is to carry it out in a cell so that it releases electrical energy and sets up an e.m.f. If this e.m.f. is then balanced against an almost equal external potential difference the conditions for reversibility are established. Slight increase in the external potential difference would make the reaction go in one direction; slight decrease, and the cell reaction would reverse.

b The dissipation of energy It follows from the second law of thermodynamics that the maximum amount of energy can never be obtained from an irreversible process. The extra amount of energy which would be converted into work, if the process were reversible, is said to be dissipated, and every irreversible process leads to some dissipation of energy of this sort. No energy is lost (that would be contrary to the first law of thermodynamics), but it appears as heat and not as work.

The energy available from a waterfall, for example, goes partially into raising the temperature of the water and cannot, under any circumstances, be converted completely into useful work. Similarly, the energy available from a definite amount of petrol is turned completely into heat if the petrol is burnt, in a irreversible process, in a closed container. If the same amount of petrol were burnt in the cylinder of a motor car, some useful work would be obtained. If the process taking place in the cylinder of a motor car could be regarded as reversible (which it is not) then the useful work would be a maximum.

As most natural processes are invariably irreversible, the second law of thermodynamics can be summarised in the statement that *there is a tendency in nature for energy to be dissipated in the form of heat* (see also p. 282).

9 More about entropy

Entropy has been related to disorderliness of energy distribution (p. 271). The total energy in a system is made up of translational, vibrational, rotational and electronic energy, and the various types are quantised (p. 176). For any total amount of energy, then, there are many ways of distributing it amongst the various available levels. The number of ways will depend on the total amount of energy and the number of levels.

In a simple, hypothetical example, three molecules with a total of four units of energy can be distributed between five energy levels as shown in Fig. 142. The molecules in (a), (b) and (c) can be arranged in three ways, and those in (d) in six ways, making a total of fifteen. Three molecules with a total of two units of energy can be distributed between three levels in only six ways.

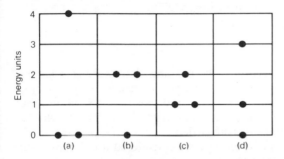

Fig. 142 Distribution of three molecules between five energy levels.

An analogy is provided by an imaginary theatre with four types of seating. One type cost £10, £20, £30, etc.; another cost £1, £2, £3, etc.; the third cost 10p, 20p, 30p etc.; and the fourth cost 1p, 2p, 3p, etc. If only 3p is taken at the box office (the total energy), only the penny seats could be occupied, and they could be occupied in three different ways. There could be one person in a 3p seat, three people in 1p seats, or one person in a 2p seat and one person in a 1p seat. If £50 is taken, all the types of seat could be occupied in a very large number of different ways. The entropy would be much higher.

When there is no energy there can be no entropy, and this is so in a crystalline solid at 0 K. At room temperature the solids have some vibrational and rotational energy, but only very little translational energy. Gases, by comparison, have translational, vibrational and rotational energy so that their entropies are higher. A complicated molecule, or a mixture of molecules, will have more energy levels than a simple molecule.

a Measure of entropy change Input of energy leads to an increase in entropy, and, for a reversible change at a fixed temperature, the increase in entropy is equal to the heat absorbed divided by the temperature. For an evolution of heat, there is a decrease in entropy. In general

$$\Delta S = \frac{\text{heat absorbed or evolved}}{\text{temperature/K}} = \frac{q}{T}$$

The simplest examples of reversible, isothermal changes are those of melting and vaporisation. The molar enthalpy of fusion of ice, for example, is $6.025\,\text{kJ}\,\text{mol}^{-1}$, giving an entropy increase of 6025/273, i.e. $22.07\,\text{J}\,\text{K}^{-1}\,\text{mol}^{-1}$ for the change from ice to water at $0\,°\text{C}$. The corresponding value for the change from water to steam at $100\,°\text{C}$ is $109\,\text{J}\,\text{K}^{-1}\,\text{mol}^{-1}$. The larger value reflects the fact that a liquid is more disorderly than a solid, but that a gas is very much more disorderly than a liquid.

b The entropy of the universe tends to a maximum An absorption of heat, q, at a temperature, T, causes an increase in entropy of q/T. For a reversible reaction, at a constant pressure (p. 279):

$$\Delta H = q_p + \Delta G$$

so that $\quad \Delta G = \Delta H - T\Delta S$

This relationship can be interpreted as meaning that *there is a natural tendency in spontaneous processes towards minimum enthalpy and maximum entropy*, for it is negative ΔH and positive ΔS values that favour negative ΔG values. That is only true, however, if what is happening in the surroundings is disregarded.

If both a system and its surroundings are considered a different result arises. An enthalpy change of ΔH_{sys} for the system, causes a corresponding enthalpy change for the surroundings of $-\Delta H_{sys}$ and this gives an entropy change for the surroundings, ΔS_{sur}, equal to $-\Delta H_{sys}/T$. As

$$\Delta G_{sys}/T = \Delta H_{sys}/T - \Delta S_{sys}$$

it follows that

$$\Delta G_{sys}/T = -\Delta S_{sur} - \Delta S_{sys} = -(\text{total entropy change})$$

It is, therefore, only when the total entropy change is positive, i.e. when the total entropy increases, that ΔG will be negative and the process will be feasible.

The second law of thermodyanmics can, therefore, be expressed as: '*the total entropy must increase in any spontaneous process*'. Both the first and second laws are described in Clausius's pithy statement: '*the energy of the universe is constant; the entropy tends always to a maximum*'.

c Entropy and probability Spontaneous changes occur when there is an increase in total entropy, which can be thought of as an increase in total disorder; they also occur when there is an increase in probability. It is the most probable changes that take place spontaneously. Dealing a hand of thirteen spades, the unmixing of two gases in a container, or the rising of this book from the table are not statistically impossible events;

they are simply highly unlikely. They all demand an increase in order. The failure of the book to rise, for example, is due to the fact that its energy is highly disordered or randomised or dispersed; it has high entropy. It is only if all the molecules of which the book is made happened to move upwards at the same moment of time that the book would rise, and the probability of this is minute.

It is not surprising, then, that there is a relationship between entropy and probability and disorder. The mathematical expression of this is

$$S = k \ln W$$

Where S is the entropy, k is Boltzmann's constant, and W is the number of ways in which the total energy can be distributed between the available energy levels (the probability). For a change from a system with a probability of W_1 to one of W_2

$$\Delta S = k \ln W_2/W_1$$

and ΔS will only be positive when W_2 is greater than W_1. Entropy will only increase if the number of ways of distributing energy increases.

Entropy values can be calculated from probability values, at least in some simple cases; this is done in statistical mechanics.

Questions on Chapter 24

1 Account for the following (a) the entropy of diamond is smaller than that of graphite (b) the entropy of a polyatomic molecule is greater than that of a monatomic molecule (c) the entropy of steam is greater than that of water at 100 °C.

2 The maximum e.m.f. of a Daniell cell in which the reaction,

$$Zn(s) + Cu^{2+}(aq) \longrightarrow Zn^{2+}(aq) + Cu(s) \qquad \Delta H_m = -210 \, kJ \, mol^{-1}$$

takes place is 1.107 V at 25 °C. Calculate the entropy change.

3 Explain what is meant by the terms (a) internal energy (b) free energy (c) unavailable energy (d) maximum work (e) net work.

4 Taking the specific latent heat of evaporation of water as 2259 J g^{-1}, calculate the molar entropy change at 100 °C for the change from water to steam.

5 The evaporation of water and the solution of ammonium chloride in water both take place spontaneously at room temperature. Both processes absorb heat. Why, then, do they take place? Under what conditions can they be reversed?

6 Explain why it is that most reactions which take place readily at room temperatures are exothermic, whilst endothermic reactions are favoured at temperatures of about 3000 °C.

7 Explain why the denser form of an element which exhibits allotropy is expected to have smaller heat of combustion.

8 In the processes of melting, evaporation and dissolution the large gain in entropy on the formation of the more disordered state offsets the positive sign of ΔH. What does this mean?

9 To what extent are the tables of electrode potentials also tables of free energy values?

10 Taking the standard electrode potential of zinc as -0.761 V, calculate the free energy change in the reaction of zinc with hydrochloric acid. Would you expect the enthalpy change to be bigger or smaller than the free energy change?

11 In any system undergoing a spontaneous change there must be an increase in entropy. Discuss that statement.

12 Explain the terms enthalpy change and entropy change, illustrating your answer with suitable examples. Show how, under constant conditions of temperature, enthalpy change and entropy change influence the direction of chemical change. For liquid methanol in equilibrium with methanol vapour at 338 K, the boiling point of methanol,

$$CH_3OH(l) \longrightarrow CH_3OH(g) \qquad \Delta H_m = +35.3 \, kJ \, mol^{-1}$$

Calculate the entropy change when methanol is evaporated. (LS)

13 Explain what is meant by 'enthalpy change of reaction' and 'entropy change of a reaction'. Discuss, with examples, how these two factors, separately and together, enable the direction of a chemical change to be predicted. What other factor or factors determine whether or not a given change will actually take place? (LS)

14* Use values of standard entropies and enthalpies of formation to calculate the free energy changes in the following reactions.

$$C(graphite) + \tfrac{1}{2}O_2(g) \longrightarrow CO(g) \qquad CO(g) + \tfrac{1}{2}O_2(g) \longrightarrow CO_2(g)$$
$$C_2H_4(g) + H_2(g) \longrightarrow C_2H_6(g) \qquad NO(g) + \tfrac{1}{2}O_2(g) \longrightarrow NO_2(g)$$

Compare the values you obtain with the quoted values for free energies of formation.

15* Calculate the equilibrium constant at 25 °C and 101.325 kPa for the change from butane to 2-methylpropane. What will be the partial pressure of butane in the equilibrium mixture?

16* Record the standard enthalpies of formation and the standard free energies of formation for CH_4, C_2H_6, C_3H_8, C_4H_{10}, C_5H_{12} and C_6H_{14}. Comment on the figures.

17* List the standard enthalpies and free energies of formation of the hydrogen halides. Comment on the figures.

18* Calculate the equilibrium constants, at 25 °C, for the following changes.

(a) The conversion of MnO_4^- into Mn^{2+} in acid solution.

(b) $Fe^{2+}(aq) + Ag^+(aq) \longrightarrow Fe^{3+}(aq) + Ag(s)$

(c) $H^+(aq) + OH^-(aq) \longrightarrow H_2O(l)$

(d) $Fe^{3+}(aq) + Ag(s) \longrightarrow Ag^+(aq) + Fe^{2+}(aq)$

(e) $CaCO_3(s) \longrightarrow CaO(s) + CO_2(g)$

19* ΔG for a change only becomes zero whan $T\Delta S$ balances ΔH. Estimate the temperature at which this happens for the following reactions. Assume that ΔH and ΔS do not change with temperature:

(a) $MgCO_3(s) \longrightarrow MgO(s) + CO_2(g)$

(b) $CaCO_3(s) \longrightarrow CaO(s) + CO_2(g)$

(c) $N_2(g) + 3H_2(g) \longrightarrow 2NH_3(g)$

(d) $SiO_2(s) + 2C(s) \longrightarrow Si(s) + 2CO(g)$

20* What are Ellingham diagrams? Quote some figures from them to illustrate the processes used in the manufacture of iron and aluminium.

21* Record some standard entropy values for chemicals in different states. What general conclusion can you draw?

25
Chemical equilibrium

1 Introduction

An equilibrium exists when two opposing forces or rates balance each other out. In everyday life such equilibria exist in a balanced seesaw or when a man walks at the right speed up a downward-moving escalator. Similar situations are widespread in chemistry.

a Phase equilibria In a saturated solution containing some undissolved solid, X, the system is in a state of equilibrium at a given temperature, because the rate at which solid X is dissolving is just balanced by the rate at which crystallisation from the solution is taking place. The equilibrium appears to be static; once it has been established no bulk or macroscopic changes seem to take place. But the equilibrium on a microscopic scale is *dynamic*, with counterbalancing processes still taking place at equal rates. This can readily be shown by adding some radioactive solid X to the mixture; radioactivity soon shows up in the solution. Or add some radioactive saturated solution and radioactivity will soon show up in the solid.

Such an equilibrium is known as a phase equilibrium. Other examples are represented by the equilibria existing between a gas and its saturated solution (p. 206), a liquid and its vapour (p. 215), a solute partitioned between two immiscible solvents (p. 199) and mixtures of two liquids (p. 218).

b Chemical equilibria Many reactions do not go to completion and then stop, for the products of the reaction themselves react to form the original reactants. The reaction may, then, provide a mixture of reactants and products, and a dynamic equilibrium will be established when the reactants and products are both reacting at the same rates.

If the products preponderate greatly in the equilibrium mixture, the reaction may be regarded, for practical purposes, as going completely. If the reactants preponderate, the reaction has hardly taken place at all. When reactants and products are both present in the equilibrium mixture in similar amounts the reaction is said to be *reversible*.

A knowledge of the composition of the equilibrium mixture resulting from any chemical reaction is important in determining *how far* the reaction actually goes, and how much product can be formed. This is related to the enthalpy, and other energy, changes associated with the reaction and is the province of *chemical thermodynamics* (p. 277).

Chemical kinetics (p. 303) is concerned with *how fast* the reaction is, and the detailed atomic rearrangements occurring in a reaction are studied under the general heading of *reaction mechanisms* (p. 318).

c Types of chemical equilibria When all the reactions and products of a reaction are all in the same phase, i.e. all gases, all liquids or all in solution, the equilibrium is known as *homogeneous equilibrium*, e.g.

$$H_2(g) + I_2(g) \rightleftharpoons 2HI(g) \qquad\qquad N_2O_4(g) \rightleftharpoons 2NO_2(g)$$

$$CH_3COOH(l) + C_2H_5OH(l) \rightleftharpoons CH_3COOC_2H_5(l) + H_2O(l)$$

$$I_2(aq) + I^-(aq) \rightleftharpoons I_3^-(aq)$$

If two or more phases are involved the equilibrium is said to be *heterogeneous*, e.g.

$$CaCO_3(s) \rightleftharpoons CaO(s) + CO_2(g)$$

$$3Fe(s) + 4H_2O(g) \rightleftharpoons Fe_3O_4(s) + 4H_2(g)$$

$$BiCl_3(aq) + H_2O(l) \rightleftharpoons BiOCl(s) + 2HCl(aq)$$

If ions are involved in an equilibrium it is generally referred to as *ionic equilibria* (p. 356).

2 The equilibrium law

It can be shown, experimentally (p. 288) and by a thermodynamic argument, that for a homogeneous equilibrium represented as

$$xA + yB \rightleftharpoons wC + zD$$

$$\frac{[C]_{eq}^w[D]_{eq}^z}{[A]_{eq}^x[B]_{eq}^y} = \text{a constant at a given temperature} = K_c$$

This is the mathematical expression of the equilibrium law, and K_c is known as the *equilibrium constant*.

[A] is a conventional way of writing the concentration* of A in $mol\,dm^{-3}$. It must be emphasised that the concentrations used shown as $[A]_{eq}$ are those existing in the *equilibrium mixture*; they are not the initial concentrations before the reaction has taken place. It is conventional to put the products of the reaction in the numerator so that high values for K_c mean that the products are present in higher concentrations than the reactants. In simple language, the reaction has 'gone well', though the value of K_c gives *no indication of the rate* of any reaction.

For reactions involving gases it is often more convenient to measure and express the gas concentrations as partial pressures (p. 10) in atmospheres. If this is done the equilibrium constant is known as K_p. For example,

$$aX(g) + bY(g) \rightleftharpoons rC(g)$$

$$\frac{(p_C)_{eq}^r}{(p_X)_{eq}^a(p_Y)_{eq}^b} = \text{a constant at a given temperature} = K_p$$

---

* Ideally the activity (p. 356) should be used, but in simple cases this can be taken as approximately equal to the concentration in $mol\,dm^{-3}$.

3 The units of K_c and K_p

The definitions of K_c and K_p, given in section 2, show that both values will be dimensionless (without units) if the total number of moles of reactants equals the total number of moles of products, i.e. if $(x + y) = (w + z)$, or $(a + b) = r$. In these cases, the concentration or pressure terms in the numerator and denominator exactly cancel out.

Otherwise, K_c will have units of $(mol\,dm^{-3})^n$, and K_p units of $(atmosphere)^n$, where n is equal to the total number of moles of products minus the total number of moles of reactants.

For example,

equilibrium	K_p expression	units of K_p
$H_2(g) + I_2(g) \rightleftharpoons 2HI(g)$	$\dfrac{(p_{HI})^2_{eq}}{(p_{H_2})_{eq}(p_{I_2})_{eq}}$	no units
$N_2(g) + 3H_2(g) \rightleftharpoons 2NH_3(g)$	$\dfrac{(p_{NH_3})^2_{eq}}{(p_{N_2})_{eq}(p_{H_2})^3_{eq}}$	atm^{-2}
$N_2O_4(g) \rightleftharpoons 2NO_2(g)$	$\dfrac{(p_{NO_2})^2_{eq}}{(p_{N_2O_4})_{eq}}$	atm

A slightly different form of equilibrium constant, giving apparently dimensionless values of K_c and K_p, is, however, sometimes used. Each pressure term, in the K_p expression, is taken as the *ratio* of the actual pressure, p atm, to an arbitrarily chosen standard pressure, $p^\ominus$, of 1 atmosphere. The dimensionless ratio, $p/p^\ominus$, is written into the K_p expression instead of the actual value of p in atmospheres. Similarly, when K_c is involved, a ratio of the actual concentration relative to a chosen standard of $1\,mol\,dm^{-3}$ is used.

The resulting values of K_c and K_p are dimensionless, but particular units have, in fact, been specified in the choice of the standard states used. It is the dimensionless form of equilibrium constant that is referred to in the relationship $\Delta G_m^\ominus(298\,K) = -RT\ln K$ (p. 277), and the use of a dimensionless K avoids the embarrassment of taking the logarithm of a quantity with units.

Investigation of Equilibrium Mixtures

To investigate an equilibrium mixture and measure K_c or K_p values, it is necessary to establish the equilibrium at a particular temperature and then find the composition of the mixture by some method which does not change the composition. Alternatively, it is possible to obtain the values by electrical methods for reactions that can be carried out in cells (p. 272).

4 Reaction between an acid and an alcohol

Acids and alcohols react to form esters and water, e.g.

$$CH_3COOH(l) + C_2H_5OH(l) \rightleftharpoons CH_3COOC_2H_5(l) + H_2O(l)$$

To find the equilibrium constant for such a reaction,

$$K_c = \frac{[CH_3COOC_2H_5]_{eq}[H_2O]_{eq}}{[CH_3COOH]_{eq}[C_2H_5OH]_{eq}}$$

it is necessary to measure the concentrations of the substances present *in the equilibrium mixture* at a particular temperature.

This is done by sealing known masses of ethanoic acid (a mol) and ethanol (b mol) in a glass tube at a definite temperature, say 50 °C. After some hours the acid and the alcohol will have reacted to form the equilibrium mixture. The tube is then cooled, and the amount of acid it contains is found by titration with standard alkali. The purpose of the cooling is to 'freeze' the equilibrium. It is essential that the equilibrium should not shift whilst the equilibrium concentrations are being measured, and 'freezing' the equilibrium means cooling to a sufficiently low temperature that the equilibrium will not shift appreciably during the measurement of the concentrations.

If there are x mol of acid present at equilibrium, it follows that $(a - x)$ mol of acid have been converted into ester and water. Since 1 mol of ester and 1 mol of water are formed from 1 mol of acid, there must be $(a - x)$ mol of ester and of water present in the equilibrium mixture. Moreover $(a - x)$ mol of alcohol must have been used up, so that the amount of alcohol present at equilibrium will be $(b - (a - x))$ mol.

For a sealed tube of volume V dm^3 the equilibrium concentrations, in mol dm^{-3}, would be as follows,

acid	+	alcohol		ester	+	water
$\dfrac{x}{V}$		$\dfrac{b - (a - x)}{V}$		$\dfrac{a - x}{V}$		$\dfrac{a - x}{V}$

and the equilibrium constant will be given by

$$K_c = \frac{\left(\dfrac{a - x}{V}\right)\left(\dfrac{a - x}{V}\right)}{\left(\dfrac{x}{V}\right)\left(\dfrac{b - (a - x)}{V}\right)} = \frac{(a - x)^2}{x(b - (a - x))}$$

In this reaction, with an equal number of molecules on each side of the equation, the volume of the container cancels out in the equilibrium expression and need not be known. For a reaction in which the number of molecules involved changes, the values of V must be known (see p. 295).

The equilibrium can also be established by starting with known amounts of ester and water. This takes a week or so even in the presence of a measured amount of concentrated hydrochloric acid to act as a catalyst. When equilibrium is established, the extra acid (the ethanoic acid) in the mixture can be measured by titration with a standard alkali.

5 Reaction between hydrogen and iodine

Hydrogen and iodine react, reversibly, to form hydrogen iodide:

$$H_2(g) + I_2(g) \rightleftharpoons 2HI(g)$$

To find the equilibrium constant, known masses of hydrogen and iodine, or a known mass of hydrogen iodide, are kept in a sealed tube at, say, $450°C$, until the equilibrium is established. The equilibrium mixture is then 'frozen' by rapid cooling; rapid so that the equilibrium has not time to shift and adjust itself to the equilibrium which would, given time, exist at the lower temperature. The amount of iodine in the equilibrium mixture at $450°C$ is then measured by titration with standard sodium thiosulphate solution.

If the original amounts of hydrogen and iodine were a mol and b mol respectively, and if x mol of iodine is converted into hydrogen iodide, then $2x$ mol of hydrogen iodide will be formed, and the equilibrium concentrations in $mol\,dm^{-3}$, in a vessel of volume $V\,dm^3$, will be

hydrogen, $\dfrac{a-x}{V}$ iodine, $\dfrac{b-x}{V}$ hydrogen iodide, $\dfrac{2x}{V}$

The equilibrium constant will be given by,

$$K_c = \frac{[HI]_{eq}^2}{[H_2]_{eq}[I_2]_{eq}} = \frac{\left(\dfrac{2x}{V}\right)^2}{\left(\dfrac{a-x}{V}\right)\left(\dfrac{b-x}{V}\right)} = \frac{4x^2}{(a-x)(b-x)}$$

V, once again, cancelling out.

6 Use of flow method

The equilibrium mixture resulting from gas reactions can be obtained by passing the reactant gases through a tube, kept at a steady temperature, at such a rate that equilibrium is established. The equilibrium gas mixture coming out of the tube can then be 'frozen', and analysed. If necessary the attainment of equilibrium within the hot region of the tube can be expedited by including a catalyst.

Such a flow method was used, for example, by Haber in his investigation of the formation of ammonia from nitrogen and hydrogen.

7 The reaction $I_2(aq) + I^-(aq) \rightleftharpoons I_3^-(aq)$

This is typical of a reversible reaction whose equilibrium constant is measured by a partition method. The equilibrium constant is given by

$$K_c = \frac{[I_3^-(aq)]_{eq}}{[I_2(aq)]_{eq}[I^-(aq)]_{eq}}$$

and it is necessary to measure, or calculate, the various concentrations in an equilibrium mixture.

This can be done by dissolving iodine and a known mass of potassium iodide, i.e. I^- ions, in a known volume of water and adding some tri- or tetra-chloromethane or benzene. Free iodine molecules distribute themselves between the aqueous and non-aqueous solvents, but I^- and I_3^- do not do this as the ions are insoluble in the non-aqueous solvent.

The concentrations of iodine in the non-aqueous layer can be measured by titrating a portion of it with a standard sodium thiosulphate solution. $[I_2(aq)]$ in the aqueous layer can be calculated from this value if the partition coefficient of iodine between water and the non-aqueous solvent is known, or is separately measured.

Titration of the aqueous layer gives the total iodine concentration in that layer. This is equal to $[I_2(aq)] + [I_3^-(aq)]$. As $[I_2(aq)]$ is known, $[I_3^-(aq)]$ can be found.

The total iodide ion concentration in the aqueous layer can be calculated from the original mass of potassium iodide added. This total iodide ion concentration is equal to $[I^-(aq)]$ plus $[I_3^-(aq)]$. As $[I_3^-(aq)]$ is known, $[I^-(aq)]$ can be found.

The three equilibrium concentrations, $[I_2(aq)]$, $[I_3^-(aq)]$ and $[I^-(aq)]$ give the value for K_c, and the same value will be obtained starting with different amounts of iodine, potassium iodide and water.

8 Measurement of gas pressure

For equilibrium mixtures in which the number of gaseous molecules changes as the equilibrium shifts, it is possible to measure equilibrium constants by measuring the pressure exerted by the equilibrium mixture. This is particularly convenient when it is required to measure equilibrium constants at different temperatures. The equilibrium constant for the dissociation of iodine molecules into iodine atoms provides a simple example.

If 1 mol of iodine is taken, in a container of volume, V, the pressure, p, exerted at a temperature, T, at which the iodine is in the vapour state, will be given by RT/V if the iodine did not dissociate. As the iodine does, in fact, dissociate, the measured pressure, P, at T, when the equilibrium has been established, will be equal to $p(1 + \alpha)$, where α is the degree of dissociation. This is because the dissociation gives an increased number of particles.

$$I_2(g) \rightleftharpoons 2I(g)$$
$$(1 - \alpha) \qquad 2\alpha$$

The total number of particles becomes $(1 + \alpha)$, so that the partial pressure of I_2 molecules will be $P(1 - \alpha)/(1 + \alpha)$ whilst the partial pressure of I atoms will be $P2\alpha/(1 + \alpha)$. The equilibrium constant is, therefore,

$$K_p = \frac{(p_I)^2_{eq}}{(p_{I_2})_{eq}} = \frac{(P2\alpha)^2(1 + \alpha)}{(1 + \alpha)^2(1 - \alpha)P} = \frac{4\alpha^2 P}{1 - \alpha^2} = \frac{4\alpha^2 p}{1 - \alpha}$$

9 Heterogeneous reactions

The application of the law of equilibrium to heterogeneous reactions, such as those between gases and solids, is difficult because of the problem of interpreting the meaning of the concentration (activity) of a solid. In some heterogeneous reactions, the law cannot be applied, as the surface area and nature of the solid seem to be the predominant factors. There are, however, some selected heterogeneous reactions for which the experimental results can be related to the law of equilibrium by assuming that the concentrations of the solids involved remain constant throughout the reaction.

a Thermal dissociation of calcium carbonate Solid calcium carbonate dissociates on heating,

$$CaCO_3(s) \rightleftharpoons CaO(s) + CO_2(g)$$

At any fixed temperature

$$K_c = \frac{[CaO(s)]_{eq}[CO_2(g)]_{eq}}{[CaCO_3(s)]_{eq}}$$

and, if $[CaO(s)]_{eq}$ and $[CaCO_3(s)]_{eq}$ are both taken as being constant,

$$[CO_2(g)]_{eq} = \text{a constant} \quad \text{and} \quad (p_{CO_2})_{eq} = \text{a constant}.$$

The pressure above hot calcium carbonate, which can be measured on a manometer, is therefore constant at any given temperature. It is sometimes referred to as the *dissociation pressure*. Its value increases with temperature because K_c does.

At 900 °C, the dissociation pressure is 101 kPa. If carbon dioxide is removed from an equilibrium mixture at that temperature the calcium carbonate will decompose to form more carbon dioxide to maintain the equilibrium pressure. That is why calcium carbonate begins to decompose quite rapidly when heated in an open container above 900 °C. The rate of dissociation is increased if the carbon dioxide is removed by a current of air. If carbon dioxide is pumped into an equilibrium mixture at 900 °C it will react with calcium oxide to form more carbonate to re-establish the equilibrium pressure.

b Reaction of steam and iron

$$3Fe(s) + 4H_2O(g) \rightleftharpoons Fe_3O_4(s) + 4H_2(g)$$

If the concentrations of the solids concerned are taken as constant

$$K_c = \frac{[H_2(g)]^4}{[H_2O(g)]^4} \quad \text{or} \quad K_c{}' = \frac{[H_2(g)]}{[H_2O(g)]} \quad \text{or} \quad K_p = \frac{(p_{H_2})^4}{(p_{H_2O})^4}$$

Such relationships can be tested by measuring the partial pressures of hydrogen and steam in equilibrium with different mixtures of iron and iron oxide. It is found that the ratio p_{H_2}/p_{H_2O} is constant at any one temperature and independent of the amounts of iron and oxide present.

If hydrogen is added to an equilibrium mixture it reduces the oxide and forms steam to re-establish the p_{H_2}/p_{H_2O} ratio. Similarly, addition of more steam produces more hydrogen.

Effects of change of conditions on equilibrium mixtures and constants

10 The effect of concentration change

For any reaction

$$xA + yB \rightleftharpoons wC + zD$$

$$K_c = \frac{[C]_{eq}^w [D]_{eq}^z}{[A]_{eq}^x [B]_{eq}^y}$$

and K_c is constant at any fixed temperature. If either A or B are added to the equilibrium mixture, more C and D will be formed so that the new concentration terms in the equilibrium expression will still equal K_c. The equilibrium is said to shift from 'left to right'. Addition of more C or D causes a shift from 'right to left'. Such changes can be seen if reactions with visible changes are used.

a Hydrolysis of bismuth(III) chloride A clear, colourless solution of bismuth(III) chloride is hydrolysed by water to form white, insoluble bismuth(III) chloride oxide,

$$BiCl_3(aq) + H_2O(l) \rightleftharpoons BiOCl(s) + 2HCl(aq)$$

Addition of water to the colourless solution produces more bismuth(III) chloride oxide and the mixture becomes opalescent; the equilibrium shifts from left to right. Addition of acid makes the mixture clear again as the equilibrium shifts from right to left.

b The colour change of an indicator (p. 368) Many indicators are complex, weak acids, generalised as HIn, in which the HIn molecules and the

293

In$^-$ ions have different colours, e.g.

$$\text{HIn} + \text{H}_2\text{O} \rightleftharpoons \text{H}_3\text{O}^+ + \text{In}^-$$
(one colour)   (different colour)

In any solution, an equilibrium mixture containing some HIn and some In$^-$ exists. Addition of acid (H$_3$O$^+$ ions) converts In$^-$ into HIn, and addition of alkali (removal of H$_3$O$^+$ ions) converts HIn into In$^-$, with corresponding colour changes.

c Acidification of potassium chromate Yellow solutions of potassium chromate turn orange on adding dilute acids as the CrO_4^{2-} ions are converted into orange $Cr_2O_7^{2-}$ ions. Adding of alkali (removal of H$^+$ ions) restores the yellow colour.

$$2CrO_4^{2-}(aq) + 2H^+(aq) \rightleftharpoons Cr_2O_7^{2-}(aq) + H_2O(l)$$
(yellow)   (orange)

d The formation of Fe(CNS)$^{2+}$ This deep red ion is formed on mixing Fe^{3+} and CNS$^-$ ions, e.g.

$$Fe^{3+}(aq) + NH_4CNS(aq) \rightleftharpoons Fe(CNS)^{2+}(aq) + NH_4^+(aq)$$

The reaction is reversible and the equilibrium position can be estimated by the intensity of the red colour of the Fe(CNS)$^{2+}$ ions. Addition of Fe^{3+} or CNS$^-$ ions increases the colour; addition of NH$_4^+$ ions decreases it.

11 Effect of catalysts

A catalyst does not alter the compostion of an equilibrium mixture and has no effect on the value of an equilibrium constant. A catalyst may, however, speed up both forward and back reactions in a reversible reaction, to the same extent, so that the same equilibrium mixture is obtained more rapidly in the presence of a catalyst.

If a catalyst did affect an equilibrium constant it would defy the law of conservation of energy. A catalyst, for example, which speeded up the change from A to B more than the reversed change from B to A could be used to obtain energy from nowhere. For, in a reversible reaction

$$A \rightleftharpoons B + x \text{ joule}$$

addition of the catalyst would cause a shift in the equilibrium from left to right. An equilibrium mixture of A and B without catalyst could, then, be made to evolve heat simply by adding the catalyst.

12 Effect of pressure

A change of pressure does not change equilibrium constants but it does change the composition of the equilibrium mixture in any reaction in which there is a volume change.

Consider a reaction in which there is an increase in volume, e.g.

$$N_2O_4(g) \rightleftharpoons 2NO_2(g)$$

no. of moles at equilibrium	a mol	b mol
partial pressures at equilibrium	$\dfrac{ap}{(a+b)}$	$\dfrac{bp}{(a+b)}$

and let there be a mol of N_2O_4 and b mol of NO_2 in the equilibrium mixture, i.e. a total of $(a+b)$ mol. If the total pressure at equilibrium is p, then the partial pressures will be as shown above and

$$K_p = \frac{(p_{NO_2})^2_{eq}}{(p_{N_2O_4})_{eq}} = \frac{b^2 p^2(a+b)}{(a+b)^2 ap} = \frac{b^2 p}{a(a+b)}$$

As K_p is constant, an increase in pressure will lead to a decrease in b and an increase in a; the equilibrium will shift from right to left. As NO_2 is brown and N_2O_4 almost colourless such a shift can be observed by compressing an N_2O_4/NO_2 mixture in a stout tube with a rubber plunger; the intensity of the colour decreases.

For a reaction in which there is a decrease in volume, e.g.

$$N_2(g) + 3H_2(g) \rightleftharpoons 2NH_3(g)$$

the effect of pressure is reversed. The higher the pressure the more ammonia there will be in the equilibrium mixture and vice versa.

In a reaction where the volume of the products is the same as that of the reactants, pressure has no effect on the composition of the equilibrium mixture, though increase in pressure will enable the equilibrium mixture to be attained more quickly in a gaseous reaction.

13 Effect of temperature

In a reversible reaction, change of temperature will affect the rates of both forward and back reactions, but not necessarily to the same extent. Change of temperature, therefore, causes a shift of equilibrium and a change in equilibrium constant. The change depends on whether exothermic or endothermic reactions are concerned.

The way in which equilibrium constants change with temperature is found, both theoretically and experimentally (p. 319), to be governed by the relationship

$$\frac{d(\ln K)}{dT} = \frac{\Delta H_m}{RT^2}$$

where ΔH_m is the molar enthalpy change for the reaction. On integrating, the expression gives

$$\ln K = -\Delta H_m/RT + c$$

where c is a constant, if it is assumed that ΔH_m does not change with temperature.

For a reversible reaction which is exothermic in the forward direction, e.g.

$$N_2(g) + 3H(g) \rightleftharpoons 2NH_3(g) \qquad \Delta H_m = -92.37 \, kJ$$

the value of ΔH_m is negative. The value of $-\Delta H_m/RT$ will, therefore, be positive, and the greater T is the smaller will be its value. K will, therefore, be smaller the greater T is and less ammonia will be found in the equilibrium mixture the higher the temperature is. In other words, increase in temperature shifts the equilibrium from right to left. Conversely a lowering of temperature will shift the equilibrium from left to right to give an equilibrium mixture containing more ammonia.

For a reversible reaction which is endothermic in the forward direction, e.g.

$$N_2(g) + O_2(g) \rightleftharpoons 2NO(g) \qquad \Delta H_m = +179.9 \, kJ$$

the value of ΔH_m will be positive, and the value of $-\Delta H_m/RT$ will be negative. Increase in T will make $-\Delta H_m/RT$ smaller, so that K will be bigger. Increase in T therefore causes a shift of equilibrium from left to right. Decrease in T causes a shift from right to left.

The expression relating ΔH_m and T can also be written as:

$$\ln K_2 - \ln K_1 = -\frac{\Delta H_m}{R}\left(\frac{1}{T_2} - \frac{1}{T_1}\right)$$

or

$$\lg K_2 - \lg K_1 = -\frac{\Delta H_m}{2.303R}\left(\frac{1}{T_2} - \frac{1}{T_1}\right)$$

Values of ΔH_m can be calculated if K values are known at any two temperatures. Alternatively, values of K at one temperature can be converted into values at another.

14 Le Chatelier's Principle

Many of the conclusions reached in the preceding sections can be deduced from the widely applicable principle of Le Chatelier. This states that '*a system in equilibrium, subjected to any change, will shift the equilibrium, if possible, in the direction which causes an opposite change*'.

Increasing the concentration of any reagent in an equilibrium mixture shifts the equilibrium in the direction which converts some of that reagent into other products.

Increasing the pressure on a system in equilibrium causes the equilibrium to shift in the direction which will bring about a lowering of pressure. In a chemical reaction, this means that the equilibrium will shift in the direction which produces the smaller number of gas molecules, i.e. smaller volume.

Increasing the temperature of a system in equilibrium causes a shift in the direction which absorbs heat; that lowers the temperature. Decreasing the temperature will shift the equilibrium in the direction evolving heat; that raises the temperature.

In summary,

$$A + B \rightleftharpoons X + Y + \text{heat evolved } (\Delta H \text{ is negative})$$
(smaller vol)　　　　　　　　　　　　　(larger vol)

⟵ addition of X or Y ⟶
⟵ increase in temperature ⟶
⟵ increase in pressure ⟶

Changes in concentration or pressure may cause an equilibrium to shift, but they do not change the value of the equilibrium constant. Changes in temperature change the equilibrium constant value, and the equilibrium shifts to adjust to the new value.

Industrial processes

15 The Haber process

The conversion of hydrogen and nitrogen into ammonia is exothermic and results in a decrease in gas volume,

$$N_2(g) + 3H_2(g) \rightleftharpoons 2NH_3(g) \qquad \Delta H_m^{\ominus}(298\,K) = -92.4\,kJ\,mol^{-1}$$

Le Chatelier's principle shows, therefore, that the yield of ammonia will be greatest at low temperature and high pressure (Fig. 143).

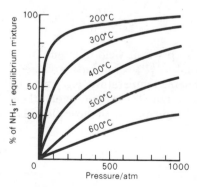

Fig. 143 The percentage of ammonia in the equilibrium mixture obtained from a 3:1 mixture of hydrogen and nitrogen at different temperatures and pressures.

At low temperature, however, equilibrium is reached slowly: at high pressure, the cost of the equipment and the running costs are high. In practice a compromise has to be struck. Pressures between 150 and 350 atm are used at a temperature of about 500 °C. The rate of reaction is increased by using a catalyst based on iron oxide.

a Effect of temperature The approximate variation of K_p with temperature can be calculated using the relationship given on page 296. Taking a value of 6.73×10^5 for K_p at 298 K, and -92.4 for ΔH_m, gives

$$\lg K_{p,T} = -10.39 + 4833/T$$

This assumes that ΔH_m does not change with temperature so that the calculated and observed values of K_p do not agree, particularly at high temperatures, as the following figures show:

temp./K	298	400	600	800
$\Delta H_m/kJ$	-92.4	-96.9	-105.8	-114.6
K_p/atm^{-2}	6.73×10^5	40.7	1.66×10^{-3}	6.92×10^{-6}
K_p/atm^{-2} (calculated)	6.73×10^5	49.3	4.62×10^{-3}	4.48×10^{-5}

b Effect of pressure By assuming that the gases behave as ideal gases and by considering an initial 3:1 mixture of hydrogen and nitrogen it is also possible to calculate approximate values for the yield of ammonia at different pressures.

K_p is taken as being independent of temperature, though it does change slightly, particularly at high pressures, due to the non-ideal behaviour of the gases. If the partial pressures in the equilibrium mixtures are p_{H_2}, p_{N_2} and p_{NH_3} then

$$p_{N_2} + p_{H_2} + p_{NH_3} = P(\text{total pressure}) \quad \text{and} \quad K_p = \frac{(p_{NH_3})^2}{(p_{N_2})(p_{H_2})^3}$$

As p_{H_2} is three times as big as p_{N_2} it follows that

$$p_{N_2} = \tfrac{1}{4}(P - p_{NH_3}) \qquad p_{H_2} = \tfrac{3}{4}(P - p_{NH_3}) \qquad K_p = \frac{9.48(p_{NH_3})^2}{(P - p_{NH_3})^4}$$

At 700 K, K_p is 7.76×10^{-5} and for a 50 per cent yield of ammonia in the equilibrium mixture p_{NH_3} must be equal to $\tfrac{1}{2}P$; P will have a value of 700 atm. For a yield of 20 per cent at the same temperature, p_{NH_3} must be equal to $\tfrac{1}{5}P$ and a pressure of 110 atm is required. These figures and others which can be obtained similarly are in good agreement with those plotted in Fig. 143.

16 The contact process

The conversion of sulphur dioxide and oxygen into sulphur trioxide is an exothermic reaction with a decrease in gas volume,

$$2SO_2(g) + O_2(g) \rightleftharpoons 2SO_3(g) \qquad \Delta H_m^{\ominus}(298\,\text{K}) = -197.6\,\text{kJ mol}^{-1}$$

Increase in pressure would give an increased yield of sulphur trioxide but the effect is small and the yield is good even at low pressures just above atmospheric which are, therefore, used in practice.

The greatest yield of sulphur trioxide would be obtained at low temperature, but the attainment of the equilibrium would be very slow. A compromise temperature, initially about 420 °C, is used, together with a catalyst of vanadium(V) oxide with potassium sulphate as a promoter, on a silica support.

Yields above 95.5 per cent are obtained but only by carrying out the process in stages. A mixture of dry air and dry sulphur dioxide containing approximately 10 per cent of sulphur dioxide, 11 per cent of oxygen and 79 per cent of nitrogen is passed into the first converter at 420 °C; a 63 per cent conversion is achieved but the temperature rises as the reaction is exothermic. Above 600 °C the catalyst deteriorates and the sulphur trioxide begins to dissociate. The partially converted gas mixture is, therefore, cooled before passing into a second converter, and then likewise into a third. The sulphur trioxide in the mixture is now removed by absorption in 98 per cent sulphuric acid and the unreacted gas mixture (containing only about 0.07 per cent sulphur dioxide) is passed into a fourth converter and the extra sulphur trioxide is again absorbed. In this way all but about 0.05 per cent of the original sulphur dioxide is converted into the trioxide.

Questions on Chapter 25

1 When 60 g of ethanoic acid were heated with 46 g of ethanol until equilibrium was reached, 12 g of water and 58.7 g of ethyl ethanoate were formed. What is the equilibrium constant for the reaction between acid and alcohol? What mass of ester would be formed under the same conditions starting from 90 g of ethanoic acid and 92 g of ethanol?

2 When 1 mol of ethanol reacts with 1 mol of ethanoic acid until an equilibrium is obtained, there is present in the mixture 0.333 mol of alcohol and acid, and 0.666 mol of ester and water. Calculate the amount of ester present at equilibrium when (a) 3 mol of ethanol react with 1 mol of acid (b) 92 g of ethanol react with 60 g of ethanoic acid (c) 1 mol of ethanol reacts with 1 mol of acid in the presence of 1 mol of water.

3 State the law of mass action and show how you would use it to derive the equilibrium constant of a reaction $2A + B = C + 2D + Q$ joule in which the reactants and resultants are all gases. What steps could be taken (a) to obtain as large a yield as possible of C from a given amount of B (b) to obtain the yield as quickly as possible? (OC)

4 Describe how the equilibrium between hydrogen, iodine and hydrogen iodide could be measured at, say, 300 °C. How would you expect the equilibrium to vary with (a) increasing temperature (b) increasing pressure? Give your reasons. The reaction at 300 °C is slightly exothermic. (OCS)

5 Starting with equimolecular quantities of ethanol and ethanoic acid, the position at equilibrium consists (in molecular quantities) of $\frac{1}{3}$ ethanol, $\frac{1}{3}$ acid, $\frac{2}{3}$ ester and $\frac{2}{3}$ water. In what molecular proportions must the acid and alcohol be mixed in order to obtain a 90 per cent yield of ester from the quantity of ethanol used?

6 What do you understand by chemical equilibrium? Suggest methods for investigating it in two of the systems:
(a) $CaCO_3 \rightleftharpoons CaO + CO_2$,
(b) $I_2 + I^- \rightleftharpoons I_3^-$,
(c) $2NH_3 \rightleftharpoons N_2 + 3H_2$.
What generalisations would you expect to be able to draw from these results?

7 Derive expressions for the equilibrium constant for the reaction represented by the equation $2A + B = C + 2D$ if the original concentrations of A and B are a and $b \, mol \, dm^{-3}$ respectively and if (i) $x \, mol \, dm^{-3}$ of C are found in the equilibrium mixture (ii) $y \, mol \, dm^{-3}$ of A are converted into C and D (iii) the equilibrium concentration of B is $a/2$.

8 Describe the effect on the equilibrium mixture arising from the following reaction

$$CO(g) + 2H_2(g) \rightleftharpoons CH_3OH(g) \qquad \Delta H_m = -92.05 \, kJ \, mol^{-1}$$

of (a) increased pressure (b) increased temperature (c) increased partial pressure of hydrogen (d) the presence of a catalyst.

9 Derive an expression for the equilibrium constant for the reaction

$$3H_2(g) + N_2(g) \rightleftharpoons 2NH_3(g)$$

in terms of a, b, x and P, where a and b are the original numbers of moles of hydrogen and nitrogen, respectively, and x is the number of moles of nitrogen which react at a total pressure of $P \, Pa$.

10 Give an account of some uses of Le Chatelier's principle.

11 Use Le Chatelier's principle to predict the effect of (i) increased pressure (ii) decreased temperature on (a) the melting of ice (b) the solution of sodium chloride in water (c) the solution of ammonia in water (d) the formation of nitrogen oxide from nitrogen and oxygen (e) the formation of sulphur trioxide from sulphur dioxide and oxygen.

12 4 g of potassium iodide was dissolved in $500 \, cm^3$ of water, and about 1 g of iodine was dissolved in $100 \, cm^3$ of benzene. The two solutions were then mixed and allowed to stand. Subsequent titrations showed that $10 \, cm^3$ of the benzene layer was equivalent to $5.1 \, cm^3$ of M/10 sodium thiosulphate solution, whilst $50 \, cm^3$ of the aqueous layer was equivalent to $2.9 \, cm^3$ of M/10 thiosulphate. The distribution coefficient of iodine between benzene and water is 130. Calculate the value of the equilibrium constant for the equilibrium,

$$I_2(aq) + I^-(aq) \rightleftharpoons I_3^-(aq)$$

13 Calculate the percentage by volume of oxygen converted into nitrogen oxide when air (79 per cent N_2 and 21 per cent O_2 by volume) is heated to $2000 \, °C$, given that at this temperature the equilibrium constant, K, for the reaction

$$\tfrac{1}{2}N_2(g) + \tfrac{1}{2}O_2(g) \rightleftharpoons NO(g)$$

is 2.9×10^{-2}. (WS)

14 Distinguish between thermal dissociation and thermal decomposition. Formulate the equilibrium constant in terms of the degree of dissociation and the total pressure for the gaseous system:

$$2NO_2(g) \rightleftharpoons 2NO(g) + O_2(g)$$

Nitrogen dioxide is dissociated to the extent of 56.6 per cent at 494 °C and 99 kPa pressure. At what pressure will the dissociation be 80 per cent at 494 °C? (WS)

15 Discuss the effects of temperature and pressure on the composition of the equilibrium mixture in a reversible reaction. The total gas pressure in a system originally consisting of a sample of solid ammonium hydrogensulphide (NH_4HS) at 25 °C is 0.66 atmospheres. Calculate the value of the equilibrium constant, K, in terms of partial pressures, explaining the reasoning underlying any assumptions you may make. What would be the partial pressure of ammonia in the mixture if 0.1 atmospheres of this gas were added to the system? (OC)

16 A solution containing 12.7 g of iodine and 166.1 g of potassium iodide in 1 dm³ of water is shaken up with 1 dm³ of benzene. Given that the partition coefficient for iodine between benzene and water is 400, and that the equilibrium constant for the reaction $I_2 + I^- = I_3^-$ is 730 (concentrations expressed in mol dm⁻³), calculate (a) the concentration of iodine molecules in the benzene (b) the concentration of I_3^- ion in the water. (Neglect the solubilities of water and potassium iodide in benzene and of benzene in water. Be careful to state the units in which concentrations are expressed.) (OCS)

17 1 g of hydrogen and 127 g of iodine are allowed to attain equilibrium in an evacuated container of volume 10 dm³ at a temperature of 450 °C. At this temperature K_c is equal to 50. What is (a) the value of K_p (b) the total pressure in the container (c) the partial pressure of hydrogen in the container?

18 The densities of diamond and graphite are 3.5 and 2.3 g cm⁻³ respectively, and the change from graphite to diamond is represented by the following equation,

$$C(graphite) \rightleftharpoons C(diamond) \qquad \Delta H = +1.883 \, kJ \, mol^{-1}$$

Is the formation of diamond from graphite favoured by (a) high or low temperature (b) high or low pressure? Explain your answers.

19 Two solid compounds, A and B, dissociate at 15 °C into gaseous products as follows:

$$A \rightleftharpoons A' + H_2S \qquad B \rightleftharpoons B' + H_2S$$

The pressure at 15 °C over excess solid A was 50 mm of mercury and over excess solid B was 68 mm of mercury. Find (a) the dissociation constant of A and of B (units, mm²) (b) the relative number of molecules of A' and of B' in the vapour over a mixture of the solids (c) the total pressure of gas over the mixture.

20 If the degree of dissociation of water is 1.93×10^{-7} at $42\,°C$ and 0.39×10^{-7} at $2\,°C$, calculate the enthalpy of reaction for the reaction

$$H_2O(l) \rightleftharpoons H^+(aq) + OH^-(aq)$$

21 If the equilibrium constant for a reaction is 2.9×10^{-5} at $947\,°C$ and 10.4×10^{-5} at $1047\,°C$ calculate the approximate enthalpy of the reaction.

22 In the thermal dissociation of solid ammonium hydrogensulphide

$$NH_4HS(s) \rightleftharpoons H_2S(g) + NH_3(g)$$

the total pressure at equilibrium was found to be $23.33\,kPa$ at $9.5\,°C$ and $66.79\,kPa$ at $25.1\,°C$. Calculate the enthalpy change of the reaction.

23 The equilibrium constants for the dissociation of oxygen molecules (O_2) into oxygen atoms ($2O$) are 9.2×10^{-27} at $527\,°C$ and 8.0×10^{-16} at $927\,°C$. Calculate the heat of dissociation.

24 Compare the value of the equilibrium constant for the reaction $H_2(g) + I_2(g) \rightarrow 2HI(g)$ with that for $\frac{1}{2}H_2(g) + \frac{1}{2}I_2(g) \rightarrow HI(g)$. Why should there be any difference?

25 How would the equilibrium constant for a reaction using concentrations expressed as mol fractions differ from that using $mol\,dm^{-3}$?

26 The K_p value (using P in atmospheres) for the reaction $N_2 + 3H_2 \rightarrow 2NH_3$ is 1.5×10^{-5} at $500\,°C$. What percentage of a $1:3$ mixture of nitrogen and hydrogen would be converted into ammonia at $500\,°C$ if the total pressure was   (a) 1 atmosphere   (b) 500 atmospheres?

27 For a gas reaction, $A(g) \rightleftharpoons 2B(g)$, plot the fraction of A which is converted into B at a fixed temperature if (a) $K_p' = 100$ (b) $K_p = 10$.

28* Look up the values for the ionic product of water at any two temperatures and use them to calculate the enthalpy change in the formation of water from H^+ and OH^-.

26
Rates of reaction. Chemical kinetics

1 Introduction

A chemical reaction is simply a rearrangement of atoms or ions, but the rearrangement may take place in a simple or complex way and at rates varying from the immeasurably slow to the explosive.

When the rate can be measured it is often found to depend on the concentration of the reagents and the temperature, and sometimes on the pressure and catalysis. In heterogeneous reactions, the state of sub-division of a solid is an important factor.

The knowledge gained from a study of reaction rates may enable the rate of any particular reaction to be controlled advantageously in a laboratory, industrial or biological process. It can also throw light on the reaction mechanism, differentiating, for instance, between reactions which involve only one simple step and others which may pass through a number of intermediate steps each having a different rate.

a The rate of a reaction This is generally measured and expressed either as the fall in concentration of the reactants, or the rise in concentration of the products, with time. For a reaction

$$A + B \longrightarrow 2C$$

one mole of A and one of B are used up for every two moles of C that are formed, so that the rate of increase in the concentration of C is twice that of the decrease in A and B. Hence

$$\text{rate} = -d[A]_t/dt = -d[B]_t/dt = 0.5\,d[C]_t/dt$$

Different units can be used but concentrations are generally given in $mol\,dm^{-3}$ and time in seconds. The units for the rate of a reaction, on this basis, are $mol\,dm^{-3}\,s^{-1}$.

As the rate of the reaction is not the same when expressed in terms of A or B as it is in terms of C it is important to specify which is being chosen when referring to the rate of the reaction.

b Rate constants The experimental relationship between a reaction rate and the concentration of the reactants is known as the *rate law* or the *rate expression* for that reaction.

For the reaction

$$xA + yB \longrightarrow \text{products}$$

it would be of the form

$$\text{rate of reaction} = -d[A]_t/dt = k[A]^m[B]^n$$

with k being known as the *rate constant*. m may be equal to x, and n to y, but there is no general relationship between the stoichiometric equation for a reaction and the rate expression. Indeed, m or n may be either whole numbers or fractions or zero.

If the rate expression and the rate constant for any reaction is known, the actual rate for any concentrations can be calculated.

As A and B react and their concentrations decrease, the rate of reaction falls, but the rate constant remains constant.

c The order of a reaction A reaction with a rate expression as in **b** is said to have a total order of $(m + n)$, the order with respect to A being m, and with respect to B, n.

This enables reactions to be classified as either zero, first, second etc. order; reactions may also be of fractional order, but the order cannot be given as a simple number for complex reactions with complicated rate expressions.

Acceptable mechanisms for any reaction must account for the experimentally measured rate expression and the related orders (p. 315).

d Half-life The half-life of a reaction, $t_{\frac{1}{2}}$, is the time taken for the concentration of a reagent to fall to half its initial value. It gives a simple indication of the rate of any change. It is used particularly in considering radioactivity (p. 78).

The way in which the half-life for a reaction depends on concentration depends on the order of the reaction (p. 315).

Experimental methods

To measure the rate of a reaction it is necessary to measure the way in which the concentration of one of the reactants or one of the products changes with time, at a fixed temperature. Any property proportional to the concentration can be measured so that many methods are available, as indicated below.

From rate measurements it is possible to derive a rate expression and to find the value of the rate constant and the half-life. For most reactions it will also be possible to find the order and to suggest a mechanism for the reaction (p. 325). Measurement of rate constants at different temperatures gives values of activation energies (p. 318).

2 Measurement of concentrations by withdrawal of samples and titration

e.g. $2H_2O_2(aq) \longrightarrow 2H_2O(l) + O_2(g)$.

The initial concentration $[A]_0$ of a solution of hydrogen peroxide is measured by titration with standard potassium manganate(VII). A catalyst is then added, at a known time, to another portion of the same

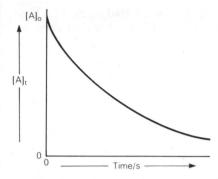

Fig. 144 Plot of $[A]_t$ against t. The slope of the curve at any time, t, gives the rate, $-d[A]_t/dt$, at that time.

solution and samples are withdrawn, in a pipette, at measured time intervals. These samples are immediately added to a measured excess of potassium manganate(VII) solution which prevents any further decomposition of the hydrogen peroxide. By back titration, the concentration of hydrogen peroxide in the sample at time t, $[A]_t$, can be calculated.

The change in concentration of the hydrogen peroxide in t seconds will be $([A]_t - [A]_0) \, mol \, dm^{-3}$ and the average rate of change will be $([A]_t - [A]_0)/t \, mol \, dm^{-3} \, s^{-1}$. To find the particular rate at any particular time a graphical method (which is rather long and inaccurate) can be used. Alternatively, the values of $[A]_0$, $[A]_t$ and t can be treated mathematically as explained on pages 310–12.

In the graphical method, $[A]_t$ is plotted against t as in Fig. 144. The slope of the curve at any time, t, which can be found by drawing the tangent, will be equal to the rate $-d[A]_t/dt$, at time t. A plot of the rate against $[A]_t$ (Fig. 145) will then give a straight line showing that the reaction is first order, i.e.

$$\text{rate} = -d[A]_t/dt = k[H_2O_2]$$

The slope of the line will equal k.

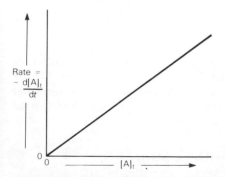

Fig. 145 Plot of rates (obtained from Fig. 144) against $[A]_t$. The straight line plot shows that the reaction is first order, and the slope of the line is equal to k.

It may well have been that the rate had to be plotted against $[A]_t^2$ to give a straight line; if so the reaction would have been of second order.

The rate of hydrolysis of an ester, e.g.

$$H.COOCH_3(l) + H_2O(l) \longrightarrow H.COOH(l) + CH_3OH(l)$$

can be measured in the same sort of way by titration of samples of the mixture with standard alkali. This measures the concentration of acid at time t, i.e. c_t. When the reaction is completed, the acid concentration, c_∞, will be equal to the initial concentration of ester, $[A]_0$. The concentration of ester at time t, i.e. $[A]_t$, will be given by $(c_\infty - c_t)$.

3 Measurement of volume of gas evolved

The decomposition of hydrogen peroxide solutions can also be followed by measuring the rate of formation of oxygen. A sample of the solution in a conical flask in a thermostat is connected to a gas burette or syringe (p. 8). A catalyst is added at a known time, and the volume of gas collected after various time intervals is measured. The final volume recorded, V_∞, will be proportional to the initial concentration of hydrogen peroxide, $[A]_0$. $(V_\infty - V_t)$, where V_t is the volume of gas at time t, will be proportional to the concentration of hydrogen peroxide at time t, i.e. $[A]_t$.

Other reactions producing gases can be monitored in the same way.

4 Measurement of gas pressure

Instead of measuring the gas volume in **3**, the reaction could have been carried out at constant volume and the increase in pressure with time measured. This is done by connecting the reaction flask to a manometer. Any reaction producing an increased gas volume can be investigated in this way.

5 Use of colour changes

The time taken for the colour of a reactant to disappear, or the colour of a product to appear, can be measured. Or, with a colorimeter (p. 187) or spectrophotometer (p. 180), the change in colour can be recorded continuously. Typical examples where such procedures can be followed are:

a $S_2O_3^{2-}(aq) + 2H^+(aq) \longrightarrow H_2O(l) + SO_2(g) + S(s)$

The time of appearance of the sulphur precipitate is inversely proportional to the rate of the reaction, and, for fixed concentrations of acid, it will be found that the rate is proportional to $[S_2O_3^{2-}(aq)]$. The reaction can also be studied at different temperatures and with different acid concentrations.

b $\quad BrO_3^-(aq) + 6I^-(aq) + 6H^+(aq) \longrightarrow 3H_2O(l) + 3I_2(aq) + Br^-(aq)$

If a little starch is added to the mixture the iodine is detected by the appearance of a blue-black colour, and the development of the colour to match a standard can be used to measure the reaction rate. It will be found that

$$\text{rate} = -d[BrO_3^-]/dt = k[BrO_3^-][I^-][H^+]^2$$

c $\quad CH_3COCH_3(aq) + I_2(aq) \longrightarrow CH_2ICOCH_3(aq) + I^-(aq) + H^+(aq)$

The rate of iodination of propanone can be measured by timing the disappearance of the iodine colour. The reaction is catalysed by hydrogen ions and it is found that

$$\text{rate} = -d[I_2]/dt = k[CH_3COCH_3][H^+]$$

Surprisingly, the rate is independent of the $[I^-]$ (p. 325).

d $\quad H_2O_2(aq) + 2H^+(aq) + 2I^-(aq) \longrightarrow 2H_2O(l) + I_2(aq)$

This reaction, in acid solution, is known as the Harcourt–Esson reaction. It can be investigated very easily by carrying out the reaction in the presence of a known amount ($x \, cm^3$) of standard sodium thiosulphate solution and a little starch.

The first iodine to be formed reacts with the sodium thiosulphate present, but when this has all been used up the mixture quite suddenly turns blue. The time is recorded and a further $x \, cm^3$ of thiosulphate are added. When the mixture once more turns blue the time is recorded again, more thiosulphate is added, and so on. The time for the reappearance of the blue colour lengthens as the rate of formation of iodine gets slower. It is found that, under these conditions (p. 314),

$$\text{rate} = -d[H_2O_2]/dt = k[H_2O_2]$$

e Clock reactions Many mixtures with colour changes can be made to function so as to give good indications of the time; they are known as clock reactions.

6 Use of a polarimeter. The inversion of sucrose

Cane sugar, or sucrose, is dextrorotatory, but it is hydrolysed in acid solution to fructose, which is laevorotatory, and to glucose, which is dextrorotatory. As the optical activity of fructose is greater than that of glucose, the mixture is, in fact, laevorotatory. As the reaction proceeds, the polarised light is rotated less and less to the right, and eventually is rotated to the left. That is why the process is referred to as the inversion of cane sugar, and the mixture of fructose and glucose as invert sugar,

$$C_{12}H_{22}O_{11}(aq) + H_2O(l) \longrightarrow C_6H_{12}O_6(aq) + C_6H_{12}O_6(aq)$$

sucrose	fructose	glucose
(dextro)	(laevo)	(dextro)

invert sugar (laevo)

The rate of the reaction is measured by taking readings on a polarimeter. If p_0, p_t and p_∞ are the polarimeter readings at the start, after time t, and at the end of the reaction respectively, then $(p_0 - p_\infty)$ is proportional to the original concentration of cane sugar, $[A]_0$, and $(p_t - p_\infty)$ is proportional to the concentration of cane sugar after time t, $[A]_t$. The $+$ or $-$ values of the polarimeter readings must, of course, be taken into account.

For dilute solutions of sucrose, with the water present in large excess (p. 314),

$$\text{rate} = -d[C_{12}H_{22}O_{11}]/dt = k[C_{12}H_{22}O_{11}]$$

7 Use of other physical methods

Rates can also be measured by using ultraviolet or infrared spectrophotometers, conductivity measurements (for reactions in which the number of ions changes), dilatometers (for measuring small volume changes in reactions in solution) and refractometers (when there are changes in refractive index).

Methods using mass spectrometers and chromatography can also be employed.

8 Investigation of very fast reactions

In recent years kinetic studies have been extended to include reactions previously regarded as instantaneous. Special techniques are necessary which enable reactions between ions, and many of biological interest, to be investigated, some of them with half-lives as low as 10^{-11} s.

a Flow methods Two reacting solutions can be forced into a mixing chamber under pressure and the resulting, reacting mixture passed along a tube. The composition of the mixture at different points along the tube, i.e. at different times, can be measured by, for example, using a movable spectrometer. Alternatively the composition at a fixed point in the tube can be measured for different flow rates. If there is a colour change in the reaction the colour intensity will vary along the tube.

b Relaxation methods This type of method has been applied to measure the backward and forward rates in the reaction

$$H^+(aq) + OH^-(aq) \rightleftharpoons H_2O(l) \qquad \Delta H^\ominus(298\,K) = -56.9\,kJ\,mol^{-1}$$

The ions and molecules are in equilibrium at any particular temperature and the equilibrium constant has a fixed value at that temperature. If, however, the temperature is quite suddenly increased by discharging a bank of condensers through the water, the equilibrium constant will be lowered (p. 295) so that the equilibrium will shift from right to left. The rate at which this happens can be measured electrically, as more ions will be formed; the time taken for a new equilibrium to be established (the *relaxation time*) can be measured similarly. From these measurements it is possible to calculate the rate constant for the back reaction $(2.4 \times 10^{-5}\,s^{-1})$ and for the forward reaction $(1.4 \times 10^{11}\,mol^{-1}\,dm^3\,s^{-1})$. The reaction between $H^+(aq)$ and $OH^-(aq)$ is the fastest reaction known.

The method can be applied more generally by using any other rapid change, e.g. pressure, to disturb an established equilibrium.

c Flash photolysis In this method a very powerful flash of light of short duration (obtained by discharging a bank of condensers through a gas discharge tube), or a laser beam, is passed through a gas mixture. Atoms and/or free radicals (p. 329) are formed, and they can be identified by passing a second flash of light through the mixture immediately after the first flash. Examination of the resulting absorption spectrum (p. 178) indicates which atoms and/or free radicals are present. The rate at which they recombine with each other, or with themselves, can be measured by photographing the absorption spectrum at a series of short time intervals after the initial flash. Changes in the absorption spectrum with time give the rates of the various processes taking place.

Passage of an initial flash through pure azomethane (diluted with argon), for example, produces methyl radicals

$$CH_3-N{=}N-CH_3(g) \longrightarrow 2{\cdot}CH_3 + N_2(g)$$

which can be detected in the resulting absorption spectrum. Measurement of the decrease in their concentration with time enables the rate of the reaction

$$2{\cdot}CH_3 \longrightarrow C_2H_6(g)$$

to be obtained. Similarly, an initial flash through a mixture of oxygen and chlorine produces $ClO{\cdot}$ free radicals, and their life can be measured.

The order of reactions

9 The kinetics of first-order reactions

Typical first-order reactions are summarised below.

$2N_2O_5(g) \rightarrow 4NO_2(g) + O_2(g)$ $-d[N_2O_5]/dt = k[N_2O_5]$

$2H_2O_2(aq) \rightarrow 2H_2O(l) + O_2(g)$ $-d[H_2O_2]/dt = k[H_2O_2]$

$(CH_3)_3.C.Cl + OH^- \rightarrow (CH_3)_3.C.OH + Cl^-$ $-d[(CH_3)_3.C.Cl]/dt = k[(CH_3)_3.C.Cl]$

a The rate expression A typical first-order reaction may be represented as

$A \longrightarrow$ products of reaction

and the rate will be given by

rate $= -d[A]_t/dt = k[A]_t$

which gives, on integration,

$-\ln[A]_t = kt + C$(a constant)

When t is zero $[A]_t$ is $[A]_0$ so that the value of C is $-\ln[A]_0$. Therefore

$\ln[A]_t = \ln[A]_0 - kt$ $\ln([A]_0/[A]_t) = kt$ $[A]_t = [A]_0 e^{-kt}$

$\lg[A]_t = \lg[A]_0 - kt/2.303$ $\lg([A]_0/[A]_t) = kt/2.303$

If time is measured in seconds the units of the rate constant for a first-order reaction will be s^{-1}.

b Half-life The half-life, $t_{\frac{1}{2}}$, is the time for $[A]_t = \frac{1}{2}[A]_0$, i.e.

$\lg 2 = kt_{\frac{1}{2}}/2.303$ or $t_{\frac{1}{2}} = 0.69/k$

The half-life for a first-order reaction is, therefore independent of the initial concentration.

c Graphical representation A plot of $[A]_t$ against t for a first-order reaction gives a typical curve as shown in Fig. 146.

But whether or not any particular set of measurements on the rate of a reaction will fit the typical rate expression is best tested by plotting $\lg[A]_t$ against t. The plot will be a straight line if the reaction is of the first-order, and the value of k can be obtained because the gradient of the line will be equal to $-k/2.303$. Titration values proportional to $[A]_t$ can be used.

The following simplified figures for a reaction, using arbitrary units of concentration, plotted in Figs. 146 and 147, illustrate the procedure.

time/s	0	200	400	600
$[A]_t$	40	20	10	5
$\lg[A]_t$	1.60	1.30	1.0	0.70

The gradient of the line in Fig. 147 is $-0.9/600$ giving a value for k of $0.00345\,s^{-1}$. The half-life is $200\,s$, i.e. $0.69/k$. It is independent of the initial concentration.

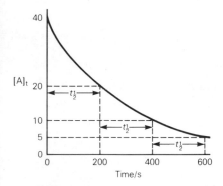

Fig. 146 Plot of $[A]_t$ against time for a first order reaction, showing that the half-life is independent of the initial concentration. Compare with Fig. 27, page 79.

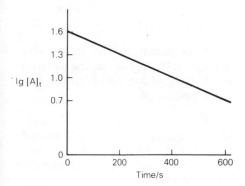

Fig. 147 Plot of $\lg[A]_t$ against time for a first order reaction. The slope of the line is equal to $-k/2.303$.

10 The kinetics of second-order reactions

Typical second-order reactions are summarised below.

$$H_2(g) + I_2(g) \rightarrow 2HI(g) \qquad -d[I_2]/dt = k[H_2][I_2]$$

$$S_2O_8{}^{2-}(aq) + 2I^-(aq) \rightarrow 2SO_4{}^{2-}(aq) + I_2(aq) \quad -d[I_2]/dt = k[S_2O_8{}^{2-}][I^-]$$

$$CH_3Br + OH^- \rightarrow CH_3OH + Br^- \qquad -d[CH_3Br]/dt = k[CH_3Br][OH^-]$$

Introduction to Physical Chemistry

a The rate expression For a simple second-order reaction

$$2A \longrightarrow \text{products of reaction}$$

the rate will be given by

$$\text{rate} = -d[A]_t/dt = k[A]_t{}^2$$

which gives, on integration,

$$1/[A]_t = 1/[A]_0 + kt$$

The units of k are $\text{mol}^{-1}\,\text{dm}^3\,\text{s}^{-1}$.

In the more general case

$$A + B \longrightarrow \text{products of reaction.}$$

When the initial concentrations of A and B are not alike,

$$\text{rate} = -d[A]_t/dt = k[A][B]$$

and the more difficult integration gives

$$kt = \frac{1}{[A]_0 - [B]_0} \ln \frac{[A]_t[B]_0}{[A]_0[B]_t}$$

b Half-life For the simpler reaction in **a** the half-life is the time taken for $[A]_t = \frac{1}{2}[A]_0$, i.e.

$$2/[A]_0 = 1/[A]_0 + kt_{\frac{1}{2}} \qquad \text{or} \qquad t_{\frac{1}{2}} = 1/k[A]_0$$

The half-life is, therefore inversely proportional to the initial concentration. This means that the half-life gets greater as the reaction proceeds. If the time taken for the first halving of concentration is t, then the time for the second halving will be $2t$, and so on.

c Graphical representation A plot of $[A]_t$ against t for a simple second-order reaction is shown in Fig. 148. A plot of $1/[A]_t$ against t will be a straight line and the gradient of the line will equal k. This is illustrated by the following figures, plotted in Figs. 148 and 149.

time/s	0	200	600	1400
$[A]_t$	40	20	10	5
$1/[A]_t$	0.025	0.05	0.1	0.2

The value of k is $125 \times 10^{-6}\,\text{mol}^{-1}\,\text{dm}^3\,\text{s}^{-1}$. The value for the first half-life is $10^6/(40 \times 125)$, i.e. $200\,\text{s}$; the value for the second half-life is $10^6/(20 \times 125)$, i.e. $400\,\text{s}$.

For a second order reaction when the initial concentration of A and B are not alike, it is a plot of $\lg([A]_t/[B]_t)$ against t that gives a straight line.

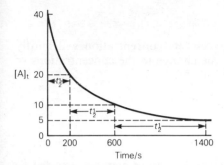

Fig. 148 Plot of $[A]_t$ against time for a second-order reaction, $2A \rightarrow$ products, showing that the half-life is inversely proportional to the concentration. Compare with Fig. 146, p. 311.

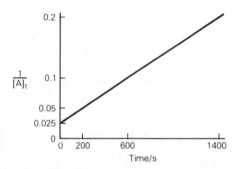

Fig. 149 Plot of $1/[A]_t$ against time for a second-order reaction, $2A \rightarrow$ products. The slope of the line equals k.

11 Zero-order reactions

The decomposition of ammonia on a hot tungsten wire, or of hydrogen iodide on a hot gold wire, are found to be of zero-order so long as the gas pressure is not too low. This means that the rate of the decompositions are constant and independent of the gas concentrations (pressures). This arises because the metal surface at which the reaction is taking place is fully saturated with adsorbed gas, and the rate of decomposition is constant so long as the gas pressure remains high enough to keep the surface saturated.

The rate expression for a typical zero-order reaction

$$A \longrightarrow \text{products} \qquad \text{is} \qquad \text{rate} = -d[A]_t/dt = k$$

which gives, on integration,

$$[A]_t = [A]_0 - kt$$

A plot of $[A]_t$ against t gives a straight line with a slope equal to $-k$. The half-life is $[A]_0/2k$.

313

12 Pseudo orders

If a reactant is present in a large excess, its concentration will hardly change during a reaction relative to the change in the concentrations of other reactants.

For a second-order reaction

$$A + B \longrightarrow products$$

$$rate = -d[A]_t/dt = k[A][B]$$

but if B is present in large excess its concentration may be regarded as constant so that

$$rate = -d[A]_t/dt = k'[A]$$

The hydrolysis of esters (p. 306) or of sucrose (p. 307), for example, can be made to exhibit first-order reaction kinetics by using excess water, even though the reactions are second-order if the water is not present in excess.

Similarly, the oxidation of iodide ions by hydrogen peroxide in acid solution is normally first-order with respect to both hydrogen peroxide and iodide ion,

$$H_2O_2(aq) + 2H^+(aq) + 2I^-(aq) \longrightarrow 2H_2O(l) + I_2(aq)$$

$$rate = -d[H_2O_2]/dt = k[H_2O_2][I^-](1 + k'[H^+])$$

but it is independent of $[I^-]$ if carried out in the presence of sodium thiosulphate (p. 307). As the iodine is formed it is converted back into $I^-(aq)$ again so that $[I^-]$ remains constant.

$$I_2(aq) + 2S_2O_3^{2-}(aq) \longrightarrow S_4O_6^{2-}(aq) + 2I^-(aq)$$

The order of a reaction when carried out under special conditions is sometimes referred to as a pseudo-order. But under the conditions, it is, nevertheless, a genuine experimentally measured order.

13 Fractional orders

The order of a reaction need not be a whole number as shown by the following examples:

$$2SO_2(g) + O_2(g) \rightarrow 2SO_3(g) \qquad d[SO_3]/dt = k[SO_2][SO_3]^{-0.5}$$

$$CH_3CHO(g) \rightarrow CH_4(g) + CO(g) \qquad -d[CH_3CHO]/dt = k[CH_3CHO]^{1.5}$$

$$H_2(g) + Br_2(g) \rightarrow 2HBr(g) \qquad -d[Br_2]/dt = \frac{k[H_2][Br_2]^{1.5}}{[Br_2] + k'[HBr]}$$

In the last example, which is a complex reaction (p. 328), the total order cannot really be expressed as a simple number.

14 To find the order of a reaction

The rate expressions and integrated rate expressions are different for different orders of reaction as summarised in the simplest cases below.

reaction	A → products	A → products	2A → products
order	zero	first	second
rate expression	$-d[A]_t/dt = k$	$-d[A]_t/dt = k[A]$	$-d[A]_t/dt = k[A]^2$
integrated rate expression	$[A]_t = [A]_0 - kt$	$\lg[A]_t = \lg[A]_0 - kt/2.303$	$1/[A]_t = 1/[A]_0 + kt$
straight line plot against t	$[A]_t$	$\lg[A]_t$	$1/[A]_t$

a Graphical method The order of a reaction can be obtained by finding out which plot of the experimental data against t, as summarised above, gives a straight line.

b Half-life method The half-life is proportional to $[A]_0$ for a zero-order reaction, independent of $[A]_0$ for a first-order reaction, and proportional to $1/[A]_0$ for a second-order reaction with equal initial concentrations.

c Ostwald's isolation method If all the reactants in a reaction except one are present in very large excess, the order with respect to that one reactant can be measured as in **a** or **b** above. By taking each reactant in turn, in this way, the total order can be obtained.

Questions on Chapter 26

1 What is meant by the rate of a chemical reaction? Describe a method of measurement of the rate of one chemical reaction, and explain to what use the result may be put. (OS)

2 The polarimeter readings in an experiment to measure the rate of inversion of cane sugar were as follows:

time/min	0	10	20	30	40	80	∞
angle	32.4	28.8	25.5	22.4	19.6	10.3	−14.1

What is the order of the reaction?

3 The rate of inversion of cane sugar is of the first order. If 25 per cent of a sample of cane sugar is hydrolysed in 60 seconds, how long will it take for 50 per cent to be hydrolysed?

4 A solution of hydrogen peroxide was titrated against an M/50 solution of potassium manganate(VII) and $25\,cm^3$ of the peroxide solution were found to be equivalent to $46.1\,cm^3$ of the manganate(VII) solution. On adding colloidal platinum to a sample of the peroxide solution decomposition began. At intervals, $25\,cm^3$ portions were withdrawn and titrated, rapidly, against the same manganate(VII) solution. The results obtained were as follows:

time/min	5	10	20	30	50
$KMnO_4/cm^3$	37.1	29.8	19.6	12.3	5.0

Find, either graphically or mathematically, the value of the velocity constant for the decomposition of the peroxide solution.

5 The rate constant of a unimolecular reaction $AB \rightarrow A + B$ is $1.4 \times 10^{-4}\,s^{-1}$. Explain what this statement means. In what time is half the compound AB changed into A and B? (OS)

6 Methoxymethane decomposes, under certain conditions, into methane, carbon monoxide and hydrogen.

$$(CH_3)_2O(g) \longrightarrow CH_4(g) + CO(g) + H_2(g)$$

A sample of the ether was found to exert an initial pressure of $40\,kPa$. After 10 seconds, the pressure had risen by $1.08\,kPa$. How long will it take for the pressure to reach $80\,kPa$?

7 In the reaction between nitrogen oxide and hydrogen to form nitrogen and water, the times for half change are 140 and 102 seconds at initial gas pressures of 38.4 and $45.39\,kPa$ respectively. What is the order of the reaction? Suggest a possible mechanism.

8 In the thermal decomposition of dinitrogen oxide the half-life in seconds rises from 255 to 470 to 860 as the initial pressure is changed from 39.47 to 18.53 to $7\,kPa$. What is the order of the reaction?

9 Explain how you would follow the decomposition of diazoethanoic ester which decomposes in aqueous solution with the evolution of nitrogen according to the following equation:

$$N_2CH.COOEt + H_2O \longrightarrow HOCH_2.COOEt + N_2$$

Describe the apparatus that you would use and outline the sort of measurements you would make. What form of rate equation would you expect the decomposition to follow? (OS)

10 What factors may influence the rate of a chemical reaction? Illustrate your answer with reference to   (a) the saponification of ethyl ethanoate (b) the reaction between sulphur dioxide and oxygen   (c) the preparation of oxygen from potassium chlorate(V). For any one of these reactions describe how the rate could be measured experimentally.

11 Some chemical reactions take place at great speed; others proceed very slowly. What are the reasons for these differences? Illustrate your answer with examples. (OS)

12 What is the order of a reaction in which half the reagents react in half an hour, three-quarters in one hour and seven-eighths in one and a half hours?

13 Define the terms molecularity and reaction order as used in chemical kinetics. Give one example of a reaction for which the two are numerically equal, and one for which they differ.

Equal volumes of two solutions, one containing a certain concentration of substance A and the other the same concentration of substance B, are mixed and the reaction $A + B \rightarrow C$ ensues. At the end of 100 s, one quarter of the orginal A remains unreacted. How much A (expressed as a fraction of the amount of A originally present) will be left unreacted at the end of 200 s if the reaction is (i) first order in A and zero order in B (ii) first order in both A and B (iii) zero order in both A and B? (CC)

14 Discuss and explain fully the methods that are available for increasing the velocity of chemical changes. Illustrate the use of the methods you mention. (OLS).

15* Using the information from your data book, calculate the standard enthalpy change at 298 K for the reaction

$$N_2O_4(g) \longrightarrow 2NO_2(g)$$

Using this reaction as an example, and giving equations where appropriate, distinguish carefully between (a) the rate of reaction and the rate constant (b) the position of equilibrium and the equilibrium constant. Explain (giving reasons) how each of these four quantities would be affected by an increase in temperature. (OC)

16* Give the rate expressions for any six reactions not mentioned in this chapter.

27
Reaction mechanisms

1 Molecularity

Once the experimental rate law or expression for a reaction is established, attempts are made to account for the experimental facts in terms of a specific reaction mechanism. What actually happens to the atoms, molecules or ions in the course of a reaction?

For simple, one-step reactions the mechanisms are, at least qualitatively, simple. If only one species is involved, i.e.

$$AB \longrightarrow A + B \qquad radium \longrightarrow other\ atoms$$

the reaction is said to be *unimolecular*. Single atoms or molecules with sufficient energy decompose. When two species are involved, e.g.

$$A + B \longrightarrow AB \qquad AB + X \longrightarrow AX + B$$

the reaction is said to be *bimolecular*. The general idea is that the two species react together when they collide so long as they have sufficient energy.

The molecularity is a theoretical postulate and not an experimentally measured quantity. It must be a whole number because fractions of atoms or molecules cannot be involved in reactions; it may or may not be equal to the order of the reaction.

Many chemical reactions are not, however, simple one-step processes. The products may react to reform the reactants. The products may partially react to form some other products, thus giving rise to consecutive reactions. The reactants may react together in more than one way, so that side reactions exist. Chain reactions may build up. Elucidation of reaction mechanisms in such cases can be very difficult. The overall reaction is generally broken down into a series of simple reactions with different molecularities and different rates. The overall molecularity of a complex reaction is generally taken as the molecularity of the rate-determining step, but the term is used most clearly when it is limited to simple, one-step processes.

2 Activation energy

The rate of a reaction increases as the temperature rises, and the rate constant is only constant at a fixed temperature. This would be expected from kinetic theory considerations, for increase in temperature will increase molecular motion and raise the rate of intermolecular collisions.

Calculations based on the total number of collisions, however, give reaction rates which are higher than those found experimentally (p. 320). They also suggest that rates of reactions, around room temperature, should rise by about 2 per cent for each 1 K rise in temperature. In reality, the rise is generally about 10 per cent, i.e. the reaction rate about doubles for a rise of 10 K.

To account for these facts, Arrhenius suggested, in 1899, that a molecule would only react if it had higher than average energy, i.e. if it was activated, the necessary energy for reaction to occur being known as the activation energy. It is generally quoted as a molar activation energy in $kJ\,mol^{-1}$.

He found, moreover, that the experimental relationship between the rate constant for a reaction, k, and the temperature was given by

$$k = Ae^{-E/RT} \qquad \text{or} \qquad d\ln k/dT = E/RT^2$$

$$\ln k = \ln A - E/RT \qquad \text{or} \qquad \lg k = \lg A - E/2.303RT$$

where A is a constant (sometimes known as the pre-exponential constant) and E is the activation energy for the reaction. Neither A nor E are, in fact, independent of temperature, but it is the much greater dependence of $e^{-E/RT}$ that predominates.

These relationships can be proved by showing that plots of $\ln k$ or $\lg k$ against $1/T$ give straight lines. The slopes of the lines, or the expressions, can be used to obtain values of E.

In the thermal decomposition of dinitrogen oxide at high temperatures, for example,

$$2N_2O(g) \longrightarrow 2N_2(g) + O_2(g)$$

the values of $\lg(k \times 1000)$ were found to be 4.064 at 1125 K and 3.575 at 1085 K. As

$$\lg k_1 - \lg k_2 = -\frac{E}{R \times 2.303}\left[\frac{1}{T_1} - \frac{1}{T_2}\right]$$

it can be calculated that the energy of activation is $285.6\,kJ\,mol^{-1}$.

Similarly, the rate constant at one temperature can be found from that at another if the value of E is known.

3 Simple collision theory

The simplest and earliest interpretation of Arrhenius's experimental results involves collision theory, the basic idea being that reaction rates are equal to the frequency of collision between activated molecules.

The total number of collisions (Z) between molecules in a gas per unit volume per unit time can be calculated* from kinetic theory considerations by making certain assumptions about molecular diameters, and it is normally expressed in molecules $s^{-1} cm^{-3}$. As the fraction of molecules which are activated is given by $e^{-E/RT}$ (p. 319) it follows that

$$\text{frequency of collision between activated molecules} = \text{rate of reaction in molecules } s^{-1} cm^{-3} = Ze^{-E/RT}$$

On this basis, reaction rates would be expected to depend on the values of Z, E and T.

a The value of Z Collision frequencies in a gas at room temperature and pressure are of the order of $10^{28} s^{-1} cm^{-3}$. As the total number of molecules in $1 cm^3$ of gas under these conditions is about 2.7×10^{19} it follows that all gas reactions would be over in about $10^{-9} s$ if all collisions led to reaction. Most gas reactions are slower than that, because it is only activated collisions that count.

b The value of E At a given temperature, the value of $e^{-E/RT}$, i.e. the fraction of molecules that are activated, will be higher the lower E is. At $27\,°C$, for example,

$E/kJ\,mol^{-1}$	50	100	200
fraction of molecules that are activated	1.97×10^{-9}	3.98×10^{-18}	1.5×10^{-35}

Low activation energies, therefore, mean higher reaction rates. Halving the activation energy from 100 to 50 increases the rate by a factor of about 5×10^8.

c The value of T Change of temperature affects both the value of Z and the value of $e^{-E/RT}$ but the effect on the latter is very much greater than on the former. For an activation energy of $50\,kJ\,mol^{-1}$

T/K	300	310	1000	2000
fraction of molecules that are activated	1.97×10^{-9}	3.76×10^{-9}	2.45×10^{-3}	4.95×10^{-2}

---

* The total number of collisions $cm^{-3} s^{-1}$ for like molecules, A, (Z_{AA}) and for unlike molecules, A and B, (Z_{AB}) are

$$Z_{AA} = 2N_A{}^2 d_A{}^2 \left\{ \frac{\pi RT}{M_A} \right\}^{\frac{1}{2}} \qquad Z_{AB} = N_A N_B d_{AB}{}^2 \left\{ 8\pi RT \left(\frac{1}{M_A} + \frac{1}{M_B} \right) \right\}^{\frac{1}{2}}$$

where N_A and N_B are the number of molecules per cm^3, M_A and M_B are the relative molecular masses, d_A is the diameter of A and d_{AB} is the mean diameter of A and B.

For temperatures around room temperature, an increase of about $10\,K$ doubles the value of $e^{-E/RT}$. That is why reaction rates of many reactions about double for a temperature rise of $10\,K$.

In passing from 1000 to $2000\,K$ the fraction of activated molecules increases by a factor of about 20. The average energy (p. 26) increases only two-fold.

d Calculation of rate constant For a second-order gas reaction

$$A(g) + B(g) \longrightarrow products$$

the total number of collisions, Z_{AB}, is proportional (p. 320) to N_A and N_B, i.e.

$$Z_{AB} = YN_AN_B$$

The number of activated collisions will be $Z_{AB}e^{-E/RT}$ and this will be equal to the rate of the reaction if it is expressed in molecules $cm^{-3}\,s^{-1}$. For equal initial amounts of A and B:

$$\text{rate (in molecules cm}^{-3}\,s^{-1}) = -\frac{dN_A}{dt} = -\frac{dN_B}{dt}$$

$$= kN_AN_B = YN_AN_Be^{-E/RT}$$

$$k(\text{in cm}^3\,\text{molecules}^{-1}\,s^{-1}) = Ye^{-E/RT}$$

To get the value of k in its more usual units of $dm^3\,mol^{-1}\,s^{-1}$ it is necessary to multiply by $L/10^3$ where L is the Avogadro constant. It follows then that

$$k(\text{in dm}^3\,\text{mol}^{-1}\,s^{-1}) = 10^{-3}LYe^{-E/RT}$$

This theoretical expression is very similar to the experimental expression of Arrhenius (p. 319),

$$k = Ae^{-E/RT}$$

but the theoretical expression does make it clear that A, equal to $10^{-3}LY$ varies with the square root of the temperature. (See footnote on page 320, for value of Y).

For bimolecular gas reactions, and for some reactions in solution when simple molecules are involved, the theoretical and experimental values of k agree within a factor of 10 or so, but large discrepancies arise in other cases.

e Limitations of collision theory The quantitative failure of simple collision theory suggest that it is oversimplified. Attempts have been made to make it more successful by introducing arbitrary correction factors, known as *steric* or *probability factors*, p, so that the basic rate equation is written as

$$\text{rate} = pZe^{-E/RT}$$

Values of p vary very greatly from reaction to reaction and its significance is not well understood. It may reflect the necessity for activated molecules to be correctly orientated with respect to each other for reaction to occur, but many other factors are probably involved.

To account more satisfactorily for the facts, it is necessary to treat the idea of collision between molecules in a more sophisticated way. This is done in transition state or activated complex theory, and more accurately still, in molecular reaction dynamics.

4 Transition state or activated complex theory

This theory considers the nature of molecular collisions in more detail than simple collision theory. In a reaction

$$AB + X \longrightarrow A + BX$$

for example, AB and X will have a certain energy when remote from each other. If X approaches AB with sufficient energy its electrons will begin to overlap with those of AB so that some bonding between B and X will begin to form. At the same time, the AB bond will lengthen and weaken. As the BX bond strengthens and the AB bond weakens an activated complex or transition state, written as [A⋯B⋯X] will be formed, and this will then decompose into A + BX. In summary:

$$AB + X \longrightarrow [A\cdots B\cdots X] \longrightarrow A + BX$$

a Reaction profile for one-step reaction The energy changes taking place in a simple reaction as above can be calculated (p. 324) and are conveniently summarised in diagrams as shown in Fig. 150. The diagrams are referred to as reaction, or energy profiles or pathways. The progress of the reaction is represented as moving from left to right; the horizontal, x-axis is sometimes referred to as the *reaction coordinate*. The activation energy for the reaction is E_I; that for the reverse reaction is

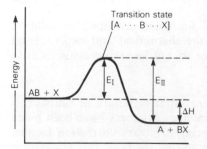
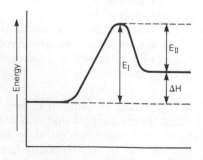

Fig. 150 Reaction profiles for a one-step reaction between AB and X to form A and BX via a transition state, [A⋯B⋯X]. The profile on the left shows an exothermic reaction; ΔH will be negative. That on the right is for an endothermic reaction with ΔH positive.

E_{II}. It will be seen that it is the difference between E_I and E_{II} that determines whether the reaction is exothermic or endothermic.

The transition state or activated complex is the arrangement of atoms of maximum energy (least stability) through which the reagents must pass before the products can be formed. It is not a discrete molecule that can be isolated.

b Activation energy and enthalpy of reaction In the relationship between rate constant and temperature the activation energy is the important factor (p. 319). The relationship between equilibrium constant and temperature is very similar in form (p. 295) but it is the enthalpy of reaction that is the important factor:

$$\frac{d(\ln k)}{dT} = E/RT^2 \qquad \frac{d(\ln K)}{dT} = \Delta H_m/RT^2$$

The close similarity between the two expressions is well understood by considering the application of Fig. 150 to a simple reversible reaction

$$AB + X \rightleftharpoons A + BX \qquad \Delta H_m = Q \, kJ$$

At equilibrium, the rate of the forward reaction will be equal to the rate of the backward reaction, so that

$$k_1[AB]_{eq}[X]_{eq} = k_2[A]_{eq}[BX]_{eq}$$

As $\quad K = \dfrac{[A]_{eq}[BX]_{eq}}{[AB]_{eq}[X]_{eq}} \quad$ it follows that $K = k_1/k_2$

For the forward and back reactions

$$\frac{d(\ln k_1)}{dT} = E_I/RT^2 \qquad \text{and} \qquad \frac{d(\ln k_2)}{dT} = E_{II}/RT^2$$

so that $\qquad \dfrac{d(\ln K)}{dT} = \dfrac{d(\ln k_1)}{dT} - \dfrac{d(\ln k_2)}{dT}$

or $\qquad \dfrac{d(\ln K)}{dT} = (E_I - E_{II})/RT^2 = \Delta H_m/RT^2$

c Reaction paths for two-step reactions In a two-step reaction the first step gives a product known as the intermediate. If this is of low enough energy it may be stable enough to be isolated, or it may be so unstable that it only has a transient existence. The intermediate takes part in the second step of the reaction, e.g.

overall reaction	$AB + X \longrightarrow A + BX$
1st step (slow)	$AB \longrightarrow (A \cdots B) \longrightarrow A + B$
2nd step (fast)	$B + X \longrightarrow (B \cdots X) \longrightarrow BX$

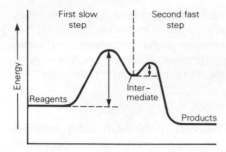

Fig. 151 Reaction profile for a two-step reaction, a slow first step with a high activation energy, being followed by a fast second step with a lower activation energy.

The overall rate of the reaction will be fixed by the rate of the slower, rate-determining step with the higher activation energy (the first step in the example given). The reaction profile (Fig. 151) links together the profiles of the two elementary reactions involved. It shows two transition states and one intermediate.

d Advantages of transition state theory Transition state theory is an advance on collision theory. In the first place, it enables activation energies to be calculated, at least in principle. The free energy of a system AB + X can be calculated for any A—B and B—X distances, and the results are generally plotted on a 'contour diagram' (Fig. 152) in which the lines show equal energy values in just the same way as contours on a map show equal heights.

The 'valley' at R represents the energy for the initial state of the system, i.e. AB + X, whilst the 'valley' at P represents the final state, i.e. A + BX. The energy contours rise in all directions from the valleys at R and P, but the 'easiest' path is shown by the bold line, with T representing the transition state. The point T is rather like a col in a mountain

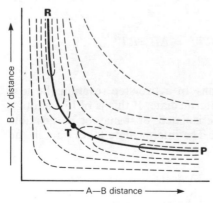

Fig. 152 The variation of energy for different BX and AB distances in the system A—B—X.

region. The reaction path is along the line RTP and it is this line which is represented in Fig. 150. The transition state is in an unusual position at T. It can either form $A + BX$, or revert to $AB + X$, by moving along RTP; in either case it will lower its energy. But it cannot move in any other direction except by raising its energy. That is why the reaction path is specific.

Transition state theory can also be used to show that the value of the pre-exponential constant, A, in the Arrhenius equation (p. 319) depends on the entropy of formation of the transition state. Values of A can, therefore, be calculated (at least in principle) without making any assumptions to estimate collision frequencies or without invoking any mysterious steric constants (p. 321).

5 Some examples of reaction mechanisms

To build up a reaction mechanism from a series of one-step reactions, and to account for the experimental order of the reaction, may not be easy because the transition states and intermediates cannot always be isolated. It may be possible, however, to discover their nature by spectroscopic methods (p. 184). Isotopic tracer techniques (p. 92) and stereochemical changes may also provide useful guidelines. Even so, very detailed work is required and some inspired guesswork may be necessary.

a The thermal decomposition of dinitrogen pentoxide This was one of the first first-order reactions to be investigated. The stoichiometric equation is

$$2N_2O_5(g) \longrightarrow 4NO_2(g) + O_2(g)$$

and the rate is proportional to $[N_2O_5]$. The suggested stages are as follows:

$$N_2O_5(g) \rightleftharpoons NO_2(g) + NO_3(g)$$

$$NO_2(g) + NO_3(g) \xrightarrow{\text{SLOW}} NO_2(g) + O_2(g) + NO(g)$$

$$NO(g) + NO_3(g) \xrightarrow{\text{FAST}} 2NO_2(g)$$

with the first stage being known as a pre-equilibrium. Support for these ideas comes from the fact that radioactive $^{15}N_2O_5$ is formed if radioactive $^{15}NO_2$ is added to the reaction mixture.

b The halogenation of propanone Propanone, and other ketones, undergo halogenation as typified with iodine

$$CH_3.CO.CH_3(aq) + I_2(aq) \rightarrow CH_3.CO.CH_2I(aq) + H^+(aq) + I^-(aq)$$

Introduction to Physical Chemistry

The rate is found to be proportional to the propanone concentration, to the H^+ ion concentration if carried out in acid solution, and to the OH^- ion concentration if carried out in alkaline solution. It is clear, then, that the mechanism must involve a slow, rate-determining step in which iodine does not take part. This is thought to be the conversion of the ketone into its enol-form, i.e.

$$CH_3-\overset{\overset{\textstyle O}{\|}}{C}-CH_3 \xrightarrow[\text{FAST}]{H^+} CH_3-\overset{\overset{\textstyle \overset{\textstyle H}{\overset{|}{O^+}}}{\|}}{C}-CH_3 \xrightarrow[\text{SLOW}]{H_2O} CH_3-\overset{\overset{\textstyle \overset{\textstyle H}{\overset{|}{O}}}{}}{C}=CH_2$$
$$+ H_3O^+$$

$$CH_3-\overset{\overset{\textstyle O}{\|}}{C}-CH_3 \xrightarrow[\text{SLOW}]{OH^-} CH_3-\overset{\overset{\textstyle O}{\|}}{C}-CH_2^- \xrightarrow[\text{FAST}]{H^+} CH_3-\overset{\overset{\textstyle \overset{\textstyle H}{\overset{|}{O}}}{}}{C}=CH_2$$
$$+ H_2O \qquad\qquad \text{enol-form}$$

The enol-form then reacts rapidly with iodine,

$$CH_3-\overset{\overset{\textstyle \overset{\textstyle H}{\overset{|}{O}}}{}}{C}=CH_2 + I_2 \longrightarrow CH_3-\overset{\overset{\textstyle O}{\|}}{C}-CH_2I + H^+ + I^-$$

c The dissociation of hydrogen iodide The reaction

$$2HI(g) \rightleftharpoons H_2(g) + I_2(g)$$

has been very extensively studied and the rate expressions are simple,

$$\text{forward rate} = k[HI]^2 \qquad \text{backward rate} = k'[H_2][I_2]$$

The forward reaction is thought to be bimolecular, collision between activated hydrogen iodide molecules leading to reaction. The rate constant calculated on this basis is in agreement with the experimental value.

For a long time, the back reaction was also thought to be similar, with reaction between colliding activated molecules of hydrogen and iodine. More recent work, however, suggests the following mechanism.

$$I_2 \rightleftharpoons 2I \qquad I + H_2 \xrightarrow{\text{FAST}} H_2I \qquad H_2I + I \xrightarrow{\text{SLOW}} 2HI$$

d The oxidation of I^- by hydrogen peroxide in acid solution The overall equation for the reaction is

$$2I^-(aq) + 2H^+(aq) + H_2O_2(aq) \longrightarrow I_2(aq) + 2H_2O(l)$$

and the rate of the reaction (p. 314) is given by

$$-d[H_2O_2]/dt = k[H_2O_2][I^-] + k'[H_2O_2][I^-][H^+]$$

The first term in the rate equation does not involve H^+ ions. This is accounted for by the following steps,

$$H_2O_2 + I^- \xrightarrow{\text{SLOW}} HOI + OH^-$$

$$HOI + I^- \xrightarrow{\text{FAST}} I_2 + OH^-$$

$$2OH^- + 2H^+ \xrightarrow{\text{FAST}} 2H_2O$$

The dependence on H^+ arises because of a parallel mechanism,

$$H_2O_2 + H^+ \rightleftharpoons H_3O_2^+$$

$$H_3O_2^+ + I^- \xrightarrow{\text{SLOW}} HOI + H_2O$$

with the HOI then undergoing the same fast reaction with I^- as previously.

e The alkaline hydrolysis of esters These reactions, e.g.

$$RCOOR'(aq) + OH^-(aq) \longrightarrow RCOO^-(aq) + R'OH(aq)$$

are second-order with

$$\text{rate} = k[\text{Ester}][OH^-]$$

When water containing $H_2{}^{18}O$ is used ${}^{18}OH^-$ is formed and the ${}^{18}O$ atom is eventually found in the resulting acid and not in the alcohol. This suggests that the bond cleavage is between the acyl group and the oxygen atom, i.e.

$$\begin{array}{c} O \\ \parallel \\ R-C \!\mid\! O- R' \end{array}$$

The suggested mechanism is as follows:

$$\begin{array}{c} O \\ \parallel \\ C-OR' \\ \mid \\ R \end{array} \xrightarrow[\text{SLOW}]{OH^-} \begin{array}{c} O^- \\ \mid \\ HO-C-OR' \\ \mid \\ R \end{array} \xrightarrow{\text{FAST}} \begin{array}{c} O \\ \parallel \\ HO-C-R \\ + \\ (OR')^- \end{array} \xrightarrow{\text{FAST}} \begin{array}{c} RCOO^- \\ + \\ R'OH \end{array}$$

f The hydrolysis of alkyl halides The hydrolysis of bromomethane by a dilute solution of potassium hydroxide in a solvent of 80 per cent ethanol and 20 per cent water is found to be of second order, and of first order with respect to both the bromomethane and OH^- ion concentrations.

$$\text{rate} = k[CH_3Br][OH^-]$$

The reaction proceeds via the formation of a transition state,

$$HO^- + CH_3Br \longrightarrow [HO\cdots CH_3\cdots Br]^- \longrightarrow HO-CH_3 + Br^-$$

Such a process is bimolecular, and a similar mechanism accounts for the hydrolysis of bromoethane under similar conditions.

But for higher halides such as 2-bromo-2-methyl propane the kinetics of the hydrolysis are first order. The rate is independent of the OH^- ion concentration, the mechanism postulated being,

$$BuBr \xrightarrow{\text{SLOW}} Bu^+ + Br^- \qquad Bu^+ + OH^- \xrightarrow{\text{FAST}} BuOH$$

This is a unimolecular process.

The bimolecular process is referred to as an S_N2 reaction: S for substitution, $_N$ for nucleophilic (the OH^- ion is a nucleophilic reagent) and 2 for bimolecular. The unimolecular process is an S_N1 reaction.

The way in which a halide will undergo hydrolysis depends mainly on the nature of the halide and how readily it can ionise, and on the ionising power of the solvent. Thus the S_N2 hydrolysis of bromomethane in ethanol–water solution is replaced by an essentially S_N1 hydrolysis in methanoic acid solution, i.e. in a better ionising medium. In ethanol–water solution, 2-iodopropane undergoes both S_N1 and S_N2 hydrolysis side by side.

The mechanism by which a halide undergoes hydrolysis under different conditions can be accounted for by electronic theories of organic chemistry.

6 Chain reactions

Many gas reactions have complex rate expressions because they proceed through a series of steps, involving highly reactive free atoms or radicals.

a Reaction of hydrogen with bromine or chlorine The rate expression for the reaction of bromine with hydrogen to form hydrogen bromide is

$$\text{rate} = k[H_2][Br_2]^{1.5}/([Br_2] + k'[HBr])$$

where k and k' are constants. Clearly the idea of a simple numerical value for the order is not valid.

The theoretical interpretation involves a chain reaction which can be summarised in the following steps:

initiating	Br_2	$\longrightarrow 2Br\cdot$	i
propagating	$Br\cdot + H_2$	$\longrightarrow HBr + H\cdot$	ii
	$H\cdot + Br_2$	$\longrightarrow HBr + Br\cdot$	iii
inhibiting	$H\cdot + HBr$	$\longrightarrow H_2 + Br\cdot$	iv
terminating	$Br\cdot + Br\cdot$	$\longrightarrow Br_2$	v

The initiation (the formation of bromine atoms) occurs by the thermal dissociation of bromine molecules. Heat, or ultraviolet light, is needed to start the process, but it need only take place to a small extent because the two rapid propagating steps form two molecules of hydrogen bromide and regenerate Br and H atoms to carry on the chain. Reaction *iv* inhibits the process by using up both H atoms and HBr molecules, and reaction *v* terminates the chain by removing the Br atom chain carriers.

The whole series of processes can be interpreted theoretically by assuming that a stationary state is set up soon after the reaction is started. The rate of formation of any atom will then be equal to its rate of removal by further reaction. By equating the necessary rate expressions it is possible to obtain an overall rate expression in agreement with that found experimentally.

In the formation of hydrogen bromide by the suggested series of reactions, the reaction *ii* is endothermic and relatively slow. With chlorine however, it is rapid, so that the chain reaction can take place very easily, and possibly explosively. With iodine, reaction *ii* is very slow so that hydrogen iodide is not formed by a chain reaction (p. 326).

b The reaction of chlorine and methane Chlorine and methane will not react in the dark, but the reaction can be initiated by heating or by ultraviolet light; it may be explosive. The chain reaction is outlined as follows:

initiating $\qquad Cl_2 \qquad\qquad \longrightarrow 2Cl\cdot$

propagating $\qquad Cl\cdot + CH_4 \longrightarrow HCl + CH_3\cdot$

$\qquad\qquad\qquad CH_3\cdot + Cl_2 \longrightarrow CH_3Cl + Cl\cdot$

terminating $\qquad 2Cl\cdot \qquad\quad \longrightarrow Cl_2$

$\qquad\qquad\qquad 2CH_3\cdot \qquad\; \longrightarrow C_2H_6$

$CH_3\cdot$, the methyl radical, is an example of an organic free radical.

c The existence of free radicals The actual transitory existence of free radicals was first shown by Paneth in 1929. A stream of pure nitrogen or helium was saturated with tetramethyl lead(IV), $Pb(CH_3)_4$, by passing it over liquid tetramethyl lead(IV). The mixture was then passed into a quartz tube heated at a point A to 500–600 °C. The thermal decomposition of the tetramethyl lead(IV) was shown by the formation of a deposit of lead at point A.

The gaseous products passing along the tube were found to react with cold metallic deposits of lead and other metals, placed in the tube at B, beyond A. The products of these reactions were identified as metallic methyl compounds, indicating that the initial tetramethyl lead(IV) had decomposed, at 500–600 °C, into lead and free methyl radicals. By varying the distance between A and B and using gas flows of different speeds it was possible to show that the methyl radical had a half-life of about 10^{-2} s.

Nowadays, flash photolysis (p. 309) can be used to show the existence and identity of free radicals, and to measure their half-lives.

d Characteristics of chain reactions The majority of chain reactions have two characteristics which may enable their existence to be recognised.

In the first place, if a chain reaction depends on free radicals then the introduction of such radicals ought to expedite the reaction. There are many examples where this is found to be so. Ethane and propane can be chlorinated almost completely in the dark at 150 °C if a trace of tetraethyl lead(IV) is present to provide ethyl radicals. The use of tetraethyl lead(IV) as an antiknock in petrol is almost certainly connected with its ability to facilitate the decomposition of hydrocarbons. Many polymerisation processes are initiated by free radicals from organic peroxides or azo-compounds.

Similarly, chain reactions can be inhibited by adding anything which will react with the necessary free radicals. Nitrogen oxide, NO, which is itself a molecule with an odd electron structure, is particularly effective. Because of the possibilities of chains being broken, chain reactions are generally very sensitive to impurities. As free radicals may be removed by reaction with the walls of a reaction vessel, chain reactions may be susceptible to surface effects.

7 Photochemistry

The chemical effects of visible, infrared and ultraviolet light are studied in photochemistry. It is important in photosynthesis, photography and in many photochemical, free-radical chain reactions.

a The laws of photochemistry *Only light which is absorbed can give rise to chemical effects*, and this is sometimes known as the first law of photochemistry. It was first suggested as early as 1818 by Grotthus and Draper. Einstein suggested, in 1908, that *the absorption of one quantum of light activates one atom or molecule*, and this statement is known as the second law.

The energy of one quantum of light is equal to hv, where h is Planck's constant (p. 60) and v the frequency of the light used. On a mole basis, the energy is Nhv where N is the Avogadro constant, and this amount of energy is known as an *einstein*. For visible light of wavelength 600 nm, its value is $199.3 \, \text{kJ mol}^{-1}$; for ultraviolet light of wavelength 200 nm it is $598 \, \text{kJ mol}^{-1}$.

b Quantum yield The initial absorption of one quantum of light might lead to a chain reaction, and the quantum yield or efficiency is defined as the number of molecules decomposed or formed per quantum absorbed. In the photochemical reaction between hydrogen and chlorine, for example, the initiating step is the splitting of chlorine molecules into Cl atoms. The quantum yield of the resulting chain reaction (p. 329) is between 10^4 and 10^6.

By comparison, the quantum yield for the decomposition of ethane-dioic acid in the presence of uranyl, UO_2^{2+}, salts, which is initiated by the activation by light of the uranyl ion, is 0.57 for light of wavelength 300 nm. This is so constant that the amount of light being absorbed can be discovered by measuring the amount of ethanedioic acid decomposed, by titration with standard potassium manganate(VII) solution.

8 Radiation chemistry

High energy radiation such as X-rays and γ-rays or beams of protons or neutrons or α- or β-particles can cause ionisation in the matter by which they are absorbed; they are called *ionising radiations*. The electrons released also have high energy and may bring about further ionisation.

Widespread, but complex, changes can ensue. α-particles, for example, can convert water into hydrogen peroxide, e.g.

$$H_2O \longrightarrow H_2O^+ + e^- \longrightarrow H + OH \qquad 2OH \longrightarrow H_2O_2$$

Radiation chemistry is still new. It is of great interest, partially because of the known biological effects of radiation and the connection with such diseases as cancer.

Questions on Chapter 27

1 The values of $\lg(k \times 1000)$ for the reaction $2N_2O \rightarrow 2N_2 + O_2$ are 4.064 and 3.575 at 1125 and 1085 K respectively. Calculate the energy of activation for the reaction.

2 The fraction of molecules with a higher energy than E is given by $e^{-E/RT}$. Use this relationship to plot a graph of the fraction of molecules having greater energy than E against E.

3 If the activation energy for a reaction is 83.14 kJ what will be the approximate ratio of the rate constant of the reaction at 27 °C to that at 37 °C? What will the ratio be for a reaction with an activation energy of 53.59 kJ?

4 Calculate the activation energy for a reaction which doubles in rate between 27 °C and 37 °C.

5 A very large number of reactions have activation energies between about 60 and 250 kJ mol^{-1}. Why is this?

6 5cm^3 portions of M/15 sodium thiosulphate solution and M/10 hydrochloric acid were mixed at different temperatures, the time for the appearance of a sulphur precipitate being recorded. The results obtained were:

temp/K	298	305	308	317	322	334	341
time/s	60	43	25	16	11	6	4

Plot a graph of $\lg(1/\text{time})$ against $1/\text{temperature}$ and comment on its significance.

7 The rate constants for the decomposition of dinitrogen pentoxide at different temperatures are given below:

temp/K	298	308	318	328	338
k/min^{-1}	0.00203	0.00808	0.0299	0.0900	0.292

Plot $\lg k$ against the reciprocal of the temperature, and use your graph to find the average activation energy for the decomposition.

8 Distinguish carefully between the meaning of activation energy and free energy. Would you expect there to be any relationship between the numerical values of these quantities for a given reaction?

9 What do you consider to be the important factors which influence the rates of chemical reactions? Indicate underlying physical principles. Over the temperature range 291K to 324K the rate at which nitrogen pentoxide decomposes in the gas phase increases by a factor of 100. By what further factor would you expect it to increase on heating to 343K? Explain your reasons. (OJSE)

10 Describe, with experimental details, how you would determine the rate of the acid-catalysed hydrolysis of methyl ethanoate (acetate) at $40 \,^\circ\text{C}$. Account for each of the following observations.   (a) The reaction $S_2O_8^{2-}(aq) + 2I^-(aq) \longrightarrow 2SO_4^{2-}(aq) + I_2(aq)$ is first order with respect to $S_2O_8^{2-}$ and also with respect to I^-.   (b) At room temperature, a mixture of hydrogen and chlorine will not react until irradiated with ultraviolet light.   (c) The calculated collision frequency between gas molecules is usually some 10^{10} times greater than the measured rate of reaction between the same molecules.   (d) The collision frequency between gas molecules is proportional to the square root of the thermodynamic temperature but a rise in temperature of 10K often increases the rate of a reaction between gases by a factor of about two. (OC)

11 Distinguish carefully between the rate constant for a chemical reaction and the equilibrium constant for a reversible process. Discuss how each of these quantities might be expected to vary with temperature. The equilibrium constant for the reversible reaction

$$H + OH \rightleftharpoons H_2 + O$$

is 1.2 at 1000K, but 0.9 at 1500K. The rate constant for the forward reaction is $2 \times 10^8 \text{dm}^3 \text{mol}^{-1}\text{s}^{-1}$ at 1000K and the rate constant for the reverse reaction is $7 \times 10^8 \text{dm}^3 \text{mol}^{-1}\text{s}^{-1}$ at 1500K. What can you deduce from these data? (OJSE)

12 The hydrolysis of methyl methanoate in dilute aqueous solution is catalysed by hydrogen ions. Describe the experiments you would make to verify this statement, and to discover how the rate of hydrolysis depends on the concentration of the ester and of the hydrogen ion. The reaction between propanone and iodine (to give iodopropanone) is catalysed by hydrogen ions. The rate of the reaction is proportional to the product of the propanone concentration and of the hydrogen ion concentration, but is independent of the iodine concentration. What can you infer from this information? (OS)

13 The thermal decomposition of phosphine into phosphorus and hydrogen is found to be a first-order reaction. Suggest a possible mechanism.

14 The rate of reaction between potassium iodate(V) and sulphurous acid solutions is found to be proportional to the product of the concentrations of potassium iodate(V) and sulphurous acid. The overall reaction is represented by the equation

$$KIO_3 + 3H_2SO_3 \longrightarrow KI + 3H_2SO_4$$

Suggest a likely mechanism for the reaction.

15 The rate of reaction between phenylamine (aniline) and iodine to form 3-iodophenylamine in the presence of potassium iodide and a dilute buffer solution is found to decrease with increasing iodide concentration and increase with increasing pH. Suggest a possible mechanism of the reaction.

16 Organic esters can be hydrolysed in both acid and alkaline solutions. Compare the two reaction mechanisms.

17 What is the value of an einstein, in $kJ\,mol^{-1}$, (a) for red light of frequency $4 \times 10^{14}\,Hz$ (b) radiation of wavenumber $10^4\,cm^{-1}$?

18* How are rate constants affected by change in temperature and change in activation energy? Illustrate your answer by quoting some numerical values.

19* Record the rate constants at various temperatures for a particular reaction and use them to calculate the activation energy for the reaction.

20* How does a catalyst affect the rate of a reaction? Illustrate your answer with some quoted values for activation energies.

28
Catalysis

1 Characteristics and examples of catalysis

A catalyst is a substance which alters the rate of a chemical reaction, itself remaining chemically unchanged at the end of the reaction. There are so many different examples, with transition metals and their compounds playing a very large part, that classification is not easy. In the widest sense, however, catalysts can be regarded as homogeneous or heterogeneous. In *homogeneous catalysis*, the reaction takes place in one phase and the catalyst is uniformly distributed in the same phase; it is generally found in reactions in aqueous solution. In *heterogeneous* or *surface catalysis*, the catalyst is not in the same phase as the reacting mixture. This type commonly occurs in gas reactions catalysed by a solid. Some of the general characteristics of catalytic action are summarised below.

a The amount of catalyst needed As a catalyst is not used up it will go on functioning for a considerable time; theoretically, but not practically, for ever. Moreover, in many reactions, only a minute trace of catalyst is needed. A concentration of 2 g of colloidal platinum in 10^6 dm^3 will catalyse the decomposition of hydrogen peroxide, and Fe^{2+} ions at a concentration of 0.00003 M catalyse the oxidation of I^- ions by $S_2O_8^{2-}$ ions.

In other reactions, however, the rate of the reaction is proportional to the concentration of catalyst. Catalysis by H^+ or OH^- ions (p. 336), as in the halogenation of propanone (p. 325), the hydrolysis of sucrose (p. 308) and esters, or the reduction of hydrogen peroxide (p. 326), is usually of this type.

In heterogeneous catalysis, large amounts of a solid may be needed. Considerable care is taken to get the solid into a form with a large surface area. Wire may be woven into a gauze, or the catalyst mixture may be spread out on a support of a ceramic material, silica gel or asbestos.

b Specificity One catalyst will alter the rate of one reaction without necessarily having any effect at all on other reactions. Different catalysts can, moreover, bring about completely different reactions. Ethanol vapour, for example, is dehydrated by passing over hot aluminium oxide; ethene or ethoxyethane are formed depending on the temperature. Hot copper causes dehydrogenation to ethanal.

Similarly, methanoic acid vapour is dehydrated by hot aluminium oxide, but dehydrogenated by hot zinc oxide.

c Promoters The activity of a catalyst is sometimes very greatly enhanced by the addition of a small amount of other substances, which may have no catalytic activity on their own; they are known as promoters.

In the Haber process, the catalyst is made by fusing magnetite, Fe_3O_4, with small quantities of aluminium oxide and potassium hydroxide. In the course of the reaction, the iron oxide is reduced to iron, which is the main catalyst; the other materials act as promoters. The catalytic effect of Fe^{2+} ions on the oxidation of I^- by $S_2O_8^{2-}$ is also promoted by traces of Cu^{2+} ions which have no catalytic action by themselves.

d Poisoning The efficiency of a catalyst may be lowered by impurities, referred to as poisons. Platinum, for example, catalyses the oxidation of sulphur dioxide to sulphur trioxide by oxygen, but arsenic impurities in the sulphur dioxide poison the catalyst. That is why platinum was replaced by vanadium pentoxide in the contact process.

Poisoning can be put to good use. In the Rosenmund reaction, for example, an acid chloride is reduced to an aldehyde, by hydrogen, using a catalyst of palladium deposited on barium sulphate containing some sulphur as a poison. The sulphur lowers the catalytic activity for the second-stage reduction of the aldehyde to the alcohol and improves the yield of aldehyde. The reduction of alkynes, by hydrogen, to alkanes can also be stopped at the alkene stage by using a catalyst of palladium on calcium carbonate poisoned by quinoline and sulphur (Lindlar's catalyst).

e Negative catalysts Some catalysts, known as negative catalysts or *inhibitors*, slow down reactions. The decomposition of hydrogen peroxide, for example, is retarded by dilute acids or by propane-1,2,3-triol (glycerol). Other everyday examples of the use of negative catalysts are provided by antioxidants in rubber and plastics to retard atmospheric oxidation which causes perishing, stabilisers in rocket fuels, and rust inhibitors in antifreeze and other mixtures.

f Autocatalysis This occurs if one of the products of a reaction acts catalytically. The reaction rate at first rises, as the catalyst is formed, instead of decreasing steadily. Three examples, with the product acting as the catalyst underlined, are represented by the following equations.

$$CH_3.COOCH_3(l) + H_2O(l) \rightarrow \underline{CH_3.COOH}(aq) + CH_3OH(aq)$$

$$2MnO_4^-(aq) + 5C_2O_4^{2-}(aq) + 16H^+(aq) \rightarrow \underline{2Mn^{2+}}(aq) + 10CO_2(g) + 8H_2O(l)$$

$$ClO_2^-(aq) + 4I^-(aq) + 4H^+(aq) \rightarrow \underline{2I_2}(aq) + 2H_2O(l) + Cl^-(aq)$$

2 The action of homogeneous catalysts

A homogeneous catalyst lowers the activation energy of the rate-determining step of a reaction by forming an intermediate. A reaction

represented as

$$A + B \longrightarrow X$$

can then take place in the presence of a catalyst, C, via the quicker steps

$$A + C \longrightarrow AC \qquad AC + B \longrightarrow X + C$$

The catalyst does play a part in the reaction but is reformed; AC is an intermediate. The reaction pathway can be summarised as in Fig. 153. In simple language, a big hill is surmounted by taking an alternative, two-stage route involving two smaller hills. Lowering the activation energy from 100 to 50 kJ at 300 K increases the rate by a factor of about 5×10^8 (p. 320).

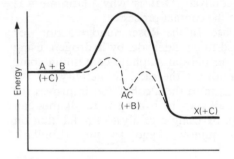

Fig. 153 Reaction profile showing how a catalyst, C, might provide an alternative path for a reaction between A and B to form X.

a Acid-base catalysis Many of the commoner examples of homogeneous catalysis involve *specific acid-base catalysis*. Acids can form intermediates by adding a proton to a molecule of one of the reactants, as in the acid catalysis of the halogenation of propanone (p. 325), or in the reaction of hydrogen peroxide and iodide ions in acid solution (p. 326). In general,

$$\text{reagent (X)} + H^+ \longrightarrow XH^+ \text{ (intermediate)}$$

and such processes are particularly easy when X includes polar bonds, such as O—H and C=O, with negatively charged O atoms.

Bases can form intermediates by extracting protons from the molecules of one of the reagents as in the base catalysed halogenation of ketones (p. 325). In general

$$\text{reagent (HX)} + OH^- \longrightarrow H_2O + X^- \text{ (intermediate)}$$

Such processes are common when a compound contains C—H bonds adjacent to C=O bonds.

When Lewis acids or bases (p. 389) are involved, as in the use of

$FeCl_3$, $AlCl_3$ or BF_3 in the Friedel–Crafts reaction or in the halogenation of benzene, the processes are referred to as *general acid-base catalysis*.

These catalytic reactions are affected by change of pH, i.e., by the strength of the acids and bases involved or by the addition of salts. The kinetics can be complicated for, particularly in aqueous solution, many different bases and acids may be present.

b The action of d-block ions As these ions can readily be converted from one oxidation state to another, it seems likely that their catalytic activity may depend on such changes. A reaction

$$A + B \longrightarrow R + S$$

may take place, in the presence of an M^{x+} catalyst, via the stages

$$A + M^{x+} \longrightarrow M^{y+} + R \qquad M^{y+} + B \longrightarrow M^{x+} + S$$

The oxidation of vanadium from $+3$ to $+4$, by Fe^{3+} is catalysed by Cu^{2+} ions in this way. The stages are

$$V^{3+}(aq) + Cu^{2+}(aq) \longrightarrow Cu^{+}(aq) + V^{4+}(aq)$$

$$Cu^{+}(aq) + Fe^{3+}(aq) \longrightarrow Fe^{2+}(aq) + Cu^{2+}(aq)$$

3 The action of heterogeneous catalysts

A very large number of important industrial processes depend on the use of heterogeneous catalysts, particularly at high temperatures, but their precise action is not known in any detail. Metals or metallic oxides are widely used, and the range of processes includes hydrogenation and dehydrogenation, oxidation and reduction, dehydration, cracking, reforming and polymerisation. There is some correlation between the type of catalyst and the nature of the process, e.g. metals are particularly effective for hydrogenation and dehydrogenation, but it is not clear-cut. As solid catalysts function best when their surface area is large it is clear that surface effects are involved.

The surface of a substance is different from the interior (p. 450). This shows up particularly well in the surface tension of liquids but there is a similar tension at the surface of solids. Each atom at the surface of a solid will, also, have free valencies and these enable the solid to form chemical bonds with other molecules or atoms in what is known as *chemisorption*. The name is chosen to distinguish it from *physical adsorption* in which molecules are held at a surface simply by van der Waals' forces (p. 151).

In a typical example of heterogeneous catalysis, such as the hydrogenation of alkenes in the presence of nickel, it is thought that both the hydrogen and the alkene are chemisorbed at the nickel surface. The bonds in the alkene and hydrogen are, therefore, weakened; so much so in the hydrogen that the molecule splits up into atoms. These H atoms can then react with an adjacent reactive alkene molecule in a two-step process. Hydrogenation can be summarised as in Fig. 154. Similar

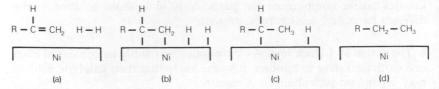

Fig. 154 The mechanism of hydrogenation of an alkene at a nickel surface. (a) Shows the approach of an alkene and a hydrogen molecule to the surface. (b) The molecules are chemisorbed and the H_2 bond has broken. (c) and (d) show the two-step addition of H atoms to the alkene molecule.

mechanisms are thought to operate in other examples of heterogeneous catalysis. The poisoning of a catalyst may be due to the preferential adsorption of the poison on the surface of the catalyst.

4 Enzymes as catalysts

Enzymes, made up of large protein molecules and colloidal in nature, provide examples of catalysis which are midway between homogeneous and heterogeneous. They are outstandingly effective, particularly specific and highly sensitive.

Urease (extracted from soya bean) will catalyse the hydrolysis of carbamide at a concentration of 1 part in 10 million, but it has no effect on the hydrolysis of methylcarbamide. Lactic dehydrogenase will oxidise L-lactic acid but not the D-acid. Other remarkable changes catalysed by enzymes include the conversion of starch into maltose by diastase, maltose into glucose by maltase, glucose into ethanol and carbon dioxide by zymase, and starch into sugars by enzymes such as pepsin, present in saliva juices. 'Natural' catalysts can, in fact, facilitate many complex reactions which simply cannot be brought about in any other way.

The functioning of an enzyme is very sensitive to temperature change, and biological enzymes generally function best at body temperatures. Enzymes are also greatly affected by changes in pH. Many of them, too, will only act as catalysts if linked to some other, simpler molecule known as a *coenzyme*. Sometimes, a metallic ion, known as an *activator*, must also be present.

All the evidence suggests that enzymes form intermediate compounds with other substances, referred to as the *substrate*, and the geometrical shape of the enzyme molecule in relation to that of the substrate seems to be an important factor. The analogy of a key fitting only one particular lock is widely used to illustrate the very specific nature of some enzymes.

Questions on Chapter 28

1 Describe and illustrate, with examples, the essential features of a catalyst. (OC)

2 What do you understand by the term catalyst? Give three examples of catalysis, and state how you would show experimentally that manganese(IV) oxide catalyses the decomposition of heated potassium chlorate(V).

3 What are the essential characteristics of a catalyst? What general explanations can you give of catalytic activity? How would you find out whether copper(II) oxide catalysed the decomposition of potassium chlorate(V)? (OC)

4 State the characteristic features of catalysis. Define the following terms, quoting one example of each: (a) homogeneous catalysis (b) heterogeneous catalysis (c) catalyst poison (d) autocatalysis. What influence has a catalyst on the composition of the final equilibrium mixture of a reaction? Mentioning essential experimental details, describe concisely how you would find out whether a given mineral acid acted as a catalyst for the hydrolysis of ethyl ethanoate by water. (JMB)

5 What effect can a catalyst have on (a) the velocity of a reaction (b) the equilibrium position in a reversible reaction? What types of catalyst are recognised and what theories have been advanced to explain their action? Name three industrial processes in which catalysts are used, stating the catalyst in each case. (OC)

6 A catalyst is something which speeds up a reaction but does not start it. Criticise this definition of a catalyst.

7 How would you demonstrate, experimentally, that the reaction between ethanedioic (oxalic) acid and potassium manganate(VII) was autocatalytic and that Mn^{2+} ions were responsible for this?

8 Draw the general shape of the rate of reaction–time graph for an autocatalytic reaction.

9 Explain why copper dipped into concentrated nitric acid reacts slowly at first and then with increasing speed. Explain, also, why the copper will not react if some carbamide is added to the nitric acid. What is the effect of (a) hydrogen peroxide (b) sodium nitrite, on the reaction between copper and nitric acid?

10 Give examples, with equations where possible, of chemical reactions which are catalysed by (a) a metallic surface (b) a colloidal metal (c) hydrogen ions.

11 Comment on the following. (a) The reaction between solutions of sodium chloride and silver nitrate is said to be instantaneous. (b) Hydrogen and oxygen normally need heating before they will combine, but on a palladium surface they will combine at ordinary temperatures. (c) Although a catalyst affects the speed of a reaction it does not alter the position of equilibrium. (d) Several reactions of chlorine are accelerated by light. (OCS)

12 Write an essay on 'Enzymes'.

13 Describe four features which are characteristic of catalytic activity. Comment on the following observations. (a) There is a marked increase with time in the rate of oxidation of copper by concentrated nitric acid. However, there is no reaction at all in the presence of excess carbamide, $(NH_2)_2CO$. (b) A mixture of hydrogen and chlorine explodes when exposed to bright sunlight but mixtures of hydrogen and bromine and of hydrogen and iodine do not. (c) The rate constant for the hydrolysis of methyl methanoate (methyl formate) in aqueous solution appears to increase with time, whereas it remains constant, but is larger, in the presence of an excess of hydrochloric acid. (OCS)

29
Ionic theory and electrolytic conductivity

Faraday's quantitative discoveries, in 1834 (p. 44), on the passage of electricity through solutions, together with many other observations to be described in this group of chapters, are accounted for by the ionic theory, first suggested by Arrhenius in 1887.

This supposes, in its broadest outline, that a molten electrolyte, or a solution of an electrolyte in a suitable solvent, contains free ions. These ions are responsible for the electrical conductivity of the liquid or solution, and for many of its physical and chemical properties. So far as conductivity is concerned, it is the movement of the positive ions (*cations*) to the cathode and of the negative ions (*anions*) to the anode, under an applied potential difference, that is important. It depends, as will be seen, on the concentration of ions present and on their speeds.

1 Electrolytes

Any compound which, in solution in a suitable solvent (most commonly water), or in the fused (molten) state, will conduct an electric current, and be decomposed by it, is called an electrolyte. Other compounds are non-electrolytes.

Electrolytes may be ionic or covalent, and they may be strong or weak. They are generally salts, acids or bases, e.g.

ionic electrolytes	covalent electrolytes	
NaCl (strong)	HCl or HNO_3 (strong)	
NaOH (strong)	CH_3COOH	(weak)
	NH_3	(weak)

a Ionic electrolytes Ionic electrolytes are made up of ions in the solid state, e.g. Na^+Cl^-, but the ions are not free to move. On heating to the melting point, sufficient energy is put in to break down the crystal structure so that free ions can move within the melt. On dissolving in water (or some other ionising solvent), the crystal structure is also broken up but the difference is that the ions formed are hydrated (solvated). If this were not so the energy changes (p. 265) would inhibit solution. The ionisation of sodium chloride in water is, then, best shown as

$$NaCl(s) + aq \longrightarrow Na^+(aq) + Cl^-(aq)$$

b Covalent electrolytes These do not contain ions so that they do not conduct electricity in the liquid state as ionic electrolytes do. Covalent substances with polar bonds, e.g. O—H or H—Hal, may, however, react with water (or some other ionising solvent) to form hydrated (solvated) ions, e.g.

$$HCl(g) + aq \longrightarrow H^+(aq) + Cl^-(aq)$$

A more specific degree of hydration for the H^+ ion is shown by writing equations involving H_3O^+, which is known as the *oxonium ion*, e.g.

$$HCl(g) + H_2O(l) \longrightarrow H_3O^+ + Cl^-(aq)$$

There is, however, some evidence that the degree of hydration is higher, e.g. $H^+(H_2O)_4$ or $H^+(H_2O)_x$.

c The role of the solvent Only so-called ionising solvents (p. 387) can bring about the processes described in **a** and **b**. Water is the commonest such solvent, and the treatment in this chapter will be limited to aqueous solutions. Some non-aqueous solvents are described on page 388.

The formation of free, gaseous ions from solid NaCl or gaseous HCl are both highly endothermic processes (pp. 116 and 264), and it is only the liberation of energy in the hydration of the ions that makes the processes in **a** and **b** feasible.

Methylbenzene, a typical non-ionising solvent, will not dissolve sodium chloride. It will dissolve hydrogen chloride, but the solution is non-conducting as no free ions are formed.

When the hydration of an ion is not of great significance it may be written as H^+ or Na^+ or Cl^-, but it must be realised that such unhydrated ions never occur in aqueous solution.

2 Weak and strong electrolytes. The degree of ionisation

a Weak electrolytes Weak electrolytes, e.g. most organic acids and bases and ammonia, form aqueous solutions with comparatively low conductivity. This is due to the fact that only a fraction of their molecules split up into ions in normal solutions. The solution formed contains some ions, but these are in equilibrium with unionised molecules, e.g.

$$\begin{array}{ll} HEt & + aq \rightleftharpoons H^+(aq) + Et^-(aq) \\ \text{ethanoic acid} \end{array}$$

$$NH_3(g) + H_2O(l) \rightleftharpoons NH_4^+(aq) + OH^-(aq)$$

The lack of complete ionisation accounts for the low conductivity.

The fraction, or percentage, of molecules which are ionised is known as the degree of ionisation. In 0.1 M ethanoic acid solution, for example, the degree of ionisation is 0.0134, or 1.34 per cent. Such a solution contains many ethanoic acid molecules, together with relatively few hydrated hydrogen and ethanoate ions. The figures for 0.1 M ammonia solution are similar.

One of the most important contributions of Arrhenius was his suggestion that the degree of ionisation of a weak electrolyte increases as the solution containing the ions is diluted, until at infinite dilution (zero concentration) the degree of ionisation rises to 1, or 100 per cent. In 0.001 M ethanoic acid solution, for example, the degree of ionisation is 0.134, or 13.4 per cent.

The actual concentration of ions in a solution of a weak electrolyte depends, therefore, on (i) the total concentration of the solution, i.e. on the number of molecules which might provide ions, and (ii) the degree of ionisation, i.e. the extent to which any molecules present are split up into ions.

b Strong electrolytes Strong electrolytes, e.g. mineral acids, alkalis and most salts, give solutions with much higher conductivities than weak electrolytes. This is because strong electrolytes are more or less completely ionised when in the molten state or in solution. In other words, their degree of ionisation is 1, or 100 per cent, or nearly so. Strong electrolytes, in fact, often exist as ions in the solid form, e.g. sodium chloride (p. 109).

If it is assumed that a strong electrolyte is fully ionised in a solution of any dilution, the concentration of ions present in the solution will depend only on the concentration of the solution.

3 The speed of ions. Ionic interference

The ionic theory assumes that the passage of an electric current through a solution of an electrolyte depends both on the concentration of ions present in the solution and on their speed.

For a strong electrolyte, fully ionised, the concentration of ions present will be proportional to the concentration of the solution, but the electrical conductivity of such solutions is not proportional to their concentrations.

The experimental effect of concentration on the conductivity of a solution of a strong electrolyte is now regarded as being due to a variation in ionic speed. In a solution containing a high concentration of ions, a high degree of ionic interference (p. 351) is envisaged, which effectively decreases ionic speed. As a solution is diluted, such ionic interference gets smaller and smaller, and it is this change in the extent of ionic interference which is used to account for the way in which the conductivity of a solution of a strong electrolyte changes with concentration (p. 347).

This idea of ionic interference is a valuable addition to the original ideas of the ionic theory.

4 Summary of ionic theory

The important fundamental ideas of the ionic theory may be summarised as follows.

a A solution of an electrolyte contains free, solvated ions.

b Passage of a current through a solution of an electrolyte depends on (i) the concentration, (ii) the speed of the ions present.

c In a solution of a weak electrolyte the degree of ionisation increases as the dilution increases until, at infinite dilution (zero concentration), there is complete ionisation.

d In a solution of a strong electrolyte there is always complete ionisation, i.e. the degree of ionisation is approximately 1, or 100 per cent, but ionic interference limits the movement of the ions. This ionic interference becomes less as the dilution increases.

Conductivity of solutions of electrolytes

5 Conductivity

Solutions of electrolytes, like metallic conductors, obey Ohm's law except under abnormal conditions. The current flowing through a given solution under given conditions is, therefore, proportional to the reciprocal of the resistance of the solution. This quantity is known as the conductance of the solution. The SI unit of resistance is the ohm (Ω). Conductance used to be measured in reciprocal ohms (Ω^{-1}), but the modern SI unit is called the siemens ($1\,S = 1\,\Omega^{-1}$). A solution with a resistance of $10\,\Omega$ has a conductance of $0.1\,S$. High resistance means low conductance, and vice versa.

To compare the resistances of different substances the idea of resistivity is used. Similarly, for conductance, conductivity is used. The resistivity of a conductor is defined as the resistance between opposite faces of a unit cube of the conducting material. As resistance is proportional to length and inversely proportional to cross-sectional area, it follows that

$$R = \rho l/a \qquad \text{or} \qquad \rho = Ra/l$$

where R is the resistance, l is the length, a is the area, and ρ is the resistivity of the material concerned.

If R is measured in ohms, l in metres, and a in square metres, ρ will be in ohm metre ($\Omega\,\text{m}$) units. These are the basic SI units but $\Omega\,\text{cm}$ are also commonly used.

The reciprocal of the resistivity ($1/\rho$) is known as the conductivity (κ). The basic SI units of conductivity are siemen metre^{-1} ($\text{S}\,\text{m}^{-1}$) but $\text{S}\,\text{cm}^{-1}$ units are very common. $1\,\text{S}\,\text{cm}^{-1} = 100\,\text{S}\,\text{m}^{-1}$.

The importance of measurement of the conductivities of solutions of electrolytes is concerned with the way in which they vary with the concentration of the solution as described on p. 340.

6 Measurement of conductivity of solutions

The conductivity of a solution is obtained by measuring its resistance using a modified Wheatstone bridge circuit. Because direct current causes a back e.m.f. (p. 439) due to polarisation, a rapidly alternating current must be used. A typical arrangement is shown in Fig. 155. The variable resistance and the position of X along a wire, AB, are changed until no current passes through the detector. At this balance point,

$$\frac{\text{resistance of conductivity cell}}{\text{resistance of variable resistance}} = \frac{\text{AX}}{\text{XB}}$$

and the value of the resistance of the conductivity cell is obtained.

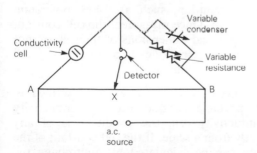

Fig. 155 Wheatstone bridge circuit for measuring conductivity.

The a.c. source can be an induction coil or a vacuum tube oscillator, and a telephone, oscilloscope or some other a.c. detector is required to find the balance point. For accurate work, specially purified water (p. 357) must be used in making the solutions, and the capacitance of the conductivity cell is balanced by having a variable condenser in parallel

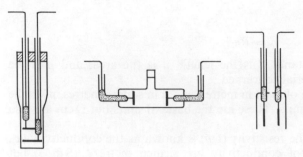

Fig. 156 Conductivity cells. The right-hand one is a dipping electrode.

with the variable resistance. The conductivity cell, containing the solution, is made of insoluble glass or fused silica. The shape of the cell varies as shown in Fig. 156. The electrodes are generally of thick platinum foil, firmly fixed in position and coated with a layer of platinum black to decrease polarisation effects.

a The cell constant From the measured resistance of a solution in a given cell, the resistivity, and hence the conductivity, could be obtained if the cross-sectional area of the electrodes, a, and the distance between them, l, were known. Such measurements would, however, be very difficult to make, and they can be avoided by making use of what is known as the cell constant. For a given cell, both l and a are constant and the ratio l/a, is called the cell constant.

As R is equal to $l/\kappa a$ it follows that

   conductivity = cell constant/measured resistance

A cell constant can be obtained from the cell dimensions, but it is more commonly measured by using a solution, such as 0.1 M potassium chloride whose conductivity is known. Once measured, the cell constant for a cell is fixed so long as the physical dimensions of the cell are not altered in any way. To ensure this, the electrodes in a conductivity cell must be rigidly fixed in their relative positions.

b Conductance or conductivity meters The circuitry just described is, nowadays, incorporated into portable conductance or conductivity meters. Using a dipping conductivity cell, the conductance of any solution, in S, can be read directly from a scale. If the cell constant of the cell is known, the conductivity can be calculated by multiplying the measured conductance by the cell constant.

7 Variation of conductivity with concentration

The conductivity of a solution varies with its concentration, c, as shown in Fig. 157. The full curve, with the maximum, is only obtained when a

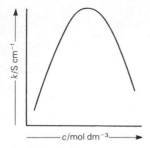

Fig. 157 The general shape of a conductivity-concentration curve for both weak and strong electrolytes. The curves for both types of electrolyte show a maximum, but it is well nigh impossible to plot both types of curve on the same scale. If judged on the same scale, the curve for a strong electrolyte lies well above that for a weak electrolyte.

wide range of concentrations is possible. In many cases a saturated solution is obtained before the maximum is reached so that only the left hand part of the curve is obtainable.

Some substances, such as potassium chloride, give solutions with much higher conductivities than others, such as ethanoic acid. Some typical figures are given below, concentration being expressed in $mol\,dm^{-3}$ and conductivity in $S\,cm^{-1} \times 10^3$.

concentration	0.0001	0.001	0.01	0.1	1	2	3
conductivity KCl	0.013	0.12	1.2	11.2	98.2	185.2	264.9
CH₃COOH	0.0107	0.041	0.143	0.46	1.32	1.60	1.62

Using $mol\,m^{-3}$ units the given concentration figures would have to be multiplied by 10^3; using $S\,m^{-1}$ units the given conductivity figures would have to be multiplied by 10^2.

For equal concentrations the differences between the two sets of figures is so marked that plotting the figures for each type of substance on conductivity–concentration graphs of the same scale is well nigh impossible.

Those substances with high conductivities are known as strong electrolytes; they include most mineral acids, alkalis and most salts. Substances with comparatively small conductivities are known as weak electrolytes; they include most organic bases and acids. There is no absolutely sharp line of demarcation between strong and weak electrolytes, for a few substances exhibit intermediate behaviour.

The shape of the conductivity–concentration curves is explained differently for weak and strong electrolytes. In both cases, an increase in concentration gives an increased total number of solute particles in a given volume of solution, and this might well be expected to give increased conductivities. It is, however, offset by a decrease in the degree of ionisation in a weak electrolyte, or by an increase in ionic interference in a strong electrolyte (Fig. 158). The more molecules there are of a weak

347

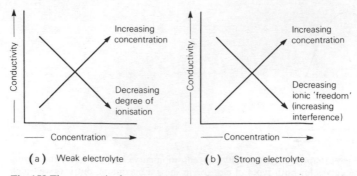

Fig. 158 The composite factors which give rise to the maxima in conductivity–concentration curves. **(a)** Weak electrolytes. **(b)** Strong electrolytes.

electrolyte, the fewer of them that ionise; the more ions there are of a strong electrolyte, the more they 'get in each others way'.

Both the change in the degree of ionisation with concentration for a weak electrolyte, and the change in ionic interference for a strong electrolyte can be investigated by making use of the concept of molar conductivity.

8 Molar conductivity

Comparison of the conductivities of say 1 M and 0.1 M acid solutions is not a fair comparison for M acid contains 1 mol of acid per cubic decimetre whereas 0.1 M acid only contains 0.1 mol. To compare them on equal terms, volumes of solutions containing equal amounts of solute ought to be considered, and this is done by comparing molar conductivities instead of conductivities.

The molar conductivity* of a solution is the electrolytic conductivity, κ, of the solution divided by its concentration, i.e.

$$\text{molar conductivity } \Lambda_c = \frac{\text{electrolytic conductivity, } \kappa}{\text{concentration, } c}$$

The symbol Λ_c is used to denote the molar conductivity of a solution at a concentration, c.

In basic SI units, concentration is expressed in $mol\,m^{-3}$ and electrolytic conductivity in $S\,m^{-1}$ giving $S\,m^2\,mol^{-1}$ units for molar conductivity. If concentration is expressed in $mol\,cm^{-3}$ units and conductivity in $S\,cm^{-1}$ units then the units of molar conductivity are $S\,cm^2\,mol^{-1}$.

---

* The use of the word 'molar' is not strictly correct. In the SI system of units 'molar' means 'divided by amount of substance', but in the term molar conductivity it means 'divided by concentration' (p. 470).

9 Variation of molar conductivity with concentration

The following figures illustrate the general way in which molar conductivity values vary with concentration; the molar conductivities are expressed in $S\,cm^2\,mol^{-1}$ and the concentrations in $mol\,dm^{-3}$.

concentration	0.0001	0.001	0.01	0.1	1	2	3
Λ_c KCl	129.1	127.3	122.4	112.0	98.3	92.6	88.3
CH$_3$COOH	107.0	41.0	14.3	4.6	1.32	0.80	0.54

Using $mol\,m^{-3}$ units the given concentrations have to be multiplied by 10^3; using $mol\,cm^{-3}$ units the multiplication factor is 10^{-3}. To convert the Λ_c values from $S\,cm^2\,mol^{-1}$ units into $S\,m^2\,mol^{-1}$ units it is necessary to multiply by 10^{-4}.

The figures are best summarised graphically, by plotting the molar conductivity against the square root of the concentration (Fig. 159). Alternatively, molar conductivity can be plotted against the reciprocal of the concentration, sometimes referred to as the dilution of the solution (Fig. 159).

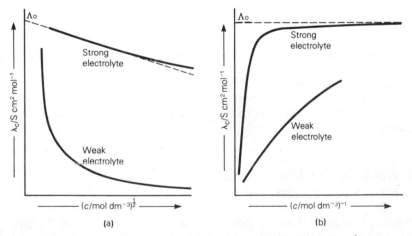

Fig. 159 Showing how molar conductivity changes with **(a)** $\sqrt{}$(concentration) and **(b)** (concentration)$^{-1}$.

It will be seen that the Λ_c values for both weak and strong electrolytes approach a maximum value at zero concentration. This is known as the molar conductivity at zero concentration, Λ_0, or the molar conductivity at infinite dilution, Λ_∞.

a Weak electrolytes Fig. 159 shows the way in which the degree of ionisation of a weak electrolyte increases as the concentration decreases. Arrhenius's original suggestion that the degree of ionisation, α, at any concentration, c, would be equal to the ratio Λ_c/Λ_∞ was confirmed when the values so obtained were found to be equal to values obtained from freezing point, boiling point and osmotic pressure measurements (p. 236). For a weak electrolyte, at a concentration, c

degree of ionisation, $\alpha = \Lambda_c/\Lambda_0$

At zero concentration, or infinite dilution, the α value for a weak electrolyte is 1, or 100 per cent. At higher concentrations, or lower dilutions, it is less than 1 or 100 per cent.

The Λ_c value for any solution can be obtained by measuring its conductivity at a concentration, c. The Λ_0 value for a weak electrolyte can, however, only be obtained indirectly, by measurements on related strong electrolytes as explained on page 351.

b Strong electrolytes Fig. 159 shows the way in which the ionic interference, for a strong electrolyte, decreases as the concentration decreases. At zero concentration, or infinite dilution, the ionic interference is negligible and Λ_c reaches its limiting value of Λ_0.

Kohlrausch, who did much of the early experimental work on measuring the conductivities of solutions, found that the data fitted reasonably well the expression

$$\Lambda_c = \Lambda_0 - kc^{\frac{1}{2}}$$

where k is a constant. That is why the plot of Λ_c against $c^{\frac{1}{2}}$ (Fig. 159), for a strong electrolyte is almost a straight line, and why values of Λ_0 can be obtained by extrapolation.

Such values are useful for obtaining the Λ_0 values for weak electrolytes (which cannot be obtained by extrapolation) as explained in the following section.

10 Kohlrausch's law. Measurement of Λ_0 for a weak electrolyte

Kohlrausch noticed that the difference between the Λ_0 values for two salts, which were strong electrolytes and had the same anion or the same cation, was always constant. For example, using Λ_0 values in $S\,cm^2\,mol^{-1}$

$$\Lambda_0(NaCl) - \Lambda_0(NaNO_3) = 3.7 \qquad \Lambda_0(KCl) - \Lambda_0(NaCl) = 21.1$$
$$\quad(108.9) \qquad (105.2) \qquad\qquad\quad (130.0) \qquad (108.9)$$

$$\Lambda_0(KCl) - \Lambda_0(KNO_3) = 3.7 \qquad \Lambda_0(KNO_3) - \Lambda_0(NaNO_3) = 21.1$$
$$\quad(130.0) \qquad (126.3) \qquad\qquad\quad (126.3) \qquad (105.2)$$

Such results can be accounted for only by assuming that the Λ_0 value of an electrolyte is the sum of two terms, one for the anion and another for the cation. These terms are known as the molar conductivities of the ions concerned (they were called ionic mobilities by Kohlrausch), and Kohlrausch's law (1876) states that *the molar conductivity at zero concentration of an electrolyte is equal to the sum of the molar conductivities at zero concentration of the ions produced by the electrolyte.* Thus,

$$\Lambda_0(\text{NaCl}) = \Lambda_0(\text{Na}^+) + \Lambda_0(\text{Cl}^-) \qquad \Lambda_0(\text{KCl}) = \Lambda_0(\text{K}^+) + \Lambda_0(\text{Cl}^-)$$

This is why $\Lambda_0(\text{NaX}) - \Lambda_0(\text{KX})$, which is equal to $\Lambda_0(\text{Na}^+) - \Lambda_0(\text{K}^+)$ is independent of the nature of X.

The Λ_0 value for a weak electrolyte, e.g. ethanoic acid, can, then, be obtained either from the values of the molar conductivities of the individual ions, i.e.

$$\Lambda_0(\text{HEt}) = \Lambda_0(\text{H}^+) + \Lambda_0(\text{Et}^-)$$

or from the Λ_0 values for selected strong electrolytes, e.g.

$$\Lambda_0(\text{HEt}) = \Lambda_0(\text{HCl}) + \Lambda_0(\text{NaEt}) - \Lambda_0(\text{NaCl})$$

Λ_0 values for individual ions can be obtained from measurements of their mobilities (p. 445).

11 Ionic interference

The theoretical treatment of ionic interference, undertaken originally by Debye and Hückel, is based on the idea that the electrical attraction between positive and negative ions in a solution prevents the ions acting as entirely isolated, single particles, except in very dilute solutions.

In an ionic crystal, a positive ion is surrounded by negative ions, and vice versa (p. 110). Debye and Hückel suggested that, in a solution, a positive ion is surrounded by an ionic atmosphere of negative ions, whilst a negative ion is surrounded by an ionic atmosphere of positive ions.

Ionisation is complete, in so far as there are no individual molecules (or only very few) of a strong electrolyte in a solution, and in so far as the attractive forces between ions in a solution are not so strong as they are in an ionic crystal. But random distribution of ions throughout a solution is not complete.

The chemical functioning, and the freedom of movement, of any ion surrounded by an ionic atmosphere with opposite charge, will clearly be different from that of a free ion. So far as motion under an applied e.m.f., i.e. conductivity, is concerned, two electrical effects will be important.

First, a positive ion moving towards the cathode will tend to drag its ionic atmosphere with it. This will result in an asymmetric ionic atmosphere with fewer negative ions in front of the positive ion, and more behind. Such an asymmetric ionic atmosphere will exert a retarding force on the positive ion and reduce its velocity.

Secondly, a negative ionic atmosphere will, as a whole, move towards the anode, so that the central positive ion will, in effect, be 'moving against the stream'.

Debye, Hückel and Onsager treated such considerations mathematically and derived what is generally known as the Onsager equation. As applied to an electrolyte giving two univalent ions, the equation is

$$\Lambda_0 - \Lambda_c = \frac{82.48}{\eta\sqrt{DT}} + \left\{\frac{8.20 \times 10^5 \times \Lambda_0}{(DT)^{3/2}}\right\}\sqrt{c}$$

where η is the viscosity of the medium, D is its dielectric constant, T the absolute temperature, and c the concentration.

Such an equation gives some idea of the complexity of the matter. The Onsager equation, nevertheless, provides good agreement with experimental data, so long as the solutions considered are dilute enough. The equation, in fact, is of the same general type as the empirical equation put forward by Kohlrausch (p. 350).

Questions on Chapter 29

1 What are the essential differences between a sodium atom and a sodium ion?

2 What is meant by the term cell constant, and for what purpose is a cell constant used? (a) The resistivity of 0.02 M potassium chloride solution is 361 Ω cm, and a conductivity cell containing such a solution was found to have a resistance of 550 Ω. What is the cell constant? (b) The same cell filled with 0.1 M zinc sulphate solution had a resistance of 72 Ω. What is the conductivity of 0.1 M zinc sulphate solution?

3 A conductivity cell with electrodes 2 cm² in area and 1 cm apart has a resistance of 7.25 ohm when filled with 5 per cent potassium chloride solution. What is (a) the cell constant (b) the conductivity of the potassium chloride solution? If the cell was filled with 0.02 M potassium chloride, with a resistivity of 361 Ω cm, what would the cell resistance be?

4 What would be the resistance of a conductivity cell with electrodes of cross-sectional area 4 cm² and 2 cm apart when the cell is filled with pure water of conductivity $0.8 \times 10^{-6}\,S\,cm^{-1}$? What current would flow through the cell under an applied potential difference of 10 volt?

5 What is meant by the term conductivity of a solution? How can conductivity be measured? How does it vary with the concentration of a solution, and how is the variation accounted for in terms of the ionic theory?

6 Using the data given on p. 341, plot the conductivity of (a) potassium chloride (b) ethanoic acid against concentration.

7 If the conductivity of 0.0005 M sulphyric acid is $x\,S\,cm^{-1}$ what is (a) the resistivity (b) the molar conductivity?

8 Why was the concept of molar conductivity introduced when that of conductivity was already established? How can the molar conductivity of 0.01 M ethanoic acid be measured?

9 Using the figures given on p. 349, plot the molar conductivity of (a) potassium chloride (b) ethanoic acid against (i) the reciprocal of concentration and (ii) the square root of concentration. From your graphs read off the Λ_0 value for potassium chloride and the apparent degree of ionisation in 0.02 M potassium chloride solution.

10 If the current carrying capacity of ions in a solution depends on their total number and their speed, how would you expect the conductivity of (a) ethanoic acid (b) sodium chloride to vary with (i) the dilution of the solution (ii) temperature?

11 How did Arrhenius account for the change in the molar conductivity of an electrolyte with dilution, and how have his ideas had to be modified?

12 Explain, briefly, how the degree of ionisation of a weak electrolyte might be obtained by measurements of (a) osmotic pressure (b) freezing point (c) boiling point (d) vapour pressure (e) conductivity.

13 Calculate the Λ_0 values for (a) methanoic acid (b) ammonia solution from the Λ_0 values given below in $S\,cm^2\,mol^{-1}$:

$NaCl = 113$ $NH_4Cl = 134.1$
$NaOH = 225.2$ Na methanoate $= 101.2$ $HCl = 397.8$

14 What would you write to convince a fifth-former that the degree of ionisation of a weak electrolyte increased as the concentration of its solution got less and less?

15 The conductivity of dichloroethanoic acid at a dilution of $8\,dm^3$ is $0.0238\,S\,cm^{-1}$. The Λ_0 value for this acid is $385\,S\,cm^2\,mol^{-1}$. Calculate the degree of ionisation at a concentration of $0.125\,mol\,dm^{-3}$.

16 The molar conductivities of Na^+, K^+ and Cl^- ions are 43.4, 64.6 and $65.5\,S\,cm^2\,mol^{-1}$ respectively. What approximate ratio of sodium chloride and potassium chloride is required to give a solution with a Λ_0 value of (i) 119.5 (ii) 124.8?

17 If the conductivity of an M solution of sodium chloride is $x\,S\,cm^{-1}$ what is (a) the conductivity of the solution in $S\,m^{-1}$ (b) the molar conductivity of the solution in $S\,cm^2\,mol^{-1}$ (c) the molar conductivity of the solution in $S\,m^2\,mol^{-1}$?

18 What are the advantages and disadvantages of expressing molar conductivities of solutions in (a) $S\,cm^2\,mol^{-1}$ units (b) $S\,m^2\,mol^{-1}$ units?

19 Express the data given in the tables on pages 347 and 349 in concentration units of $mol\,m^{-3}$, conductivity units of $S\,m^{-1}$ and molar conductivity units of $S\,m^2\,mol^{-1}$.

20 The conductivity of 0.01 M ethanoic acid is $0.143 \times 10^{-3}\,S\,cm^{-1}$, and the molar conductivities of H^+ and Et^- are 349.8 and $40.9\,S\,cm^2\,mol^{-1}$ respectively. What value for the K_a of ethanoic acid do these figures give?

21 Discuss the factors that affect the conductivity of aqueous solutions of electrolytes. When there is a known impurity present in water a measurement of the conductivity may be used to estimate its concentration. An alternative method is to weigh the residue left after evaporation. Calculate the ratio of the conductivity to the mass of the residue from $1 \, dm^3$ of water containing   (a) only Na^+ and Cl^- ions, and (b) only Ca^{2+} and HCO_3^- ions. The molar conductances are Na^+ 50, Cl^- 76, Ca^{2+} 119, and HCO_3^- $45 \, cm^2 \, ohm^{-1} \, mol^{-1}$. You may assume that the impurities are at infinite dilution. (OJSE)

22* A conductivity cell filled with 0.1 M KCl at 25 °C was found to have a resistance of $77.5 \, \Omega$. What is the cell constant? The same cell, under the same conditions, but filled with 0.1 M ethanoic acid, had a resistance of $2000 \, \Omega$. What is the percentage degree of ionisation of ethanoic acid in 0.1 M solution?

23* The resistance between two parallel electrodes, x cm apart, immersed in 0.1 M ethanoic acid at 25 °C was found to be $1000 \, \Omega$. How far apart would the electrodes have to be to give the same resistance using 0.01 M ethanoic acid?

24* List the values of the molar conductivities at zero concentration of the following ions: Li^+, Na^+, K^+, Mg^{2+}, Ca^{2+}, Sr^{2+}, Ba^{2+}, Al^{3+}, F^-, Cl^-, Br^- and I^-. Comment on the values.

25* Record the values of the molar conductivities at zero concentration of (a) H^+ (b) OH^- (c) HCl (d) NaOH (e) NaCl. How are they interrelated? Why are the values for H^+ and OH^- so much higher than those for Na^+ and Cl^-?

30
Ostwald's dilution law and ionic equilibria

1 Derivation of the law

The partial ionisation of molecules into hydrated ions when a weak electrolyte is dissolved in water produces an equilibrium mixture of molecules and hydrated ions. Ostwald (1888) applied the idea of the law of equilibrium (p. 287) to this equilibrium and arrived at an important relationship known as Ostwald's dilution law.

If the original concentration of a binary electrolyte, AB, in a solution is c mol dm^{-3}, and if the degree of ionisation is α, the concentrations of AB, $A^+(aq)$ and $B^-(aq)$, in mol dm^{-3}, at equilibrium will be

$$AB(aq) \rightleftharpoons A^+(aq) + B^-(aq)$$
$$c(1 - \alpha) \qquad c\alpha \qquad c\alpha$$

According to the equilibrium law

$$\frac{[A^+(aq)]_{eq}[B^-(aq)]_{eq}}{[AB(aq)]_{eq}} = K = \frac{c\alpha^2}{(1 - \alpha)}$$

If α is small as compared with 1, i.e. for a weak electrolyte in a solution where the degree of ionisation is low, $(1 - \alpha)$ will be approximately equal to 1 so that

$$c\alpha^2 \approx K \qquad \text{or} \qquad \alpha \approx \sqrt{K/c}$$

These two relationships, the first accurate and the second an approximation, are expressions of Ostwald's dilution law. *For a weak binary electrolyte, with a small degree of ionisation, the degree of ionisation is proportional to the square root of the reciprocal of the concentration.*

K is an equilibrium constant known as the *ionisation* or *dissociation constant*. The expression given refers only to a binary electrolyte, AB: different expressions apply for other electrolytes, e.g. AB_2, A_2B etc.

Ostwald's dilution law is only valid for weak electrolytes in dilute solutions. The fact that it does not apply to strong electrolytes was, historically, something of a mystery, but it is now realised that the concept of a degree of ionisation for a strong electrolyte is not valid. They are fully, or nearly fully, ionised, i.e., α is 1 or nearly 1, and the characteristics of solutions of strong electrolytes depends mainly on the extent of the ionic interference (p. 351).

2 Activities

The simple derivation of Ostwald's dilution law in the preceding section uses concentration in mol dm^{-3}. This is valid in very dilute solutions where ionic interference is negligible and where the solutions approach ideal behaviour. In a more accurate and broader treatment, however, it is necessary to use activities rather than concentrations. The activity is, in simple language, the effective concentration. Activities are lower than concentrations, the two being related by the activity coefficient, i.e.

activity, a = concentration × activity coefficient, f

Activity coefficients for individual ions cannot be obtained, but it is possible to get mean values for particular electrolytes. For example, that for hydrochloric acid in 0.1 M solution is 0.798, rising to 0.966 in 0.001 M solution. Electrolytes giving ions with multiple charges have lower values.

3 Types of ionic equilibria

Ostwald's argument can be applied to a number of different types of ionic equilibria. A summary is given together with the special name, if any, given to the equilibrium constant.

The various equilibrium constants, K, involved often have very low values. If so, it is convenient to express them as pK values, i.e. $-\lg K$. If K is 10^{-5}, for example, pK is 5.

a The ionic product of water, K_w (p. 357)

$$H_2O(l) \rightleftharpoons H^+(aq) + OH^-(aq) \qquad [H^+(aq)]_{eq}[OH^-(aq)]_{eq} = K_w$$

The concentration of the water is regarded as constant.

b The dissociation constant of a weak acid, K_a (p. 358)

$$HA(aq) \rightleftharpoons H^+(aq) + A^-(aq) \qquad \frac{[H^+(aq)]_{eq}[A^-(aq)]_{eq}}{[HA(aq)]_{eq}} = K_a$$

c The dissociation constant of a weak base, K_b (p. 361)

$$BOH(aq) \rightleftharpoons B^+(aq) + OH^-(aq) \qquad \frac{[B^+(aq)]_{eq}[OH^-(aq)]_{eq}}{[BOH(aq)]_{eq}} = K_b$$

d The dissociation constant of an indicator, K_{in} (p. 369)

$$HIn(aq) \rightleftharpoons H^+(aq) + In^-(aq) \qquad \frac{[H^+(aq)]_{eq}[In^-(aq)]_{eq}}{[HIn(aq)]_{eq}} = K_{in}$$

e The hydrolysis constant of a salt, K_h (p. 377)

$$BA(aq) + H_2O(l) \rightleftharpoons HA(aq) + BOH(aq) \qquad \frac{[HA(aq)]_{eq}[BOH(aq)]_{eq}}{[BA(aq)]_{eq}} = K_h$$

f The solubility product, K_s (p. 392)

$$AgCl(s) \rightleftharpoons Ag^+(aq) + Cl^-(aq) \qquad [Ag^+(aq)]_{eq}[Cl^-(aq)]_{eq} = K_s$$

g The stability constant, K_{stab} (p. 398)

$$Ag^+(aq) + 2NH_3(aq) \rightleftharpoons Ag(NH_3)_2^+ + aq \qquad \frac{[Ag(NH_3)_2{}^+(aq)]_{eq}}{[Ag^+(aq)]_{eq}[NH_3(aq)]_{eq}^2} = K_{stab}$$

4 Ionic product of water

It is sometimes said that pure water is a non-conductor. What is really meant is that pure water has such a low conductivity that, in some cases, it can be neglected. There are, however, many circumstances in which the conductivity of water cannot be neglected.

The fact that even the purest water ever obtained is a conductor (p. 395) shows that water is very feebly ionised,

$$H_2O(l) \rightleftharpoons H^+(aq) + OH^-(aq)$$

and application of the law of equilibrium gives the expression,

$$K = \frac{[H^+(aq)]_{eq}[OH^-(aq)]_{eq}}{[H_2O(l)]_{eq}}$$

In pure water, the concentration of water molecules is so very large compared with those of $H^+(aq)$ and $OH^-(aq)$ ions that it can be regarded as constant. In a mixture of 1 000 000 water molecules with one $H^+(aq)$ and one $OH^-(aq)$ ion, a four-fold change in the ion concentrations to $4H^+(aq)$ and $4OH^-(aq)$ will still leave 999 997 water molecules. The concentration of water molecules has hardly altered, and it can be taken as constant. Therefore,

$$[H^+(aq)]_{eq}[OH^-(aq)]_{eq} = \text{a constant, } K_w$$

and the constant is known as the ionic product of water. The units are $mol^2\,dm^{-6}$.

The numerical value of this ionic product can be obtained from experimental measurements. The conductivity of the purest conductivity water ever made is $0.054 \times 10^{-6}\,S\,cm^{-1}$. The concentration of water is $18\,g$ per $18\,cm^3$ or $1\,mol$ per $18\,cm^3$ or $(1/18)\,mol\,cm^{-3}$.

The molar conductivity of water is therefore given by

$$\varLambda_c = \frac{\text{conductivity in } S\,cm^{-1}}{\text{concentration in } mol\,cm^{-3}} = 9.72 \times 10^{-7}\,S\,cm^2\,mol^{-1}$$

The Λ_0 value for water, 545, can be obtained either from the molar conductivities of H^+ and OH^- ions or from Λ_0 values for hydrochloric acid, sodium hydroxide and sodium chloride,

$$\Lambda_0(H_2O) = \Lambda_0(H^+) + \Lambda_0(OH^-)$$
$$\Lambda_0(H_2O) = \Lambda_0(HCl) + \Lambda_0(NaOH) - \Lambda_0(NaCl)$$

The degree of ionisation of water, Λ_c/Λ_0, is therefore $(9.72 \times 10^{-7})/545$ or approximately 18×10^{-10}. This very low value (0.000 000 18 per cent) shows how very feebly water is ionised.

Water has a concentration of $1000/18$ mol dm^{-3} and if it was fully ionised the resulting concentrations of $H^+(aq)$ and $OH^-(aq)$ would be $1000/18$ mol dm^{-3}. But, as the degree of ionisation is only 18×10^{-10} it follows that

$$[H^+(aq)]_{eq} = [OH^-(aq)]_{eq} = \frac{1000}{18} \times 18 \times 10^{-10} = 10^{-7} \text{ mol dm}^{-3}$$

$$K_w = [H^+(aq)]_{eq}[OH^-(aq)]_{eq} = 10^{-7} \times 10^{-7} = 10^{-14} \text{ mol}^2 \text{ dm}^{-6}$$

5 Ionic product and temperature

The figure of 10^{-14} for the ionic product of water is a convenient one for general use, but the accurate value depends on the temperature as is shown in the following table.

$T/°C$	18	25	40	75
$K_w/\text{mol}^2\,\text{dm}^{-6}$	0.61×10^{-14}	1×10^{-14}	2.92×10^{-14}	16.9×10^{-14}

The equilibrium constant for a reaction is related to temperature by the expression (p. 295),

$$\frac{d(\ln K)}{dT} = \frac{\Delta H_m}{RT^2}$$

By taking the value of K_w, which is a multiple of K, at two different temperatures it is possible to calculate ΔH_m for the reaction

$$H^+(aq) + OH^-(aq) \rightleftharpoons H_2O(l)$$

The calculated value is found to be in close agreement with the experimentally measured value (p. 252).

6 Dissociation constants for weak monobasic acids

In an aqueous solution of ethanoic acid, HEt, of concentration c mol dm^{-3} and with a degree of ionisation, α, the equilibrium situation can be summarised as follows (p. 355),

$$HEt(aq) \rightleftharpoons H^+(aq) + Et^-(aq)$$
$$c(1 - \alpha) \qquad c\alpha \qquad c\alpha$$

all concentrations being given in $mol\,dm^{-3}$.

The dissociation constant, K_a, is therefore, given by

$$K_a = \frac{[H^+(aq)]_{eq}[Et^-(aq)]_{eq}}{[HEt(aq)]_{eq}} = \frac{c\alpha^2}{(1 - \alpha)}$$

or, if α is small as compared with 1,

$$K_a \approx c\alpha^2 \qquad or \qquad \alpha \approx \sqrt{K_a/c}$$

The degree of ionisation, α, of a weak acid in a solution of concentration, c, can be measured by conductivity (p. 349) or other methods (p. 236), and, from such measurements, the acid dissociation constants can be obtained; they are often quoted as pK_a values (where $pK_a = -\lg K_a$). Typical examples, at 298 K, are

	$K_a/mol\,dm^{-3}$	pK_a		$K_a/mol\,dm^{-3}$	pK_a
HCOOH	1.8×10^{-4}	3.8	HOCl	3.7×10^{-8}	7.4
CH_3COOH	1.8×10^{-5}	4.8	HCN	4.9×10^{-10}	9.3
$CH_2ClCOOH$	1.4×10^{-3}	2.9	HNO_2	4.6×10^{-4}	3.34

The strongest acids have the highest K_a values, but the lowest pK_a values.

7 Typical calculations

The expressions given in the preceding section relate α, c and K so that any one of these can easily be calculated if the other two are known. The calculation of the degree of ionisation, and the resulting $H^+(aq)$ ion concentration, in a solution of known concentration, is very common, as in the following examples.

a 0.1 M ethanoic acid The dissociation constant of ethanoic acid is $1.8 \times 10^{-5}\,mol\,dm^{-3}$. In 0.1 M solution, the concentration is $0.1\,mol\,dm^{-3}$. Using the approximate form of Ostwald's dilution law,

$$\alpha^2 \approx 10 \times 1.8 \times 10^{-5} \qquad or \qquad \alpha \approx 1.34 \times 10^{-2}$$

The degree of dissociation of ethanoic acid in 0.1 M solution is, therefore, 1.34×10^{-2} or 1.34 per cent.

Complete ionisation of ethanoic acid, at a concentration of $0.1\,mol\,dm^{-3}$ would give $0.1\,mol\,dm^{-3}$ of $H^+(aq)$ ion. Ionisation of a fraction, 0.0134, of the ethanoic acid would give a $H^+(aq)$ ion concentration of 0.1×0.0134, i.e. $0.00134 \, mol \, dm^{-3}$. This is therefore the concentration of $H^+(aq)$ ion in 0.1 M ethanoic acid.

Notice that the approximate form of Ostwald's dilution law has been used in this calculation because the α value is small as compared with 1.

b 0.001 M ethanoic acid Here, c is 0.001 so that

$$\alpha^2 \approx 1000 \times 1.8 \times 10^{-5} \quad \text{or} \quad \alpha \approx 0.134$$

The degree of ionisation in 0.001 M ethanoic acid is, therefore, approximately 0.134 or 13.4 per cent, and the resulting $H^+(aq)$ ion concentration is 0.001×0.134 or $0.000134 \, mol \, dm^{-3}$.

The α value in this example is not very small and the use of the approximate form of Ostwald's dilution law is inaccurate. Using the accurate form

$$\alpha^2 = 1000 \times 1.8 \times 10^{-5}(1 - \alpha)$$

$$\alpha^2 + (1.8 \times 10^{-2})\alpha - (1.8 \times 10^{-2}) = 0$$

gives a value of 0.125 for α.

8 Dissociation constants for weak polybasic acids

Polybasic acids ionise in stages and have more than one dissociation constant. Carbonic acid, which is dibasic, has two dissociation constants, outlined as follows.

$$H_2CO_3(aq) \rightleftharpoons HCO_3^-(aq) + H^+(aq) \qquad K_1 = 3 \times 10^{-7} \, mol \, dm^{-3}$$

$$HCO_3^-(aq) \rightleftharpoons CO_3^{2-}(aq) + H^+(aq) \qquad K_2 = 6 \times 10^{-11} \, mol \, dm^{-3}$$

$$\text{where} \quad K_1 = \frac{[H^+(aq)]_{eq}[HCO_3^-(aq)]_{eq}}{[H_2CO_3(aq)]_{eq}}$$

$$\text{and} \quad K_2 = \frac{[H^+(aq)]_{eq}[CO_3^{2-}(aq)]_{eq}}{[HCO_3^-(aq)]_{eq}}$$

The second dissociation constant for H_2CO_3 is of course the first dissociation constant for HCO_3^-.

Other typical values at 298 K are:

	$K_a/mol \, dm^{-3}$	pK_a		$K_a/mol \, dm^{-3}$	pK_a
H_3PO_4	$K_1 = 7.5 \times 10^{-3}$	$pK_1 = 2.1$	H_2SO_3	$K_1 = 1.3 \times 10^{-2}$	$pK_1 = 1.9$
	$K_2 = 6.2 \times 10^{-8}$	$pK_2 = 7.2$		$K_2 = 6 \times 10^{-8}$	$pK_2 = 7.2$
	$K_3 = 2.2 \times 10^{-13}$	$pK_3 = 12.7$			

For many inorganic oxoacids, it is a fairly general rule that successive dissociation constants, $K_1, K_2, K_3 \ldots$ are in the approximate ratio of $1:10^{-5}:10^{-10} \ldots$

9 Dissociation constant for weak bases

The argument applied to weak acids in section 6 can also be applied to dilute solutions of weak bases, e.g. ammonia, NH_3. The equilibrium involved is

$$NH_3(g) + H_2O(l) \rightleftharpoons NH_4^+(aq) + OH^-(aq)$$

The equilibrium constant, known as the dissociation constant, K_b, of the base is given by

$$K_b = \frac{[NH_4^+(aq)]_{eq}[OH^-(aq)]_{eq}}{[NH_3(aq)]_{eq}} = \frac{c\alpha^2}{(1-\alpha)} \approx c\alpha^2$$

using the symbols as in section 6.

The degree of ionisation, α, of a weak base in a solution of known concentration, c, can be measured by conductivity or other methods, and the dissociation constant of the base can be obtained from the results. It is expressed as K_b or pK_b, as in the following examples.

	K_b/mol dm^{-3}	pK_b		K_b/mol dm^{-3}	pK_b
NH_3	1.8×10^{-5}	4.8	$C_2H_5NH_2$	6.5×10^{-4}	3.2
CH_3NH_2	4.0×10^{-4}	3.4	$C_6H_5NH_2$	4.3×10^{-10}	9.4

From the dissociation constant of a weak base it is a simple matter to calculate the degree of ionisation and the corresponding $OH^-(aq)$ ion concentration in any solution.

The dissociation constant for ammonia in aqueous solution, for example, is 1.8×10^{-5} mol dm^{-3}. In 0.1 M solution, therefore,

$$\alpha^2 \approx 10 \times 1.8 \times 10^{-5} \qquad \text{or} \qquad \alpha \approx 1.342 \times 10^{-2}$$

The degree of ionisation in 0.1 M ammonia solution is, therefore, 1.342×10^{-2} or 1.342 per cent.

Complete ionisation into OH^- ions at a concentration of 0.1 mol dm^{-3} would give 0.1 mol dm^{-3} of $OH^-(aq)$ ion. Ionisation of a fraction 0.013 42 would give an $OH^-(aq)$ ion concentration of $0.1 \times 0.013\,42$, i.e. 0.001 342 mol dm^{-3}. This is, therefore, the concentration of $OH^-(aq)$ ions in 0.1 M aqueous ammonia.

Questions on Chapter 30

1 The molar conductivity of a M/32 solution of a weak acid is 9.2 S cm^2 mol^{-1}. If the molar conductivity at zero concentration is 389 S cm^2 mol^{-1}, what is the dissociation constant of the acid?

2 The dissociation constant of ethanoic acid is 1.8×10^{-5} mol dm^{-3}. Calculate the values of the degree of ionisation at dilutions of 10, 100, 1000, 10 000 and 100 000 dm^3, and plot the degree of ionisation against the dilution.

3 The dissociation constant of benzenecarboxylic acid is $6.7 \times 10^{-5} \, \text{mol dm}^{-3}$. What are the concentrations of the solutions, in mol dm^{-3}, in which benzenecarboxylic acid is   (a) 10 per cent   (b) 25 per cent   (c) 50 per cent   (d) 90 per cent ionised?

4 The molar conductivities of ethanoic acid in various solutions (V is the volume containing 1 mol of solute) are given below:

V/dm^3	13.57	54.28	434.2	1737	6948	∞
$\Lambda_0/\text{S cm}^2 \text{mol}^{-1}$	6.09	12.09	33.22	63.60	116.8	387.9

Do these figures agree with Ostwald's dilution law? What conclusion would you draw about the value of the dissociation constant of ethanoic acid?

5 On what basic principles is the derivation of Ostwald's law, and its testing by the use of Λ_c/Λ_0 values, based? Which principle is it that is false so far as strong electrolytes are concerned?

6 The dissociation constant for ammonia in aqueous solutions is $1.99 \times 10^{-5} \, \text{mol dm}^{-3}$. What is the concentration of $OH^-(aq)$ ions in a 0.01 M solution?

7 The dissociation constant of methanoic acid is $2.1 \times 10^{-4} \, \text{mol dm}^{-3}$. Using the relationship $c\alpha^2 \approx K$, calculate the percentage degree of ionisation of methanoic acid in 0.00001 M solution. Why is the answer you get absurd? What is the correct answer?

8 A dibasic acid has two dissociation constants, usually written as K_1 and K_2 and referred to as the first and secondary dissociation constants. For carbonic acid, for instance, K_1 is 3×10^{-7} and K_2 is $6 \times 10^{-11} \, \text{mol dm}^{-3}$. Why is K_1 usually bigger than K_2?

9 Compare the OH^- ion concentration in 0.01 M, 0.001 M, 0.0001 M and 0.00001 M solutions of   (i) ammonia solution   (ii) potassium hydroxide. The dissociation constant of ammonia is $1.99 \times 10^{-5} \, \text{mol dm}^{-3}$; potassium hydroxide may be regarded as fully ionised.

10 Define molar conductivity of a solution of an electrolyte. The molar conductance of a solution of ethanoic acid containing $0.03 \, \text{mol dm}^{-3}$ is 8.50 units at $18\,^{\circ}\text{C}$, and 14.7 units at $100\,^{\circ}\text{C}$. The molar conductivity of ethanoic acid at zero concentration at $18\,^{\circ}\text{C}$ is 347 units and at $100\,^{\circ}\text{C}$ is 773 units. Calculate   (a) the degree of dissociation of ethanoic acid at $18\,^{\circ}\text{C}$ and $100\,^{\circ}\text{C}$ when the concentration is $0.03 \, \text{mol dm}^{-3}$ (b) the dissociation constants of ethanoic acid at these two temperatures. From your results deduce whether the dissociation of ethanoic acid is exothermic or endothermic. (CS)

11 Describe the experiments you would make in order to determine the relation between the degree of dilution of a solution of ethanoic acid and its molar conductivity. How are the results of such experiments explained? The molar conductivity of ethanoic acid at zero concentration is $388 \, \text{S cm}^2 \text{mol}^{-1}$. The molar conductivity of a solution containing 0.3 g of ethanoic acid in $50 \, \text{cm}^3$ of water is $4.6 \, \text{S cm}^2 \text{mol}^{-1}$. Calculate the freezing point of this solution. (A solution containing 34.2 g of cane sugar, $C_{12}H_{22}O_{11}$, in 100 g of water freezes at $-1.85\,^{\circ}\text{C}$.) (CS)

12 What do you understand by (a) the law of mass action (b) an equilibrium constant? If the dissociation constant for ethanoic acid is $1.8 \times 10^{-5} \, \text{mol dm}^{-3}$, calculate the concentration of $H^+(aq)$ ions in 0.1 M ethanoic acid.

13 Derive the mathematical expression for Ostwald's dilution law for an electrolyte, one molecule of which ionises to give n cations and m anions.

14 The conductivity of a 0.05 M solution of ethanoic acid is $4.4 \times 10^{-4} \, \text{S cm}^{-1}$. The molar conductivities of H^+ and Et^- ions are 310 and $77 \, \text{S cm}^{-1}$ respectively. Calculate the ionisation constant of ethanoic acid.

15* Record the ionisation constants of some typical carboxylic and halogenated carboxylic acids, and comment on the values.

16* Record the pK values of some oxoacids and use the values to show how the strength of such an acid, when expressed by the formula $XO_m(OH)_n$, depends on the electronegativity of X and the value of m.

17* Compare the pK values of (a) H_2SO_3 and HSO_3^- (b) H_2CO_3 and HCO_3^- (c) H_3PO_4, $H_2PO_4^-$ and HPO_4^{2-}. Comment on the values.

18* Record how, and explain why, the ionic product of water varies with temperature.

19* Plot values of $\lg(K_w \times 10^{14})$ against $T^{-1}/K \times 10^3$. Use the graph to find the value for the enthalpy of the reaction between $H^+(aq)$ and $OH^-(aq)$.

31

pH values, indicators and hydrolysis

1 pH values

In any aqueous solution, the equilibrium

$$H_2O(l) \rightleftharpoons H^+(aq) + OH^-(aq)$$

may shift according to the conditions, but the equilibrium constant will remain constant, at a fixed temperature, so that the ionic product of water (p. 357)

$$[H^+(aq)]_{eq}[OH^-(aq)]_{eq} = 10^{-14}\,mol^2\,dm^{-6} \quad \text{(at 25°C)}$$

will also remain constant.

In pure water $[H^+(aq)]_{eq}$ equals $[OH^-(aq)]_{eq}$, and this equality means that pure water is neutral. In an acidic solution $[H^+(aq)]_{eq}$ will be greater than $[OH^-(aq)]_{eq}$: in an alkaline solution $[OH^-(aq)]_{eq}$ will be greater than $[H^+(aq)]_{eq}$.

neutral solution $[H^+(aq)]_{eq} = [OH^-(aq)]_{eq} = 10^{-7}$

acidic solution $[H^+(aq)]_{eq} > [OH^-(aq)]_{eq}$

alkaline solution $[OH^-(aq)]_{eq} > [H^+(aq)]_{eq}$

the ionic concentrations being expressed in $mol\,dm^{-3}$ at 25°C.

The very wide range, from a strongly acidic solution with $H^+(aq)$ concentration of about $10^{-1}\,mol\,dm^{-3}$ to a strongly alkaline solution with $H^+(aq)$ concentration about $10^{-13}\,mol\,dm^{-3}$, made a more convenient scale for expressing concentration desirable. Such a scale was suggested by Sørensen, in 1909. On this scale, *the negative logarithm of the $H^+(aq)$ concentration, in $mol\,dm^{-3}$, in a solution is called the pH of the solution*, (p for potenz, meaning strength)*. Thus

$$pH = -\lg[H^+(aq)] = \lg 1/[H^+(aq)]$$

For a neutral solution, with $[H^+(aq)] = [OH^-(aq)] = 10^{-7}\,mol\,dm^{-3}$ the pH is 7. For an acid solution with $[H^+(aq)] > 10^{-7}\,mol\,dm^{-3}$, the pH is less than 7. For an alkaline solution, with $[H^+(aq)] < 10^{-7}\,mol\,dm^{-3}$ the pH is greater than 7.

---

* A more precise, operational definition is given in terms of the e.m.f. of a specified cell, containing a standard solution, compared with the e.m.f. of the same cell containing a solution of unknown pH. If the e.m.f. and pH values concerned are $E(S)$ and pH(S), for the standard, and $E(X)$ and pH(X), for the unknown, then

$$pH(X) = pH(S) + \frac{E(X) - E(S)}{(RT\ln 10)/F}$$

A number of standard solutions, e.g. potassium hydrogen phthalate, are assigned pH(S) values at different temperatures.

$[H^+(aq)]$ 10^{-1} |_____|_____| 10^{-13}

10^{-7}

```
        N
        E
ACIDIC  U  ALKALINE
        T
        R
        A
        L
```

pH |_____|_____|
    1   7   13

2 pH changes during acid–alkali titrations

As an alkaline solution is run into an acidic solution, during a titration, there is a change in the pH value and, at the end point, the change may be a sharp one which can be detected by using an indicator (p. 367). The way in which the pH changes during the course of a titration depends upon the nature of the acid and alkali used.

a Titration of strong alkali against strong acid e.g. 0.1 M sodium hydroxide against 25 cm³ of 0.1 M hydrochloric acid.

Before starting the titration, the pH of the 0.1 M acid (fully ionised to give $[H^+(aq)]$ equal to 10^{-1} mol dm⁻³) will be 1. There will be 0.0025 mol of HCl in the 25 cm³ of solution.

After the addition of 5 cm³ of 0.1 M NaOH (0.0005 mol), there will be 0.002 mol of HCl in 30 cm³ of solution. The concentration of HCl and $H^+(aq)$ will be 0.0667 mol dm⁻³. The pH is calculated as follows:

$$pH = -\lg 0.0667 = -(\bar{2}.8240) = 1.176$$

When 25 cm³ of 0.1 M alkali have been added to the acid, the mixture will be just neutral with a pH of 7.

With 26 cm³ of 0.1 M NaOH (0.0026 mol) added, there will be 0.0001 mol of NaOH in 51 cm³ of solution. The NaOH and the OH⁻(aq) concentrations will be 0.001 96 mol dm⁻³. The $H^+(aq)$ ion concentration will be $10^{-14}/0.001\,96$ mol dm⁻³ so that the pH is given by

$$pH = -\lg(10^{-14}/0.001\,96) = -(\overline{12}.7077) = 11.292$$

Further figures, given below, are plotted in Fig. 160(a).

0.1 M NaOH added/cm³	0	5	10	20	24	24.9	25	25.1	26
pH	1	1.176	1.367	1.954	2.690	3.155	7	10.683	11.292

The graph shows that there is a quite sudden change in pH, from 4 to 10, near the end point, and when equivalent amounts of acid and alkali are present the solution is exactly neutral, with a pH of 7.

365

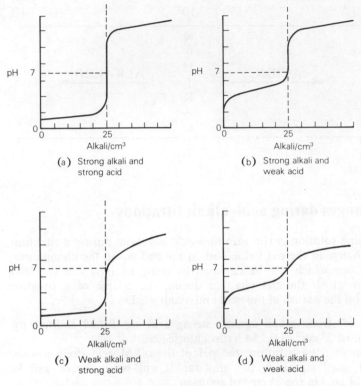

Fig. 160 Titration curves showing the pH changes as a 0.1 M solution of a strong or weak alkali is added to 25 cm³ of a 0.1 M solution of a strong or weak acid. The precise shape of the curves depends on the actual strengths of the alkalis and acids concerned. Slightly different curves are obtained, too, if solutions other than 0.1 M are used.

b Titration of a strong alkali against a weak acid e.g. 0.1 M sodium hydroxide against 25 cm³ of 0.1 M ethanoic acid. The pH at the end point, when equivalent amounts of alkali and acid are present, is greater than 7, because a mixture of equal amounts of a strong alkali and a weak acid forms an alkaline solution. This is because the salt formed is hydrolysed, i.e. interacts with water (p. 375). Sodium ethanoate, for example, is fully ionised, but the $Na^+(aq)$ and $Et^-(aq)$ ions interact with $OH^-(aq)$ and $H^+(aq)$ ions from water

$$NaEt(aq) \longrightarrow Na^+(aq) + Et^-(aq)$$

$$H_2O(l) \rightleftharpoons OH^-(aq) \quad H^+(aq)$$

NaOH   HEt
(strong)   (weak)

The interaction of the ions is not, however, equal. Formation of molecules of weakly ionised ethanoic acid removes some H^+(aq) ions from the solution, but there is no compensating removal of OH^-(aq) ions, because sodium hydroxide is a strong electrolyte, fully ionised. The solution, therefore, contains excess OH^-(aq) ions and is alkaline.

The titration curve (Fig. 160(b)) shows a marked change in pH between 7.5 and 10.5. At the precise end point, when the solution is 0.1 M sodium ethanoate, the pH is 8.87 (p. 377).

c Titration of a weak alkali against a strong acid e.g. 0.1 M ammonia solution against 0.1 M hydrochloric acid. In the titration of a weak alkali against a strong acid, the sudden change in pH at the end point occurs between about 3.5 and 6.5 as shown in Fig. 160(c). The pH at the end point, when equivalent amounts of alkali and acid are present, is 5.13 (p. 378), i.e. the mixture is acidic. This is accounted for by hydrolysis as follows.

$$NH_4Cl(aq) \longrightarrow NH_4^+(aq) + Cl^-(aq)$$
$$+ \qquad +$$
$$H_2O(aq) \rightleftharpoons OH^-(aq) \quad H^+(aq)$$

$$NH_3 + H_2O \qquad HCl$$
$$\text{(weak)} \qquad \text{(strong)}$$

Removal of OH^-(aq) ions from the solution in the formation of un-ionised ammonia is not balanced by an equal removal of H^+(aq) ions, so that the concentration of H^+(aq) ions is greater than that of OH^-(aq) ions.

d Titration of a weak alkali against a weak acid e.g. 0.1 M ammonia solution against 0.1 M ethanoic acid. There is only a gradual change in pH throughout this type of titration of the kind shown in Fig. 160(d).

3 Neutralisation indicators

Indicators are substances which vary in colour according to the H^+(aq) ion concentration of the solution to which they are added.

Such indicators can be used to measure H^+(aq) ion concentrations (p. 375), or to detect changes in H^+(aq) ion concentration or pH. Different indicators change colour over different ranges of pH, and the most useful are those having a distinct colour change over a narrow range of pH, e.g.

indicator	'acid' colour	'alkaline' colour	pH range
thymol blue	red	yellow	1.2–2.8
screened methyl orange	red	green	3.1–4.4
methyl red	red	yellow	4.4–6.3
azolitmin (in litmus)	red	blue	5.0–8.0
thymol blue	yellow	blue	8.0–9.6
phenolphthalein	colourless	red	8.3–10

It will be seen that thymol blue undergoes two colour changes at different pH values, and this is not unusual. Mixtures of selected indicators give a gradual change of colour over a wide range of pH; such mixtures are known as *universal indicators*.

In an acid–alkali titration, it is important to pick an indicator which changes colour over the range in which there is a marked pH change during the titration (see Fig. 160). The ideal indicator would change colour over a range whose midpoint was the midpoint of the marked pH change occurring during the titration. This means that the choice of common indicator is limited as follows:

titration	marked pH change	indicator
strong acid and strong alkali	4–10	litmus (5–8) or almost any indicator
weak acid and strong alkali	7.5–10.5	phenolphthalein (8.3–10)
strong acid and weak alkali	3.5–6.5	methyl red (4.4–6.3)
weak acid and weak alkali	no marked change	end point cannot be detected accurately by any indicator

4 The functioning of an indicator

The most commonly used indicators are weak acids, the un-ionised molecule being one colour whilst the anion is a different colour, e.g.

$$HIn(aq) \rightleftharpoons H^+(aq) + In^-(aq)$$
(one colour)   (different colour)

In an acidic solution, where the concentration of $H^+(aq)$ ions is high, the indicator will be mainly present as HIn molecules. In alkaline solution, where $H^+(aq)$ ions are removed by combination with $OH^-(aq)$ ions, the indicator will be mainly present as In^- ions.

The equilibrium constant for the ionising indicator is given by

$$K_{in} = \frac{[H^+(aq)]_{eq}[In^-(aq)]_{eq}}{[HIn(aq)]_{eq}}$$

where K_{in} is the ionisation constant of the indicator. When the indicator is at the midpoint of its colour change there will be as many HIn molecules present as $In^-(aq)$ ions, i.e. $[HIn(aq)]$ will be equal to $[In^-(aq)]$, and, at this stage, K_{in} will be equal to $[H^+(aq)]$. Thus *the midpoint of an indicator's colour change occurs when the hydrogen ion concentration is equal to the ionisation constant of the indicator.* The corresponding pH value is the pK_{in} value for the indicator. Some typical numerical values are given below:

thymol blue	1.51	methyl red	5.0	thymol blue	8.9
methyl orange	3.6	azolitmin	7.9	phenolphthalein	9.6

The pH at which an indicator is midway through its colour change, i.e. the pK_{in} of the indicator, is of value, but the range over which the colour change may be said to be completed is also useful. To decide this theoretically, it is necessary to assume the conditions under which the colour change might be said to be complete. The assumption most generally made is that the colour change from the HIn colour to that of $In^-(aq)$ ions will be complete when $[In^-(aq)]$ is equal to $10[HIn(aq)]$. Similarly, the colour change, in the other direction, will be complete when $[HIn(aq)]$ is equal to $10[In^-(aq)]$.

When $[In^-(aq)]$ is equal to $10[HIn^-(aq)]$,

$$K_{in} = 10[H^+(aq)] \quad \text{or} \quad [H^+(aq)] = K_{in}/10$$

and when $[HIn(aq)]$ is equal to $10[In^-(aq)]$,

$$K_{in} = [H^+(aq)]/10 \quad \text{or} \quad [H^+(aq)] = 10K_{in}$$

A $H^+(aq)$ ion concentration of K_{in} corresponds to a pH equal to pK_{in}; $[H^+(aq)]$ of $0.1 K_{in}$ corresponds to a pH of $pK_{in}+1$; $[H^+(aq)]$ of $10K_{in}$ corresponds to a pH of $pK_{in}-1$.

The colour change of an indicator can therefore be summarised as:

	$H^+(aq)$	pH
first change of colour	$0.1K_{in}$	$pK_{in}+1$
midpoint of change	K_{in}	pK_{in}
colour change complete	$10K_{in}$	$pK_{in}-1$

The $H^+(aq)$ ion concentration has to change 100-fold to complete the colour change, or in other words, the range over which an indicator changes colour completely is 1 unit of pH on each side of the pK_{in} value for the indicator.

These theoretical figures for the range of an indicator do not accurately coincide with the experimentally measured range, as shown in the following table. This is due to the arbitrary assumption made about the completion of the colour change in the theoretical considerations.

indicator	pK_{in} value	'theoretical' range	'experimental' range
methyl orange	3.6	2.6–4.6	3.1–4.4
methyl red	5.0	4.0–6.0	4.4–6.3
phenolphthalein	9.6	8.6–10.6	8.3–10.0

5 Uses of indicators in special titrations

It is very important to realise that the terms acidic, alkaline and neutral can be very misleading unless carefully understood in their particular context. A solution of a weak acid, e.g. carbonic acid, may have a pH of 5. Phenolphthalein would indicate that such a solution was acidic, but with screened methyl orange, the solution would appear to be 'alkaline'.

True neutrality occurs when the pH is 7, but as has been seen (p. 366), the pH of mixtures formed when equivalent quantities of acids and alkalies are mixed may be less than or greater than 7 if the acid and alkali are not of equal strength. The end point of an acid–alkali titration is not always at a pH of 7.

a Titration of sodium carbonate Sodium carbonate is a salt of a strong alkali and a weak acid, and its solution in water is alkaline, because of hydrolysis. It will react with solutions of strong acids, e.g.

$$Na_2CO_3(aq) + 2HCl(aq) \longrightarrow 2NaCl(aq) + H_2CO_3(aq)$$

and the detection of the end point of such a reaction is affected by the fact that one of the products of the reaction is itself an acid. It is, however, a very weak acid and the pH of the carbonic acid solutions obtained in a typical carbonate–acid titration is about 5. Such a solution would be 'acid' to phenolphthalein (range 8.3–10) and turn it colourless, but it would not be acid to screened methyl orange (range 3.1–4.4), and would not turn it red.

The pH of 0.025 M sodium carbonate solution is 11.5, so that screened methyl orange will be green in this solution. At the end point, in titration with 0.1 M hydrochloric acid, the colour will change to red. This colour change will not have been affected by the carbonic acid formed as carbonic acid, at this concentration, will not affect screened methyl orange.

Carbonates can, therefore, be titrated against strong acids by using screened methyl orange as indicator.

b Titration of phosphoric(V) acid Phosphoric(V) acid is tribasic so that it can react with an alkali such as sodium hydroxide in three stages,

i $\quad H_3PO_4(aq) \quad + NaOH(aq) \longrightarrow NaH_2PO_4(aq) + H_2O(l)$

ii $\quad NaH_2PO_4(aq) + NaOH(aq) \longrightarrow Na_2HPO_4(aq) + H_2O(l)$

iii $\quad Na_2HPO_4(aq) + NaOH(aq) \longrightarrow Na_3PO_4(aq) \quad + H_2O(l)$

to form two acid salts and one normal one.

The pH value of the various products, all in 0.1 M solution are

H_3PO_4	NaH_2PO_4	Na_2HPO_4	Na_3PO_4
1.5	4.4	9.6	≈ 12

Addition of sodium hydroxide solution to phosphoric(V) acid solution containing screened methyl orange, until the colour changes completely to green, will give a solution of sodium dihydrogenphosphate(V). The amount of sodium hydroxide required will be equivalent to one-third of the phosphoric(V) acid present.

Using phenolphthalein, addition of sodium hydroxide will give a deep magenta colour when all the phosphoric(V) acid has been converted into disodium hydrogenphosphate(V) (pH = 9.6). But before then the phenolphthalein will become pale rose when the pH reaches 8.3, i.e. when about 93 per cent of disodium hydrogenphosphate(V) has been formed. It is, therefore, not easy to obtain an accurate end point using phenolphthalein as an indicator. The measured reading is however, approximately equivalent to two-thirds of the total phosphoric(V) acid present (Fig. 161).

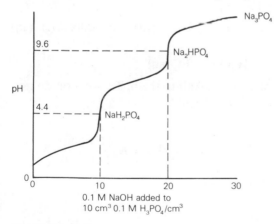

Fig. 161 The pH changes in a titration of 0.1 M sodium hydroxide solution against 0.1 M phosphoric(V) acid.

6 Buffer solutions

A buffer solution, or a solution of reserve acidity or alkalinity, is one which maintains a fairly constant pH even when small amounts of acid or alkali are added to it. Water, or simple aqueous solutions, do not maintain their pH value at all well because of the marked effect of impurities, such as dissolved carbon dioxide from the air or silicates from a glass vessel.

Acid buffer solutions can be made by mixing a weak acid with a salt of the same weak acid; alkaline buffer solutions, by mixing a weak base with a salt of the weak base.

It is so useful to be able to make and keep buffer solutions that many proprietary mixtures are marketed. They are commonly provided in tablet form to be made into solution as required.

a Acidic buffer solutions A simple acid buffer solution can be made from ethanoic acid and sodium ethanoate,

$$HEt(aq) \rightleftharpoons H^+(aq) + Et^-(aq)$$

$$NaEt(aq) \longrightarrow Na^+(aq) + Et^-(aq)$$

Sodium ethanoate is fully ionised, and the ethanoate ions it produces suppress the ionisation of the ethanoic acid so that the mixture contains more acid molecules and more ethanoate ions than a simple solution of ethanoic acid.

The excess ethanoate ions react with any $H^+(aq)$ ions which might be added,

$$Et^-(aq) + H^+(aq) \longrightarrow HEt(aq)$$

whilst the excess ethanoic acid molecules react with any added $OH^-(aq)$ ions,

$$HEt(aq) + OH^-(aq) \longrightarrow H_2O(l) + Et^-(aq)$$

Addition of small amounts of acid or alkali to the buffer solution do not, therefore, greatly affect the pH of the mixture.

In the buffer solution

$$K_a = \frac{[H^+(aq)]_{eq}[Et^-(aq)]_{eq}}{[HEt(aq)]_{eq}} \qquad [H^+(aq)]_{eq} = K_a \frac{[HEt(aq)]_{eq}}{[Et^-(aq)]_{eq}}$$

and, by changing signs and taking logarithms,

$$pH = pK_a + \lg \frac{[Et^-(aq)]_{eq}}{[HEt(aq)]_{eq}}$$

The pH of the solution depends, then, on the K_a value of the acid and the ratio of the concentrations of $Et^-(aq)$ and $HEt(aq)$. As this ratio will not change on dilution, the pH of a buffer solution is not affected by adding water.

The value of the ratio can be calculated by making two approximations. First, it can be assumed that all the $Et^-(aq)$ comes from the fully ionised salt, i.e. the very small amount of $Et^-(aq)$ from the weakly ionised acid can be neglected. Secondly, it can be assumed that the concentration of HEt molecules is approximately equal to the total acid concentration, the weak ionisation of the acid again being neglected.

K_a for ethanoic acid is $1.8 \times 10^{-5}\,mol\,dm^{-3}$, giving a pK_a value of 4.74. The pH of a mixture of equal volumes of ethanoic acid and sodium ethanoate solutions of equal concentrations will, therefore, be 4.74. When the ratio of the salt to the acid is 2, the pH will be 5.04; when the ratio is 0.5, the pH will be 4.44.

In the general case of any weak acid with one of its salts,

$$[H^+(aq)]_{eq} = K_a \frac{[acid]_{eq}}{[salt]_{eq}} \qquad pH = pK_a + lg\frac{[salt]_{eq}}{[acid]_{eq}}$$

Measurement of the pH of a solution with a known salt/acid ratio can be used to find K_a values.

b Alkaline buffer solutions An alkaline buffer solution contains a weak alkali together with one of its salts. A mixture of ammonia solution and ammonium chloride provides an example. The $NH_4^+(aq)$ ions from the fully ionised ammonium chloride suppress the ionisation of the ammonia solution

$$NH_3(aq) + H_2O(l) \longrightarrow NH_4^+(aq) + OH^-(aq)$$

so that the mixture contains a greater concentration of both ammonia molecules and ammonium ions than ammonia solution by itself.

Added $OH^-(aq)$ and $H^+(aq)$ ions are 'taken up' by the reactions

$$NH_4^+(aq) + OH^-(aq) \longrightarrow NH_4OH(aq)$$
$$NH_3(aq) + H^+(aq) \longrightarrow NH_4^+(aq)$$

In the buffer solution

$$K_b = \frac{[NH_4^+(aq)]_{eq}[OH^-(aq)]_{eq}}{[NH_3(aq)]_{eq}}$$

$$[OH^-(aq)]_{eq} = K_b\frac{[NH_3(aq)]_{eq}}{[NH_4^+(aq)]_{eq}}$$

As

$$[OH^-(aq)]_{eq}[H^+(aq)]_{eq} = K_w$$

$$\frac{K_w}{[H^+(aq)]_{eq}} = K_b\frac{[NH_3(aq)]_{eq}}{[NH_4^+(aq)]_{eq}}$$

and taking logarithms,

$$pH = pK_w - pK_b + lg\frac{[NH_3(aq)]_{eq}}{[NH_4^+(aq)]_{eq}}$$

In the more general case, and making the same approximations as in **a**,

$$pH = pK_w - pK_b + \lg \frac{[base]_{eq}}{[salt]_{eq}}$$

For a mixture of equal volumes of ammonia and ammonium chloride solutions, of equal concentrations,

$$pH = pK_w - pK_b = 14 - 4.74 = 9.26$$

c The effectiveness of a buffer solution The pH of a buffer solution made from equal volumes of 0.1 M ethanoic acid and 0.1 M sodium ethanoate is 4.74. If $10 \, cm^3$ of M hydrochloric acid, i.e. 0.01 mol of $H^+(aq)$, is added to $1 \, dm^3$ of the mixture, it will react with $Et^-(aq)$ ions to form HEt molecules. 0.01 mol of $Et^-(aq)$ will be used up, and 0.01 mol of HEt will be formed. The concentrations of $Et^-(aq)$ ions and HEt(aq) molecules (in $mol \, dm^{-3}$), after the addition of the hydrochloric acid, will be

$$[Et^-(aq)] = 0.1 - 0.01 = 0.09 \qquad [HEt(aq)] = 0.1 + 0.01 = 0.11$$

The new pH will be given by

$$pH = 4.74 + \lg(0.09/0.11) = 4.65$$

Addition of $10 \, cm^3$ of M hydrochloric acid to $1 \, dm^3$ of this buffer solution lowers the pH from 4.74 to 4.65. Addition of $10 \, cm^3$ of M hydrochloric acid to $1 \, dm^3$ of water would produce 0.01 M hydrochloric acid; the pH would be changed from 7 to 2.

d To make a buffer solution of known pH The expressions given in **a** and **b** enable the acid/salt, or base/salt, ratio for any particular pH to be calculated.

To obtain a buffer solution with pH of 5 from ethanoic acid and sodium ethanoate solutions it is necessary that

$$5 = 4.74 + \lg \frac{[salt]_{eq}}{[acid]_{eq}} \qquad \text{or} \qquad \frac{[Et^-(aq)]_{eq}}{[HEt(aq)]_{eq}} = 1.82$$

This can be achieved by adding $182 \, cm^3$ of a sodium ethanoate solution to $100 \, cm^3$ of an ethanoic acid solution of the same concentration.

7 Measurement of pH

Many industrial and biological processes depend on $H^+(aq)$ concentration so that the measurement and control of pH can be important. The simplest method is to use a *pH meter*, as described on page 412, but indicators can also be used, for if the same amount of the same indicator gives the same colour in the same amount of two different solutions, then the solutions must have the same pH.

$H^+(aq)$ ion concentrations in unknown solutions can, therefore, be measured if a set of buffer solutions of known pH is available. The approximate pH of an acid solution is first measured using universal indicator (p. 368). If its value is approximately 4, say, then methyl orange (range 3.1–4.4) is a suitable indicator to choose for further investigation. Two drops of methyl orange are added to $10\,cm^3$ of the acid solution and the colour obtained is matched against that given by 2 drops of methyl orange in $10\,cm^3$ portions of solutions of known pH varying from 3.1 to 4.4. By matching the colour of the unknown solution against that of the known solutions, the pH of the unknown solution can be found. The colour matching can be done by eye or by using a colorimeter (p. 187). The set of buffer solutions, each with a slightly different colour, can be replaced by a coloured chart or by a set of coloured glasses. The use of a colorimeter with a set of coloured glasses provides a quick and simple method.

Alternatively, wide- and narrow-range pH papers can be used in the same sort of way. The method is quicker but less accurate.

8 Hydrolysis of salts

A salt may be regarded as formed, together with water, by the reaction of an acid and a base, e.g.

$$HA(aq) + BOH(aq) \longrightarrow BA(aq) + H_2O(l)$$
$$\text{acid} \qquad \text{base} \qquad \text{salt}$$

where A is an anion and B a cation.

If such a reaction is reversible, a solution of a salt in water will contain some acid and some base, and if the acid and the base are not of equal strengths the solution will be either acidic or basic, the equilibrium between $H^+(aq)$ and $OH^-(aq)$ ions in water being disturbed.

When this happens, salt hydrolysis is said to have taken place, and the nature and extent of the hydrolysis depends on the strengths of the acid and base from which the hydrolysed salt can be formed.

a Salt of strong acid and strong base Such salts, e.g. sodium chloride, sodium sulphate and potassium nitrate are not hydrolysed, and their aqueous solutions are neutral, with a pH of 7.

In a solution of sodium chloride in water, for example, there are $Na^+(aq)$, $Cl^-(aq)$, $H^+(aq)$ and $OH^-(aq)$ ions present, but, as sodium hydroxide and hydrochloric acid are both fully ionised there is no tendency for reaction between $Na^+(aq)$ ions and $OH^-(aq)$ ions, or between $H^+(aq)$ and $Cl^-(aq)$ ions. The $H^+(aq)$ and $OH^-(aq)$ ion concentrations remain as they are in pure water.

b Salt of weak acid and strong base e.g. sodium ethanoate. The interaction between the ions, giving an alkaline solution, is shown on page 366. The hydrolysis may be referred to as *anion hydrolysis*,

$$Et^-(aq) + H_2O(l) \rightleftharpoons HEt(aq) + OH^-(aq)$$

It is the formation of $OH^-(aq)$ that makes the solution alkaline.

c Salt of strong acid and weak base e.g. ammonium chloride. The interaction between the ions, giving an acidic solution, is shown on page 367. This is an example of cation hydrolysis,

$$NH_4^+(aq) + H_2O(l) \rightleftharpoons NH_3(aq) + H_3O^+(aq)$$

with the $H_3O^+(aq)$ ions causing the acidity.

The acidity of a copper(II) sulphate solution may be explained in a similar way. An over-simplified summary shows

$$
\begin{array}{ccc}
CuSO_4(aq) \longrightarrow & Cu^{2+}(aq) & + SO_4^{2-}(aq) \\
& + & + \\
2H_2O(l) \rightleftharpoons & 2OH^-(aq) + & 2H^+(aq) \\
& \downarrow & \uparrow \\
& Cu(OH)_2 & H_2SO_4 \\
& \text{(weak and} & \text{(strong)} \\
& \text{insoluble)} &
\end{array}
$$

the acidity being attributed to the removal of $OH^-(aq)$ ions in the formation of copper(II) hydroxide,

$$Cu^{2+}(aq) + 2H_2O(l) \longrightarrow 2H^+(aq) + Cu(OH)_2$$

Alternatively, hydrated metal cations can be regarded as acids, dissociating in stages, e.g.

$$Al(H_2O)_6^{3+}(aq) + H_2O(l) \rightarrow Al(H_2O)_5OH^{2+}(aq) + H_3O^+(aq)$$

$$Al(H_2O)_5OH^{2+}(aq) + H_2O(l) \rightarrow Al(H_2O)_4(OH)_2^+(aq) + H_3O^+(aq)$$

$$Al(H_2O)_4(OH)_2^+(aq) + H_2O(l) \rightarrow Al(H_2O)_3(OH)_3(s) + H_3O^+(aq)$$

The final dissociation, resulting in the formation of hydrated aluminium hydroxide will only take place if basic anions are present to lower the $H_3O^+(aq)$ ion concentration; $OH^-(aq)$, $CO_3^{2-}(aq)$ and $S^{2-}(aq)$ ions all serve this purpose. As a result, addition of these ions to a solution of aluminium salts precipitates the hydroxide.

Other hydrated trivalent metal ions, and Cu^{2+}, Zn^{2+}, Cd^{2+}, Hg^{2+} and Pb^{2+} act similarly.

d Salt of weak acid and weak base Both anion and cation hydrolysis will occur together, and the resulting solution may be acidic, neutral or alkaline depending on the relative extent to which they do occur.

A solution of ammonium ethanoate is approximately neutral (pH = 7), hydrolysis of $Et^-(aq)$ ions being just about balanced by hydrolysis of $NH_4^+(aq)$ ions. Ammonium carbonate is alkaline, and ammonium nitrite, acidic.

9 The pH of salt solutions

a Salt of a weak acid and strong base. Anion hydrolysis If a salt, BA, in a solution of concentration $c \, mol \, dm^{-3}$, has a degree of hydrolysis, h, the resulting equilibrium concentrations, in $mol \, dm^{-3}$, will be

$$A^-(aq) + H_2O(l) \rightleftharpoons HA(aq) + OH^-(aq)$$
$$c(1-h) \qquad\qquad\quad ch \qquad\quad ch$$

and the equilibrium constant, known as the hydrolysis constant, K_h, will be given by

$$K_h = \frac{[HA(aq)]_{eq}[OH^-(aq)]_{eq}}{[A^-(aq)]_{eq}} = \frac{ch^2}{(1-h)} \approx ch^2 \text{ (if } h \text{ is small}$$
$$\text{compared with 1)}$$

In the solution,

$$[H^+(aq)]_{eq}[OH^-(aq)]_{eq} = K_w$$

and $$\frac{[H^+(aq)]_{eq}[A^-(aq)]_{eq}}{[HA(aq)]_{eq}} = K_a$$

so that $$K_h = K_w/K_a \qquad \text{and} \qquad h \approx \sqrt{K_w/K_a c}$$

It is, then, very easy to calculate the degree of hydrolysis in a solution of any concentration from values of K_w and K_a.

The $OH^-(aq)$ ion concentration is equal to $ch \, mol \, dm^{-3}$. The $H^+(aq)$ ion concentration is, therefore, K_w/ch, i.e.

$$[H^+(aq)]_{eq} = K_w/ch \approx \sqrt{K_w K_a/c}$$

The pH of the solution will be given by

$$pH = -lg[H^+(aq)] = -0.5 \lg K_w - 0.5 \lg K_a + 0.5 \lg c$$

$$pH = 0.5(pK_w + pK_a + \lg c)$$

The pH of 0.1 M sodium ethanoate, for example, will be

$$-0.5 \lg 10^{-14} - 0.5 \lg(1.8 \times 10^{-5}) + 0.5 \lg 0.1$$

i.e. $(7 + 2.37 - 0.5)$ or 8.87.

b Salt of strong acid and weak base. Cation hydrolysis

$$B^+(aq) + H_2O(l) \rightleftharpoons H^+(aq) + BOH(aq)$$
$$c(1 - h) \qquad\qquad ch \qquad ch$$

$$K_h = ch^2/(1 - h) \approx ch^2$$

$$K_h = K_w/K_b \qquad \text{and} \qquad h \approx \sqrt{K_w/K_b c}$$

The $H^+(aq)$ ion concentration is ch, or

$$[H^+(aq)] = ch = \sqrt{K_w c/K_b}$$

so that

$$pH = -\lg[H^+(aq)] = -0.5\lg K_w + 0.5\lg K_b - 0.5\lg c$$
$$pH = 0.5(pK_w - pK_b - \lg c)$$

For 0.1 M ammonium chloride

$$pH = -0.5\lg 10^{-14} + 0.5\lg(1.8 \times 10^{-5}) - 0.5\lg 0.1$$

i.e. $(7 - 2.37 + 0.5)$ or 5.13.

c Salt of weak acid and weak base

$$B^+(aq) + A^-(aq) + H_2O(l) \rightleftharpoons HA(aq) + BOH(aq)$$
$$c(1 - h) \quad c(1 - h) \qquad\qquad ch \qquad ch$$

The concentrations, at equilibrium, will be as shown and

$$K_h = \frac{[HA(aq)]_{eq}[BOH(aq)]_{eq}}{[B^+(aq)]_{eq}[A^-(aq)]_{eq}} = \frac{h^2}{(1 - h)^2}$$

Here, the relationship between K_h and h does not involve c, and the degree of hydrolysis is independent of the concentration of the solution.

The $H^+(aq)$ ion concentration is given by

$$[H^+(aq)] = \sqrt{K_w K_a/K_b}$$

and the pH by

$$pH = -\lg(H^+(aq)) = -0.5\lg K_w - 0.5\lg K_a + 0.5\lg K_b$$
$$pH = 0.5(pK_w + pK_a - pK_b)$$

d Hydrolysis of acid salts The pH of a solution of an acid salt is determined by the extent of the hydrolysis together with the extent of the dissociation of the hydrogen-containing anion.

For sodium hydrogencarbonate, $NaHCO_3$, anion hydrolysis leads to

$$HCO_3^-(aq) + H_2O(l) \longrightarrow H_2CO_3 + OH^-(aq)$$

whereas dissociation of the anion leads to

$$HCO_3^-(aq) \longrightarrow CO_3^{2-}(aq) + H^+(aq)$$

The hydrolysis predominates, as HCO_3^- is such a very, very weak acid ($K_a = 6 \times 10^{-11}$, p. 360), so that a solution of sodium hydrogencarbonate is alkaline, with a pH of 8.3 at a concentration of $0.1\,mol\,dm^{-3}$.

With sodium hydrogensulphite,

$$HSO_3^-(aq) + H_2O(l) \longrightarrow H_2SO_3 + OH^-(aq)$$

$$HSO_3^-(aq) \longrightarrow SO_3^{2-}(aq) + H^+(aq)$$

The HSO_3^- ion is a stronger acid ($K_a = 1 \times 10^{-7}$) than HCO_3^- and it dissociates more than it is hydrolysed. A solution of sodium hydrogensulphite is, consequently, acidic with a pH of about 5.

Questions on Chapter 31

1 What is the approximate percentage decrease in $H^+(aq)$ ion concentration corresponding to an increase of 0.1 in pH?

2 What is the hydrogen ion concentration in solutions with pH of   (a) 4 (b) 3.6?

3 Calculate the pH of (a) 0.1 M ethanoic acid solution if the dissociation constant of the acid is taken as $1.75 \times 10^{-5}\,mol\,dm^{-3}$ (b) 0.01 M hydrochloric acid   (c) 0.01 M sodium hydroxide solution (d) a solution with a hydrogen ion concentration of $2 \times 10^{-6}\,mol\,dm^{-3}$.

4 The logarithmic nature of the pH scale means that the mean hydrogen ion concentration between 4 and 5 is not 4.5 but 4.3. Explain what this means.

5 Prove that the pH of an acid with dissociation constant, K, in a solution in which the degree of dissociation is x, is equal to $\lg 1/K + \lg (x/(1 - x))$.

6 Calculate the approximate pH values of the solutions obtained by adding 20, 24, 24.9 and $25\,cm^3$ of 0.1 M sodium hydroxide to $25\,cm^3$ of 0.1 M hydrochloric acid. Why would the values be different if 0.1 M ethanoic acid were used instead of hydrochloric acid, and what would be the effect in each case if ammonia solution were substituted for sodium hydroxide? (OCS)

7 Draw and explain the curves showing approximately the change of pH when $50\,cm^3$ of 0.1 M sodium hydroxide solution is run slowly with stirring into   (a) $25\,cm^3$ of 0.1 M hydrochloric acid   (b) $25\,cm^3$ of 0.1 M ethanoic acid. What light do these curves throw on the indicators which may be used for the titration of sodium hydroxide with these acids? Why is it not possible to titrate ethanoic acid with ammonia solution, using ordinary indicators? (OC)

8 Show how the formula expressing Ostwald's dilution law is derived. To what extent is it generally applicable? The dissociation constant of ethanoic acid in aqueous solution at 25°C is $1.75 \times 10^{-5}\,mol\,dm^{-3}$. Calculate its degree of ionisation in 0.1 M solution at 25°C, and the pH of the solution.

9 What is the pH of 0.01 M potassium hydroxide solution at 25 °C if the ionic product of water at that temperature is $1 \times 10^{-14} \, \text{mol}^2 \, \text{dm}^{-6}$? What is the hydrogen ion concentration (in $\text{mol} \, \text{dm}^{-3}$) of a solution which has a pH of 5.5?

10 Explain the meaning of the term dissociation constant as applied to acids and bases in solution. Calculate (a) the mass of hydrogen chloride required per dm^3 to give a pH of 3 (b) the pH of a 0.01 M solution of propanoic acid $(K_a = 1.45 \times 10^{-5} \, \text{mol} \, \text{dm}^{-3})$ (c) the concentration in $\text{mol} \, \text{dm}^{-3}$ of an aqueous ammonia solution whose pH is 10.3 $(K_b = 1.74 \times 10^{-5} \, \text{mol} \, \text{dm}^{-3})$.

11 A 0.025 M solution in water of a monobasic acid was found to have a freezing point of $-0.06 \, °\text{C}$. What is the dissociation constant and pK value of the acid?

12 What methods are used for determining the 'end points' of titrations? Outline the principles involved. (OS)

13 What is the evidence that when ammonia dissolves in water ammonium ions are formed? Will the pH of a solution of ammonium chloride in water be greater or less than 7? Give reasons. (OS)

14 What is a buffer solution, and why are such solutions important? Given that the dissociation constant of ethanoic acid is $1.8 \times 10^{-5} \, \text{mol} \, \text{dm}^{-3}$, describe how you would prepare a buffer solution of pH 5. (OS)

15 What is a buffer solution? Explain its mode of action. Give two instances of the use of ammonia solution in the presence of ammonium chloride. (OC)

16 What is the pH of a buffer solution made by dissolving 0.005 mol of methanoic acid $(K_a = 1.8 \times 10^{-4} \, \text{mol} \, \text{dm}^{-3})$ and 0.007 mol of sodium methanoate in $1 \, \text{dm}^3$ of aqueous solution? What is the effect on the pH of a tenfold dilution?

17 Calculate the pH of a mixture of $50 \, \text{cm}^3$ of M ethanoic acid with $20 \, \text{cm}^3$ of M sodium hydroxide. Point out clearly what assumptions you make in the calculation.

18 The K_{in} values for methyl red and methyl orange are 1×10^{-5} and $4 \times 10^{-4} \, \text{mol} \, \text{dm}^{-3}$ respectively. They both change colour from red in acid solution, through orange, to yellow in alkaline solution. Draw up a table to show their probable colours in solutions of pH 2, 3, 4, 5 and 6.

19 The acid dissociation constant of methyl orange is $4 \times 10^{-4} \, \text{mol} \, \text{dm}^{-3}$. The solubility product of magnesium hydroxide is $2 \times 10^{-11} \, \text{mol}^3 \, \text{dm}^{-9}$. Describe what will happen when molar sodium hydroxide solution is added dropwise to $25 \, \text{cm}^3$ of 0.5 M sulphuric acid in which 0.1 g of magnesium has been dissolved and containing a few drops of methyl orange. $(K_w = 10^{-14}; \, Mg = 25.) \, (OS)$

20 Explain the distinction between the terms 'the molarity of an acid' and 'the strength of an acid'. Discuss the way in which knowledge of the strength of an acid would affect your choice of the indicator to be used for determining its molarity. (OS)

21 What is meant by the term hydrolysis? Discuss briefly three examples from inorganic chemistry, and three from organic chemistry.

22 What do you understand by (a) the hydrolysis of a salt (b) the hydrolysis of an ester (c) the hydrolysis of an acid chloride? Give two examples of (a) and one each of (b) and (c). What tests would you apply in each case to show that hydrolysis has taken place? (OC)

23 Give illustrative examples to show the difference between (a) anion hydrolysis (b) cation hydrolysis (c) hydration (d) dehydration (e) dehydrogenation.

24 Hydrolysis is a term used very widely in chemistry. Illustrate its various usages.

25 Explain why solutions of aluminium salts and of sodium hydrogensulphate are acidic whilst solutions of sodium sulphide, sodium phosphate(V) and sodium hydrogencarbonate are alkaline.

26 Explain why there is a large evolution of heat when anhydrous aluminium chloride is dissolved in water and why the resulting solution is acidic.

27 Explain why (a) a solution of sodium dihydrogenphosphate(V) is slightly acidic (b) a solution of disodium hydrogenphosphate(V) is distinctly alkaline (c) a solution of sodium phosphate(V) is still more strongly alkaline.

28 Write equations to show the various stages in the hydrolysis of hydrated aluminium salts containing $Al(H_2O)_6^{3+}$ ions and of hydrated iron(III) salts containing $Fe(H_2O)_6^{3+}$ ions.

29 Solid, hydrated iron(III) alum is pale violet in colour. Solutions of iron(III) salts in water are generally yellow-brown, but become violet in colour if strongly acidified. Account for these facts in terms of the extent of hydrolysis into hydroxy-complexes.

30 What is the pH of a 0.01 M solution of sodium ethanoate if the dissociation constant of ethanoic acid is $1.8 \times 10^{-5} \, mol \, dm^{-3}$?

31 Calculate the percentage degree of hydrolysis in a 0.1 M solution of ammonium chloride if the ionisation constant of ammonia is $1.8 \times 10^{-5} \, mol \, dm^{-3}$.

32 What is the pH of the solution obtained when a 0.1 M solution of an acid with dissociation constant $1.8 \times 10^{-5} \, mol \, dm^{-3}$ is just neutralised by a strong base? What indicator would you suggest might be best used in a titration of this sort?

33 How many grams of sodium ethanoate must be added to $500 \, cm^3$ of water to give a solution with a pH of 8.52? The dissociation constant of ethanoic acid is $1.75 \times 10^{-5} \, mol \, dm^{-3}$.

34 If the dissociation constants of hydrocyanic acid and ammonia are 7.2×10^{-10} and $1.8 \times 10^{-5} \, mol \, dm^{-3}$ respectively, what is the pH of a 0.05 M solution of ammonium cyanide?

35 Would it be possible to estimate sodium hydroxide and ammonia in a solution of the two in water by direct titration with hydrochloric acid using two indicators, one of pK 11 and the other of pK 4?

36 Prove that, for a mixture of a weak acid and one of its salts,

$$pH = pK_a + lg\,[\text{salt}]/[\text{acid}]$$

Use the relationship to plot a graph showing how the pH of a mixture of a weak acid (ionisation constant $= 10^{-5}$) and one of its salts varies with the salt/acid concentration ratio.

37* What is the CN^- ion concentration and the pH in a 0.1 M solution of HCN?

38* Use varied examples to explain the difference between the theoretical and experimental range of an indicator. Why is there any difference?

39* What solutions are generally used to give standard pH values at 25 °C of (a) 3.557 (b) 4.008 (c) 6.865 (d) 7.413 (e) 9.180?

40* Calculate the pH of 1 M solutions of (a) KCN (b) NaOCl (c) HF.

32
Acids and bases

1 Early definitions

Acids were long ago recognised, by Boyle and others, as a group of hydrogen-containing compounds with characteristic properties such as sour taste, ability to change the colour of indicators, ability to react with many metals and carbonates, and ability to neutralise metallic oxides and hydroxides to form salts. These metallic oxides and hydroxides were called bases (Raoult, 1754). A base metal was one which would react with mineral acids, as opposed to a noble metal which would not react.

The term alkali is much more ancient, and was originally associated with the ashes from burnt wood. As chemistry developed, a distinction was made between mild alkalis, such as sodium and potassium carbonate, and caustic alkalis, such as sodium and potassium hydroxide. Eventually the word alkali was most commonly used to denote soluble bases, such as sodium and potassium hydroxides.

In the early days of the ionic theory, acids were regarded as compounds which could provide H^+ ions when dissolved in water, e.g.

$$H_2SO_4 \longrightarrow H^+ + HSO_4^- \qquad H_2SO_4 \longrightarrow 2H^+ + SO_4^{2-}$$

and bases were substances which would react with H^+ ions from acids to form a salt and water, e.g.

$$PbO + 2H^+ \longrightarrow Pb^{2+} + H_2O$$

$$NaOH + H^+ \longrightarrow Na^+ + H_2O$$

Alkalis were also defined as substances which, on dissolving in water, dissociated to give OH^- ions, e.g.

$$NaOH \longrightarrow Na^+ + OH^-$$

2 The oxonium ion

Once it was realised that H^+ and other ions are hydrated in aqueous solution (p. 263), acidity was related to H_3O^+ (oxonium) or $H(H_2O)_x^+$ or $H^+(aq)$ ions, and the ionisation of water and of typical acids came to be written in terms of H_3O^+ ions, e.g.

$$H_2O(l) \quad + H_2O(l) \longrightarrow H_3O^+ + OH^-(aq)$$

$$HCl(g) \quad + H_2O(l) \longrightarrow H_3O^+ + Cl^-(aq)$$

$$H_2SO_4(l) + 2H_2O(l) \longrightarrow 2H_3O^+ + SO_4^{2-}(aq)$$

Introduction to Physical Chemistry

Such equations show that the solvent (water) for an acid has a role to play in the ionisation process (p. 342), and this is an important part of modern views, as will be seen.

3 Brønsted–Lowry theory

The definition given in the preceding sections are not entirely satisfactory, and they certainly only apply to aqueous solutions. If solvents other than water are used, it is not possible to define acids and bases solely in terms of H_3O^+ and OH^-(aq) ions. An acid dissociates in liquid ammonia, for example, to produce NH_4^+ ions, and, in this solvent, the NH_2^- ion is responsible for basicity (p. 388). Nitric acid, a typical acid in aqueous solution,

$$HNO_3(l) + H_2O(l) \longrightarrow H_3O^+ + NO_3^-(aq)$$

can ionise to give OH^- ions in other solvents and can, therefore, act as a base,

$$HNO_3 \longrightarrow OH^- + NO_2^+$$

Broader definitions of acids and bases are, therefore, needed.

Brønsted and Lowry provided such definitions in 1922.

An acid is a substance that will give up protons to a base.

A base is a substance that will accept protons from an acid.

An acid is a proton-donor and a base is a proton-acceptor.

The relationship between an acid and a base is then summarised by

$$\begin{array}{ccc} A & \rightleftharpoons H^+ + & B \\ \text{acid} & & \text{base} \\ \text{(proton-donor)} & & \text{(proton-acceptor)} \end{array}$$

or, in aqueous solution,

$$\begin{array}{cc} A + H_2O(l) \rightleftharpoons H_3O^+ + & B \\ \text{(acid)} & \text{(base)} \end{array}$$

A and B are said to be *conjugate* or a *conjugate* pair. B is the conjugate base of acid A, and A the conjugate acid of base, B. Typical conjugate pairs are shown below.

acid	base	acid	base	acid	base
HCl	Cl^-	H_3O^+	H_2O	HSO_4^-	SO_4^{2-}
H_2SO_4	HSO_4^-	NH_4^+	NH_3	$H_2PO_4^-$	HPO_4^{2-}
H_2O	OH^-	$C_6H_5NH_3^+$	$C_6H_5NH_2$	HPO_4^{2-}	PO_4^{3-}
H_3PO_4	$H_2PO_4^-$	$Al(H_2O)_6^{3+}$	$Al(H_2O)_5OH^{2+}$	OH^-	O^{2-}

In the left hand group, molecular acids are conjugate with anion bases; in the central group, cation acids are conjugate with molecular or cation bases; in the right hand group, anion acids are conjugate with anion bases of higher charge.

The examples given above include many species not originally regarded as acids and bases and show how the concept of acids and bases is widened by the Brønsted–Lowry theory. It will also be seen that some species, e.g. H_2O, OH^- and HSO_4^- can function either as acids or bases.

4 The strengths of conjugate pairs

A is a strong acid if the equilibrium

$$A \rightleftharpoons H^+ + B^-$$

lies well to the right; B^- is a strong base if the equilibrium is well to the left. Thus the conjugate base of a strong acid is a weak base and vice versa. Similarly, the conjugate acid of a strong base is a weak acid, and vice versa.

In aqueous solution, the equilibria involved for an acid, HA, and its conjugate base, A^-, are

$$HA(aq) + H_2O(l) \rightleftharpoons H_3O^+ + A^-(aq)$$
(acid)   (base)

$$A^-(aq) + H_2O(l) \rightleftharpoons HA(aq) + OH^-(aq)$$
(base)   (acid)

and the acid and basic strengths are given by K_a or pK_a and by K_b or pK_b (pp. 358–61), i.e.

$$K_a = \frac{[H_3O^+]_{eq}[A^-(aq)]_{eq}}{[HA(aq)]_{eq}} \qquad K_b = \frac{[HA(aq)]_{eq}[OH^-(aq)]_{eq}}{[A^-(aq)]_{eq}}$$

It follows that

$$K_a \times K_b = [H_3O^+]_{eq}[OH^-(aq)]_{eq} = K_w \quad (\text{i.e. } 10^{-14})$$

or $pK_a + pK_b = 14$

When an acid reacts with a base, by donating a proton, a second acid and base are formed. The change can be generalised as

$$acid(I) + base(II) \rightleftharpoons acid(II) + base(I)$$

Acid(I) is conjugate with base(I) and acid(II) is conjugate with base(II).

The equilibrium position in such reactions will depend on the relative strengths of the acids and bases. If acid(I) is stronger than acid(II), then base(II) will be stronger than base(I). If so, the equilibrium will lie to the right, e.g.,

$$HCl(g) + H_2O(l) \rightleftharpoons H_3O^+ + Cl^-(aq)$$

When acid(I) is weaker than acid(II) the equilibrium lies to the left, e.g.

$$H_2O(l) + H_2O(l) \rightleftharpoons H_3O^+ + OH^-(aq)$$

and when the acids and bases are of similar strength the equilibrium is evenly balanced, e.g.

$$H_2CO_3(aq) + HS^-(aq) \rightleftharpoons H_2S(g) + HCO_3^-(aq)$$

The situation is very much like that existing in redox reactions (p. 428).

5 The strengths of oxoacids in aqueous solution

Oxoacids contain one or more O—H bonds and can be represented by the general formula $XO_m(OH)_n$. They can be molecules or ions, e.g.

sulphuric acid	$SO_2(OH)_2$	nitric acid	$NO_2(OH)$
hydrogensulphate ion	$SO_3(OH)^-$	nitrous acid	$NO(OH)$

Their strengths, in aqueous solution, depends on the position of the equilibrium

$$-O-H + H_2O(l) \rightleftharpoons H_3O^+(aq) + -O^-(aq)$$

which is measured by the K_a or pK_a value for the acid.

Anything which favours the withdrawal of H^+ from the O—H bond will increase the acid strength, and the polarity of the $-O^{\delta-}\!\!=\!\!H^{\delta+}$ bond is an important factor.

The higher the electronegativity of X, in $XO_m(OH)_n$, the stronger the acid will be, because elements with high electronegativity will withdraw electrons from the OH bond and build up the $-O^{\delta-}\!\!=\!\!H^{\delta+}$ polarity. For example,

$B(OH)_3$	$pK_a = 9.2$			
$Al(OH)_3$	$pK_a = 11.2$	$Cl(OH)$	$pK_a = 7.2$	
		$Br(OH)$	$pK_a = 8.7$	
$PO(OH)_3$	$pK_a = 2.1$	$I(OH)$	$pK_a = 10.6$	
$AsO(OH)_3$	$pK_a = 2.3$			

The atom, X, will withdraw electrons even more readily if it is linked to oxygen atoms because of the high electronegativity of the O atom. That is why higher values of m give stronger acids. For example,

$Cl(OH)$	$pK_a = 7.2$		$NO(OH)$	$pK_a = 3.3$
$ClO(OH)$	$pK_a = 2$		$NO_2(OH)$	$pK_a = -1.4$
$ClO_2(OH)$	$pK_a = -1$			
$ClO_3(OH)$	$pK_a = c. -10$		$SO(OH)_2$	$pK_a = 1.9$
			$SO_2(OH)_2$	$pK_a = -3$

When m is 0, 1, 2 or 3, the approximate values for K_a are less than 10^{-7}, 10^{-2}, 10^3 and 10^8 respectively.

The higher the negative charge on an acid the weaker it will be because it is increasingly difficult to withdraw H^+ from a negatively charged species. Thus

H_2SO_3	$pK_a = 1.9$	H_3PO_4	$pK_a = 2.1$
HSO_3^-	$pK_a = 7.2$	$H_2PO_4^-$	$pK_a = 7.2$
		HPO_4^{2-}	$pK_a = 12.4$

6 Effects of solvent on the strength of acids and bases

Substances can act as acids in aqueous solution if they can donate protons and, because the water can accept them, it is said to be a *protophilic* solvent. Other solvents are *protogenic* (capable of donating protons) or *aprotic* (neither accepting nor donating protons).

a Protophilic solvents In an aqueous solution of hydrogen chloride, the equilibrium

$$HCl(g) + H_2O(l) \rightleftharpoons H_3O^+ + Cl^-(aq)$$

lies far to the right because of the protophilic nature of the water and the high polarity of the $H^{\delta+}$—$Cl^{\delta-}$ bond.

In a solvent of glacial ethanoic acid, which is much less protophilic, the equilibrium

$$HCl(g) + CH_3COOH(l) \rightleftharpoons CH_3COOH_2^+ + Cl^-$$

will be more to the left so that hydrogen chloride is, here, a weaker acid than it is in aqueous solution. Differences in acid strength, not readily apparent in aqueous solution, become noticeable in glacial ethanoic acid solution, the order of decreasing strength for some common acids being $HClO_4 > H_2SO_4 > HCl > HNO_3$. In 0.005 M solution in glacial ethanoic acid, for instance, the conductivity of chloric(VII) acid is more than fifty times that of nitric acid.

Conversely, in a solution of liquid ammonia, which is more strongly protophilic than water, an acid such as ethanoic acid, regarded as weak in aqueous solution, ionises much more fully and becomes stronger,

$$HEt(l) + NH_3(l) \rightleftharpoons NH_4^+ + Et^-$$

$$HEt(l) + H_2O(l) \rightleftharpoons H_3O^+ + OH^-$$

b Protogenic solvents Substances such as liquid hydrogen chloride, liquid hydrogen fluoride and concentrated sulphuric acid are themselves proton donors. They are said to be protogenic. Compared with water, they decrease the strength of acids and increase that of bases.

The strength of an acid may, in fact, be so greatly decreased that it functions as a base. Nitric acid, for example, acts as a base in liquid hydrogen fluoride solution,

$$HF(l) + HNO_3(l) \rightleftharpoons H_2NO_3^+ + F^-$$
$$\text{(acid)} \quad \text{(base)} \qquad \text{(acid)} \qquad \text{(base)}$$

and in a solution of concentrated sulphuric acid,

$$HNO_3(l) \rightleftharpoons NO_2^+ + OH^-$$
$$2H_2SO_4(l) \rightleftharpoons 2HSO_4^- + 2H^+$$
$$HNO_3(l) + 2H_2SO_4(l) \rightleftharpoons NO_2^+ + 2HSO_4^- + H_3O^+$$

The NO_2^+ (the nitryl cation) is the electrophile in the nitration of benzene.

c Aprotic solvents Neither acids nor bases can function as such in aprotic solvents, e.g. methylbenzene and trichloromethane, which can neither donate nor accept protons. Hydrogen chloride, for example, will dissolve in methylbenzene but the solution is neither acidic nor conducting; the hydrogen chloride does not ionise.

d Summary The effect of the solvent is summarised as follows:

nature of solvent	protophilic (proton accepting)	aprotic	protogenic (proton donating)
strength of acid	increased	nil	decreased
strength of base	decreased	nil	increased

e Cady and Elsey's definitions The part played by the solvent in acid-base considerations is emphasised in Cady and Elsey's definitions which may be summarised as follows.

An acid is a solute which increases the concentration of the cation characteristic of the pure solvent.

A base is a solute which increases the concentration of the anion characteristic of the pure solvent.

7 Reactions in non-aqueous solvents

Chemical reactions in liquid ammonia solution have been extensively studied, following the original investigations of Franklin, and more recently many other non-aqueous solvents have been used. Many of the reactions in such non-aqueous solvents are closely related to similar reactions in water. Liquid ammonia, for example, is slightly ionised, and can act as an acid or a base just as water can,

$$\underset{\text{acid}}{NH_3(l)} + \underset{\text{base}}{NH_3(l)} \rightleftharpoons \underset{\text{acid}}{NH_4^+} + \underset{\text{base}}{NH_2^-}$$
$$\underset{\text{acid}}{H_2O(l)} + \underset{\text{base}}{H_2O(l)} \rightleftharpoons \underset{\text{acid}}{H_3O^+} + \underset{\text{base}}{OH^-(aq)}$$

The NH_4^+ and NH_2^- ions are related in just the same way as H_3O^+ and OH^- are.

A solution of ammonium chloride in liquid ammonia is acidic, just as a solution of hydrogen chloride in water is,

$$HCl \longrightarrow H^+ + Cl^- \qquad NH_4Cl \longrightarrow NH_4^+ + Cl^-$$

Similarly, a solution of sodamide, $NaNH_2$, in liquid ammonia is basic, like a solution of sodium hydroxide in water,

$$NaOH \longrightarrow Na^+ + OH^- \qquad NaNH_2 \longrightarrow Na^+ + NH_2^-$$

Moreover, solutions of ammonium chloride and sodamide in liquid ammonia can be titrated using phenolphthalein as an indicator, just as aqueous solutions of hydrogen chloride and sodium hydroxide can.

The small ionisation of liquid sulphur dioxide is represented by the equation,

$$2SO_2(l) \rightleftharpoons SO^{2+} + SO_3^{2-}$$

so that compounds, such as sulphur dichlorideoxide (thionyl chloride), which dissolve in this solvent to give thionyl, SO^{2+}, ions are acidic, whilst compounds, e.g. caesium sulphite, which dissolve to give sulphite ions, SO_3^{2-}, are basic,

$$SOCl_2 \rightleftharpoons SO^{2+} + 2Cl^- \qquad CsSO_3 \rightleftharpoons 2Cs^+ + SO_3^{2-}$$

Analogous situations are found with other non-aqueous solvents which ionise slightly. Examples are provided by liquid hydrogen fluoride, liquid hydrogen sulphide, liquid hydrogen cyanide, and bromine trifluoride,

$$2HF(l) \rightleftharpoons H_2F^+ + F^- \qquad 2HCN(l) \rightleftharpoons H_2CN^+ + CN^-$$

$$2H_2S(l) \rightleftharpoons H_3S^+ + SH^- \qquad 2BrF_3 \rightleftharpoons BrF_2^+ + BrF_4^-$$

8 Lewis acids and bases

The idea of acids and bases has been extended still further, by Lewis, using the following definitions:

An acid is a substance that can accept a pair of electrons.

A base is a substance that can donate a pair of electrons.

For example,

$$H^+ \quad + \quad :\overset{..}{O}-H^- \quad \rightleftharpoons \quad H:\overset{..}{O}-H$$

$$Cl:\overset{x}{B}\,{:}\,Cl \quad + \quad :\overset{x}{N}_x H \quad \longrightarrow \quad Cl\,\overset{x}{:}\,B\,{:}\,N\,_x H$$

$$AlCl_3 \quad + \quad Cl^- \quad \longrightarrow \quad AlCl_4^-$$

acid   base

It will be seen that these definitions do not necessarily involve any ions at all. Nevertheless a typical Lewis acid, BCl_3, can be titrated against a typical Lewis base, $(C_2H_5)_3N$, in chlorobenzene solution, using crystal violet as an indicator.

This very broad generalisation links Lewis acids (electron-pair acceptors) with oxidising agents (electron acceptors) and electrophiles (electron-seekers). It links Lewis bases (electron-pair donors) with reducing agents (electron donors) and nucleophiles (electron-donors or nucleus-seekers).

There are different types of Lewis acid and Lewis base, as summarised below.

Lewis acids	Lewis bases
a Positive ions, e.g. H^+, Al^{3+}, Fe^{3+}	a Negative ions, e.g. OH^-, Cl^-, CN^-
b Molecules with incomplete octets, e.g. BF_3, $BeCl_2$	b Molecules with lone pairs e.g. H_2O, NH_3, ROH
c Molecules in which the central atom can expand its octet, e.g. $SiCl_4$, PCl_5	c Molecules with multiple bonds between C atoms, e.g. C_2H_4, C_2H_2

Questions on Chapter 32

1 Explain the meaning of the following terms as applied to an acid or its aqueous solution; concentration, strength, basicity. What experiments would you make to show that (a) sulphuric acid is dibasic (b) hydrochloric acid is stronger than sulphuric acid? (OCS)

2 The term 'strong acid' is very commonly used. What, precisely, does it mean?

3 Choose any two chemicals and explain how they can be regarded as acids on the various different theories of acid character.

4 An acid, in elementary chemistry, is sometimes defined as a compound which contains hydrogen replaceable by a metal, which will turn blue litmus red, and which will react with metallic carbonates to yield carbon dioxide. To what extent is this a satisfactory definition of an acid?

5 Comment on the definition of an acid as a substance giving rise, in solution, to a cation characteristic of the solvent.

6 Comment on the following statements: (a) The carbonate ion is a fairly strong base (b) The ionisation constant of an acid, HA, in aqueous solution measures the competition for protons between water and A^- ions (c) No acid, in aqueous solution, can be stronger than H_3O^+.

7 What is an acid? Illustrate, by means of examples, the factors influencing the strength of an acid. (OS)

8 Define what you mean by the strength of an acid, and describe methods of measuring it. Discuss the reaction between sulphuric acid and sodium chloride in terms of the relative strengths of sulphuric and hydrochloric acids. (OS)

9 What do you understand by the terms acid, base and salt? Into which of these classes would you place H_2O, $NaHSO_4$, NH_4^+, CuO, C_6H_5OH and CH_3COO^-? Give your reasons.

10 The most important acid is a proton. Discuss this statement.

11 What are the conjugate bases of the following when they act as monobasic acids: HCl, HNO_3, H_3PO_4, H_2O, H_3O^+, CH_3COOH, NH_4^+, HSO_4^- and HCO_3^-? Which of these conjugate bases can still be regarded as acids?

12 Illustrate the statement that salts which contain an ion which can act as a base are more soluble in dilute acids than in water.

13 Which of the following are to be regarded as acids or bases or both: HSO_4^-, HCO_3^-, $H_2PO_4^-$, NH_4^+, OH^-, H_2O, H_3O^+, CO_3^{2-}, NH_3? Give your reasoning.

14 The greater the number of oxygen atoms in a molecule of an oxyacid of an element, the greater is the strength of the acid. Illustrate and discuss this statement.

15 (a) Define the terms 'Lewis acid' and 'Lewis base' and give one example of each.   (b) Define the terms 'acid' and 'base' on the Lowry–Brønsted theory and give one example of each   (c) Explain why (i) CCl_3COOH is a stronger acid than CH_3COOH (ii) the $[Al(H_2O)_6]^{3+}$ ion is a stronger acid than the $[Mg(H_2O)_6]^{2+}$ ion. (JMB)

16 Discuss the ways in which the concepts of 'acidity' and 'alkalinity' have developed since 1777 when Lavoisier generalized that all acids contained oxygen. You may like to illustrate your answer by reference to some or all of the following species, but equal credit will be given for using examples of your own choice: HCl, H_2SO_4, HSO_4^-, NH_3, H_2O, H_3O^+, NH_4^+. (L)

17 What factors influence the strengths of acids and bases?

18 Describe carefully what you would do to compare, experimentally, the strengths of two solid acids.

19* Choose any six weak acids and any six weak bases and illustrate your answer by quoting K_a and K_b values.

20* Record the pK values for   (a) NH_3 (b) CH_3NH_2 (c) $C_6H_5NH_2$. What are their conjugate acids, and what will their pK_a values be? Similarly, record the pK_a values for ethanoic acid and its three chlorinated derivatives. What will the corresponding conjugate bases be, and what are their pK_b values?

33
Solubility product and stability constant

1 Meaning of solubility product

In a saturated solution of a sparingly soluble electrolyte, such as silver chloride, the ions in the solution will be in equilibrium (p. 286) with the undissolved solid, i.e.

$$AgCl(s) \rightleftharpoons Ag^+(aq) + Cl^-(aq)$$

$$\frac{[Ag^+(aq)]_{eq}[Cl^-(aq)]_{eq}}{[AgCl(s)]_{eq}} = K$$

As the concentration of solid silver chloride must be constant it follows that

$$[Ag^+(aq)]_{eq}[Cl^-(aq)]_{eq} = \text{a constant, } K_s(AgCl)$$

The constant, $K_s(AgCl)$, is known as the solubility product of silver chloride. It is equal to $1 \times 10^{-10} \text{ mol}^2 \text{ dm}^{-6}$.

More generally, for an electrolyte $A_x B_y$,

$$A_x B_y(s) \rightleftharpoons xA^{y+} + yB^{x-}$$

$$K_s(A_x B_y) = [A^{y+}]_{eq}^x [B^{x-}]_{eq}^y$$

The concept is only accurately valid if the activities (p. 356) of the ions concerned are used, but for dilute solutions, i.e. solutions of sparingly soluble electrolytes, accurate results can be obtained by using ionic concentrations expressed in mol dm^{-3}. Moreover, the general·qualitative idea is applicable to more concentrated solutions.

The main use of solubility products is concerned with the conditions under which an electrolyte will dissolve, or will come out of solution as a solid, i.e. form a precipitate. It is important to remember that for an electrolyte, AB, with $K_s(AB)$ equal to $[A^+]_{eq}[B^-]_{eq}$:

i a solution in which $[A^+]_{eq}[B^-]_{eq}$ is less than $K_s(AB)$ is not saturated, and more AB can be dissolved in it;

ii a solution in which $[A^+]_{eq}[B^-]_{eq}$ is equal to $K_s(AB)$ is saturated, and

iii if anything be done to a solution which tends to make $[A^+]_{eq}[B^-]_{eq}$ greater than $K_s(AB)$ then solid AB will be precipitated. Evaporation of a saturated solution, for instance, tends to increase the ionic concentrations of the ions present. As a result, solid is deposited.

2 Some applications

It is the value of the solubility product of a substance that determines the conditions under which it will be precipitated.

a The common ion effect The equilibrium in a saturated solution of, for example, silver ethanoate, is

$$AgEt(s) \rightleftharpoons Ag^+(aq) + Et^-(aq)$$

Addition of either Ag^+ or Et^- ions will precipitate silver ethanoate. Alternatively, silver ethanoate is less soluble in solutions containing a common ion than in water (p. 395).

b Precipitation of hydroxides A hydroxide, $M^{x+}(OH)_x$, will only be precipitated if the value of the solubility product, $[M^{x+}]_{eq}[OH^-]^x_{eq}$, is exceeded. For a given $[M^{x+}]_{eq}$, then , higher OH^- concentrations will be needed the larger the solubility product.

The value for $Al(OH)_3$, for example, is $1 \times 10^{-33} \, mol^4 \, dm^{-12}$. If the Al^{3+} concentration in a solution is $10^{-3} \, mol \, dm^{-3}$, the concentration of OH^- to form a precipitate will be $10^{-10} \, mol \, dm^{-3}$. If $[Al^{3+}]$ is $10^{-21} \, mol \, dm^{-3}$, a concentration of OH^- of $10^{-4} \, mol \, dm^{-3}$ will produce a precipitate.

Under normal conditions, ammonia solution will contain a high enough concentration of OH^- to precipitate both $Al(OH)_3$, and $Ca(OH)_2$, with a much higher solubility product ($8 \times 10^{-6} \, mol^3 \, dm^{-9}$):

$$NH_3(g) + H_2O(l) \rightleftharpoons NH_4^+(aq) + OH^-(aq)$$

But if solid ammonium chloride is added, i.e. the concentration of NH_4^+ is increased, this will suppress the ionisation of the ammonia solution and limit the OH^- concentration. Only hydroxides with very low solubility products, e.g. $Al(OH)_3$, $Fe(OH)_3$ and $Cr(OH)_3$, will be precipitated under these conditions.

c Precipitation of sulphides For equal concentrations of metallic ion, sulphides with low solubility products, e.g. HgS ($3 \times 10^{-54} \, mol^2 \, dm^{-6}$) will be precipitated by lower S^{2-} ion concentrations than those with high solubility products, e.g. ZnS ($1 \times 10^{-23} \, mol^2 \, dm^{-6}$).

Passing hydrogen sulphide into solutions of the metallic salt is a convenient way of adding S^{2-} ions. But if acidic solutions are used, the H^+ ions suppress the ionisation of the hydrogen sulphide,

$$H_2S \rightleftharpoons H^+ + HS^- \qquad HS^- \rightleftharpoons H^+ + S^{2-}$$

and the sulphide ion concentration will only be high enough to precipitate sulphides with low solubility products. The S^{2-} ion concentration is greatly increased in alkaline solution so that many more sulphides can be precipitated.

d Solubility of sparingly soluble salts of weak acids in strong acids A salt of a weak acid, sparingly soluble in water, is often soluble in a strong acid. Calcium ethanedioate (oxalate), for example, is only sparingly soluble in water, but soluble in dilute hydrochloric acid. This is sometimes regarded as the 'turning out' of a weak acid by a stronger one,

$$CaOx(s) + 2HCl(aq) \longrightarrow CaCl_2(aq) + H_2Ox(aq)$$

In a saturated solution of calcium ethanedioate in water,

$$[Ca^{2+}(aq)]_{eq}[Ox^{2-}(aq)]_{eq} = K_s(CaOx)$$

but addition of a strong acid, i.e. of $H^+(aq)$ ions, reduces the ethanedioate ion concentration by the formation of ethanedioic acid,

$$Ox^{2-}(aq) + 2H^+(aq) \longrightarrow H_2Ox(aq)$$

This reduction in the ethanedioate ion concentration enables more calcium ethanedioate to go into solution.

Addition of Cl^- ions along with H^+ ions does not materially affect the position, for Ca^{2+} and Cl^- ions do not combine together for the product they would form, $CaCl_2$, is a salt which is soluble and fully ionised.

e Use of potassium chromate(VI) as an indicator In a silver nitrate–chloride titration, the silver nitrate is run into the chloride solution to which some potassium chromate(VI) solution is added to act as an indicator.

The first addition of silver nitrate precipitates silver chloride, but not silver chromate(VI). This is because silver chloride has a lower solubility than silver chromate(VI). When sufficient silver nitrate has been added to precipitate all the silver chloride, further addition precipitates brick-red silver chromate(VI) and this serves as the end-point.

Potassium chromate(VI) can only be used as an indicator in neutral solution because silver chromate(VI) is soluble in acids. Its action is not like that of the indicators used in acid–alkali titrations (p. 367) where only one or two drops are required. Sufficient potassium chromate(VI) is required to produce enough silver chromate(VI) to saturate the solution so that silver chromate(VI) can be precipitated.

3 Typical calculations

a Calculation of solubility product from solubility The solubility of, for example, silver chloride ($M_r = 143.5$), at 18 °C, is 0.001 46 g dm^{-3} or approximately 10^{-5} mol dm^{-3}.

As the solution is necessarily dilute because of the low solubility, the silver chloride can be regarded as fully ionised, so that the concentrations of $Ag^+(aq)$ and $Cl^-(aq)$ will each be 10^{-5} mol dm^{-3}. The solubility product is given by

$$K_s(AgCl) = [Ag^+(aq)]_{eq}[Cl^-(aq)]_{eq} = 10^{-10} \, mol^2 \, dm^{-6}$$

b Calculation of solubility from solubility product Silver chromate(VI), Ag_2CrO_4 ($M_r = 333$) has a solubility product of $2.5 \times 10^{-12} \, mol^3 \, dm^{-9}$ at 18 °C.

The chromate can be regarded as fully ionised in its saturated solution. If, then, the solubility of silver chromate is $x \, mol \, dm^{-3}$, the concentration of $Ag^+(aq)$ ions will be $2x \, mol \, dm^{-3}$ and that of $CrO_4{}^{2-}(aq)$ will be $x \, mol \, dm^{-3}$. Therefore

$$[Ag^+(aq)]^2_{eq}[CrO_4{}^{2-}(aq)]_{eq} = 2.5 \times 10^{-12} = (2x)^2 x$$

and the solubility is $0.855 \times 10^{-4} \, mol \, dm^{-3}$, or ($0.855 \times 10^{-4} \times 333$), i.e. $0.0284 \, g \, dm^{-3}$ at 18 °C.

c Calculation of solubility in a solvent with a common ion When silver chloride is dissolved in 0.1 M hydrochloric acid, the $Cl^-(aq)$ coming from the silver chloride can be neglected compared with that coming from the fully ionised hydrochloric acid. The $Cl^-(aq)$ concentration is, therefore, $0.1 \, mol \, dm^{-3}$.

The solubility product value remains constant, so that

$$[Ag^+(aq)]_{eq}[Cl^-(aq)]_{eq} = 10^{-10} \, mol^2 \, dm^{-3}$$

and $\qquad [Ag^+(aq)]_{eq} = 10^{-9} \, mol \, dm^{-3}$

The solubility of the silver chloride is, therefore, $10^{-9} \, mol \, dm^{-3}$, or $143.5 \times 10^{-9} \, g \, dm^{-3}$. This compares with a solubility of $146 \times 10^{-5} \, g \, dm^{-3}$ in water.

4 Measurement of the solubility of a sparingly soluble salt

This experimental measurement can be carried out by conductivity measurements.

The measured conductivity of a saturated solution of silver chloride, at 18 °C, for example, is $1.274 \times 10^{-6} \, S \, cm^{-1}$. At the same temperature, the conductivity of the conductivity water used in making the saturated solution was $0.054 \times 10^{-6} \, S \, cm^{-1}$. The conductivity due to the silver chloride, is, therefore, $1.220 \times 10^{-6} \, S \, cm^{-1}$.

If the solubility of silver chloride at 18 °C is $x \, g \, dm^{-3}$, it is $x/143.5 \, mol \, dm^{-3}$. The concentration of a saturated solution of silver chloride is, therefore, $x/143.5 \, mol \, dm^{-3}$ or $10^{-3}x/143.5 \, mol \, cm^{-3}$.

The molar conductivity, Λ_c, for the saturated solution is given by

$$\Lambda_c = \frac{\text{conductivity in } S \, cm^{-1}}{\text{concentration in } mol \, cm^{-3}} = 1.22 \times 10^{-3} \times \frac{143.5}{x}$$

and, as the solution is very dilute, this can be taken as being equal to Λ_0, i.e.

$$\Lambda_0 = 1.22 \times 10^{-3} \times 143.5/x \, \text{S cm}^2 \, \text{mol}^{-1}$$

The Λ_0 value for silver chloride can also be obtained from the molar conductivities of Ag^+ and Cl^- ions, or from the Λ_0 values for silver nitrate, sodium chloride and sodium nitrate (p. 350). Thus

$$\Lambda_0(\text{AgCl}) = \Lambda_0(\text{Ag}^+) + \Lambda_0(\text{Cl}^-)$$

$$\Lambda_0(\text{AgCl}) = \Lambda_0(\text{AgNO}_3) + \Lambda_0(\text{NaCl}) - \Lambda_0(\text{NaNO}_3)$$

Such measurements give a value for $\Lambda_0(\text{AgCl})$ of $120 \, \text{S cm}^2 \, \text{mol}^{-1}$. Therefore,

$$1.22 \times 10^{-3} \times 143.5/x = 120 \quad \text{or} \quad x = 0.001\,46$$

The solubility of silver chloride is $0.001\,46 \, \text{g dm}^{-3}$.

5 Solubility product and solubility from electrode potentials

The $E^{\ominus}$ value for the $Ag^+(aq)/Ag$ electrode is $+0.799 \, \text{V}$, but the value if the electrode is immersed in a saturated solution of silver chloride is $0.504 \, \text{V}$. It follows (p. 411) that

$$0.504 = 0.799 + 0.059 \lg [\text{Ag}^+(\text{aq})]$$

$$[\text{Ag}^+(\text{aq})] = 10^{-5} \, \text{mol dm}^{-3}$$

where $[\text{Ag}^+(\text{aq})]$ is the silver ion concentration in the saturated solution. As the $Cl^-(aq)$ ion concentration must be equal to it, then

$$[\text{Ag}^+(\text{aq})][\text{Cl}^-(\text{aq})] = K_s(\text{AgCl}) = 10^{-10} \, \text{mol}^2 \, \text{dm}^{-6}$$

6 The effect of complex formation on solubility

Silver chloride is, generally, regarded as insoluble in water, and it is even less soluble in solvents with a common ion. It will, however, dissolve in ammonia solution because the formation of the $Ag(NH_3)_2^+$ complex ion lowers the concentration of Ag^+ ion,

$$\text{AgCl(s)} \rightleftharpoons \text{Ag}^+(\text{aq}) + \text{Cl}^-(\text{aq})$$

$$\text{Ag}^+(\text{aq}) + 2\text{NH}_3(\text{aq}) \rightleftharpoons \text{Ag}(\text{NH}_3)_2^+(\text{aq})$$

In other words, $Ag(NH_3)_2Cl$ has a much higher solubility product than AgCl and is soluble.

Other common examples of a substance insoluble in water being converted into a soluble complex are summarised opposite:

complex formation		use
$Ag^+(aq) + 2CN^-$	$\rightleftharpoons Ag(CN)_2^-$	in photography
$Ag^+(aq) + 2S_2O_3^{2-}(aq)$	$\rightleftharpoons Ag(S_2O_3)_2^{3-}$	in photography
$I_2(s) + I^-(aq)$	$\rightleftharpoons I_3^-(aq)$	making 'iodine' solutions
$HgI_2(s) + 2I^-(aq)$	$\rightleftharpoons HgI_4^{2-}$	Nessler's solution
$\left.\begin{array}{l} Au \\ Pt \end{array}\right. + \left\{\begin{array}{l} \text{aqua regia} \\ \text{(c. HCl + c. HNO}_3\text{)} \end{array}\right.$	$\begin{array}{l} \rightleftharpoons AuCl_4^- \\ \rightleftharpoons PtCl_6^{2-} \end{array}$	$\left.\right\}$ dissolving Au and Pt
$\begin{array}{l} Au + 4CN^-(aq) \\ Ag + 2CN^-(aq) \end{array}$	$\begin{array}{l} \rightleftharpoons AuCN)_4^- \\ \rightleftharpoons Ag(CN)_2^- \end{array}$	$\left.\right\}$ extraction of Au and Ag

In some cases, apparently odd solubilities are found because both the common ion effect and complex formation play a part. The solubility of lead chloride, for example, in solutions of other chlorides, depends on the chloride ion concentration. When this is low, the lead chloride is less soluble than in water due to the common ion effect. As the chloride ion concentration increases, the solubility rises due to the formation of a soluble $PbCl_4^{2-}$ complex. It can, therefore, be difficult to interpret solubility measurements.

7 Chelation

A complex consists of a central atom or ion linked to other molecules, atoms or ions, known as *ligands*. Simple, or monodentate, ligands are linked to the central atom or ion by only one atom of the ligand. They are, mainly, neutral molecules, e.g. H_2O, NH_3, or anions, e.g. CN^- OH^-, Cl^-, which have lone pairs. The number of ligands is, generally, 2, 4 or 6, but may be 8 or an odd number in rare cases. The complex may be a cation, anion or neutral molecules as illustrated below.

complex cations	$Cu(H_2O)_4^{2+}$	tetraaquocopper(II)
	$Ag(NH_3)_2^+$	diammine silver(I)
	$CrCl_2(H_2O)_4^+$	dichlorotetraaquochromium(III)
complex anions	$Fe(CN)_6^{4-}$	hexacyanoferrate(II)
	$Fe(CN)_6^{3-}$	hexacyanoferrate(III)
	AlF_6^{3-}	hexafluoroaluminate
	BH_4^-	tetrahydridoborate
neutral complexes	H_2SiF_6	hexafluorosilicic acid
	$Cr(NH_3)_3Cl_3$	trichlorotriamminechromium(III)
	$Ni(CO)_4$	tetracarbonylnickel(0)

Other ligands make use of two or more atoms to form more than one bond to the central ion. Such ligands are called chelate groups, and the compounds formed, *chelate compounds*. A group capable of forming two bonds is called a bidentate group, one forming three groups is a tridentate group, and so on.

Ethylene diamine (ethane-1,2-diamine), usually abbreviated to en, is a typical bidentate group giving typical chelate compounds as shown below.

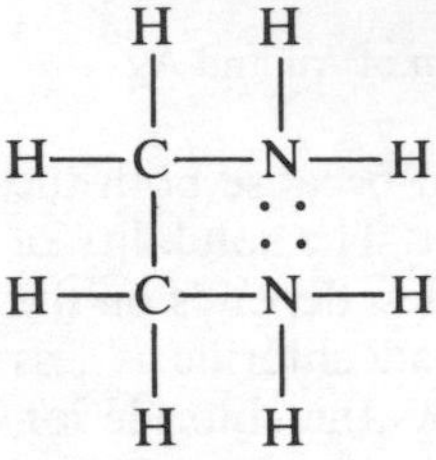

ethylene diamine (ethane-1,2-diamine)

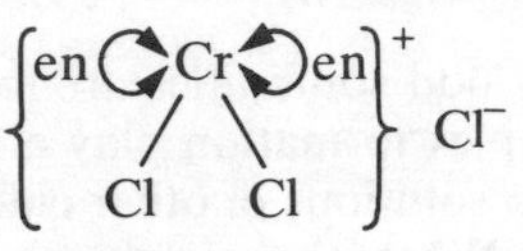

dichlorobis (ethylenediamine) chromium (III) chloride

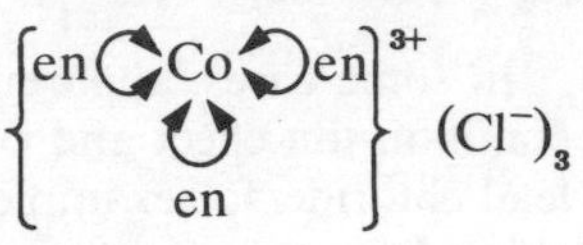

tris (ethylenediamine) cobalt (III) chloride

Ethylenediaminetetra-acetic acid, EDTA, bis[di(carboxymethyl)amino]-ethane, is an important hexadentate ligand, bonding through two N atoms and four O atoms.

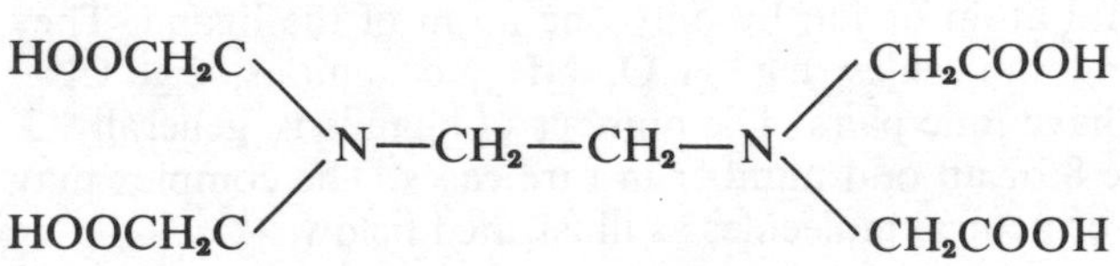

EDTA

8 Stability constants

The stability constant of a complex ion is a measure of its stability so far as dissociation into its component species is concerned. The $Ag(NH_3)_2^+$ ion, for example, is in equilibrium in aqueous solution with $Ag^+(aq)$ ions and ammonia, i.e.

$$Ag^+(aq) + 2NH_3(aq) \rightleftharpoons Ag(NH_3)_2^+(aq) + aq$$

and the equilibrium constant, taking the concentration of water as constant, is known as the stability constant of $Ag(NH_3)_2^+$, i.e.

$$\frac{[Ag(NH_3)_2^+(aq)]_{eq}}{[Ag^+(aq)]_{eq}[NH_3(aq)]_{eq}^2} = \text{stability constant, } K_{stab}$$

The value, for $Ag(NH_3)_2^+$ is $1.7 \times 10^7 \, mol^{-2} \, dm^6$; it is commonly quoted as a $\lg K_{stab}$ value, i.e. 7.23.

The higher the K_{stab} or $\lg K_{stab}$ value, the more stable the complex in aqueous solution. Some typical $\lg K_{stab}$ values are given below.

$Ag(S_2O_3)_2^{3-}$	13.0	$Fe(CN)_6^{4-}$	8.3	AlF_6^{3-}	19.8
$Ag(CN)_2^-$	21.0	$Fe(CN)_6^{3-}$	31.0	I_3^-	2.9

The overall stability constant may be made up of a number of terms if the dissociation can take place in stages, and many different ionic species may co-exist in equilibrium. For $Ag(NH_3)_2^+$,

$$Ag^+(aq) + NH_3(aq) \rightleftharpoons Ag(NH_3)^+(aq) \qquad \lg K_{stab} = 3.32$$
$$Ag(NH_3)^+(aq) + NH_3(aq) \rightleftharpoons Ag(NH_3)_2^+(aq) \qquad \lg K_{stab} = 3.91$$

The logarithm of the overall stability constant will be the sum of the logarithms of the individual stability constants for each stage of the dissociation. The overall stability constant is the product of the stepwise values.

Measurement of the values involves the determination of the equilibrium concentrations of, perhaps, many ions, and methods which do not upset the equilibrium must be used. In a solution of copper(II) sulphate and ammonia, for example, $Cu(NH_3)_{6-x}(H_2O)_x^{2+}$ ions exist, with x having a value of 6 when the ammonia concentration is very low but rising to 1 as the ammonia concentration is increased. The proportion of each ion in a particular mixture can be measured by investigating the absorption spectrum, for each ion absorbs differently.

9 Factors affecting the stability of complexes

Considerable work has been done on trying to relate the stability of a complex ion to its structure, but it has only led to some broad generalisations.

The greater the electrical fields of the central ion and the ligand, i.e. the smaller the size and the higher the charge, the more stable the complex ion tends to be. $Fe(CN)_6^{3-}$, for example, is much more stable than $Fe(CN)_6^{4-}$, and fluoro-complexes tend to be stabler than the corresponding chloro-complexes.

The electron-donating (basic) properties of the ligand are also very important, and many common ligands which give stable complexes are, indeed, strong bases. The formation of a complex and the formation of an acid are similar processes, i.e.

$$H^+ + base \rightleftharpoons acid \qquad M^+ + ligand \rightleftharpoons complex$$

As a broad generalisation, too, multidentate ligands (unless they are very bulky) form stabler complexes than monodentate ligands. The EDTA complex, for example, is much more stable than the corresponding ammine.

10 The use of stability constants

Many chemical facts can be accounted for in terms of stability constant values, particularly by comparing the value of one complex with that of another.

CN^- ions, for example, are more effective in dissolving silver chloride than $S_2O_3^{2-}$ ions; in turn, $S_2O_3^{2-}$ ions are more effective than ammonia solution.

Addition of ammonia to copper(II) sulphate solution produces a deep blue solution containing, mainly, $Cu(NH_3)_4(H_2O)_2^{2+}$ ions. Further addition of EDTA solution converts some of these ions into the lighter-coloured, more stable $Cu(EDTA)^{2+}$ ions, and the colour is removed altogether if CN^- ions are added, because $Cu(CN)_4^{2-}$ ions are colourless and very stable.

Similarly, iron(III) chloride solution gives a deep red colour with SCN^- ions due to the formation of $Fe(H_2O)_5(SCN)^{2+}$ ions. Addition of F^- or $C_2O_4^{2-}$ ions removes the colour as the colourless, and very stable, FeF_6^{3-} and $Fe(C_2O_4)_3^{3-}$ ions are formed.

Kinetic effects may, however, prevent reactions, and ligands are classified as *labile* (easily replaced) or *inert*.

11 Calculation of ionic concentrations

If the stability constant of a complex is known, the ionic concentrations in an aqueous solution of the ion under different conditions can be calculated. If solutions of 0.02 M silver nitrate and 2 M ammonia are mixed in equal volumes, for example, the initial concentration of $Ag^+(aq)$ will be $0.01\ mol\ dm^{-3}$ and that of $NH_3(aq)$ will be $1\ mol\ dm^{-3}$. The state of affairs at equilibrium will be

$$Ag^+(aq) + 2NH_3(aq) \rightleftharpoons Ag(NH_3)_2^+(aq)$$
$$(0.01 - x) \quad (1 - 2x) \qquad\qquad x$$

so that $\quad x/(0.01 - x)(1 - 2x)^2 = K_{stab} = 1.7 \times 10^7$

from which the concentrations of each species can be obtained.

As an approximation, it can be assumed that almost all the silver is present as the complex ion, i.e. that x is approximately 0.01. Therefore,

$$0.01/[Ag^+(aq)] \approx 1.7 \times 10^7 \qquad or \qquad [Ag^+(aq)] \approx 5.88 \times 10^{-10}$$

The $Ag^+(aq)$ ion concentration in the mixture is, therefore, $5.88 \times 10^{-10}\,mol\,dm^{-3}$. As the solubility product of silver chloride is $10^{-10}\,mol^2\,dm^6$ it will require a $Cl^-(aq)$ concentration of $0.17\,mol\,dm^{-3}$ to precipitate silver chloride.

12 Complexometric titrations

In a complexometric titration the concentration of a metallic ion is estimated by direct titration with a standard solution of a complexing agent. The determination of the hardness of water provides a typical example.

Both Ca^{2+} and Mg^{2+} ions form stable complexes with EDTA, one mole of the simple ion combining with one mole of the EDTA anion,

$$Mg^{2+}(aq) + EDTA^{4-}(aq) \longrightarrow Mg(EDTA)^{2-}(aq)$$

The calcium complex is more stable than the magnesium one, so that the calcium complex is first formed when EDTA solution is added to water, followed by the magnesium complex.

An indicator is necessary to denote the point at which all the Ca^{2+} ion and/or all the Mg^{2+} ion is used up. The indicator is itself a complexing agent; one which forms coloured complexes with both Ca^{2+} and/or Mg^{2+} ions, the complexes being less stable than those formed with EDTA. Eriochrome Black T forms red complexes with both Ca^{2+} and Mg^{2+} but as soon as enough EDTA is added to react with all the Ca^{2+} and Mg^{2+} the red colour changes to blue, the complexes with the indicator being replaced by the more stable complexes with EDTA. Using murexide, which forms a complex only with Ca^{2+} and not with Mg^{2+}, it is possible to determine the Ca^{2+} ion concentration.

The titrations are carried out in an alkaline buffer solution which favours the quantitative formation of the complexes, and a solution of the sodium salt of EDTA is used as it is more soluble than the acid.

Questions on Chapter 33

1 Account for the following (a) Zinc sulphide alone is precipitated when hydrogen sulphide is passed into an aqueous solution containing Zn^{2+} and Mn^{2+} ions and dilute ethanoic acid. (b) Magnesium hydroxide is not precipitated from solutions of magnesium salts by ammonia solution in the presence of ammonium chloride. (c) A concentrated solution of calcium chloride gives no precipitate with ammonia solution, but with sodium hydroxide solution a precipitate forms immediately.

2 The solubility product of magnesium hydroxide is 3.4×10^{-11} $mol^3 dm^{-9}$. Calculate its solubility in $g dm^{-3}$.

3 If the solubility of silver chloride in water is $1.3 \times 10^{-5} mol dm^{-3}$, what will it be in 0.1 M sodium chloride solution?

4 If one molecule of bismuth(III) sulphide ionises to give two Bi^{3+} ions and three S^{2-} ions, what is the numerical relationship between the solubility of bismuth(III) sulphide expressed as $x g dm^{-3}$ and the solubility product, K, of bismuth(III) sulphide?

5 If the solubility of potassium chlorate(VII) is $x mol dm^{-3}$ in water containing $y mol dm^{-3}$ of potassium chloride, what will the solubility be in pure water?

6 (a) The dissociation constant for hydrogen sulphide into hydrogen and sulphide ions is $10^{-22} mol^2 dm^{-6}$, and the concentration of a saturated solution of the gas in water is approximately M/10. What will be the S^{2-} ion concentration in a saturated solution of hydrogen sulphide in (i) 1 M hydrochloric acid (ii) M/10 hydrochloric acid? (b) Cadmium sulphide is first precipitated by passing hydrogen sulphide into an acidic solution when the acid is 2 M and the concentration of Cd^{2+} ions is $0.01 mol dm^{-3}$. What is the solubility product of cadmium sulphide?

7 Explain what is meant by the term solubility product and why it can be given an approximately constant value for a given salt, solvent and temperature. The solubility products of silver chloride and iodide are approximately 10^{-10} and $10^{-16} mol^2 dm^{-6}$ respectively. What will be the effect of (a) adding sodium chloride solution to a saturated solution of silver chloride (b) shaking up solid silver chloride with a solution of potassium iodide and (c) shaking up solid silver iodide with a solution of hydrochloric acid? (OS)

8 What is meant by the term solubility product? Give two examples of the application of this concept to qualitative analysis. The solubility of lead sulphate in water at 17°C is $0.035 g dm^{-3}$. Calculate (a) the solubility product of lead sulphate (b) the solubility of lead sulphate (in $g dm^{-3}$) in a 0.01 M solution of sodium sulphate at the same temperature. (Assume complete dissociation of both solutes). (OCS)

9 Explain why calcium carbonate is insoluble in water but soluble in dilute hydrochloric acid, whilst lead carbonate is insoluble in both.

10 Explain why (a) zinc sulphide is insoluble in water but soluble in dilute hydrochloric acid (b) mercury(II) sulphide is insoluble in water and dilute hydrochloric acid (c) copper(II) sulphide is insoluble in water and dilute hydrochloric acid, but soluble in aqua regia.

11 Why is it that lead chloride is less soluble in dilute hydrochloric acid than in water, and why does the solubility rise as the acid becomes more concentrated?

12 Account for the following facts: (a) Silver iodide is not appreciably soluble in concentrated ammonia solution, though silver chloride is. (b) Addition of potassium bromide solution to a dilute ammoniacal solution of silver chloride precipitates silver bromide.

13 What experimental evidence can be advanced in support of the following statements? (a) When carbon dioxide dissolves in water, chemical reaction takes place between the two substances. (b) The solubility of silver chloride in aqueous ammonia is due to the formation of a complex ion $[Ag(NH_3)_2]^+$. (OS)

14 What is a complex ion? What experimental observations suggest that such ions exist in certain solutions? (OS)

15 Zinc will replace silver from a solution of potassium dicyanoargentate(I) but will not replace copper from an equivalent solution of potassium tetracyanocopper(I). Zinc will also replace copper from a solution of tetraamminecopper(II) sulphate but not from an equivalent solution of potassium tetracyanocopper(I). Account for these facts.

16 What tests do you know in organic qualitative analysis which depend on complex formation? Wherever possible, give the formula of the complex formed.

17 Explain why copper(II) hydroxide will dissolve both in dilute sulphuric acid, which is acidic, and in ammonia solution, which is alkaline. Why will silver chloride dissolve in the latter, but not in the former?

18 $25 \, cm^3$ (excess) of ammonia solution were added to $25 \, cm^3$ of 0.1 M copper(II) sulphate and the resulting deep-blue solution was shaken with trichloromethane (chloroform). After the layers had been allowed to settle, $50 \, cm^3$ of the trichloromethane layer required $25.5 \, cm^3$ of 0.05 M hydrochloric acid for neutralisation. $20 \, cm^3$ of the blue aqueous layer were neutralised by 33.3 cm of 0.5 M hydrochloric acid. If the partition coefficient of ammonia between water and trichloromethane is 25, calculate the probable formula of the 'cuprammonium ion'.

19 The freezing point of a solution containing 3 g of potassium iodide in 100 g of water was $-0.619\,°C$. When 4.11 g of mercury(II) iodide were dissolved in the same solution the freezing point rose to $-0.504\,°C$. Calculate (a) the degree of dissociation of the potassium iodide before the addition of the mercury(II) iodide (b) the formula of the complex ion produced when the mercury(II) iodide is added. (Freezing point constant per $100 \, g = 18.6\,°C$.)

20 How would you demonstrate that copper was present in a complex anion in Fehling's solution?

21* Record the stability constants of (a) CdI_4^{2-} (b) $Ni(NH_3)_6^{2+}$ (c) $Zn(NH_3)_4^{2+}$. What do the values represent?

22* Compare the stability constants of some hydroxy-complexes, and comment on the values.

23* Quote some stability constants to illustrate (a) that the stability of a complex increases as the charge on the central ion increases (b) that the stability of a complex of a simple ion is greater for polydentate ligands than for monodentate ones.

34
Cells and corrosion

1 A simple cell

Chemical reactions often release energy, most commonly as heat, but in electrical cells a chemical reaction takes place and releases some of its energy in the form of electricity. A typical, simple cell consists of a rod of zinc immersed in zinc sulphate solution and a rod of copper immersed in copper(II) sulphate solution, the two solutions being kept apart, though in electrical contact, by a porous partition (Fig. 162).

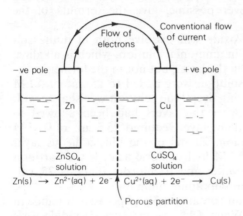

Fig. 162 A simple cell.

If the zinc and copper rods are connected by a wire, a current will flow. The copper is referred to as the positive pole, and the zinc as the negative pole. The current is said to flow, conventionally, from the copper to the zinc in the external circuit. This corresponds, as will be seen, to a flow of electrons from zinc to copper in the external circuit.

How does the current originate? The zinc rod is in contact with $Zn^{2+}(aq)$ ions, and some zinc atoms from the rod form $Zn^{2+}(aq)$ ions by the loss of two electrons,

$$Zn(s) \longrightarrow Zn^{2+}(aq) + 2e^-$$

The zinc ions pass into the solution whilst the liberated electrons are free to flow away from the zinc rod. The copper rod is in contact with $Cu^{2+}(aq)$ ions. Here, some $Cu^{2+}(aq)$ ions deposit on the copper rod and are discharged (converted into atoms) by taking up electrons,

$$Cu^{2+}(aq) + 2e^- \longrightarrow Cu(s)$$

The flow of electrons from the zinc rod to the copper rod constitutes the current.

The two electrode processes, each comprising what is known as a *half-cell*, taken together make up the reaction,

$$Zn(s) + Cu^{2+}(aq) \longrightarrow Zn^{2+}(aq) + Cu(s)$$

and this is the chemical reaction taking place in the cell, liberating energy as electricity. As the reaction proceeds, the zinc rod decreases in mass whilst the copper rod increases. The concentration of $Zn^{2+}(aq)$ ions increases, whilst that of $Cu^{2+}(aq)$ ions decreases.

The direction of the current is controlled by the relative ease of ionisation of zinc and copper atoms. Zinc will form ions more readily than copper and this is why electrons flow from the zinc to the copper. If a rod of magnesium in magnesium sulphate solution replaced the copper rod in copper(II) sulphate solution, the flow of current would be reversed, i.e. the zinc rod would be the positive pole of such a cell.

2 Electric potentials

The conversion of zinc atoms into zinc ions in one half of the cell means that the zinc rod becomes negatively charged with respect to the solution surrounding it. If the cell is disconnected in the external circuit, this negative charge will build up making the loss of positively charged zinc ions more and more difficult, as the zinc rod will increasingly attract zinc ions rather than release them. Eventually, an equilibrium position will be established.

The situation is conveniently represented by what is known as a Helmholtz double layer (Fig. 163) and the zinc rod is said to have a negative electric potential. The tendency for a metal to lose ions to a surrounding solution is known as its *electrolytic solution pressure*, whilst the opposite tendency for ions to deposit on a metal from a solution is known as the *deposition* pressure. For a zinc rod in contact with a M solution of zinc ions, the electrolytic solution pressure is greater than the deposition pressure. This gives rise to the negative electric potential.

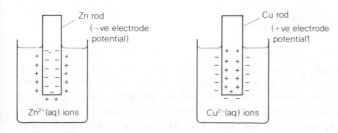

Zn rod
(−ve electrode potential)

$Zn^{2+}(aq)$ ions

Cu rod
(+ve electrode potential)

$Cu^{2+}(aq)$ ions

Fig. 163 Helmholtz double layers.

For a copper rod immersed in copper(II) sulphate solution of M concentration, the deposition pressure is greater than the electrolytic solution pressure. $Cu^{2+}(aq)$ ions therefore deposit on the copper rod, giving it a positive charge in respect to the solution. As this positive charge builds up on the rod the deposition of further ions is hindered by repulsion, until, in the end, an equilibrium position is established. The copper rod is said to have a positive electric potential (Fig. 163).

The electric potential of a metal depends on the concentration of ions with which it is in contact. The electrolytic solution pressure of the metal remains constant but the deposition pressure decreases as the concentration of the ions decreases. Electric potential values, therefore, fall as the concentration of the ions is lowered, and rise as the concentration is increased.

3 Measurement of standard electrode potentials

A *single* electric potential of a metal cannot be measured absolutely. To measure the potential difference between a metal and a solution necessitates the making of electrical connection between the metal and the solution. The connection to the solution could only be made by immersing some other metal in the solution, and this would introduce another electric potential.

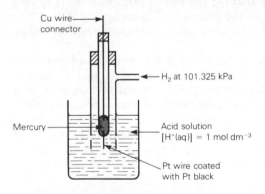

Fig. 164 A simple standard hydrogen electrode.

The difficulty is overcome by using an arbitrary, relative scale of measurement based on a *standard hydrogen electrode* (Fig. 164). Hydrogen gas, at 101.325 kPa, is bubbled over a platinum electrode coated with platinum black and immersed in an acid solution with a $H^+(aq)$ ion concentration of $1 \, mol \, dm^{-3}$ (strictly, unit activity, p. 356; to achieve this it is necessary to use 1.18 M hydrochloric acid). The hydrogen is adsorbed on the platinum black and an equilibrium is set up between the adsorbed layer of hydrogen and the hydrogen ions in the solution,

$$\tfrac{1}{2}H_2(g) \rightleftharpoons H^+(aq) + e^-$$

The platinum black catalyses the setting up of this equilibrium. There will, in fact, be an electric potential between the adsorbed hydrogen and the solution but this is arbitrarily taken as zero, at 25 °C, for the standard electrode as described.

This fixes a scale against which other relative electrode potentials, or electrode potentials, can be measured. The values depend on the metal used, on the concentration of metallic ions with which it is in contact, and on the temperature. When the ionic concentration is $1 \, mol \, dm^{-3}$ (strictly, unit activity) the electrode potential value is known as the *standard electrode potential*. Values are commonly quoted at 25 °C and signified as $E^\ominus$.

The standard electrode potential of zinc, for example, is the e.m.f. of a cell consisting of a standard hydrogen electrode and a standard zinc electrode. The two electrodes, or half-cells, are connected by a 'bridge' (Fig. 165) so that the overall e.m.f. can be measured. The 'bridge' can consist of a bent glass tube containing some inactive solution such as saturated KCl, or a folded filter paper saturated with KCl, or some type of porous partition as shown in Fig. 162.

The e.m.f. of the cell must be measured without drawing any current so that the composition of the chemicals in the cells is not changed. A valve voltmeter, or some other high resistance instrument, can do this. Alternatively, a potentiometer circuit is used. The measured value, at 25 °C, for the cell shown in Fig. 165 is 0.76 volt, and the zinc is the negative pole. The standard electrode potential of zinc is, therefore, $-0.76 \, V$ (p. 409).

Other standard electrode potentials can be measured in the same way.

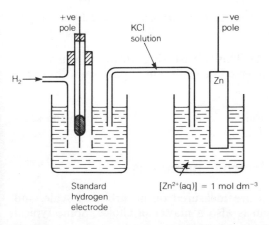

Fig. 165

4 Other reference electrodes

A standard hydrogen electrode is not easy to use; it is bulky, slow to reach equilibrium and susceptible to the presence of impurities. Other reference electrodes are therefore commonly used. Their electrode potentials in relation to the standard hydrogen electrode must be known.

a Calomel electrode This consists (Fig. 166(a)) of a platinum wire dipping into mercury in contact with a solution of potassium chloride saturated with mercury(I) chloride (calomel). The equilibrium established is

$$\tfrac{1}{2}Hg_2Cl_2 + e^- \rightleftharpoons Hg + Cl^-$$

so that the potential of the electrode relative to a standard hydrogen electrode depends on the Cl^- ion concentration. For 0.1 M KCl the value is $+0.334\,V$ at 25°C; for saturated KCl it is $+0.24\,V$. The electrode potential of zinc against a 0.1 M calomel electrode is therefore $-1.094\,V$; against a saturated calomel electrode it is $-1.00\,V$.

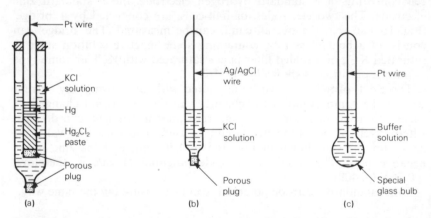

Fig. 166 Reference electrodes. **(a)** A calomel electrode. **(b)** A silver/silver chloride electrode. **(c)** A glass electrode.

b Silver/silver chloride electrode This consists (Fig. 166(b)) of a silver wire coated with silver chloride immersed in a solution of potassium chloride,

$$AgCl + e^- \rightleftharpoons Ag + Cl^-$$

For saturated KCl solution the electrode potential relative to the standard hydrogen electrode is $+0.203\,V$ at 25°C.

5 Values of standard electrode potentials

Standard electrode potentials are measured on an arbitrary scale, and the sign to be alloted to them is also a matter of choice. Some typical values, in volts at 25°C, are given opposite:

$Li^+(aq) + e^- \longrightarrow Li(s)$	-3.04	$Fe^{2+}(aq) + 2e^- \longrightarrow Fe(s)$	-0.44
$K^+(aq) + e^- \longrightarrow K(s)$	-2.92	$Co^{2+}(aq) + 2e^- \longrightarrow Co(s)$	-0.28
$Ba^{2+}(aq) + 2e^- \longrightarrow Ba(s)$	-2.90	$Ni^{2+}(aq) + 2e^- \longrightarrow Ni(s)$	-0.25
$Ca^{2+}(aq) + 2e^- \longrightarrow Ca(s)$	-2.87	$Sn^{2+}(aq) + 2e^- \longrightarrow Sn(s)$	-0.14
$Na^+(aq) + e^- \longrightarrow Na(s)$	-2.71	$Pb^{2+}(aq) + 2e^- \longrightarrow Pb(s)$	-0.13
$Mg^{2+}(aq) + 2e^- \longrightarrow Mg(s)$	-2.37	$H^+(aq) + e^- \longrightarrow \frac{1}{2}H_2(g)$	0.00
$Al^{3+}(aq) + 3e^- \longrightarrow Al(s)$	-1.66	$Cu^{2+}(aq) + 2e^- \longrightarrow Cu(s)$	$+0.34$
$Zn^{2+}(aq) + 2e^- \longrightarrow Zn(s)$	-0.76	$Ag^+(aq) + e^- \longrightarrow Ag(s)$	$+0.80$

The sign convention adopted* is that recommended by the IUPAC, but as the opposite signs are sometimes used it is essential to be sure which convention is being used when interpreting quoted data.

The electrode written in the table above as $Li^+(aq) + e^- \rightarrow Li(s)$ may also be written as $Li^+(aq)\,|\,Li(s)$, but on the convention being used the more oxidised state is always written first, i.e. the electrode process is taken as a reduction and the electrode potential is, really, a reduction potential. If quoted with reversed sign it is an oxidation potential. Both values are known as redox potentials (p. 427).

The order of standard electrode potentials is the order of the electrochemical (reactivity) series, with the most reactive metals having the highest negative values. The more negative the value, the easier it is for the oxidation process ($M \rightarrow M^{n+} + ne^-$) to take place and conversely, the more difficult it is for the reduction process ($M^{n+} + ne \rightarrow M$) to take place.

6 The e.m.f. of cells

The standard electrode potential of zinc is $-0.76\,V$. When combined with a standard hydrogen electrode, as in Fig. 165, the cell is written, conventionally, as follows:

$$H_2(g)/Pt \,|\, H^+(aq)\,(1\,mol\,dm^{-3}) \,|\, Zn^{2+}(aq)\,(1\,mol\,dm^{-3}) \,|\, Zn(s)$$
$+$ ve pole $\qquad\qquad\qquad\qquad\qquad\qquad\qquad\qquad\qquad$ $-$ ve pole

The zinc ionises more readily than the hydrogen so that it is the negative pole, and the flow of electrons is from the negative pole to the positive

---

* This convention has the advantage that an electrode with a positive electrode potential is positively charged with respect to the solution with which it is in contact; such an electrode also forms the positive pole when combined with a standard hydrogen electrode. Moreover, in any cell, the positive pole will be the electrode with the highest positive electrode potential. The convention has the disadvantage that the most electropositive metals are alloted high negative standard electrode potentials. (IUPAC = International Union of Pure and Applied Chemistry.)

pole in the external circuit. The cell e.m.f. is conventionally quoted as $-0.76\,V$, i.e.

$$E^{\ominus}(\text{cell}) = \begin{array}{c} E^{\ominus} \text{ of the right-} \\ \text{hand electrode} \end{array} - \begin{array}{c} E^{\ominus} \text{ of the left-} \\ \text{hand electrode} \end{array}$$

The negative value of the e.m.f. indicates that the actual reaction taking place in the cell is

$$Zn(s) + 2H^+(aq) \longrightarrow Zn^{2+}(aq) + H_2(g)$$

A cell made up of a standard copper electrode and a standard hydrogen electrode is written as follows,

$$H_2(g)/Pt \mid H^+(aq)\,(1\,mol\,dm^{-3}) \mid Cu^{2+}(aq)\,(1\,mol\,dm^{-3}) \mid Cu(s)$$
$-$ve pole $\qquad\qquad\qquad\qquad\qquad\qquad\qquad\qquad$ $+$ve pole

It will have an e.m.f. of $+0.34\,V$, and this positive value indicates that the cell reaction is

$$H_2(g) + Cu^{2+}(aq) \longrightarrow 2H^+(aq) + Cu(s)$$

In a cell using standard zinc and copper electrodes, the arrangement is written as

$$Zn(s) \mid Zn^{2+}(aq)(1\,mol\,dm^{-3}) \mid Cu^{2+}(aq)(1\,mol\,dm^{-3}) \mid Cu(s)$$
$-$ve pole $\qquad\qquad\qquad\qquad\qquad\qquad\qquad\qquad$ $+$ve pole

The cell e.m.f. is $+0.34\ -(-0.76)$, i.e. $+1.1\,V$. The positive value indicates that the cell reaction is

$$Zn(s) + Cu^{2+}(aq) \longrightarrow Zn^{2+}(aq) + Cu(s)$$

7 The use of $E^{\ominus}$ values

a To calculate the e.m.f. of a cell This is described in the preceding section.

b To predict the feasibility of a reaction A reaction generalised as

$$A(s) + B^{n+}(aq) \longrightarrow B(s) + A^{n+}(aq)$$

where A and B are metals and A^{n+} and B^{n+} are the cations, will only be feasible when the e.m.f. of the cell

$$A(s) \mid A^{n+}(aq) \mid B^{n+}(aq) \mid B(s)$$

is positive. This means that the $E^{\ominus}$ value for the cell reaction, calculated from the two $E^{\ominus}$ values concerned, must be positive.

For example

$Cu^{2+}(aq) + 2e^- \longrightarrow Cu(s)$	$E^{\ominus} = +0.337\,V$
$Zn^{2+}(aq) + 2e^- \longrightarrow Zn(s)$	$E^{\ominus} = -0.763\,V$
$Cu^{2+}(aq) + Zn(s) \longrightarrow Zn^{2+}(aq) + Cu(s)$	$E^{\ominus} = +1.1\,V$

The reaction given is well known to be feasible (p. 431); the combined

$E^{\ominus}$ value is $+1.1$ V. The reverse reaction, with E equal to -1.1 V, is not feasible. Remember, always, that such considerations give no indication about the rate of any reaction.

c To find free energy changes See pages 272 and 429.

d To find equilibrium constants See page 430.

e To measure solubility products and solubilities See page 396.

f To calculate electrode potentials under non-standard conditions As described in the following section.

8 The Nernst equation

Standard electrode potentials are measured when an electrode is in contact with a solution of its ions at a concentration of $1 \, mol \, dm^{-3}$, or, more accurately, unit activity. To find electrode potentials for other ion concentrations it is necessary to use the Nernst equation (p. 431). It shows that the electrode potential of a metal, E, in contact with its ion at a concentration of $[M^{z+}]$ is related to its standard electrode potential, $E^{\ominus}$, by the relationship,

$$E = E^{\ominus} + (RT/zF)\ln[M^{z+}]$$

where R is the gas constant, F the Faraday constant, T the absolute temperature and z the number of positive charges on the metal ion. Using numerical values of R and F at $25\,°C$ gives

$$E = E^{\ominus} + (0.059/z)\lg[M^{z+}]$$

The $E^{\ominus}$ value for copper is $+0.344$ V; it follows that the E value when copper is in contact with 0.01 M $Cu^{2+}(aq)$ ions will be $+0.344 - 0.059$, i.e. 0.285 V. The $E^{\ominus}$ value for zinc is -0.76 V; in 0.01 M solution it will be $-0.076 - 0.059$, i.e. -0.819 V. Each ten-fold lowering of the ion concentration lowers the electrode potential by $0.059/z$ volts.

9 Concentration cells

The difference in the electrode potential for different ion concentrations means that concentration cells can be set up, as shown below:

$-$ve pole		$+$ve pole
$Cu(s) \mid Cu^{2+}(aq)\,(0.01\,M)$	$\mid$	$Cu^{2+}(aq)\,(1\,M) \mid Cu(s)$
$(Cu(s) \longrightarrow Cu^{2+}(aq) + 2e^-)$		$(Cu^{2+}(aq) + 2e^- \longrightarrow Cu(s))$

Such a cell would give an initial e.m.f. of $0.344 - 0.285$, i.e. 0.059 volt. $Cu^{2+}(aq)$ ions will deposit on the right-hand electrode from the more concentrated solution more readily than on the left-hand electrode from the weaker solution. Copper atoms, therefore, form $Cu^{2+}(aq)$ ions at the left-hand electrode, whereas Cu^{2+}(aq) ions form copper atoms at the right hand. Eventually, the two solutions will have equal Cu^{2+}(aq) ion concentrations, and no e.m.f. will be recorded.

10 Measurement of H^+(aq) ion concentration. The pH meter

The e.m.f. of the concentration cell between a standard and a non-standard hydrogen electrode,

H_2(g)/Pt | H^+(aq) (1 mol dm^{-3}) | H^+(aq) (x mol dm^{-3}) | (H_2(g)/Pt
standard H electrode   non-standard H electrode

will be equal to $0.059 \lg x$, where x is the concentration of H^+(aq) in the unknown solution. Measurement of the e.m.f. will, therefore give the value of x, or, indirectly, the pH of the solution.

Such a cell would involve two hydrogen electrodes and would be very inconvenient to use. The standard hydrogen electrode is generally replaced by some other reference electrode (p. 408), and the non-standard one by a *glass electrode*. This glass electrode (Fig. 166(c)) consists of a platinum wire immersed in a buffer solution in a sealed, thin-walled bulb made of special glass. Such an electrode develops a potential depending on the H^+(aq) ion concentration in which it is immersed. Measurement of the potential, against a reference electrode, enables the particular H^+(aq) concentration, or the pH, to be found.

Instruments designed to do this are known as pH meters, and a variety of portable ones are available. They involve, in general, a cell summarised as follows,

reference electrode	solution of unknown pH	glass electrode

though both electrodes may be incorporated into one unit (Fig. 167)

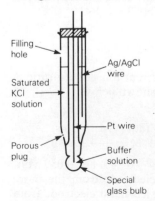

Filling hole

Saturated KCl solution

Porous plug

Ag/AgCl wire

Pt wire

Buffer solution

Special glass bulb

Fig. 167 A combined glass and silver/silver chloride electrode for use in a pH meter.

which can be dipped into the solution under test. The e.m.f. is recorded on a valve voltmeter, which is calibrated in pH units. Before use on an unknown solution, the pH meter is calibrated against standard buffer solutions (p. 372).

The glass electrode, which is sensitive to change in H^+(aq) ion concentration, is known as an ion selective electrode. Other electrodes sensitive to other specific ions are also available, and can be used for measuring the concentration of the particular ion.

11 Potentiometric, electrometric or conductimetric titrations

The H^+(aq) ion concentration, or pH, of a solution changes during an acid–alkali titration, and the end point can be conveniently obtained by measuring the pH as the titration is carried out, using a pH meter. Alternatively, a hydrogen electrode can be immersed in the solution being titrated and connected to a reference electrode. The change in pH during the titration can be followed by measuring the change in e.m.f. between the two electrodes. pH curves, as shown on page 366, can be obtained experimentally in these ways.

Such a procedure is known as potentiometric or electrometric titration. One of the advantages of the method is that it can be used with coloured solutions where indicators are not suitable.

In a similar way, the conductivity of a solution being titrated can be measured. At the end point there may be a sharp change of conductivity. Such a titration is known as a conductimetric titration (Fig. 168).

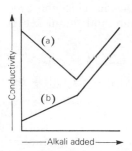

Fig. 168 The change in conductivity as a strong alkali is added to **(a)** a strong acid and **(b)** a weak acid.

12 Useful cells

A cell is, generally, only useful as a source of electricity if it can produce both a reasonably high current and a reasonably high voltage, i.e. it is the power (the product of current and voltage, p. 473) that matters. There are three main types of practical cell.

a Simple cells A simple cell is an arrangement in which a chemical reaction takes place to release electrical energy. It serves as a compact, portable supply of electricity. The *Mallory cell* is typical. Zinc and a compressed pellet of mercury(II) oxide and graphite are separated by a porous material in a solution of potassium hydroxide. The cell can be represented as

$$Zn(s) \mid ZnO(s) \mid NaOH(aq) \mid HgO(s) \mid Hg(l)$$

and the overall reaction within the cell is

$$HgO(s) + Zn(s) \longrightarrow Hg(l) + ZnO(s)$$

b Storage cells A storage cell is simply a reversible simple cell. Once the chemicals in the cell have been used up they can be re-formed by charging the cell. In the lead/acid battery or accumulator there is a lead cathode and a lead(IV) oxide anode in sulphuric acid. The electrode processes, during discharge, are

cathode $\qquad Pb(s) \longrightarrow Pb^{2+}(aq) + 2e^-$

anode $\qquad PbO_2(s) + 4H^+(aq) + 2e^- \longrightarrow Pb^{2+}(aq) + 2H_2O(l)$

The reverse processes take place on charging. The overall change is

$$Pb(s) + PbO_2(s) + 2H_2SO_4(aq) \rightleftharpoons 2PbSO_4(s) + 2H_2O(l)$$

The lead/acid battery is heavy but it is widely used in motor cars because it can provide the very large currents required to start them.

For smaller currents, a nickel/cadmium cell may be used. The electrodes are made of cadmium and nickel(II) hydroxide pastes, and they are immersed in potassium hydroxide solution. On charging, the cadmium hydroxide cathode is converted into cadmium, and the nickel(II) hydroxide anode into hydrated nickel(III) oxide, i.e.

cathode $\qquad Cd(OH)_2(s) + 2e^- \longrightarrow Cd(s) + 2OH^-(aq)$

anode $\qquad 2Ni(OH)_2(s) + 2OH^-(aq) \longrightarrow Ni_2O_3.H_2O(s) + 2H_2O(l) + 2e^-$

The reverse processes take place on discharge.

c Fuel cells In these cells, fuel is supplied to the cell during its operation. The hydrogen/oxygen cell is typical. Hydrogen is fed in at the anode and oxygen at the cathode. In an acid electrolyte, the electrode processes are

cathode $\qquad O_2(g) + 4H^+(aq) + 4e^- \longrightarrow 2H_2O(l)$

anode $\qquad 2H_2(g) \longrightarrow 4H^+(aq) + 4e^-$

giving a theoretical voltage under standard conditions of 1.23 V. Other similar cells use methanol or hydrocarbons at the anode and air at the cathode.

Fuel cells are used to provide small currents for long periods in, for example, satellites, but the successful development of cells for large-scale use, though attractive, lies in the future.

13 Metallic couples

If zinc and copper are immersed, apart from each other, in dilute sulphuric acid they will form the cell

$-$ ve pole $\qquad\qquad$ $+$ ve pole

$Zn(s) \mid$ dil. $H_2SO_4 \mid$ $Cu(s)$

$$Zn(s) \longrightarrow Zn^{2+}(aq) + 2e^- \qquad\qquad H^+(aq) + e^- \longrightarrow \tfrac{1}{2}H_2(g)$$

If the zinc and copper are connected internally, by allowing them to touch within the acid, a similar cell, constantly short-circuited, will be set up, and hydrogen gas will bubble off from the copper. Such an arrangement is known as a galvanic or metallic couple.

A piece of pure zinc will react only very, very slowly with dilute sulphuric acid, because the conversion of $H^+(aq)$ ions into hydrogen gas is difficult at a zinc surface; hydrogen has a high overvoltage on zinc (p. 437). If the pure zinc is touched, within the dilute sulphuric acid, by a copper or platinum wire, evolution of hydrogen will quickly take place from the copper or platinum surface. Similarly, evolution of hydrogen can be expedited by adding a little copper(II) sulphate or platinum(IV) chloride solution. This causes a deposit of copper or platinum on the zinc surface, and the establishment of a lot of galvanic or metallic couples.

14 Galvanised iron

Coating iron with zinc to make galvanised iron is a very common method of rust prevention, but a metallic couple of zinc and iron is immediately set up if the zinc coating becomes imperfect and, if a solution of an electrolyte, or even water, is present, corrosion will take place. The cell set up will be

$-$ ve pole $\qquad\qquad$ $+$ ve pole

$Zn(s) \mid H_2O(l) \mid Fe(s)$

$$Zn(s) \longrightarrow Zn^{2+}(aq) + 2e^- \qquad\qquad H^+(aq) + e^- \longrightarrow \tfrac{1}{2}H_2(g)$$

and zinc ions will be formed at the zinc surface whilst hydrogen will form at the iron surface. The process can be summarised as in Fig. 169 (p. 416).

The zinc will slowly corrode away, being converted into zinc hydroxide, but the iron will not corrode, i.e. it will not rust. The zinc is referred to as a *sacrificial coating*, or it is said to provide *cathodic protection* as the iron is functioning as the cathode. Similar cathodic

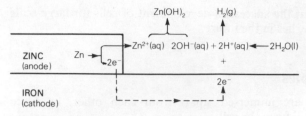

Fig. 169 Cathodic protection of iron by zinc.

protection of underground iron pipes is provided by connecting them to buried magnesium rods.

15 Tinplate

Coating steel with tin is another very common method of protecting iron, but it is very different in its effect from zinc and it is not, in general, a sacrificial coating. Once a tin coating is pierced the cell set up will be

$$- ve \ pole \qquad\qquad + ve \ pole$$

$$Fe(s) \mid H_2O(l) \mid Sn(s)$$

$$Fe(s) \longrightarrow Fe^{2+}(aq) + 2e^- \qquad H^+(aq) + e^- \longrightarrow \tfrac{1}{2}H_2(g)$$

with the iron acting as the anode (Fig. 170).

The Fe^{2+}(aq) ions and the iron(II) hydroxide formed are ultimately converted into rust (p. 417) so that, although tin plating prevents rusting so long as the plating is perfect, it encourages rusting once imperfections arise. Zinc is, therefore, a better protective material but it cannot be used where foodstuffs are concerned because zinc salts are poisonous. That is why tin cans are so widely used for preservation of food, and why zinc plating is limited to such articles as buckets and dustbins.

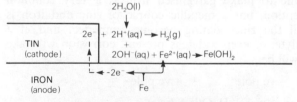

Fig. 170 Corrosion of iron at an imperfection in a tin coating.

16 Rusting of iron

It is well known that the rusting of iron requires the presence of both water and oxygen, but the formation of rust is not fully understood. No two examples of rusting are exactly alike, and rust $(Fe_2O_3.xH_2O)$ is of variable composition.

Imperfections in the iron surface and the different availability of dissolved oxygen in the solution in contact with the iron are the two main causes of rusting.

a Imperfections in iron Even very small imperfections in an iron surface, caused by impurities, or unequal strains during mechanical treatment, will give rise to a heterogeneous surface in which some points will act as anodes and others as cathodes (Fig. 171). At the points acting as anodes, iron will form Fe^{2+}(aq) ions, and at the points acting as cathodes, hydrogen will be produced.

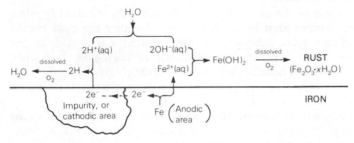

Fig. 171 The rusting of iron initiated at a surface impurity.

The hydrogen formed at the cathodic points will hinder the ionisation unless it is removed. This is where the oxygen comes in, for dissolved oxygen in any water or solution present, reacts with the hydrogen to form water.

The Fe^{2+}(aq) ions liberated at the anodic points pass into solution, where, together with the OH^-(aq) ions present, they form iron(II) hydroxide, which is oxidised by dissolved oxygen to form rust.

b Differential aeration Even a piece of iron with no surface imperfections will rust if the dissolved oxygen in contact with it is at different concentrations over the surface. This will be so when a piece of iron is half immersed in water or a solution of an electrolyte. More dissolved oxygen will be available at the surface of the water than below it. Conversion of H^+(aq) ions into water, will, therefore, be easier at the surface of the water than lower down, and the iron in the region of the water surface will be cathodic. Consequently, iron below the surface will be anodic, and it is here that rusting takes place.

This accounts for the fact that rust is known to form at points where the oxygen supply is limited. On partially painted iron, for example, rust formation is greatest underneath the paint layer. Rust also forms at the centre of a drop of water placed on an iron surface, rather than at the edges of the drop. Rusting, moreover, often causes deep pitting in iron. This is because a pit gets deeper, once it has started, as oxygen is less available at the bottom of the pit so that the main rusting occurs there.

17 Passivity

Some metals, particularly iron, nickel, cobalt and chromium can be rendered passive, i.e. unreactive, by treatment which covers them with a thin coating of oxide.

For example, an iron nail dipped into copper(II) sulphate solution is immediately coated with copper. If the same iron nail is first dipped, on a thread, into concentrated nitric acid, and then washed in water, it will not be coated with copper on dipping into copper(II) sulphate solution. As soon as the nail is touched, however, it loses its passivity due to breaking of the oxide film.

Similarly, a piece of stainless steel will corrode quite noticeably when placed in hot, concentrated hydrochloric acid because the acid breaks down the surface layer. Addition of concentrated nitric acid to the mixture results, at first, in an increased rate of corrosion, but when sufficient nitric acid has been added to give a strongly oxidising mixture, the corrosion suddenly stops as the oxide layer is once again built up on the steel surface.

Other oxidising agents, and electrolytic treatment with the metal as the anode, can also be used for rendering metals passive. This is done, particularly, in making anodised aluminium.

Questions on Chapter 34

1 What is meant by describing one metal as more electropositive than another? Briefly describe three experiments by which you could show that zinc is more electropositive than copper.

2 The statement that zinc replaces copper from solution means that at equilibrium the ratio of copper(II) ion concentration to zinc ion concentration is very small. Explain in more detail what this statement means.

3 What are the advantages and the disadvantages of the sign convention chosen in this chapter for electrode potentials?

4 The standard electrode potential of zinc referred to the standard hydrogen electrode is -0.76 V. Referred to a decimolar calomel electrode it is -1.094 V. Will it be higher or lower than -0.76 V when referred to a hydrogen electrode containing a concentration of hydrogen ions of 0.1 mol dm^{-3}, and will it be higher or lower than -1.094 V when referred to a 1 M calomel electrode? Explain your reasoning.

5 What reactions take place in each half cell, and overall, in the following cells:

(a) $H_2(g)/Pt \mid HCl(0.1 \text{ M}) \mid Cl_2(g)/graphite$
(b) $Ag/AgCl(s) \mid HCl(1 \text{ M}) \mid H_2(g)/Pt$
(c) Cd amalgam $\mid CdSO_4(aq) \mid Hg_2SO_4(s)/Hg(l)$
(d) $Fe(s)/Fe^{2+}(aq)(1 \text{ M}) \mid Sn(s)/Sn^{2+}(aq)(1 \text{ M})$

6 Summarise the cells which would have to be set up to bring about the following reactions:

(a) $2AgBr(s) + H_2(g) \longrightarrow 2Ag(s) + 2HBr(aq)$
(b) $H_2(g) + I_2(s) \longrightarrow 2HI(aq)$
(c) $Mg(s) + 2Ag^+(aq) \longrightarrow Mg^{2+}(aq) + 2Ag(s)$

7 What would you do to measure the standard electrode potential of a $Cl^-(aq)/Cl_2(g)$ electrode?

8 What is the e.m.f. of a cell made up of two hydrogen electrodes, one with a concentration of hydrogen ion of $0.025 \, mol \, dm^{-3}$ and the other with a concentration of $10^{-8} \, mol \, dm^{-3}$? Which is the negative pole of the cell?

9 Plot a graph of the e.m.f. of a cell, made up of a standard hydrogen electrode and a hydrogen electrode immersed in solution of pH equal to x, against x. What is the slope of the line?

10 Describe, carefully, how you would measure the pH of a solution using (a) a pH meter (b) indicators.

11 Compare the change in pH with the change in conductivity when $25 \, cm^3$ of 0.1 M NaOH is titrated against 0.2 M NaOH.

12 What are the advantages and disadvantages of carrying out an acid-base titration using (a) a pH meter (b) a conductance meter (c) indicators to determine the end point?

13 Describe some processes in which the control of pH is important.

14 Sketch, approximately to scale, the graphs showing the change in conductivity during the following titrations: (a) $50 \, cm^3$ of 0.1 M HCl against 1 M NaOH (b) $50 \, cm^3$ of 0.1 M ethanoic acid against 1 M NaOH (c) $50 \, cm^3$ of 1 M NaCl against 1 M $AgNO_3$ (d) $50 \, cm^3$ of 0.1 M HCl against 1 M NH_3 solution.

15 What sort of corrosion would you expect at a zinc surface imperfectly coated with a layer of tin?

16 The rusting of iron can be inhibited to some extent by (a) adding sodium sulphate (b) adding sodium borate(III) (c) adding zinc sulphate (d) adding sodium carbonate. Suggest reasons for the inhibiting effect of these substances.

17 A piece of iron immersed in some water is found to rust quite rapidly, but when another sample of the same iron is constantly whirled round in the same sort of water it does not rust. Explain why this is so.

18 Discuss the various methods which have been used to prevent rusting of iron.

19 What factors determine the nature of the reaction (if any) that takes place when a metal is immersed in an aqueous solution of an acid? Suggest explanations of the following observations: (a) impure commercial zinc dissolves more rapidly in dilute sulphuric acid than high purity zinc (b) iron is not attacked by concentrated nitric acid at room temperature (c) amalgamated aluminium is rapidly attacked by cold water, whereas the unamalgamated metal is almost unattacked. (CS)

20 Explain carefully what will happen if a strip of magnesium ribbon is cleaned with emery paper, dipped into silver nitrate solution, washed in cold water, and then dipped into warm water.

21* What will be the e.m.f. of the following cells: (a) a zinc rod immersed in $0.05\,M$ $ZnSO_4$ and a $0.1\,M$ calomel electrode (b) a standard lead electrode and a saturated calomel electrode (c) a standard zinc and a standard silver electrode (d) copper in $1\,M$ $CuSO_4$ and copper in $0.01\,M$ $CuSO_4$.

22* Arrange the following ions, when in molar solution, in the order of their strength as oxidising agents: $Cu^+(aq)$, $Ag^+(aq)$, $Pb^{2+}(aq)$, $Cr^{3+}(aq)$, $Ca^{2+}(aq)$, $Al^{3+}(aq)$, $Fe^{3+}(aq)$ and $Sn^{2+}(aq)$.

23* List the following metals in the order of their strength as reducing agents: Ni, Co, Fe, Sr, Hg, Cd, Al, Li.

24* Select any three reactions between a metal and a solution containing a metallic ion to show how standard electrode potentials can be used to indicate equilibrium positions.

35
Electronic theory of oxidation and reduction

1 Definitions

The term oxidation was first used to mean the addition of oxygen to an element or compound, or the removal of hydrogen from a compound. Reduction meant the addition of hydrogen to an element or compound, or the removal of oxygen from a compound. Such definitions have been extended and nowadays many oxidation–reduction, or *redox*, reactions are best interpreted in terms of transfer of electrons.

The conversion of iron(II) into iron(III) oxide,

$$4FeO(s) + O_2(g) \longrightarrow 2Fe_2O_3(s)$$

is clearly an oxidation of the iron(II) oxide, but the conversion of iron(II) chloride into iron(III) chloride is an analogous change.

$$2FeCl_2(aq) + Cl_2(g) \longrightarrow 2FeCl_3(aq)$$

It is, therefore, regarded as an oxidation of the iron(II) chloride, even though oxygen is not involved. Chlorine is the oxidising agent.

The change can be still further generalised into a conversion of $Fe^{2+}(aq)$ ions into $Fe^{3+}(aq)$ ions. This will, therefore, also be an oxidation, with the reverse change being a reduction,

$$Fe^{2+}(aq) \xrightarrow{\text{oxidation}} Fe^{3+}(aq) + e^- \qquad Fe^{3+}(aq) + e^- \xrightarrow{\text{reduction}} Fe^{2+}(aq)$$
$$\text{(R.A.)} \qquad\qquad\qquad\qquad\qquad \text{(O.A.)}$$

When $Fe^{2+}(aq)$ ions are being oxidised they are acting as reducing agents, and when $Fe^{3+}(aq)$ ions are being reduced they are acting as oxidising agents. In general

$$\text{O.A.} + \text{electrons} \underset{\text{oxidation}}{\overset{\text{reduction}}{\rightleftharpoons}} \text{R.A.}$$

Oxidation is the loss of electrons by an atom, ion or molecule.

Reduction is the gain of electrons by an atom, ion or molecule.

An oxidising agent takes electrons; it is an electron acceptor.

A reducing agent gives electrons; it is an electron donor.

2 Oxidation number or oxidation state

When an element is oxidised it must be acting as a reducing agent and it, therefore, loses electrons: when reduced, it gains electrons. The oxidation state or oxidation number of an element is the number of electrons it might be considered to have lost or gained.

All elements in the elementary, uncombined state are given oxidation numbers of zero. When sodium, for example, is oxidised it loses one electron, and the Na^+ ion is said to have an oxidation number of $+1$. Similarly, the Cu^{2+} and Al^{3+} ions have oxidation numbers of $+2$ and $+3$, whilst F^- and O^{2-} have oxidation numbers of -1 and -2. *For simple ions, the oxidation number is equal to the ionic charge*, e.g.

$$Na \longrightarrow Na^+ + e^- \qquad Al \longrightarrow Al^{3+} + 3e^- \qquad Cl + e^- \longrightarrow Cl^-$$
$$\;\;0 \qquad\qquad +1 \qquad\quad\; 0 \qquad\qquad +3 \qquad\qquad\; 0 \qquad\qquad\qquad -1$$

The oxidation number for an element in a covalent compound is decided, on an arbitrary basis, by taking the oxidation number to be equal to the charge that the element would carry if all the bonds in the compound were regarded as ionic instead of covalent. In doing this, a shared pair of electrons between two atoms is assigned to the atom with the greater electronegativity. Or, if the two atoms are alike, the shared pair is split between the two, one electron being assigned to each atom. The resulting charges on the various atoms when the bonding electrons are so assigned are the oxidation numbers of the atoms.

Some examples of oxidation numbers of atoms in different compounds will illustrate the usage:

$C\ H_4$	$C\ H_3\ Cl$	$C\ H_2\ Cl_2$	$C\ Cl_4$
$-4\ +1$	$-2\ +1\ -1$	$0\ +1\ -1$	$+4\ -1$

$N\ H_3$	$H_2\ O$	$H\ F$	$Cl\ F$	$I\ Cl_3$	$I\ F_5$
$-3\ +1$	$+1\ -2$	$+1\ -1$	$+1\ -1$	$+3\ -1$	$+5\ -1$

The following points should be noted.

a The algebraic sum of the oxidation numbers of all the atoms in an uncharged compound is zero. In an ion, the algebraic sum is equal to the charge on the ion, e.g.

$N\ H_4^+$	$O\ H^-$	$S\ O_4^{2-}$	$Al\ F_6^{3-}$
$-3\ +1$	$-2\ +1$	$+6\ -2$	$+3\ -1$

b All elements in the elementary state have oxidation numbers of zero, shared pairs between like atoms being split equally.

c As fluorine is the most electronegative element it always has an oxidation number of -1 in any of its compounds.

d Oxygen, second only to fluorine in electronegativity, has an oxidation number of -2 in almost all its compounds, e.g.

$Mg\ O$	$Fe_2\ O_3$	$C\ O_2$	$Mn_2\ O_7$	$Cr\ O_3$
$+2\ -2$	$+3\ -2$	$+4\ -2$	$+7\ -2$	$+6\ -2$

Exceptions are provided by oxygen difluoride and the peroxides,

$$\underset{-1\ +2}{F_2\ O}$$

$$\left[\begin{matrix} \underset{-1}{\ddot{O}} & \underset{-1}{\overset{\times\times}{\underset{\times\times}{O}}} \end{matrix}\right]^{2-}$$

the oxidation numbers in the peroxide ion being calculated by splitting the shared pair equally between the two oxygen atoms.

e In all compounds, except ionic metallic hydrides, the oxidation number of hydrogen is $+1$, e.g.

$$\underset{+1\ -1}{H\ Cl} \qquad \underset{+1\ -2}{H_2\ O} \qquad \underset{-3\ +1}{N\ H_3} \qquad \underset{+1\ -1}{Li\ H} \qquad \underset{+2\ -1}{Ca\ H_2}$$

f In compounds containing more than two elements, the oxidation number of any one of them may have to be obtained by first assigning reasonable oxidation numbers to the other elements. In sulphuric acid, H_2SO_4, for example, the most reasonable oxidation numbers for hydrogen and oxygen are $+1$ and -2, which gives sulphur an oxidation number of $+6$. Other examples are,

$$\underset{+1\ +4\ -2}{H_2\ S\ O_3} \qquad \underset{+1\ +7\ -2}{K\ Mn\ O_4} \qquad \underset{+1\ +6\ -2}{K_2\ Cr_2\ O_7} \qquad \underset{+1\ +7\ -2}{K\ Cl\ O_4}$$

g Some elements may have widely different oxidation numbers in different compounds as is shown by the following compounds of manganese, chromium, nitrogen and chlorine.

-3	-2	-1	0	1	2	3	4	5	6	7
			Mn		$MnCl_2$	$MnCl_3$ Mn_2O_3	MnO_2		MnO_4^{2-}	MnO_4^- Mn_2O_7
			Cr		$CrCl_2$	$CrCl_3$ Cr_2O_3			CrO_3 CrO_4^{2-} $Cr_2O_7^{2-}$	
NH_3 NH_4^+	N_2H_4	NH_2OH	N_2	N_2O	NO	N_2O_3 HNO_2	NO_2	N_2O_5 HNO_3		
		HCl		Cl_2	HClO		$HClO_2$ ClO_2	$HClO_3$		$HClO_4$

h When an element is oxidised its oxidation number is increased. When an element is reduced its oxidation number is decreased. Change in oxidation number can be used to decide whether an oxidation or a reduction has taken place. In the change from chloromethane to dichloromethane, for example,

$$\underset{-2\ +1\ -1}{C\ H_3\ Cl} \qquad \underset{0\ +1\ -1}{C\ H_2\ Cl_2}$$

the oxidation number of carbon is increased from -2 to 0. The carbon is therefore being oxidised.

3 Common oxidising and reducing agents

The functioning of some common oxidising and reducing agents is summarised below.

oxidising agent	effective change	decrease in oxidation number
$KMnO_4$ in acid solution	$MnO_4^- \longrightarrow Mn^{2+}$	5
$KMnO_4$ in alkaline solution	$MnO_4^- \longrightarrow MnO_2$	3
$K_2Cr_2O_7$ in acid solution	$Cr_2O_7^{2-} \longrightarrow Cr^{3+}$	3
dilute HNO_3	$NO_3^- \longrightarrow NO$	3
concentrated HNO_3	$NO_3^- \longrightarrow NO_2$	1
concentrated H_2SO_4	$SO_4^{2-} \longrightarrow SO_2$	2
manganese(IV) oxide	$MnO_2 \longrightarrow Mn^{2+}$	2
chlorine	$Cl \longrightarrow Cl^-$	1
chloric(I) acid	$ClO^- \longrightarrow Cl^-$	2
KIO_3 in dilute acid	$IO_3^- \longrightarrow I$	5
KIO_3 in concentrated acid	$IO_3^- \longrightarrow I^-$	4

reducing agent	effective change	increase in oxidation number
iron(II) salts (acid)	$Fe^{2+} \longrightarrow Fe^{3+}$	1
tin(II) salts (acid)	$Sn^{2+} \longrightarrow Sn^{4+}$	2
ethanedioates (acid)	$C_2O_4^{2-} \longrightarrow CO_2$	1
sulphites (acid)	$SO_3^{2-} \longrightarrow SO_4^{2-}$	2
hydrogen sulphide	$S^{2-} \longrightarrow S$	2
iodides (dilute acid)	$I^- \longrightarrow I$	1
iodides (concentrated acid)	$I^- \longrightarrow I^+$	2
metals, e.g. Zn	$Zn \longrightarrow Zn^{2+}$	2
hydrogen	$H \longrightarrow H^+$	1

4 Balancing redox reaction equation

The full equation for a redox reaction can be obtained by combining two half-equations. The manganate(VII) ion, for example, in acid solution, acts as an oxidising agent according to the equation

$$MnO_4^-(aq) \longrightarrow Mn^{2+}(aq)$$

To balance this, electrically and atomically, it must be written as

$$MnO_4^-(aq) + 8H^+(aq) + 5e^- \longrightarrow Mn^{2+}(aq) + 4H_2O(l) \qquad i$$

Iron(II) ions are oxidised, in acid solution, according to the equation,

$$Fe^{2+}(aq) \longrightarrow Fe^{3+}(aq) + e^- \qquad ii$$

To combine the two half-equations, i and ii, it is necessary to multiply ii by five. When this is done, and the equations are combined, the result is

$$MnO_4^-(aq) + 8H^+(aq) + 5Fe^{2+}(aq) \rightarrow Mn^{2+}(aq) + 5Fe^{3+}(aq) + 4H_2O(l)$$

Alternatively, oxidation numbers can be used. For the above reaction, the two half-equations give

$$MnO_4^-(aq) + Fe^{2+}(aq) \longrightarrow Mn^{2+}(aq) + Fe^{3+}(aq)$$
$${+7} {+2} {+2} {+3}$$

and show that the oxidation number of the oxidising agent has decreased by 5 whilst that of the reducing agent has increased by 1. The equation must, therefore, be balanced as follows:

$$MnO_4^-(aq) + 5Fe^{2+}(aq) \longrightarrow Mn^{2+}(aq) + 5Fe^{3+}(aq)$$

The oxygen atoms must now be balanced by introducing water,

$$MnO_4^-(aq) + 5Fe^{2+}(aq) \longrightarrow Mn^{2+}(aq) + 5Fe^{3+}(aq) + 4H_2O(l)$$

and a final balancing of the hydrogen, by introducing hydrogen ions, gives the final equation,

$$MnO_4^-(aq) + 5Fe^{2+}(aq) + 8H^+(aq) \rightarrow Mn^{2+}(aq) + 5Fe^{3+}(aq) + 4H_2O(l)$$

5 Electron transfer

The transfer of electrons from the reducing agent to the oxidising agent in a redox reaction can be demonstrated using the apparatus shown in Fig. 172 (overleaf). A current flows through the galvanometer because the following changes take place:

a In right-hand beaker $\qquad Fe^{3+}(aq) + e^- \xrightarrow{\text{reduction}} Fe^{2+}(aq)$
$\qquad\qquad\qquad\qquad\qquad$ (O.A.)

The formation of iron(II) ions in the beaker can be shown by adding potassium hexacyanoferrate(III). The iron(III) ions are gaining electrons and being reduced; they are taking electrons and acting as oxidising agents.

b In left-hand beaker $\qquad I^-(aq) \xrightarrow{\text{oxidation}} \frac{1}{2}I_2(aq) + e^-$
$\qquad\qquad\qquad\qquad\qquad$ (R.A.)

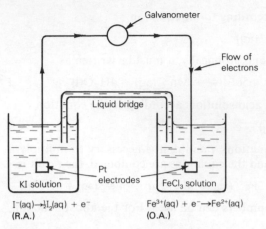

I⁻(aq)→½I₂(aq) + e⁻
(R.A.)

Fe³⁺(aq) + e⁻→Fe²⁺(aq)
(O.A.)

Fig. 172 Electron transfer in a redox reaction.

The formation of iodine in the beaker can be seen, and demonstrated by adding starch solution. The iodide ions are losing electrons and being oxidised; they are acting as reducing agents.

The overall change taking place is

$$Fe^{3+}(aq) + I^-(aq) \longrightarrow Fe^{2+}(aq) + \tfrac{1}{2}I_2(aq)$$
    (O.A.)   (R.A.)

The liquid bridge contains a solution, e.g. saturated potassium chloride, which does not react with either the potassium iodide or the iron(III) chloride. It can be contained in a tube, as shown, or a folded piece of filter paper, saturated with the solution, can be used. Alternatively, the solutions in the two beakers can be separated by a porous partition, e.g. porous pot or a plug of Polyfilla.

The bridge serves to complete the circuit internally. Electrons flow in the external wires: ions flow in the internal circuit. For every Fe^{3+}(aq) ion converted into Fe^{2+}(aq) in the right-hand beaker one Cl^-(aq) ion passes from right to left. Conversely, for every I^-(aq) ion converted into I atoms in the left-hand beaker, one K^+(aq) ion passes from left to right. This is necessary to maintain electrical neutrality.

Similar changes take place (but more rapidly) when iron(III) chloride and potassium iodide solutions are mixed in a test tube; it corresponds to the arrangement in Fig. 172 being short-circuited.

6 The strength of oxidising and reducing agents

The standard electrode, or redox, potentials for the Zn^{2+}(aq)/Zn(s) and Cu^{2+}(aq)/Cu(s) couples are $-0.76\,V$ and $+0.34\,V$ respectively (p. 409). These values imply that zinc ionises more readily than hydrogen, which ionises more readily than copper. In other words, zinc is a stronger reducing agent than hydrogen, which is stronger than copper. Alternatively, $Cu^{2+}(aq)$ ions are stronger oxidising agents than $H^+(aq)$ ions, which are stronger than $Zn^{2+}(aq)$.

The changes can be generalised as

$$\text{O.A.} + \text{electrons} \rightleftharpoons \text{R.A.}$$

oxidised form + electrons $\rightleftharpoons$ reduced form

The more positive the corresponding standard electrode, or redox, potential, the stronger the oxidising agent and the weaker the reducing agent. The more negative the potential, the stronger the reducing agent and the weaker the oxidising agent.

Standard electrode, or redox, potentials for a change of one ion into another, for example in the $Fe^{3+}(aq)/Fe^{2+}(aq)$ couple, can be measured by setting up a cell written as

$$H_2(g)/Pt \mid H^+(aq)\,(1\,mol\,dm^{-3}) \left| \begin{array}{c} Fe^{3+}(aq)\,(1\,mol\,dm^{-3}) \\ Fe^{2+}(aq)\,(1\,mol\,dm^{-3}) \end{array} \right| Pt$$

$-$ve pole $\qquad\qquad\qquad\qquad\qquad\qquad$ $+$ve pole

The platinum electrodes simply serve as conductors and/or catalysts. The cell e.m.f. shows that the standard electrode, or redox, potential for the $Fe^{3+}(aq)/Fe^{2+}(aq)$ couple is $+0.77\,V$. $Fe^{3+}(aq)$ is, therefore, a stronger oxidising agent than $H^+(aq)$, or hydrogen is a stronger reducing agent than $Fe^{2+}(aq)$.

Such electrode potential values, when listed in numerical order, as below, give the relative strengths of the oxidising and reducing agents concerned.

	oxidising agents (take electrons)		reducing agents (give electrons)	$E^{\ominus}/V$
	$Li^+(aq) + e^-$ ⟷		$Li(s)$	-3.04
	$Na^+(aq) + e^-$ ⟷		$Na(s)$	-2.71
	$Zn^{2+}(aq) + 2e^-$ ⟷		$Zn(s)$	-0.76
increasing strength as o.a.	$H^+(aq) + e^-$ ⟷	increasing strength as r.a.	$\frac{1}{2}H_2(g)$	0.00
	$Cu^{2+}(aq) + 2e^-$ ⟷		$Cu(s)$	$+0.34$
	$\frac{1}{2}I_2(aq) + e^-$ ⟷		$I^-(aq)$	$+0.54$
	$Fe^{3+}(aq) + e^-$ ⟷		$Fe^{2+}(aq)$	$+0.77$
	$\frac{1}{2}Br_2(l) + e^-$ ⟷		$Br^-(aq)$	$+1.07$
	$Cr_2O_7{}^{2-}(aq) + 14H^+(aq) + 6e^-$ ⟷		$2Cr^{3+} + 7H_2O(l)$	$+1.33$
	$\frac{1}{2}Cl_2(aq) + e^-$ ⟷		$Cl^-(aq)$	$+1.36$
	$MnO_4{}^-(aq) + 8H^+(aq) + 5e^-$ ⟷		$Mn^{2+}(aq) + 4H_2O(l)$	$+1.52$
	$\frac{1}{2}F_2(aq) + e^-$ ⟷		$F^-(aq)$	$+2.80$

7 The feasibility of a redox reaction

Lists, as in section 6, are very useful in predicting feasible reactions. Any oxidising agent on the left-hand side of the list will, theoretically, react with any reducing agent higher up on the list on the right-hand side.

Lithium is the strongest reducing agent given; it will reduce any of the oxidising agents below it, e.g.

$$Li(s) \ + Na^+(aq) \longrightarrow Li^+(aq) \ + Na(s)$$
(R.A.) (O.A.)   (weaker O.A.)   (weaker R.A.)

Fluorine is the strongest oxidising agent given and it will oxidise any of the reducing agents above it, e.g.

$$\tfrac{1}{2}F_2(g) + Cl^-(aq) \longrightarrow \tfrac{1}{2}Cl_2(g) \ + F^-(aq)$$
(O.A.)   (R.A.)   (weaker O.A.)   (weaker R.A.)

In other words, this means that a redox reaction will only be feasible if its combined $E^\ominus$ value, made up of the two redox potentials involved, is *positive*. For example

a
$$Cu^{2+}(aq) + 2e^- \longrightarrow Cu(s) \qquad E^\ominus = +0.34 \, V$$
$$Zn^{2+}(aq) + 2e^- \longrightarrow Zn(s) \qquad E^\ominus = -0.76 \, V$$
___
$$Cu^{2+}(aq) + Zn(s) \longrightarrow Zn^{2+}(aq) + Cu(s) \qquad E^\ominus = +1.01 \, V$$

The positive $E^\ominus$ value for the combined couples means that the reaction is feasible, as is well known.

b
$$Fe^{2+}(aq) + 2e^- \longrightarrow Fe(s) \qquad E^\ominus = -0.44 \, V$$
$$Sn^{4+}(aq) + 2e^- \longrightarrow Sn^{2+}(aq) \qquad E^\ominus = +0.15 \, V$$
___
$$Fe^{2+}(aq) + Sn^{2+}(aq) \longrightarrow Fe(s) + Sn^{4+}(aq) \qquad E^\ominus = -0.59 \, V$$

The negative $E^\ominus$ value means that this reaction is not feasible. It will be the reverse reaction, for which $E^\ominus$ will be $+0.59 \, V$, that is feasible.

Care must, however, be taken. Standard redox potentials refer only to standard conditions, though values at other temperatures, pressures and concentrations can be calculated as indicated in section 11 (p. 430). Moreover, redox potentials only indicate which reactions are thermodynamically feasible; they give no indication of the rate at which any reaction will take place.

8 Disproportionation

In some reactions an atom in a compound or an ion appears to undergo simultaneous oxidation and reduction, i.e. its oxidation number is both increased and decreased in the same reaction. Such a change is known as disproportionation. Typical examples are described below:

a Copper(I) compounds in aqueous solution Copper(I) ions can act either as reducing agents, as in *i*, or as oxidising agents, as in *ii*,

i	$Cu^{2+}(aq) + e^- \longrightarrow Cu^+(aq)$	$E^\ominus = +0.15\,V$
ii	$Cu^+(aq) + e^- \longrightarrow Cu(s)$	$E^\ominus = +0.52\,V$

iii	$2Cu^+(aq) \longrightarrow Cu^{2+}(aq) + Cu(s)$	$E^\ominus = +0.37\,V$
	$+1 \qquad\qquad +2 \qquad\quad 0$	

In the overall change, as in *iii*, which will take place because the $E^\ominus$ value is positive, the $Cu^+(aq)$ ions acting as oxidising agents in *ii* have oxidised those acting as reducing agents in *i*. The oxidation states are as shown.

Copper(I) compounds are, therefore, only stable in the presence of water when they are insoluble and the $Cu^+(aq)$ ion concentration is necessarily very low. Under these circumstances, the $E^\ominus$ figures quoted above are not valid.

b Iron(II) ions in aqueous solution $Fe^{2+}(aq)$ ions might be expected to disproportionate in the same way as $Cu^+(aq)$ ions

$$3Fe^{2+}(aq) \longrightarrow 2Fe^{3+}(aq) + Fe(s)$$
$$+2 \qquad\qquad +3 \qquad\qquad 0$$

Such disproportionation will not, however, take place for the change would have a negative $E^\ominus$ value:

$Fe^{2+}(aq) + 2e^- \longrightarrow Fe(s)$	$E^\ominus = -0.44\,V$
$2Fe^{3+}(aq) + 2e^- \longrightarrow 2Fe^{2+}(aq)$	$E^\ominus = +0.77\,V$

$3Fe^{2+}(aq) \longrightarrow 2Fe^{3+}(aq) + Fe(s)$	$E^\ominus = -1.21\,V$

Indeed, iron will reduce $Fe^{3+}(aq)$ to $Fe^{2+}(aq)$ ions, the $E^\ominus$ value for this change being $+1.21$ V.

9 Standard electrode potentials and free energy change

A standard electrode potential, $E^\ominus$, indicates how readily an ionisation process will take place and the free energy change for the process does likewise. It is not surprising, then, that the two are related. The expression is

$$\Delta G_m{}^\ominus \times 10^3 = -zFE^\ominus$$

where z is the number of electrons taking part in the process, $\Delta G_m{}^\ominus$ is expressed in kJ, and F is the Faraday constant ($96.485 \times 10^3\,C\,mol^{-1}$).

For the $Cu^{2+}(aq)/Cu$ couple, for example, the $E^\ominus$ values is 0.337 V and z is 2. The $\Delta G_m{}^\ominus$ value is, therefore, $-(2 \times 96\,485 \times 0.337)$, i.e. $-65\,kJ\,mol^{-1}$. For the $Zn^{2+}(aq)/Zn$ couple it is $-(2 \times 96\,485 \times (-0.763))$, i.e. $+147.2\,kJ\,mol^{-1}$.

An overall reaction will only take place if the combined $E^\ominus$ value made up of the two redox potentials involved, is *positive*. This, as can be seen in the following example, means a *negative* $\Delta G_m^{\ominus}$ value:

	$E^{\ominus}/V$	$\Delta G_m^{\ominus}/kJ\,mol^{-1}$
$Cu^{2+}(aq) + 2e^- \longrightarrow Cu(s)$	$+0.337$	-65
$Zn^{2+}(aq) + 2e^- \longrightarrow Zn(s)$	-0.763	$+147.2$
$Cu^{2+}(aq) + Zn(s) \longrightarrow Zn^{2+}(aq) + Cu(s)$	$+1.1$	-212.2

Measurement of cell e.m.f.'s provides a good way of measuring free energy changes (p. 271).

10 Standard electrode potentials and equilibrium constants

As free energy change and equilibrium constant are related by the expression (p. 277)

$$\Delta G_m^{\ominus}(298\,K) = -RT\ln K \qquad \lg K = -0.175\Delta G_m^{\ominus}(298\,K)$$

it follows (using the relationships in the previous section) that, at 25 °C,

$$\lg K = E^{\ominus}z/0.0592$$

where $E^{\ominus}$ is the cell potential for the reaction concerned, and z is the number of electrons concerned. A high positive $E^{\ominus}$ value, then, means a high equilibrium constant; a high negative $E^{\ominus}$ means a low equilibrium constant.

In general, a cell potential greater than $+0.592\,V$ means that the reaction may be regarded as going completely. A cell potential more negative than $-0.592\,V$ means that the reaction does not, for practical purposes, take place. A cell potential of zero means an equilibrium constant of 1, so that reversible reactions have cell potentials around zero. For example,

	$E^{\ominus}/V$	$\Delta G_m^{\ominus}/$ $kJ\,mol^{-1}$	K
$Zn(s) + Cu^{2+}(aq) \longrightarrow Zn^{2+}(aq) + Cu(s)$	$+1.1$	-212.2	1.4×10^{37}
$Ag^+(aq) + Fe^{2+}(aq) \longrightarrow Fe^{3+}(aq) + Ag(s)$	$+0.028$	-2.7	2.98
$Cu(s) + Pb^{2+}(aq) \longrightarrow Cu^{2+}(aq) + Pb(s)$	-0.463	$+89.3$	2.3×10^{-16}

Measurement of cell potentials provides a simple method of obtaining equilibrium constants.

11 Redox potentials under non-standard conditions

a **Effect of ionic concentration** Any redox couple can be generalised as

$$O.A. + z \text{ electrons} \longleftrightarrow R.A.$$

and the value of E for the couple for different ionic concentrations is related to the standard redox potential, $E^\ominus$, by the Nernst equation,

$$E = E^\ominus + (RT/zF)\ln([\text{O.A.}]/[\text{R.A.}])$$

where R is the gas constant $(8.313\,\text{J K}^{-1}\,\text{mol}^{-1})$, T is the temperature in K, F is the Faraday constant $(96\,485\,\text{C mol}^{-1})$, z is the number of electrons concerned, and $[\text{O.A.}]$ and $[\text{R.A.}]$ are the respective concentrations in mol dm^{-3}. Putting in the numerical values gives, at 25 °C,

$$E = E^\ominus + (0.0592/z)\lg([\text{O.A.}]/[\text{R.A.}])$$

For the $Zn^{2+}(aq)/Zn$ couple, the concentration of solid zinc is taken as unity so that

$$E = -0.76 + 0.0296\lg[Zn^{2+}(aq)]$$

It follows that the redox potential changes by 0.0296 V for every 10-fold change of $Zn^{2+}(aq)$ ion concentration. The redox potential is, therefore, $(-0.76-0.0296)\,\text{V}$ in 0.1 M solution, or $(-0.76+0.0296)\,\text{V}$ in 10 M solution. As the ionic concentration increases, the value of the redox potential gets less negative or more positive.

For the $Fe^{3+}(aq)/Fe^{2+}(aq)$ couple, the relationship is

$$E = +0.76 + 0.0592\lg([Fe^{3+}(aq)]/[Fe^{2+}(aq)])$$

As the $[Fe^{3+}(aq)]/[Fe2^{+}(aq)]$ ratio increases, the redox potential gets larger. When the $[Fe^{3+}(aq)]$ is 100 times greater than the $[Fe^{2+}(aq)]$ the redox potential is $(0.76+0.118)\,\text{V}$, i.e. 0.878 V.

b Reaction between zinc and copper(II) sulphate solution In practice, this reaction is carried out by putting some zinc into a solution of copper(II) sulphate. At the start there is no concentration of $Zn^{2+}(aq)$ ions. As $Zn^{2+}(aq)$ ions are formed and $Cu^{2+}(aq)$ ions are used up, however, the $Zn^{2+}(aq)/Zn$ potential becomes more positive, whilst the $Cu^{2+}(aq)/Cu$ potential becomes less positive. Eventually the two potentials become equal, the E value becomes zero, and an equilibrium mixture is established. At this stage

$$E_{Cu}^\ominus + 0.0296\lg[Cu^{2+}(aq)] = E_{Zn}^\ominus + 0.0296\lg[Zn^{2+}(aq)]$$

The equilibrium constant for the reaction is given by $[Zn^{2+}(aq)]$ divided by $[Cu^{2+}(aq)]$ so that

$$\lg K = \lg([Zn^{2+}(aq)]/[Cu^{2+}(aq)]) = (E_{Cu}^\ominus - E_{Zn}^\ominus)/0.0296$$

giving a value for K of 1.45×10^{37} (p. 278). This very high value means that the reaction goes almost completely.

c Effect of acidity Hydrogen ions are concerned in some redox processes, e.g.

$$MnO_4^-(aq) + 8H^+(aq) + 5e^- \longrightarrow Mn^{2+}(aq) + 4H_2O(l)$$
$$E^\ominus = +1.51\,\text{V}$$

and they must be regarded as part of the oxidising agent so that

$$E = E^{\ominus} + \frac{0.059}{5} \lg \frac{[MnO_4^-(aq)][H^+(aq)]^8}{[Mn^{2+}(aq)]}$$

For equal concentrations of $MnO_4^-(aq)$ and $Mn^{2+}(aq)$ ions the value of E changes with the pH value as follows:

pH	0	1	2	5	6
E/V	1.52	1.43	1.33	1.05	0.96

so that the oxidising power of the MnO_4^- ion is greatly affected by acidity.

Thus, at a pH of 1, potassium manganate(VII) will oxidise Cl^-, Br^- and I^- ions; at pH of 2, Br^- and I^- will be oxidised; at pH of 6, only I^- will be oxidised. The halides can be selectively oxidised in this way.

d Effect of complex formation The formation of complex ions can have a big effect on standard electrode potential values by stabilising one simple ion more than another. The standard electrode potential for the $Fe^{3+}(aq)/Fe^{2+}(aq)$ couple, for example is $+0.77\,V$, but that for $Fe(CN)_6^{3-}/(aq)Fe(CN)_6^{4-}(aq)$ is $+0.36\,V$. $Fe(CN)_6^{3-}(aq)$ ions are, therefore, weaker oxidising agents than $Fe^{3+}(aq)$ ions. In other words, the $+3$ oxidation state of iron is stabilised by complex formation with cyanide ions.

The standard electrode potential for the $Co^{3+}(aq)/Co^{2+}(aq)$ couple is $+1.82\,V$ so that the $Co^{2+}(aq)$ ion is a very weak reducing agent. It can only be oxidised by very strong oxidising agents such as fluorine or ozone. The standard electrode potential for $Co(CN)_6^{3-}(aq)/Co(CN)_6^{4-}(aq)$, however, is $-0.83\,V$, and cobalt(II) salts can be oxidised to cobalt(III), even by air, in the presence of CN^- ions.

Questions on Chapter 35

1 Write down the oxidation numbers of the atoms in the following compounds and ions (a) H_2SiO_3 (b) PO_4^{3-} (c) PH_3 (d) CrF_3 (e) Al_2O_3.
2 What is the oxidation state of the metal in the following compounds (a) $FeCl_2$ (b) FeO (c) Fe_2O_3 (d) $HgCl_2$ (e) Hg_2Cl_2 (f) $Hg_3(PO_4)_2$ (g) $CdCl_2$ (h) Co_2S_3 (i) KF?
3 Give examples of sulphur-containing compounds or ions in which the oxidation number of sulphur is (a) -2 (b) -1 (c) 4 (d) 6.
4 Arsenic, antimony and bismuth can exhibit oxidation numbers of -3, 3 and 5. Illustrate this statement.
5 What is the oxidation number of phosphorus in the following compounds (a) P_4O_{10} (b) P_4O_6 (c) H_3PO_2 (d) P_2H_4 (e) PH_3; and of nitrogen in the following compounds (f) N_2O_5 (g) NO_2 (h) N_2O_4 (i) HNO_2 (j) NO (k) N_2O (l) NH_2OH (m) N_2H_4 (n) NH_3?

6 What are the oxidation states of chlorine in Cl^-, Cl_2, $HClO_4$, HCl, $HClO_3$ and $HClO_2$?

7 Write balanced equations for the following reactions: (a) the reduction of iron(III) chloride to iron(II) chloride by aluminium (b) the addition of tin(II) ions to acidified sodium dichromate(VI) solution (c) the addition of sodium chlorate(I) to acidified iron(II) sulphate solution (d) the addition of sulphite ions to acidified potassium manganate(VII) (e) the addition of hydrogen sulphide gas to concentrated nitric acid.

8 What will happen (a) when a 1 M solution containing iron(II) and iron(III) ions is added to a 1 M solution containing tin(II) and tin(IV) ions (b) when an iron rod is placed in a solution 1 M with respect to H^+, Cu^{2+} and Zn^{2+} ions (c) when a 1 M solution of iron(II) sulphate is added to a 1 M solution of mercury(II) sulphate?

9 Explain the following facts: (a) copper(II) ions will oxidise iodide ions, but they will not do so in the presence of ethane-1,2-diamine (b) chlorine will oxidise a solution of potassium manganate(VI) to potassium manganate(VII), but iodine will not do this (c) the equilibrium pressure of hydrogen in the reaction of zinc with acid is higher than that of tin with acid (d) the reducing power of hexacyanoferrate(II) ions is greater than that of iron(II) ions.

10 Illustrate the statement that compounds containing an element in two different oxidation states, e.g. Prussian blue, are highly coloured.

11 Compare and contrast the conceptions of the valency and the oxidation number of an element by choosing selected elements to illustrate your points.

12 Explain carefully why it is that copper(I) ions disproportionate in aqueous solution whereas iron(II) ions do not.

13 Collect together some examples of disproportionation other than those given in this chapter.

14 Illustrate the statement that the strength of the oxoacids of an element, X, increases as the oxidation number of X increases.

15 What part does the liquid bridge play in the apparatus shown in Fig. 172? Would any current flow through the galvanometer if the liquid bridge was removed? What would happen if the liquid bridge was replaced by a platinum wire?

16 Illustrate the statement that oxidation may be regarded as a transfer of oxygen, a transfer of electrons, or a change in oxidation state.

17 Which of the following conversions involves oxidation? Give your reasons.

(a) $Cr(H_2O)_6^{3+} \longrightarrow CrO_4^{2-}$ (b) $Cr_2O_7^{2-} \longrightarrow CrO_4^{2-}$

(c) $CrO_3 \longrightarrow CrO_4^{2-}$ (d) $Cr(H_2O)_6^{3+} \longrightarrow Cr_2O_7^{2-}$

(e) $H_2O_2 \longrightarrow 2OH^-$ (f) $IO_3^- \longrightarrow I^-$

(g) $2H^- \longrightarrow H_2$ (h) $Fe(CN)_6^{3-} \longrightarrow Fe(CN)_6^{4-}$

(i) $MnO_2 \longrightarrow MnO_4^-$

18 How may the standard electrode potential of a metal in contact with a solution of its ions be determined? Comment on the usefulness to chemists of a knowledge of standard electrode potentials. The standard electrode potential for Zn^{2+}/Zn is $-0.76\,V$ and for Cu^{2+}/Cu it is $+0.34\,V$. Calculate the equilibrium constant for the reaction

$$Cu^{2+} + Zn \rightleftharpoons Cu + Zn^{2+} \qquad \text{at 298 K. (OJSE)}$$

19 Describe the usefulness of electrochemical data and electrochemical measurement in chemistry. (WS)

20 What would you expect the slope of a graph plotting $lg([Fe^{3+}(aq)]/[Fe^{2+}(aq)])$ against the corresponding E values to be? If the standard redox potential is $+0.76\,V$ what would be the potential when the ratio was (a) 4 and (b) 0.25?

21 The electrode potential of a metal was $-0.286\,V$ in 0.5 M solution of its ion and $-0.3155\,V$ in 0.05 M solution. What conclusions can you draw about its standard redox potential and about the charge on the metallic ion?

22* Draw a diagram of the cell you would set up to bring about the reaction $\frac{1}{2}Br_2(aq) + I^-(aq) \rightarrow \frac{1}{2}I_2(aq) + Br^-(aq)$ under standard conditions. Calculate (a) the cell e.m.f. (b) the equilibrium constant for the reaction as specified in the above equation.

23* Draw up a table of the redox potential values for the $Fe^{3+}(aq)/Fe^{2+}(aq)$ couple when $[Fe^{3+}(aq)] = n[Fe^{2+}(aq)]$ and n has values of 100, 50, 10, 5, 1, 0.2, 0.1, 0.02 and 0.01.

36
Electrolysis

1 The process of electrolysis

In an electrolytic process, a current is passed through the electrolyte to be decomposed, the electrolyte being in solution (usually in water) or in the molten state. The current is led in and out of the electrolyte through electrodes immersed in it, and electrodes which do not react with the electrolyte or any of the electrolysis products are generally chosen.

The electrode connected to the positive pole of the electricity supply is called the *anode*; that connected to the negative pole is called the *cathode*. Conventionally, the current is said to flow from the positive to the negative pole, but the actual flow of electrons is in the opposite direction, i.e. from negative to positive pole. *Electrons therefore pass into the cathode and out of the anode during electrolysis*. The source of the current is best regarded as an electron pump. One mole of electrons is 'pumped' when 96 485 C are involved, and as 1 C means 1 A flowing for 1 s, it follows that a current of 1 A means a flow of approximately 6.25×10^{18} electrons per second.

To maintain the flow of electrons through the electrolyte, the ions present in the electrolyte move towards the electrodes, the negatively charged ions (anions) to the anode and the positively charged ions (cations) to the cathode. The anions give up electrons to the anode and are discharged, whilst the cations are discharged at the cathode by taking electrons,

$$A^{n-}(aq) \longrightarrow A + ne^- \qquad C^{n+}(aq) + ne^- \longrightarrow C$$
$$\text{(anion)} \qquad\qquad\qquad \text{(cation)}$$

A and C will be the products of electrolysis, A being liberated at the anode and C at the cathode.

The giving up of electrons to the anode and the taking of electrons from the cathode maintains the flow of current, the whole process of a simple electrolysis being summarised in Fig. 173.

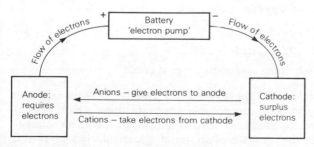

Fig. 173 The process of a simple electrolysis.

The electrode process at the anode (giving electrons) is an oxidation; that at the cathode (taking electrons) is a reduction (p. 421).

n mol of electrons, i.e. $96\,485n\,C$ will be required to discharge 1 mol of A^{n+} or C^{n-}, so that the amount of A and C produced by any amount of electricity can easily be calculated.

2 Electrode processes

When there is only one cation and one anion, e.g. in the electrolysis of molten sodium chloride, the two ions are discharged at the relevant electrodes in what are known as electrode processes. The precise nature of the electron transfer involved is not well understood but it can be thought of, in general, in terms of a Helmoltz double layer (p. 405). Movement of ions to the electrode builds up the decomposition pressure so that it becomes greater than the electrolytic solution pressure. As a result, any equilibrium is upset and atoms deposit on the electrode.

If more than one cation, or more than one anion, are present they will all move towards the relevant electrode but the particular electrode process which will take place will be the one which takes place most easily. One cation and one anion are said to be preferentially discharged.

a Preferential discharge Those ions which are most easily formed from their elements will be least easily discharged. In terms of electrode potentials (p. 409), elements with a high negative electrode potential will be least readily discharged at a cathode, whilst elements with a high positive electrode potential will be least readily discharged at an anode.

$Zn^{2+}(aq)$ ions, for example, will be preferentially discharged at a cathode, if present under comparable conditions with Na^+ ions. $Cu^{2+}(aq)$ ions will be discharged before H^+ ions. Similarly, $I^-(aq)$ ions will be discharged at an anode before $Cl^-(aq)$ ions.

On this basis, a list can be drawn up, following the electrode potential values, of the order in which common hydrated ions will, usually, be preferentially discharged.

K^+	Ca^{2+}	Na^+	Mg^{2+}	Al^{3+}	Zn^{2+}	Fe^{2+}	Pb^{2+}	H^+	Cu^{2+}	Ag^+
-2.92	-2.87	-2.71	-2.37	-1.66	-0.76	-0.44	-0.13	0	$+0.34$	$+0.80$

$\longrightarrow$ increasing ease of discharge at cathode $\longrightarrow$

OH^-	I^-	Br^-	Cl^-	SO_4^{2-}	F^-
$+0.40$	$+0.54$	$+1.07$	$+1.36$	$+2.01$	$+2.85$

$\longleftarrow$ increasing ease of discharge at anode $\longleftarrow$

These orders of preference apply only for discharge from aqueous solutions containing ions at comparable concentrations which are approximately 1 M, and for discharge involving zero, or low, overvoltages. High overvoltages, or the presence of an ion at a particularly high or low concentration, will affect the orders.

b Effect of overvoltage Depending on the nature of the electrode being used, it is found that the discharge of an ion may be more difficult than would be expected. This is particularly so using certain electrodes when gases are involved, e.g. in the discharge of H^+(aq), OH^-(aq) and Cl^-(aq) ions.

So far as electrode potentials are concerned, H^+(aq) ions should be discharged from a solution containing Zn^{2+}(aq) and H^+(aq) ions of comparable concentration. If zinc electrodes be used, however, the Zn^{2+}(aq) ions will be discharged before the H^+(aq) ions. Hydrogen is said to have a high overvoltage at a zinc electrode. At a platinum electrode, with a low hydrogen overvoltage, H^+(aq) ions would be discharged before Zn^{2+}(aq) ions.

Similarly, OH^-(aq) ions might be expected to be discharged at an anode in preference to any other anion, but the high oxygen overvoltage at most electrodes makes the discharge of OH^- ions more difficult.

Overvoltages at particular electrodes must, therefore, be taken into account in predicting the electrode process which will take place. Further details, and some numerical values, are given on page 440.

c Concentration effect An ion present at a low concentration is more difficult to discharge than the same ion at a higher concentration. By carefully balancing the concentrations of two metallic ions it is possible for them to be discharged together as an alloy. Brass, for instance, can be deposited in this way from suitable mixtures of zinc and copper(II) sulphates.

The concentration of an ion must, therefore, be taken into account in predicting whether it is likely to be discharged or not.

d Solution of anode The liberation of electrons at an anode commonly originates from the discharge of an anion. It can also originate from the ionisation of the anode with the formation of a cation.

In the electrolysis of copper(II) sulphate solution, for example, using a copper anode, there are three possible anode processes.

$$SO_4^{2-}(aq) \longrightarrow \tfrac{1}{2}S_2O_8^{2-}(aq) + e^-$$

$$2OH^-(aq) \longrightarrow \tfrac{1}{2}O_2(g) + H_2O(l) + 2e^-$$

$$Cu(s) \longrightarrow Cu^{2+}(aq) + 2e^-$$

Of these, the last takes place most easily, and the copper anode passes into solution as copper(II) ions; it is referred to as a soluble anode (p. 442).

3 Mechanism of electrolysis

The concentration and overvoltage effects, the possibility of an anode going into solution, the use of different electrode materials, and changes in temperature and current density can all affect an electrolytic process,

electrolyte	electrode	at anode	at cathode	notes
molten NaCl (p. 442)	C	$2Cl^- \longrightarrow Cl_2(g) + 2e^-$	$Na^+ + e^- \longrightarrow Na(s)$	only one anion and one cation present
dil. H_2SO_4 (p. 439)	Pt	$2OH^-(aq) \longrightarrow H_2O(l) + \frac{1}{2}O_2(g) + 2e^-$	$2H^+(aq) + 2e^- \longrightarrow H_2(g)$	a OH^-(aq) preferentially discharged at anode b 2 vol. H_2 for 1 vol. O_2 c acid concentration increases
NaOH soln. (p. 441)	Pt	$2OH^-(aq) \longrightarrow H_2O(l) + \frac{1}{2}O_2(g) + 2e^-$	$2H^+(aq) + 2e^- \longrightarrow H_2(g)$	a H^+(aq) preferentially discharged at cathode b 2 vol. H_2 for 1 vol. O_2 c NaOH concentration increases
conc. HCl (p. 441)	C	$2Cl^-(aq) \longrightarrow Cl_2(g) + 2e^-$	$2H^+(aq) + 2e^- \longrightarrow H_2(g)$	a Cl^-(aq) preferentially discharged at anode b one vol. H_2 for 1 vol. Cl_2 c acid concentration decreases
$CuSO_4$ soln. (p. 442)	Pt	$2OH^-(aq) \longrightarrow H_2O(l) + \frac{1}{2}O_2(g) + 2e^-$	$Cu^{2+}(aq) + 2e^- \longrightarrow Cu(s)$	a OH^-(aq) preferentially discharged at anode b Cu^{2+}(aq) preferentially discharged at cathode c solution converted into H_2SO_4
$CuSO_4$ soln. (p. 442)	Cu	$Cu(s) \longrightarrow Cu^{2+}(aq) + 2e^-$	$Cu^{2+}(aq) + 2e^- \longrightarrow Cu(s)$	a soluble anode b Cu^{2+}(aq) preferentially discharged at cathode c no change in solution d transfer of Cu from anode to cathode
dil. NaCl (p. 443)	Pt	$2OH^- \longrightarrow H_2O(l) + \frac{1}{2}O_2(g) + 2e^-$	$2H^+(aq) + 2e^- \longrightarrow H_2(g)$	a OH^-(aq) preferentially discharged at anode b low concentration of Cl^-(aq)
conc. NaCl (p. 443)		see section 7, p. 442		

so that varied results can be achieved. A summary of some simple processes is given opposite.

Many of the theoretical details are not yet fully understood, but a consideration of the nature of the electrolysis of 0.5 M sulphuric acid introduces the main general principles.

The overall result is that hydrogen bubbles off from the cathode, and oxygen from the anode. Two volumes of hydrogen are produced for every one volume of oxygen, and the concentration of the acid increases. Water, in effect, is being decomposed. The changes are summarised as follows:

$$H_2SO_4(l)$$
$$\downarrow$$

H^+ ions discharged	to cathode	$2H^+(aq)$ $\quad$ $SO_4^{2-}(aq)$	both ions to anode	$OH^-(aq)$ ions preferentially discharged
$2H^+(aq) + 2e^- \rightarrow H_2(g)$		$H^+(aq)$ $\quad$ $OH^-(aq)$		$2OH^-(aq) \longrightarrow$ $\frac{1}{2}O_2(g) + H_2O(l) + 2e^-$

$$\uparrow$$
$$H_2O(l)$$

a Decomposition voltage If the voltage applied between the electrodes is gradually increased from zero, the current passing through the cell will rise only very slightly, and no significant electrolysis will take place, until a certain voltage is reached. Further increase in the applied voltage above this point will lead to more rapid electrolysis and a marked rise in current (Fig. 174(a)).

The voltage at which electrolysis first begins is known as the *experimental decomposition voltage*. It can be measured by varying the voltage across an electrolytic cell and measuring the associated current so that a graph, as in Fig. 174(a), can be obtained. The value for 0.5 M sulphuric acid is 1.70 V.

This minimum voltage is necessary to bring about any electrolysis of 0.5 M sulphuric acid for two reasons. First, the formation of hydrogen and oxygen at the platinum electrodes converts them into hydrogen and oxygen electrodes (they are said to be polarised). As a result, a back e.m.f. equal to the algebraic sum of a hydrogen and an oxygen electrode,

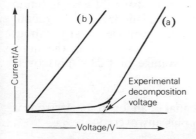

Fig. 174 Current voltage curves in electrolysis. (a) 0.5 M sulphuric acid solution using bright platinum electrodes. (b) Copper(II) sulphate solution using copper electrodes.

is produced. Such a back e.m.f. has to be overcome before electrolysis can take place.

This back e.m.f. is equal to the theoretical decomposition voltage for, once the back e.m.f. has been overcome, it might be expected that electrolysis would take place. In some cases this is so, i.e.

$$\text{theoretical decomposition voltage} = \text{electrode potential at anode} + \text{electrode potential at cathode}$$

For 0.5 M sulphuric acid, however, the experimental decomposition voltage (1.70 V) is greater than the theoretical value (1.23 V) and this is common, particularly when gases are being liberated at the electrodes. The necessity to apply a higher voltage than the theoretical decomposition voltage is due to overvoltage.

b Overvoltage The potential required to discharge a metallic ion at the electrode is generally equal, or very nearly equal, to the electrode potential of the metal concerned under the same conditions. To deposit zinc ions from a zinc sulphate solution with an ionic concentration of $1\,\text{mol dm}^{-3}$, for example, requires a cathode potential of approximately 0.76 V, the standard electrode potential for zinc.

But when gases are concerned in a discharge process, e.g. in the discharge of $H^+(aq)$, $OH^-(aq)$ and $Cl^-(aq)$ ions, the potential necessary may be considerably greater than the electrode potential concerned. The difference is known as the overvoltage. Its value depends on the particular ion, the particular electrode, and the particular way in which the electrolysis is being carried out.

The origin of overvoltage is obscure, but one likely cause is a slow rate of change from atoms of gas into gas molecules at an electrode which does not catalyse such a change. It follows that

$$\text{experimental decomposition voltage} = \text{theoretical decomposition voltage} + \text{overvoltage at anode} + \text{overvoltage at cathode}$$

and, when no overvoltage exists, the theoretical and experimental decomposition voltages are equal.

In the electrolysis of 0.5 M sulphuric acid using bright platinum electrodes, the overvoltage of hydrogen at the anode is small (about 0.03 V) but the overvoltage of oxygen at the cathode is large (about 0.44 V). The experimental decomposition voltage of 1.70 V is, therefore, made up of a theoretical decomposition voltage of 1.23 V plus overvoltages of 0.03 V and 0.44 V.

c Discharge potential The discharge potential of an ion at an electrode is equal to the electrode potential of the ion plus any overvoltage, i.e.

$$\text{discharge potential} = \text{electrode potential} + \text{overvoltage at the electrode}$$

As overvoltages for metallic ions are always small, the discharge potential for a metallic ion is approximately equal to the electrode potential of the corresponding metal. The discharge potential for a non-metallic ion varies considerably with the nature of the electrode concerned, as the overvoltage varies so much.

d Resistance of electrolyte Over and above the back e.m.f. and any overvoltages, the resistance of the electrolyte and any electrical connections have also to be overcome. Application of the minimum experimental decomposition voltage will give only a very small current (Fig. 174) and very slow electrolysis. The higher the voltage above the minimum, the more rapid the electrolysis.

4 Electrolysis of sodium hydroxide solution

a Using platinum electrodes In this process, hydrogen is produced at the cathode and oxygen at the anode. Two volumes of hydrogen are formed for each volume of oxygen. The concentration of the sodium hydroxide increases: water is, in effect, being decomposed.

The experimental decomposition voltage for a 1 M solution is approximately 1.7 V, the same value as for 0.5 M sulphuric acid. This common value is explained by the fact that both processes are essentially the same, water being decomposed with the liberation of hydrogen and oxygen at platinum electrodes.

At the anode, $OH^-(aq)$ ions are discharged, as they are the only anions present; there will be an overvoltage of 0.44 V. At the cathode there are two cations, $Na^+(aq)$, and $H^+(aq)$, and the $H^+(aq)$ ions are preferentially discharged, even though they are present in only very low concentration. The discharge potential for $Na^+(aq)$ ions from a 1 M solution is approximately 2.71 V. The concentration of $H^+(aq)$ ions in a 1 M solution of sodium hydroxide will be about $10^{-14}\,mol\,dm^{-3}$, and the discharge potential for $H^+(aq)$ ions, under such conditions will be about 0.83 V. Thus the discharge of hydrogen ions will be easier than that of sodium ions.

b Using a mercury cathode If a mercury cathode is used, $Na^+(aq)$ ions will be discharged, as hydrogen has a high overvoltage (0.78 V) at a mercury cathode. Sodium amalgam will be formed but, if the concentration of sodium in the amalgam reaches a high enough value, the hydrogen overvoltage will be sufficiently lowered for hydrogen ions to be discharged, as at a platinum electrode.

5 Electrolysis of solutions of hydrogen halides

In the electrolysis of hydrochloric acid, hydrogen is liberated at the cathode, and chlorine and/or oxygen at the anode. Whether or not chlorine or oxygen, or a mixture of the two, is produced at the anode depends on a variety of factors, the discharge potentials of Cl^-(aq) and OH^-(aq) ions being fairly close under certain conditions.

Dilute hydrochloric acid, with a low concentration of chloride ions, will tend to produce mainly oxygen: concentrated hydrochloric acid will liberate mainly chlorine. At an electrode with a high overvoltage for oxygen and a lower one for chlorine, evolution of chlorine will be favoured.

Electrolysis of solutions of hydrogen iodide will invariably produce hydrogen and iodine, because discharge of I^-(aq) ions will always be easier than discharge of OH^-(aq) ions, because of oxygen overvoltage. On the other hand, electrolysis of fluoride solutions will always give oxygen at the anode, as discharge of F^-(aq) ions is so difficult.

6 Electrolysis of copper(II) sulphate solution

a Using platinum electrodes The cathode becomes plated with copper, and oxygen bubbles off from the anode. The electrolyte slowly loses its colour and is converted into sulphuric acid.

The two cations are Cu^{2+}(aq) and H^+(aq), with the former being preferentially discharged. The two cations are SO_4^{2-}(aq) and OH^-(aq) with the latter being selectively discharged. The experimental decomposition voltage is about 1.8 V.

b Using copper electrodes The ions present are the same as in **a**. Cu^{2+}(aq) ions are preferentially discharged at the cathode, depositing copper. But with a copper anode, the easiest process at the anode is for it to form Cu^{2+}(aq) ions (p. 437).

In this electrolysis, the electrodes do not become polarised as they remain Cu/Cu^{2+} electrodes throughout. As there are also no overvoltage effects, there is no decomposition voltage. Even the smallest voltage will bring about electrolysis, and increasing the voltage will cause a steady rise in current, approximately following Ohm's law (Fig. 174b).

The theoretical decomposition voltage is also zero, the electrode potentials at the two electrodes being equal. In other words, the theoretical decomposition voltage is zero because there is really no overall decomposition taking place.

It is this process that is widely used in copper plating and in the purification (refining) of copper.

7 Electrolysis of sodium chloride

The electrolysis of *molten* sodium chloride involves only the discharge of Na^+ and Cl^- ions to form sodium and chlorine. The process is used industrially to make sodium (the Downs' process). The melting point of the sodium chloride is lowered by adding about 60 per cent of calcium chloride. The Na^+ ions are preferentially discharged at the temperature used.

The electrolysis of a *dilute solution* of sodium chloride in water (weak brine) will generally yield oxygen at the anode and hydrogen at the cathode, the solution itself becoming more concentrated. As with solutions of sulphuric acid and sodium hydroxide, this is, again, an effective decomposition of water. $H^+(aq)$ ions are preferentially discharged at the cathode (rather than $Na^+(aq)$ ions) and $OH^-(aq)$ ions at the anode (rather than $Cl^-(aq)$ ions).

Using a *concentrated solution* of sodium chloride in water (saturated brine), in special cells, it is possible to obtain sodium hydroxide, chlorine and hydrogen as the products. This is an important industrial process which is carried out in two ways.

a Use of mercury cell (See Fig. 175.) This type of cell consists of a long, rectangular, shallow trough made of steel. The sides and top are lined with rubber. The cell is slightly tilted so that a film of mercury, which acts as the cathode, can flow across the base. The anodes are titanium plates positioned just above the mercury surface.

Pure, saturated brine flows through the cell between the mercury cathode and the titanium anode. $Cl^-(aq)$ ions are preferentially discharged at the anode. This is because they are present in much greater concentration than the $OH^-(aq)$ ions and because oxygen has a higher overvoltage than chlorine. Chlorine gas is fed off from the top of the cell.

At the mercury cathode, $Na^+(aq)$ ions are discharged in preference to the $H^+(aq)$ ions, because of the high hydrogen overvoltage at a mercury surface. The sodium formed dissolves in the mercury, but the sodium content never rises above 0.5 per cent as the amalgam is running off continuously as it is formed.

The amalgam is run into a separate container fitted with activated graphite blocks, and water is added. This reacts with the sodium in the

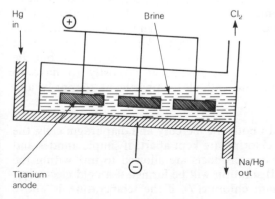

Fig. 175 The principle of a mercury cell.

amalgam to produce hydrogen, sodium hydroxide solution and mercury. The sodium hydroxide can be used as a solution or evaporated to give the solid; the mercury is recycled.

Mercury cells are operated at about 4.4 V and currents up to 400 kA can be used. They produce very pure sodium hydroxide, but they are expensive to construct, and the mercury used is both expensive and toxic.

b Use of a diaphragm cell (See Fig. 176.) In this type of cell, a titanium anode is separated from a steel cathode by a porous, asbestos diaphragm.

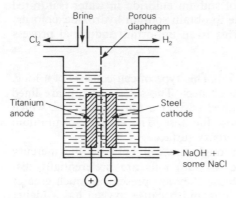

Fig. 176 The principle of a diaphragm cell.

Pure, saturated brine, containing hydrated Cl^-, Na^+, H^+ and OH^- ions, is fed into the anode compartment; the $Cl^-(aq)$ ions are preferentially discharged, and chlorine gas bubbles off. The solution, depleted in $Cl^-(aq)$ ions, passes through the diaphragm into the cathode compartment, where $H^+(aq)$ ions are preferentially discharged, and hydrogen bubbles off.

The solution remaining in the cathode compartment is a solution of sodium hydroxide with some chloride impurity. On concentration by evaporation, sodium chloride crystallises out, and a concentrated solution of fairly pure sodium hydroxide remains. A typical cell operates at 3.8 V with currents up to 150 kA. It is inexpensive to construct, but the diaphragm needs frequent replacement and it is costly to have to evaporate the solution from the cathode compartment to purify and concentrate it.

c Electrolysis with mixing In both the mercury and diaphragm cells, the sodium hydroxide and the chlorine are kept apart. If simple anodes and cathodes are used and the two products are allowed to mix within the cell, then sodium chlorate(I) solution will be formed if a cold electrolyte solution is used, and sodium chlorate(V), if the temperature is above 70 °C.

8 The movement of ions in electrolysis

The movement of cations to the cathode and anions to the anode can easily be shown by placing crystals of potassium manganate(VII) and tetraamminecopper(II) sulphate on a piece of moist filter paper. When a potential of about 20 V is applied across the paper, the movement of the highly coloured MnO_4^- and $Cu(NH_3)_4^{2+}(aq)$ ions can be seen.

Alternatively gels can be used as shown in Fig. 177. In such experiments the velocity of the ions can be measured.

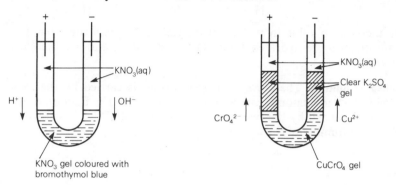

Fig. 177 Apparatus to show movement of ions in electrolysis and to measure ionic velocities.

a Mobility of ions The mobility of an ion is defined as its velocity in $m\,s^{-1}$ under a potential gradient of $1\,V\,m^{-1}$. Values can readily be obtained, for it can be shown that

$$\frac{\text{mobility of}}{\text{an ion }(u)} = \frac{\text{molar conductivity of ion }(\Lambda)}{96\,485}$$

Some typical values are given below.

ion	molar conductivity $(\Lambda)/$ $S\,cm^2\,mol^{-1}$	mobility $(u)/$ $cm\,s^{-1}\,V^{-1} \times 10^4$
H^+	349.8	36.25
Li^+	38.7	4.01
Na^+	50.1	5.19
K^+	73.5	7.62
OH^-	198.6	20.6
F^-	55.4	5.7
Cl^-	76.4	7.9
NO_3^-	71.4	7.4

It will be seen that the values for H^+ and OH^- are particularly high. Such high velocities for H^+ and OH^- ions are found only in hydroxylic solvents, such as water or ethanol, and it is suggested that the ions may, in such solvents, be passed on from one solvent molecule to another by a mechanism such as that depicted below.

$$H-O-\!\!|-\overset{\frown}{H^+} \quad + \quad O-H \quad \longrightarrow \quad H-O \quad + \quad H-O-\!\!|-\overset{\frown}{H^+}$$
$$\underset{H}{|} \qquad\qquad \underset{H}{|} \qquad\qquad \underset{H}{|} \qquad\qquad \underset{H}{|}$$

$$\overset{\frown}{O^-} \quad + \quad H-O \quad \longrightarrow \quad O-H \quad + \quad \overset{\frown}{O^-}$$
$$\underset{H}{|} \qquad\qquad \underset{H}{|} \qquad\qquad \underset{H}{|} \qquad\qquad \underset{H}{|}$$

The increase in value in passing from Li^+ to K^+ indicates that Li^+ ions are much more hydrated than Na^+ or K^+ ions, and are therefore larger and slower.

b Transport number The current, in electrolysis, is carried by the ions, with the faster moving ones carrying the larger share. The fraction of the total current carried by the two ions in a solution of C^+A^- is known as the transport number of the ions, t_C or t_A. Thus

$$\frac{t_C}{t_A} = \frac{u_C}{u_A} \qquad t_A = \frac{u_A}{u_C + u_A} \qquad t_C = \frac{u_C}{u_C + u_A}$$

The different mobilities and transport numbers of two ions means that there are different changes in the amount of electrolyte around the anode and cathode in an electrolysis.

If a current of $96\,485n$ coulomb is passed through an electrolyte, CA, using inert electrodes, n mol of C^+ ion will be discharged at the cathode, and n mol of A^- ion will be discharged at the anode. If the transport number of C^+ is t_C and of A^- is t_A, then the C^+ ions will carry $96\,485nt_C$ C towards the cathode, whilst the A^- ions will carry $96\,485nt_A$ C towards the anode. In other words, nt_C mol of C^+ ions will pass towards the cathode, whilst nt_A mol of A^- ions will pass towards the anode. This will result in a loss of nt_C mol of CA in the solution around the anode, and a loss of nt_A mol of CA in the solution around the cathode. By measuring these two losses it is possible to obtain values for t_C and t_A.

9 Polarography

Polarography, introduced by Heyrovsky in 1922, provides a rapid and accurate method of analysis requiring only small quantities of materials and simple equipment. It makes use of a dropping mercury cathode (Fig. 178). The dropping of the mercury continually renews the cathode surface, and this allows a series of cathode processes to take place at what is essentially the same cathode.

The e.m.f. applied between the cathode and anode can be steadily increased. If the mercury is dropping into a solution containing $Cu^{2+}(aq)$ ions, there will be no appreciable current until the applied e.m.f. is high

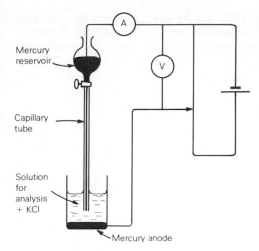

Fig. 178 Arrangement of a dropping mercury cathode for obtaining a polarogram.

enough for the Cu^{2+}(aq) ions to be discharged at the cathode. When this discharge begins there will be a marked rise in current.

As the ions are discharged other Cu^{2+}(aq) ions will diffuse towards the mercury cathode. Increase in the applied e.m.f. will cause an increase in the rate of discharge, increased current and increased rate of diffusion. There is a limit to the rate of diffusion, however, depending on the bulk concentration of Cu^{2+}(aq) ions, and when this limit is reached further increase in e.m.f. will not cause any further increase in current as ions cannot diffuse to the cathode quickly enough. The current, at this stage, is called the *limiting current*, and the way in which the current changes as the e.m.f. is increased is shown in Fig. 179. The e.m.f. value when the current is half the limiting current is known as the *half-wave potential*; it is a characteristic of a particular ion.

If other ions are present they, too, will cause a change in current, similar to that caused by Cu^{2+}(aq) ions, as they are discharged. The half-wave potentials are different for different ions so that current–voltage

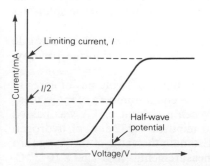

Fig. 179 Simplified change of current with voltage in polarography.

graphs, called *polarograms* (Fig. 180), can be used to detect the presence of various ions in a mixture. The method can also be used for detecting molecules which are reduced at a mercury cathode under the conditions being used. Quantative analysis can also be carried out as the limiting current for each ionic species is proportional to its concentration.

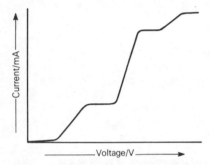

Fig. 180 A simplified polarogram showing the presence of various ions.

In practice, the resistance of the cell is high because of the small surface area of the dropping mercury cathode. Potassium chloride is, therefore, added to the electrolyte solution to increase the conductivity. Potassium ions are themselves only reduced at a mercury cathode at high voltages, and, therefore, do not interfere in the polarographic analysis for other ions.

Questions on Chapter 36

1 If the electrical cost of producing 1 kg of magnesium by an electrolytic process is £x, what will it cost to produce y kg of aluminium by a similar process? Assume that the processes are 100 per cent efficient.

2 State Faraday's laws of electrolysis. What correspondence, if any, can you find between them and the laws of chemical combination, and how far do they indicate the electrical character of atomic structure? From what sources can evidence be obtained of the existence of ions in solution? (OC)

3 Passage of a current of 1000 A through a solution of sodium chloride in a diaphragm cell is found to produce 28.58 kg of chlorine per day. What is the efficiency of the process?

4 What evidence supports the hypothesis that equal volumes of gases at the same temperature and pressure contain equal numbers of molecules? Is this hypothesis true for all gases? When a current of 1 A was passed through slightly acidified water for 30 minutes, 0.232 dm^3 of hydrogen gas was collected (at 20 °C, 0.97 atm). Estimate the value of the Avogadro constant on the basis of this measurement. ($e = 1.60 \times 10^{-19}$ C, $R = 0.082$ dm^3 atm mol^{-1}). (OJSE)

5 Explain carefully what you mean by the terms 'electrolysis', 'electrolyte' and 'electrode'. If a current passing for 1200 s deposits 0.4818 g of metallic silver, what volume of (a) hydrogen (b) oxygen would you expect it to liberate in a voltameter (in the same circuit) where dilute sulphuric acid is being electrolysed?

6 State Faraday's law of electrolysis. 1 M NaOH solution is electrolysed at 0 °C between platinum electrodes. What current would be required to produce from the cathode (a) 100 cm³ (at s.t.p.) of dry hydrogen per minute (b) 100 cm³ of hydrogen at 0 °C and 101.325 kPa containing, as it actually would, water vapour? Take the vapour pressure of M NaOH as 101.4 Pa at 0 °C. Calculate the charge in coulombs on all the ions present in 1 cm³ of 1 M NaOH. (Avogadro constant $= 6.02 \times 10^{23}$; Faraday constant $= 96\,500$ coulomb.)

7 (a) In the electrolysis of brine using a flowing mercury cathode, 1000 A liberate 30.16 kg of chlorine per day. If the working voltage of the cell is 4 volt, how many kilowatt hours of energy are required per 1016 kg of chlorine? (b) In a diaphragm cell, 2832 kilowatt hours are required per 1016 kg of chlorine. What advantages has the mercury cell over the diaphragm cell?

8 If a cylindrical rod of length 127 mm and diameter 19.05 mm is completely and evenly plated with nickel for 300 s, using a current of 2.5 A, what will be the thickness of the nickel plate? (The density of nickel is $8.9\,\mathrm{g\,cm^{-3}}$.)

9 Describe any two industrial electrolytic processes and any two industrial electrothermal processes.

10 Write down the electrode processes which you would expect to take place in the electrolysis of 1 M solutions of the following electrolytes, using platinum electrodes (a) zinc sulphate (b) sodium sulphate (c) gold(III) chloride (d) potassium nitrate (e) cadmium iodide.

11 Explain the differences you would expect in the electrolysis of an approximately 1 M solution of silver nitrate using (a) platinum (b) silver electrodes.

12 Distinguish between the meaning of the terms transport number of an ion, molar conductivity of an ion and electric mobility of an ion. Show how the terms are related.

13 Summarise the evidence which supports the suggestion that anions move towards the anode during electrolysis.

14 In a transport number measurement on a solution of silver nitrate using platinum electrodes the mass of silver in the anode compartment fell from 10.075 g to 9.420 g, whilst that in the cathode compartment fell from 8.346 g to 7.517 g. Calculate the transport number of the NO_3^- ion.

37
Surface effects and colloids

The situation at the surface of a substance is different from that in the interior, and this has important consequences. It is like being on the edge of a crowd or in the middle of it. A molecule in the interior of a substance is surrounded on all sides by similar molecules, and intermolecular forces are exerted, on average, equally in all directions. A surface molecule, however, has similar molecules on one side only so that the forces on it are different. This causes the obvious surface tension in liquids (p. 30), and similar, but less obvious, effects can occur at solid surfaces to cause adsorption. This is particularly so for large solid surface areas such as those in finely powdered solids and in some colloids (p. 454).

1 Liquid surfaces

A freely suspended liquid forms spherical drops to lower its surface energy (p. 30), and any other process which lowers the surface energy is favoured. Those substances which can lower the surface energy of a liquid are said to be *surface active* or are called *surfactants*.

Liquids which are soluble in water and have a lower surface tension, concentrate at the surface, when added to water, and lower the surface energy; simple alcohols are typical. Insoluble liquids or solids may form a film on the surface of water if they contain polar (*hydrophilic*) groups, e.g. —OH or —COOH, which can 'anchor' in the water surface. Liquids with only *hydrophobic* groups, e.g. alkyl groups, will not generally mix with, or spread on, water. They split up into lens shaped drops if they are lighter than water, or sink as globules if heavier.

a Water conservation Loss of water by evaporation from a reservoir can be reduced by spreading hexadecanol over the surface. The hydrophilic —CH_2OH groups enter the water surface, whilst the hydrophobic, —$C_{15}H_{31}$ groups, stick out of it. The rate of loss of water is lowered because the surface becomes much more like an alkane surface.

b Emulsifiers In an emulsion – e.g. ice cream, milk, margarine and many cosmetics – one phase is dispersed as small droplets in another. If the two phases are immiscible, e.g. oil and water, they will separate out, for there will be high interfacial tension particularly as the surface area between the two phases in an emulsion is high. Emulsifiers are used to stabilise such systems.

In an oil and water emulsion, for example, an emulsifier with a hydrophilic and a hydrophobic group is used. The hydrophilic group

450

attaches itself to the water drops and gives them an essentially hydrophobic surface. Likewise, the hydrophobic group attaches to the oil droplets giving them a hydrophilic surface. The emulsifier acts as a go-between, making the oil and water much more nearly miscible.

c **Detergents** Detergents are substances with both hydrophilic and hydrophobic groups. Their detailed action is complex but, essentially, they enable water and fatty or oily substances to come together into emulsions which can be washed away. Soap is the traditional detergent, but it has been partially replaced in recent years by many synthetic detergents. Their hydrophobic group is usually a large alkyl group or an alkyl substituted benzene ring. The hydrophilic group is generally $-COOH$, $-SO_3H$, $-SO_3^-$, $-OSO_3H$, $-OSO_3^-$ or $-OH$. Teepol, $C_{10}H_{21}CH(CH_3)OSO_3^- Na^+$, is typical.

d **Froth flotation** This is the main method of separating the valuable metallic parts of an ore from the unwanted rock (gangue). The powdered ore is frothed up, by air bubbles, in a foaming aqueous solution containing a chemical known as a *collector*. This is a selective surfactant which coats the metallic particles, enabling air bubbles to stick to them and repelling water. The air bubbles make the particles buoyant so that they rise to the top and are carried away in the froth, whilst the gangue gets wet and sinks to the bottom.

2 Adsorption by solids

Solids with porous structures, e.g. charcoal and silica gel, can be very effective adsorbents. Activated charcoal, i.e. charcoal which has been heated in steam, is particularly adsorptive and is used, for example, in gas masks and in decolorising many coloured solutions.

a **Physisorption and chemisorption** Benzene vapour is adsorbed rapidly by charcoal in a reversible process. Increase in pressure or decrease in temperature cause increased adsorption, but decrease in pressure or increase in temperature releases adsorbed gas. The forces between the solid and the gas or vapour are thought to be van der Waals' forces, and this type of adsorption is known as physisorption or physical adsorption. The more easily liquefiable a gas the more readily will it be adsorbed.

Chemisorption involves the formation of what might be called a surface compound. As it corresponds to something much closer to a chemical change than physisorption it is, commonly, irreversible. In many examples, efforts to free the adsorbed gas produce a definite compound. Oxygen adsorbed on carbon or tungsten, for example, is liberated as oxides of carbon or tungsten, and carbon monoxide adsorbed on tungsten is liberated as a carbonyl.

b Adsorption isotherms The extent of adsorption at a particular temperature can sometimes be related to a mathematical expression known as an adsorption isotherm. Some of these are empirical, others have a more theoretical background, but none meet all known cases.

The empirical Freundlich isotherm applies to many cases of adsorption of gases, or of solutes from solutions. If x is the mass of gas or solute adsorbed by m gram of adsorbent at a pressure, p, or a concentration, c, then

$$x/m = kp^n \qquad \text{or} \qquad x/m = kc^n$$

$$\lg(x/m) = \lg k + n\lg p \qquad \lg(x/m) = \lg k + n\lg c$$

where k and n are constants. The isotherm can be tested by plotting $\lg(x/m)$ against $\lg p$ or $\lg c$.

Langmuir argued that adsorption would only take place until the adsorbing surface was completely covered with a unimolecular layer of adsorbed gas. He considered a kinetic balance between gas molecules striking the surface, and adsorbed molecules being evaporated from the surface. The resulting Langmuir isotherm can be written as

$$pm/x = \alpha + \beta p$$

where α and β are both constants; it can be tested by plotting pm/x against p.

3 Ion-exchange resins

In recent years, many synthetic resins have been made which function as ion-exchangers. In effect, a solid releases one ion and adsorbs another. When the ions exchanged are cations, the material is known as a *base* or *cation exchanger*; when anions, as an *acid* or *anion exchanger*.

Cation exchangers are high polymers containing acidic groups such as $-HSO_3$, $-COOH$ or $-OH$; anion exchangers contain basic groups such as secondary, tertiary or quaternary amine groups.

a Purification of water The first, often naturally occurring, ion-exchange materials were used for softening water by removing $Ca^{2+}(aq)$ and/or $Mg^{2+}(aq)$ ions. The insoluble sodium salt of a cation exchanger was used, e.g.

$$2NaR(s) + Ca^{2+}(aq) \rightleftharpoons 2Na^+(aq) + CaR_2(s)$$

where R is a resin. To regenerate the sodium salt when it has been completely converted into the calcium salt it is only necessary to treat it with a solution of sodium chloride.

Water can, however, be purified much further, and all dissolved salts removed, by using both a cation and an anion exchanger. Such purified water is known as deionised or demineralised water, the process of purification being referred to as deionisation or demineralisation.

In the two-stage process, water is first passed through a column of a cation exchanger in the form of an acid; any cations in the water are replaced by $H^+(aq)$ ions.

$$HR(s) + X^+(aq) \rightleftharpoons H^+(aq) + XR(s)$$

The partially purified water is then passed through an anion exchanger in the form of a base. This replaces any anions by OH^- ions,

$$HOR(s) + Y^-(aq) \rightleftharpoons OH^-(aq) + YR(s)$$

When the resins are exhausted they can be regenerated; the cation exchanger with acid, and the anion exchanger with alkali.

A one-stage process can be used by passing water down a single column containing a mixture of cation and anion exchangers. For regeneration purposes the two exchangers have to be separated, but this is facilitated by using an anion exchanger with a lower density than the cation exchanger.

b Medical uses Removal of excess sodium salts from body fluids can be achieved by giving the patient a suitable ion-exchanger to eat. Weakly basic anion exchangers can also be used to remove excess acid in the stomach and thus relieve indigestion, and blood can be prevented from clotting by exchanging the $Ca^{2+}(aq)$ ions which it contains, and which cause it to clot, for $Na^+(aq)$ ions.

c Ion-exchange membranes Ion-exchange materials supported on paper, fibre or some other material can be used as membranes through which only anions or cations can pass. They act as 'ionic sieves'.

A typical use of such membranes is in an electrical demineralisation of water in a cell as shown in Fig. 181. Both anions and cations can pass out of the central compartment of the cell, but neither anions nor cations can pass into this compartment except by slow diffusion. Electrolytes can, therefore, be removed from the central compartment and, in this way, salts can be removed from sea-water and other solutions.

Natural membranes such as those in plant cells and nerve fibres function in the same way as ion-exchange membranes.

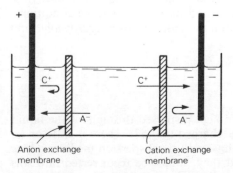

Anion exchange membrane

Cation exchange membrane

Fig. 181 Use of ion-exchange membranes in an electrolytic cell.

4 Colloidal systems

The foundations of colloidal science were laid down by Thomas Graham (1805–1869). He introduced most of the terms still used in describing a colloidal system, in which a *disperse phase* is distributed, in small particles, throughout a *dispersion medium*. Different types of colloid can be made as summarised below.

dispersion medium	disperse phase	type of colloid
gas	liquid	aerosol, fog, mist, cloud
gas	solid	aerosol, smoke, dust
liquid	gas	foam, e.g. froth on beer, soap suds, whipped cream
liquid	liquid	emulsion, e.g. milk, rubber latex, salad dressing, haircream
liquid	solid	sols or gels or pastes, e.g. some paints, fruit jellies
solid	gas	solid foam, e.g. pumice, meringues
solid	liquid	solid emulsion, e.g. butter
solid	solid	solid sol or solid gel, e.g. some coloured glasses, wings of butterflies, pearls

The small particles of the disperse phase are between 1 nm and 1000 nm in size, and they may consist of agglomerates of small particles, or of single particles such as large molecules of rubber, proteins, cellulose or synthetic polymers. It is, in the main, the large surface area of the tiny particles that accounts for the peculiar and distinctive properties of colloids.

The treatment in this chapter is mainly limited to a solid disperse phase in a liquid dispersion medium. Such systems can be subdivided into *lyophobic* (solvent-hating) colloids and *lyophilic* (solvent-loving) colloids, the latter giving rise to gels and pastes under some conditions. When water is the dispersion medium the terms *aquasol*, *hydrophobic* and *hydrophilic* may be used.

5 Lyophobic sols

The solid particles in these sols are kept dispersed throughout the liquid phase because they are electrically charged (p. 457). There is little or no interaction between the disperse phase and the dispersion medium. Once the solid particles have coagulated they cannot be reconverted into sols again simply by remixing with the dispersion medium. For this reason, lyophobic sols are sometimes called *irreversible* sols; they may also be referred to as *suspensoid* sols.

They can be made either by condensation methods in which atoms or molecules are allowed to build up into colloidal size, or by dispersion methods in which large particles are broken down.

a Condensation methods Many reactions in which one of the products is an insoluble solid can be carried out in such a way that the solid builds up into colloidal size particles. Care is necessary, but many stable sols can be made. A sulphur sol, for example, from sodium thiosulphate and acid solutions; an iron(III) hydroxide sol from iron(III) chloride solution and boiling water; an arsenic(III) sulphide sol from solutions of arsenic(III) oxide and hydrogen sulphide; a gold sol by reducing gold(III) chloride solution with methanal; and a silver sol by reducing ammoniacal silver nitrate with dextrin.

Sols may also be made by exchange of solvent. Alcoholic solutions of sulphur or resin, for example, will form sulphur or resin sols when added to water.

b Dispersion methods Colloidal particles of varying size can be made, possibly commercially, by powdering coarse particles in a *colloid mill*. Metallic sols, e.g. Pt, Ag and Au, can be made by *Bredig's* method, in which an arc is struck between two electrodes of the metal concerned immersed in the dispersion medium.

Large particles of a precipitate can sometimes be broken down into colloidal particles by adding an electrolyte. The precipitate may adsorb one of the added ions and split off electrically charged, colloidal size particles. Silver chloride sols, for example, can be made by adding silver nitrate or potassium chloride to silver chloride precipitates. Similarly, sulphide precipitates can be broken down by adding hydrogen sulphide.

The process is known as *peptisation*, the name originating from the fact that egg white, partially coagulated by heating, can be reformed as a colloidal solution by adding pepsin (extracted from pig's or sheep's stomachs) and a trace of hydrochloric acid.

6 Lyophilic sols

In lyophilic sols there is a strong interaction between the disperse phase and the dispersion medium, the latter being strongly adsorbed by the large surface area of the former. It is the adsorbed layer of the dispersion medium on the disperse phase which stabilises the sols, any electrical charge which may exist playing a smaller part. If the disperse phase is separated from the dispersion medium, the sol can be remade simply by remixing. That is why lyophilic sols are called *reversible* sols. They may also be called *emulsoid* sols because they are not unlike emulsions.

Lyophilic sols can be made from substances containing large molecules simply by mixing them with the dispersion medium under suitable conditions. Gelatine, agar-agar, starch, corn flour, and custard powder, for example, can be mixed with hot water. Rubber can be mixed with benzene to give so-called rubber solution. Nitrocellulose gives a sol with propanone.

7 Gels and pastes

A gel, like a fruit jelly, has a certain rigidity but can be deformed very easily by very weak forces. A paste is not unlike a gel but generally contains more solid particles which give it more rigidity. Examples are provided by starch paste, Plasticine, putty, wet cements, oil paints, soils, clay and mud.

Gels of gelatine, agar-agar and starchy materials can be formed by cooling the hot sols. The setting of jam involves gel formation aided by the presence of pectin which occurs in many fruit rinds.

Gels may also be made by chemical reaction under the right conditions. Mixing saturated solutions of barium thiocyanate and manganese(II) sulphate gives a barium sulphate gel; mixing saturated solutions of calcium chloride and sodium carbonate gives calcium carbonate gel; adding sulphur dichloride oxide to dry sodium 2-hydroxybenzoate gives sodium chloride gel. In all these examples ionic solids are being formed as gels.

Sodium silicate solution (diluted waterglass) reacts with dilute hydrochloric acid to form silicic acid, $SiO_2.xH_2O$, in the form of a gel; it is known as silica gel. The gel loses water on heating, and the resulting hard, brittle mass can be broken up and used as a dehydrating agent, as a catalyst, and as a catalyst support. The product is very stable to heat and to attack by most chemicals.

8 Optical properties of lyophobic sols

Particles of a certain size can scatter light, and that is why a beam of sunlight is visible in a dusty room, or the beam of a headlight is particularly visible in fog. A beam of light shows up, similarly, when passed through a lyophobic sol, the phenomenon being known as the *Tyndall effect*. The same beam is not scattered on passing through a dust-free solution (Fig. 182). This is one way of distinguishing between a true solution and a lyophobic sol.

If the scattered light is viewed from the side through a microscope, points of light originating from the individual colloidal particles can be seen. Such an arrangement is known as an ultramicroscope, and was first used in studying colloids by Zsigmondy. The ultramicroscope shows that

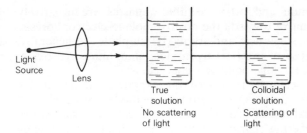

Fig. 182 The Tyndall effect.

the particles in a lyophobic sol undergo Brownian motion (p. 30), and it can be used to obtain information about the numbers of particles and about their shape and size.

9 Electrical properties of lyophobic sols

The stability of a lyophobic sol is due, predominantly, to the fact that the colloidal particles are electrically charged and therefore repel each other. The charge may be negative or positive and it is thought to originate in the adsorption of ions from the dispersion medium by the disperse phase. Many lyophobic sols are, indeed, more stable in the presence of a trace of electrolyte.

a Electrophoresis The charged particles in a lyophobic sol move under the influence of an applied potential difference, and this is known as electrophoresis. The movement can be observed by an ultramicroscope, or by observing the boundary between a sol and its dispersion medium (Fig. 183). The direction of movement can be used to discover the nature of the charge on the sol.

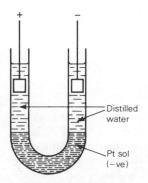

Fig. 183 Electrophoresis.

For example, platinum and other metallic aquasols are negatively charged; the particles move towards the positive pole in electrophoresis. Aquasols of starch, clay, silicic acid and arsenic(III) sulphide are, generally, negatively charged; those of iron(III) hydroxide, aluminium hydroxide, and haemoglobin are, generally, positively charged.

Tiselius, a Swedish scientist, improved the experimental techniques used in studying electrophoresis, and modern methods are capable of separating complex mixtures of proteins or amino acids. The separation is dependent on the differing charge of different proteins or amino acids in solution. The separation can be brought about between two electrodes immersed in a buffer solution on both sides of a solution of the mixed proteins or, alternatively, the solution may be subjected to an applied electrical field whilst adsorbed on paper. In the second method electrophoresis is combined with paper chromatography, a combination which has provided a very powerful means for separating amino acids.

b Electro-osmosis As a colloidal solution is, as a whole, electrically neutral, the charge on the colloidal particles must be balanced by an equal and opposite charge on the dispersion medium. Something like a Helmholtz double layer (p. 405) is probably set up.

In electrophoresis, the charged disperse phase moves whilst the dispersion medium remains stationary. If, however, the disperse phase is fixed, then the dispersion medium will move under an applied potential difference. This is known as electro-osmosis. It can be demonstrated

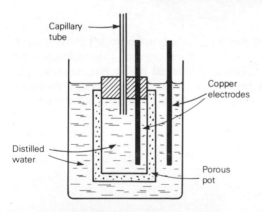

Fig. 184 Electro-osmosis.

using the arrangement shown in Fig. 184. The porous pot provides a stationary medium which is colloidal-like in nature by virtue of its large surface area. Application of a potential difference in one direction causes a flow of water from inside to outside the pot. If the potential difference is reversed, the flow is from outside to inside.

Electro-osmosis can be used to limit rising damp in old brickwork. A conductor is embedded in the bricks and a potential difference is maintained between it and the earth.

c Streaming or sedimentation potential Motion is produced in electrophoresis and electro-osmosis by the application of an external potential difference. In the reverse effects, a potential difference can be caused by motion. If a liquid is made to pass through a capillary tube or through the capillary pores in a porous pot, a potential difference, known as the *streaming potential*, is set up. It can be detected by placing electrodes at the inlet and outlet ends of the capillary. Similarly, a potential difference, known as the *sedimentation* or *centrifugation potential*, is set up if solid particles are made to pass through a liquid by sedimention or by the application of a centrifugal field.

All these effects in which motion and potential difference are related are known, collectively, as *electrokinetic* phenomena.

d Coagulating effects of electrolytes A negatively charged ion causes the coagulation of a positively charged colloidal particle, and vice versa. For equal concentrations, the coagulating effect depends on the valency of the ion causing the coagulation. This is the *Hardy–Schulze rule*. Al^{3+}(aq) ions, for instance, will coagulate negatively charged sols much more effectively than bivalent ions, and bivalent ions are much more effective than monovalent ones. Similarly SO_4^{2-}(aq) ions coagulate a positive colloid more readily than Cl^-(aq) ions. Mixing negative and positive colloids also causes mutual coagulation.

The coagulation of colloidal particles is thought to be due to adsorption of the coagulating ion. Such adsorption will decrease the electrical charge on the colloidal particles until, at the *isoelectric point*, the particles carry no charge and coagulate. That such a mechanism is likely is shown by the fact that the electrophoretic velocity of a colloidal particle decreases as an electrolyte is added until it becomes zero at the isoelectric point.

Examples in which electrolytes are used to bring about coagulation of sols are provided by:

i the use of alum (a cheap source of Al^{3+} ions) in treating sewage and dirty swimming-bath water, and in styptic pencils which are used to make blood clot,

ii the formation of deltas when the colloidal particles present in fresh river water are coagulated by the higher concentration of electrolytes in sea-water,

iii the coagulation of rubber latex by adding methanoic (formic) acid,

iv the curdling of milk by the 2-hydroxypropanoic (lactic) acid formed as the milk goes sour.

e Electrostatic precipitation of smokes or mists If a smoke or a mist is passed between highly charged plates, the colloidal particles are deposited. The process can be used industrially.

f Adsorption indicators Addition of silver nitrate solution to a sodium chloride solution produces a precipitate of silver chloride which is partly in colloidal form. As soon as any excess silver nitrate is added to the mixture, the precipitate preferentially adsorbs Ag^+ ions and becomes positively charged. If a dye, such as fluorescein, is present, the positively charged precipitate will adsorb the fluorescein anion. As a result, the precipitate becomes pink in colour, and sometimes coagulates.

Because of this adsorption, fluorescein can be used as an adsorption indicator in the titration of solutions of chlorides, bromides, iodides and thiocyanates against silver nitrate. When first added to the solution in the conical flask, the fluorescein imparts a yellowish green colour. As silver nitrate solution is run in, a white precipitate of silver salt forms and, at the end point, this precipitate turns sharply pink, and may coagulate.

Eosin, dichlorofluorescein and diphenylamine blue are also commonly used as adsorption indicators.

10 Diffusion of lyophobic sols. Dialysis

A solute will diffuse from a true solution into the pure solvent or into a weaker solution (p. 429). A layer of water, for example, can be placed, with care, on top of a copper(II) sulphate solution. On standing, the blue colour of the copper(II) sulphate will diffuse upwards into the water layer.

The rate of diffusion of colloidal particles is very, very much slower because they are bigger than true solute particles and the difference is used, in dialysis, to separate sols from true solutions.

If, for example, a mixture of starch and sodium chloride in water is placed in a dialyser (Fig. 185), the $Na^+(aq)$ and $Cl^-(aq)$ ions will pass through the membrane into the water whilst the starch will remain.

Patients suffering from kidney failure can be connected to a kidney machine in which a dialyser removes unwanted, non-colloidal substances, e.g. urea, from their blood stream.

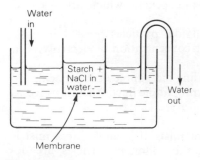

Fig. 185 Dialysis.

11 Properties of lyophilic colloids

Lyophilic colloids resemble lyophobic sols so far as diffusion and dialysis are concerned, and, if they are electrically charged, they will, also, undergo electrophoresis and electro-osmosis. They differ from lyophobic sols, however, in other ways.

a Optical properties In a lyophobic sol the scattering of light by the colloidal particles can only be observed because these particles have a different refractive index from that of the dispersion medium. A lyophilic sol is much more optically homogeneous, the refractive index of the colloidal particles with their adsorbed layer, being much the same as that of the dispersion medium. Consequently, lyophilic colloids do not exhibit the Tyndall effect at all markedly.

b Coagulation by electrolytes Low concentrations of electrolytes will not coagulate lyophilic sols, and the Hardy–Schulze rule does not apply. Addition of a lot of solid electrolyte might coagulate a lyophilic sol, probably because the hydration of the added ions lowers the number of water molecules available for adsorption on the colloidal particles. Soap, for example, can be coagulated (salted out) of a colloidal solution by adding solid salt, and ammonium sulphate is used for salting out proteins.

c Viscosity The particles in most lyophobic sols are approximately spherical in shape and there is no interaction between the particles and the dispersion medium. As a result, the viscosity of a lyophobic sol is only very slightly greater than that of the pure dispersion medium. That of a lyophilic sol is, however, much bigger than that of the dispersion medium due to the marked interaction between the disperse phase and the dispersion medium. The viscosity rises, too, as gels form.

Non-drip or thixotropic paints exhibit unusual changes in viscosity. At rest they have quite high viscosity but this is lowered by the shearing forces applied by a brush, enabling the paint to be spread easily. The shearing forces temporarily alter the structure of the colloidal system.

d The protective action of lyophilic colloids The coagulating effect of an electrolyte on a lyophobic sol is markedly decreased by the addition of a lyophilic colloid to the lyophobic sol.

It is supposed that the lyophilic material is adsorbed by the colloidal particles of the lyophobic sol, the new particle being, in effect, lyophilic. The lyophobic sol treated in this way is said to be protected, and the lyophilic colloid is known as a protective colloid.

Various methods have been used to measure the protective action of different materials quantitatively, the commonest being the use of the *gold number*, introduced by Zsigmondy. This is defined as the mass in milligram of the protective colloid which must be added to $10 \, cm^3$ of a specially prepared red gold sol just to prevent a change to violet (i.e. an increase in particle size) on the addition of 1 cm^3 of 10 per cent sodium chloride solution. The gold number of gelatine lies between about 0.005 and 0.01; that of potato starch is about 25.

The use of protective colloids to stabilise colloidal systems is widespread. Typical examples are provided by the addition of egg yolks (egg albumen) to olive oil and vinegar in making mayonnaise, and the use of gelatine in ice cream to prevent the formation of small ice crystals.

12 Sedimentation. The ultracentrifuge

Solid particles suspended in a liquid will fall or rise depending on the density of the particles. The fall of suspended particles is known as sedimentation; the rise, as creaming.

When the particles concerned are very small, as in colloidal solutions, the Brownian motion will be sufficient to prevent any great sedimentation under the influence of gravity. By using a centrifuge, however, the force exerted on the particles can be greatly increased. In order to study the sedimentation of colloidal solutions, Svedberg developed an ultracentrifuge with a rotational speed high enough to produce forces over 10^5 times greater than the force of gravity.

Svedberg used the ultracentrifuge to measure rates of sedimentation for colloidal particles and, by applying a form of Stoke's law (p. 47), developed one of the most reliable methods for measuring the relative molecular masses of colloids. The ultracentrifuge can also be used to separate colloids.

Questions on Chapter 37

1 The amount of ethanedioic (oxalic) acid, in moles, adsorbed by 5 g of activated charcoal varies with the equilibrium concentration of acid.

acid adsorbed/mol	0.29	0.60	0.75	0.90	1.05
equil. conc. acid/mol dm^{-3}	0.030	0.080	0.115	0.175	0.260

Use these figures to show the validity of the Freundlich isotherm and to find the value of the constant n. Do the figures also fit the Langmuir isotherm?

2 The volumes of nitrogen, at s.t.p. adsorbed by 1 g of activated charcoal at the same temperature but different pressures are tabulated as follows:

vol. of N_2	0.987	3.04	5.08	7.04	10.31
pressure	3.93	12.98	22.94	34.01	26.23

Do these figures agree with the Langmuir isotherm?

3 How would you attempt to establish, experimentally, that the amount of gas adsorbed by charcoal was proportional to the surface area of the charcoal?

4 $500 \, cm^2$ of a water surface is found to be covered by 0.106 mg of octadecanoic (stearic) acid, which has a relative molecular mass of 284 and a density of $0.85 \, g \, cm^{-3}$. Assuming a close-packed and unimolecular film, estimate the cross-sectional area of an octadecanoic acid molecule, and the thickness of the surface film. What estimate would you make, from this data, of the film thickness to be expected from hexacosanoic (cerotic) acid?

5 Describe the experiments you would carry out to get a numerical measurement of the relative adsorptive powers for ammonia gas of a sample of powdered charcoal and a sample of powdered sulphur.

6 Give an account of the use of adsorption indicators.

7 Distinguish between the meaning of the terms adsorption, absorption, chemisorption and physical adsorption.

8 Explain why (a) drops of water are spherical (b) alkalis are good emulsifiers for grease (c) ducks cannot float in water containing too much detergent (d) water will spread on clean glass and wet it, but it will not do so on a waxy or greasy surface.

9 What is the total surface area of a cube of side 1 cm? What is the surface area when the same cube is divided into cubes of side $10^{-6} \, cm$?

10 If the average diameter of gold particles in a sol containing 1 g of gold per dm^3 is 4 µm, and the density of gold is $20 \, g \, cm^{-3}$, calculate the approximate number of gold particles per dm^3. What osmotic pressure would you expect this sol to have, and what would be its freezing point?

11 If $1 \, cm^3$ of gold is completely subdivided into small spheres of radius 2.5 nm, how many spheres will there be and what will be the total surface area?

12 How would you prepare colloidal solutions of (a) arsenic(III) sulphide (b) a metal such as gold or platinum? Either describe a method by which the sign of the electrical charge on the particles of a colloid could be determined, or indicate briefly the part played by the electric charge in colloidal phenomena. (OCS)

13 Describe in detail how you would prepare (a) one lyophilic (b) one lyophobic colloidal solution. What experiments would you use to show three essential differences in character between the two colloidal solutions whose preparation you describe? (OCS)

14 What will be the osmotic pressure at $25 \, °C$ of a solution containing $20 \, g \, dm^{-3}$ of a protein of relative molecular mass 69×10^3?

15 What would you observe if a mixture of a solution of starch and potassium iodide was separated from chlorine water by cellophane in a dialyser?

16 Describe and explain two ways in which the nature of the electrical charge on the particles in an iron(III) hydroxide sol can be determined.

17 Coagulation of colloids is caused by heat, salt solutions, acids or alkalis, electrical precipitation and mixing with an oppositely charged colloid. Illustrate this statement, using specific examples. Is the statement invariably true?

18 Write short notes on the following: peptisation, protective action, dialysis, Brownian motion.

19 What is the Tyndall effect? Why are motor car headlights less effective in a fog than on a clear night? Why are fog lights usually yellow?

20 Compare and contrast, in three columns, the general properties of (i) a solution of sucrose in water (ii) a suspension of calcium carbonate in water   (iii) an aquasol of sulphur.

21 How would you obtain crystals of pure sodium chloride from a mixture of sodium chloride and starch?

Revision questions

1 Comment on the following statements. (a) Volatility is often said to be a criterion of covalency, but ammonium chloride volatilises and diamond does not (b) Silicates of metals form the greater part of the Earth's crust, but metals are hardly ever extracted from silicates on an industrial scale (c) Lead chloride is less soluble in normal hydrochloric acid than in water, but it is much more soluble in concentrated hydrochloric acid than in either (d) It is difficult to convert rhombic sulphur directly into monoclinic sulphur, but a sample of monoclinic sulphur changes spontaneously in the cold into rhombic sulphur. (OC)

2 Explain the meaning of the terms italicised and thereby elucidate and illustrate the following statements. (a) Solid *allotropes* can be either *monotropic* or *enantiotropic* (b) *Osmotic pressure* is the pressure necessary to prevent *osmosis* (c) A *catalyst* cannot affect the *equilibrium point* of a reaction.

3 Distinguish between the following terms, either giving examples to illustrate them or stating the units in which the quantities concerned could be measured: (a) degree of dissociation and dissociation constant (b) solubility and solubility product (c) osmosis and dialysis (d) conductivity and molar conductivity (e) drying and dehydration (f) isomer and isotope. (OC)

4 Explain clearly, with examples, the meaning of the terms diffusion, solubility product, common ion, efflorescence, stability constant.

5 Explain the meaning of the following statements and describe how you would verify any one of them experimentally. (a) Sulphuric acid is dibasic (b) Methanoic acid is stronger than ethanoic (c) Hydrogen fluoride is associated in the vapour phase.

6 What is meant by the relative molecular mass of a substance? Discuss the factors which determine it. Explain how you would investigate experimentally the values for three of the following (a) phosphorus trioxide (b) hydrogen chloride (c) ammonium chloride (d) mercury (e) benzoic acid. Mention any interesting conclusions to which your results might lead.

7 Explain briefly the difference between: (a) an atom and its ion (b) allotropes and isotopes (c) group and period (in the periodic table) (d) electrolysis and electrolytic dissociation (e) negative catalyst and catalyst poison. (JMB)

8 Suggest explanations for the following facts. (a) Very small solid particles undergo erratic motion when suspended in gases or liquids, but large ones do not (b) A wet substance dries more rapidly in a vacuum desiccator than in an ordinary one, even though the same drying agent is used (c) A gas cools on expanding adiabatically (d) A liquid cools when it evaporates freely.

465

9 Explain why sodium chloride: (a) dissolves in water (b) lowers the freezing point of water (c) is less soluble in solutions of hydrochloric acid than in water (d) coagulates a colloidal solution of iron(III) hydroxide.

10 Explain concisely the difference between three of the following (a) a weak and a strong electrolyte (b) electrovalency and covalency (c) cooling by adiabatic expansion and the Joule–Thomson effect (d) dissociation and thermal decompositions (e) monotropy and enantiotropy. (OCS)

11 Comment on, illustrate, or explain the following statements. (a) Hydrogen chloride does not obey Henry's law (b) Metals can displace hydrogen from sodium hydroxide (c) A strong electrolyte does not obey Ostwald's dilution law (d) Some allotropes differ in chemical properties, others in physical properties only. (OC)

12 Comment on, illustrate, or explain four of the following statements. (a) A catalyst does not alter the point of equilibrium in a reversible reaction (b) Combustion is not necessarily accompanied by flame (c) A colloidal particle carries an electric charge (d) Amorphous carbon absorbs gases easily, but diamond will not (e) The end point of an acid–alkali titration is not always at neutrality. (OC)

13 Define (a) atomic number (b) Avogadro constant (c) the gas constant, R (d) Faraday constant. How could the value of one of these be obtained? (OS)

14 Explain fully what you understand by three of the following (a) a negative catalyst (b) the heat of formation of a metallic oxide (c) the vapour pressure of a liquid (d) the valency of nitrogen in ammonium chloride. (OC)

15 Explain the meaning of the following terms: supercooling, superheating, supersaturation, metastable state, triple point.

16 Give examples in which (a) metals (b) oxides (c) enzymes (d) gases are used as catalysts.

17 Give an account of either clathrates, layer lattice structures, graphitic compounds, or molecular seives.

18 Ammonium sulphate can be made for use as a fertiliser by stirring a solution of ammonium carbonate with finely divided calcium sulphate. At equilibrium, the mixture is saturated with both calcium sulphate and calcium carbonate, with solubility products of 2.3×10^{-4} and 4.8×10^{-9} $mol^2\,dm^{-6}$ respectively. What is the approximate ratio of the concentration of ammonium sulphate to that of ammonium carbonate in the mixture?

19 If a solution of sodium sulphate is stirred with excess barium carbonate until equilibrium is established, what percentage of it will be converted into sodium carbonate? The solubility products of barium carbonate and barium sulphate are 8.0×10^{-9} and 1.1×10^{-10} $mol^2\,dm^{-6}$ respectively.

20 Under what circumstances are measurements of the physical properties of solutions suitable for the determination of relative molecular masses? 9.3 g of *p*-cresol in 1000 g of benzene depress the freezing point of benzene by 0.42 °C. 200 g of *p*-cresol in the same mass of benzene depress the freezing point 5.0 °C. Comment on these observations. (OE)

21 Discuss the effects of changes of temperature and pressure on the position of equilibrium in a gaseous system. How does the situation change if a heterogeneous system is considered?

22 What do you understand by a 'perfect gas'? Predict the experimental conditions required for a real gas to approach 'perfect' behaviour. (OE)

23 Explain carefully why heat is evolved when many electrolytes dissolve in water although it would be expected that the work done in separating their ions would be revealed by the absorption of heat.

24 Write short notes on any four of the following: free energy, steam distillation, constant boiling mixtures, critical solution temperatures, fractionating towers, eutectic mixtures, entropy and activation energy.

25 A solution of iodine in carbon tetrachloride gives, on boiling, a mixture of iodine vapour which is purple, and carbon tetrachloride vapour which is colourless. The amount of iodine dissolved is shown by the colour of the solution. Design an apparatus using this solution which illustrates the functioning of a fractionating tower.

26 Write short notes on the following: ionic product for water, solubility product, buffer solutions, hydrolysis of salts, radioactivity.

27 The isotopic composition of magnesium is found to be ^{24}Mg (77.4%), ^{25}Mg (11.0%) and ^{26}Mg (11.6%). Assuming the isotopic mass to be equal to the mass number, calculate the relative atomic mass of magnesium. In most cases the chemical relative atomic mass of an element is the same from whatever source it comes. What does this imply about the origin of the elements? In which element is there a variation in relative atomic mass and why?

28 In what ways does radioactive disintegration differ from ordinary chemical reactions? Explain briefly what is meant by: (a) a radioactive series (b) radioactive equilibrium (c) half-life period (d) a curie.

At each point marked by an asterisk a number has been omitted from the following nuclear equations which represent the net changes occurring in the ultimate disintegration of three radioactive elements P, Q and R. Complete the equations by inserting the appropriate numbers and assign P, Q and R to their correct groups in the periodic table, briefly explaining how you deduce the necessary information.

$$^{232}_{90}P = {}^*_*Pb + 6{}^4_2He + 4{}_{-1}^{0}e$$

$$^{234}_{92}Q = {}^{206}_{*}Pb + *{}^4_2He + *{}_{-1}^{0}e$$

$$^*_*R = {}^{207}_{*}Pb + 4{}^4_2He + 2{}_{-1}^{0}e$$

Describe one simple application of a radioactive isotope to the solution of a chemical problem. (WS)

29 Derive the mathematical expression for Ostwald's dilution law for an electrolyte, one molecule of which ionises to give *n* cations and *m* anions.

30 Explain the following. (a) The pH of 0.1 M hydrochloric acid is 1.0 but that of 0.1 M ethanoic acid is 2.75 (b) The cations discharged when separate aqueous solutions of copper sulphate and sodium sulphate are electrolysed between platinum electrodes are copper and hydrogen respectively (c) Pure water is acid to phenolphthalein but alkaline to methyl orange (d) Strong electrolytes do not obey Ostwald's dilution law.

31 Give and explain a definition of an acid (a) making use of the concept of electrons (b) without the use of this concept. Are there any substances covered by one of the definitions you give which would not be covered by the other? Which do you consider to be the better definition, and why? (OCS)

32 Using the kinetic theory and the concept of reaction velocity, explain the following facts. (a) When nitrogen oxide is heated to 1000 °C it is almost completely decomposed into nitrogen and oxygen (b) When oxygen and nitrogen are heated to 3000 °C the mixture of gases is found to contain about 40 per cent of nitrogen oxide (c) Hydrogen and oxygen combine to form water with the evolution of considerable heat, but the mixture of gases needs heating before combination takes place. (OCS)

33 Describe, with a sketch or diagram of the apparatus, how you would measure two of the following (a) the partition coefficient of iodine between potassium iodide solution and benzene (b) the molar conductivity of a 0.001 M solution of ethanoic acid (c) the relative molecular mass of urea. (OCS)

34 Describe clearly experiments you would carry out to demonstrate three of the following (a) that hydrogen diffuses through a porous wall more rapidly than oxygen (b) that the osmotic pressure of sugar increases with rise of temperature (c) that the decomposition of hydrogen peroxide in solution is a first-order reaction (d) that the constant boiling mixture of hydrogen chloride and water is a mixture and not a compound (e) that chrome alum and potash alum are isomorphous. (OS)

35 Suggest methods of investigating one of the following (a) the rate of decomposition of ammonium nitrite solution at its boiling point (b) the equilibrium between hydrogen, iodine and hydrogen iodide in the vapour phase (c) the formula of the copper complex in solution of copper(II) sulphate containing excess ammonia. (OS)

36 Explain, with examples where possible, what you understand by five of the following: Avogadro constant, active mass, molar conductivity, electrode potential, buffer solution, normal salt. (OC)

37 How would you measure (a) the degree of dissociation of ethanoic acid in water (b) the degree of association of ethanoic acid in benzene?

38 Many fundamental chemical definitions have had to be modified in the last hundred years. Give some illustrative examples.

39 Explain what is meant by the following statements. (a) The gas constant R is $8.314 \, \text{J K}^{-1} \, \text{mol}^{-1}$ (b) The ionic product of water at 25 °C is $1.0 \times 10^{-14} \, mol^2 \, dm^{-6}$ (c) The hydrolysis constant of sodium acetate is $5.5 \times 10^{-10} \, mol \, dm^{-3}$ at $25 \, °C$.

Calculate (i) the specific heat capacities of argon (relative atomic mass = 40) (ii) the dissociation constant of ethanoic acid at $25 \, °C$ (iii) the pH value of a 0.1 M solution of sodium ethanoate. (W)

40 Explain the following (a) The relative atomic mass of nickel (atomic number = 28) is less than that of cobalt (atomic number = 27) (b) The rate of simple reactions between gases increases with temperature much more rapidly than does the rate at which the molecules collide (c) Sulphur, when gently heated from room temperature, melts at $119 \, °C$, but when strongly heated melts at $113 \, °C$ (d) When hydrogen sulphide is passed into a solution of sodium arsenite a yellow precipitate is obtained, but when passed into dilute aqueous arsenious oxide a faintly opalescent yellow solution results (e) A Bunsen burner flame sometimes 'lights back' and burns at the bottom of the tube. (W)

41 Discuss critically the following statements (a) 'The heat of formation of water is $285 \, kJ$' (b) 'The properties of the elements are in periodic dependence on their relative atomic masses' (c) 'The rate of a reaction and accordingly the yield of product increases with temperature' (d) 'On electrolysing an aqueous solution containing several cations, the cation present in greatest concentration is preferentially discharged.' (W)

42 Suggest an experimental method for studying the dependence of the rate of the following reaction on the concentration of bromine in the presence of excess methanoic acid.

$$Br_2 + H.COOH = 2Br^- + 2H^+ + CO_2$$

43 Write an account of the work of any one famous chemist.

44 Do you prefer inorganic, organic or physical chemistry? Give your reasons.

45 'Ein Chemiker, der kein Physiker ist, ist gar nichts.' 'All the interesting scientific developments are now in the field of biochemistry.' Comment on one of these sayings.

Appendix I
SI Units

1 The basic SI units (p. 1)

a Length The metre is the length of 1 650 763.73 wavelengths in a vacuum of the radiation corresponding to the transition between the levels $2p_{10}$ and $5d_5$ of the krypton-86 atom.

b Mass The kilogram is the mass of the international prototype of the kilogram (a piece of platinum–iridium alloy kept at Sèvres in France).

c Time The second is the duration of 9 192 631 770 periods of the radiation corresponding to the transition between the two hyperfine levels of the ground state of the caesium-133 atom.

d Electric current The ampere is that constant current which, if maintained in two straight parallel conductors of infinite length, of negligible cross-section and placed 1 metre apart in a vacuum, would produce between these conductors a force equal to 2×10^{-7} newton per metre of length.

e Thermodynamic temperature The kelvin is the fraction 1/273.16 of the thermodynamic temperature of the triple point of water (pages 5 and 216).

f Amount of substance The mole is the amount of substance which contains as many elementary entities as there are atoms in 0.012 kilogram of carbon-12. The elementary entities must be specified and may be atoms, molecules, ions, electrons, other particles, or specified groups of such particles (p. 16).

2 The use of 'specific' and 'molar'

The word 'specific' before the name of a physical quantity means 'divided by mass'.

The word 'molar' means divided by amount of substance. Molar conductivity (p. 348) and molar extinction coefficient (p. 186) are exceptions to the rule, and the word molar should not be used to describe 1 M solutions.

The subscript m is used unless there is no ambiguity. Thus V_m is the molar volume (p. 5) and ΔH_m the molar enthalpy change (p. 246).

3 Derived quantities and units

a Volume The basic SI unit is m^3 but this is quite a large volume so that dm^3 and cm^3 are commonly used. The dm^3 is, sometimes, called a litre (l); the cm^3 is a millilitre (ml).

The temperature and pressure at which a gas volume is measured must always be quoted. It is often convenient to quote values at an arbitrarily chosen temperature and pressure (s.t.p). $273.15 \, K$ ($0 \, °C$) and $101.325 \, k \, Pa$ are chosen.

The molar volume of a gas is the volume occupied by 1 mol of the gas; it has a value of $22.413 \, 8 \, dm^3 \, mol^{-1}$ at s.t.p.

b Concentration Concentration means the amount of a solute per unit volume of solution. In a solution of A in B, for example, it is the amount of A divided by the volume of the solution; the base SI unit is $mol \, m^{-3}$. However, $mol \, dm^{-3}$ are much more commonly used and the symbol $[A]$ means the concentration of A in $mol \, dm^{-3}$. A solution with a concentration $1 \, mol \, dm^{-3}$ is said to be $1 \, M$; one with a concentration of $0.1 \, mol \, dm^{-3}$ is $0.1 \, M$; and so on.

Other methods of expressing the proportions of A and B in a mixture of the two are summarised below:

$$\text{mass concentration} = \frac{\text{mass of A}}{\text{volume of solution}} \quad (kg \, m^{-3} \text{ or } g \, dm^{-3})$$

$$\text{molality of A} = \frac{\text{amount of A}}{\text{mass of B}} \quad (mol \, kg^{-1} \text{ or } mol \, g^{-1})$$

$$\text{mol fraction of A} = \frac{\text{amount of A}}{\text{amount of A} + \text{amount of B}}$$

See also pages 195 and 205.

c Force The derived SI unit of force, the newton (N), is the force required to accelerate a mass of $1 \, kg$ at the rate of $1 \, m \, s^{-2}$. The gravitational force on a mass of $1 \, kg$ is approximately $10 \, N$ so that is the force required to hold a mass of $1 \, kg$.

d Pressure This is the force per unit area and the derived SI unit is $N \, m^{-2}$, which is called the pascal (Pa). For other units in common use, see page 4.

e Energy The derived SI unit of energy is the joule (J), with 1 joule being the energy needed to push against a force of $1 \, N$ for $1 \, m$ ($1 \, J = 1 \, Nm$). To raise a mass of $1 \, kg$ by $1 \, m$ requires approximately $10 \, J$ of energy. To raise the temperature of $250 \, cm^3$ of water by $1 \, °C$ requires about $1 \, kJ$. Molar energy, in $kJ \, mol^{-1}$ is often used.

The calorie (cal) is 4.184 J; it is the energy needed to raise the temperature of 1 g of water by 1 °C, so that 1 J raises the temperature of 1 g of water by 0.239 °C. The electron volt (eV) is the energy acquired by one electron when it is accelerated through a potential of 1 V ($1 \text{ eV} = 1.602\,189\,2 \times 10^{-19} \text{ J}$). If a mole of electrons is involved, the energy acquired is 96.486 kJ mol^{-1}. Spectral lines arise from transitions between two different energy levels; a line with a wave number of 1000 cm^{-1} corresponds to an energy change of 11.96 kJ mol^{-1} (p. 177).

4 Summary

The first six units are basic SI units. (Non SI units are marked with an asterisk.)

physical quantity	SI unit	definition of unit	other units used
amount of substance (n)	mole (mol)	mol	
electric current (I)	ampere (A)	A	
length (l)	metre (m)	m	km, cm, mm, nm, Ångström* ($1 \text{ Å} = 10^{-10} \text{ m}$)
mass (m)	kilogram (kg)	kg	$1 \text{ g} = 10^{-3} \text{ kg}$ $1 \text{ Mg} = 10^6 \text{ g} = 1 \text{ tonne*}$ atomic mass unit* (amu) = $1.660\,565 \times 10^{-27} \text{ kg}$
thermodynamic temperature (T)	kelvin (K)	K	$1 °C^* = 1 \text{ K}$ (interval)
time (t)	second (s)	s	minute*, hour*
concentration	mol m^{-3}	mol m^{-3}	mol dm^{-3} (see 3b above)
electric charge (Q)	coulomb (C)	A s	
electric conductance (G)	siemens (S)	m^{-2} kg^{-1} s^3 A^2 Ω^{-1}	
electric potential difference (V)	volt (V)	kg m^2 s^{-3} A^{-1} J A^{-1} s^{-1}	
electric resistance (R)	ohm (Ω)	m^2 kg s^{-3} A^{-2} V A^{-1}	
energy (E)	joule (J)	kg m^2 s^{-2}	1 calorie* (cal) = 4.184 J 1 electron volt* (eV) = $1.602\,189\,2 \times 10^{-19}$ J 1 erg* = 10^{-7} J 1000 cm^{-1} = 11.96 kJ mol^{-1} 1 kilowatt hour* (kW h) = 3.6×10^6 J

physical quantity	SI unit	definition of unit	other units used
force (F)	newton (N)	$\mathrm{kg\,m\,s^{-2}}$ $\mathrm{J\,m^{-1}}$	1 dyne* (dyn) $= 10^{-5}\,\mathrm{N}$
frequency (v)	hertz (Hz)	$\mathrm{s^{-1}}$	
magnetic flux (Φ)	weber (Wb)	$\mathrm{kg\,m^2\,s^{-2}\,A^{-1}}$	
magnetic flux density (B)	tesla (T)	$\mathrm{kg\,s^{-2}\,A^{-1}}$	
power (P)	watt (W)	$\mathrm{kg\,m^2\,s^{-3}}$ $\mathrm{J\,s^{-1}}$	
pressure (p)	pascal (Pa)	$\mathrm{kg\,m^{-1}\,s^{-2}}$ $\mathrm{N\,m^{-2}}$	1 atmosphere* $= 101\,325\,\mathrm{Pa}$ 1 mm of Hg* $= 1\,\mathrm{Torr*} =$ $101\,325/760\,\mathrm{Pa}$ 1 bar* $= 10^5\,\mathrm{Pa}$
volume (V)	$\mathrm{m^3}$	$\mathrm{m^3}$	$1\,\mathrm{dm^3} = 1\,\mathrm{litre\ (1\,l)}$ $1\,\mathrm{cm^3} = 1\,\mathrm{millilitre\ (1\,ml)}$

Appendix II
Physical constants

The values, in SI units (unless in brackets), of some important physical constants are summarised below, in alphabetical order.

constant	symbol	value
atomic mass constant	m_u	$1.660\,565\,5 \times 10^{-27}\,\text{kg}$
Avogadro constant	L	$6.022\,045 \times 10^{23}\,\text{mol}^{-1}$
Bohr magneton	μ_B	$9.274\,078 \times 10^{-24}\,\text{J}\,\text{T}^{-1}$
Boltzmann constant	$k\,(=R/L)$	$1.380\,662 \times 10^{-23}\,\text{J}\,\text{K}^{-1}$
charge on electron	e	$1.602\,189\,2 \times 10^{-19}\,\text{C}$
Faraday constant	F	$9.648\,456 \times 10^{4}\,\text{C}\,\text{mol}^{-1}$
ice point	T_{ice}	$273.15\,\text{K}$
mass of electron	m_e	$9.109\,534 \times 10^{-31}\,\text{kg}$
mass of neutron	m_n	$1.674\,954\,3 \times 10^{-27}\,\text{kg}$
mass of proton	m_p	$1.672\,648\,5 \times 10^{-27}\,\text{kg}$
molar gas constant	R	$8.314\,41\,\text{J}\,\text{K}^{-1}\,\text{mol}^{-1}$ $(1.987\,17\,\text{cal}\,\text{K}^{-1}\,\text{mol}^{-1})$ $(8.205\,75 \times 10^{-2}\,\text{dm}^3\,\text{atm}\,\text{K}^{-1}\,\text{mol}^{-1})$
molar volume of ideal gas at s.t.p.	V_m	$2.241\,38 \times 10^{-2}\,\text{m}^3\,\text{mol}^{-1}$
Planck constant	h	$6.626\,176 \times 10^{-34}\,\text{J}\,\text{s}$
Rydberg constant	R_∞	$1.097\,373\,177 \times 10^{7}\,\text{m}^{-1}$
triple point of water		$273.16\,\text{K}$
velocity of light	c	$2.997\,924\,580 \times 10^{8}\,\text{m}\,\text{s}^{-1}$

Appendix III
Relative atomic masses 1977

From information provided by the International Union of Pure and Applied Chemistry.

element	symbol	atomic number	relative atomic mass accurate	relative atomic mass approx.
actinium	Ac	89	227.027 8	227
aluminium	Al	13	26.981 54	27
americium	Am	95		243
antimony	Sb	51	121.75	121.5
argon	Ar	18	39.948	40
arsenic	As	33	74.921 6	75
astatine	At	85		210
barium	Ba	56	137.33	137.5
berkelium	Bk	97		249
beryllium	Be	4	9.012 18	9
bismuth	Bi	83	208.980 4	209
boron	B	5	10.81	11
bromine	Br	35	79.904	80
cadmium	Cd	48	112.41	112.5
caesium	Cs	55	132.905 4	133
calcium	Ca	20	40.08	40
californium	Cf	98		251
carbon	C	6	12.011	12
cerium	Ce	58	140.12	140
chlorine	Cl	17	35.453	35.5
chromium	Cr	24	51.996	52
cobalt	Co	27	58.933 2	59
copper	Cu	29	63.546	63.5
curium	Cm	96		247
dysprosium	Dy	66	162.50	162.5
einsteinium	Es	99		254
erbium	Er	68	167.26	167
europium	Eu	63	151.96	152
fermium	Fm	100		257
fluorine	F	9	18.998 403	19
francium	Fr	87		223
gadolinium	Gd	64	157.25	157
gallium	Ga	31	69.72	69.5
germanium	Ge	32	72.59	72.5
gold	Au	79	196.9665	197
hafnium	Hf	72	178.49	178.5

element	symbol	atomic number	relative atomic mass accurate	relative atomic mass approx.
helium	He	2	4.002 60	4
holmium	Ho	67	164.930 4	165
hydrogen	H	1	1.007 9	1
indium	In	49	114.82	115
iodine	I	53	126.904 5	127
iridium	Ir	77	192.22	192
iron	Fe	26	55.847	56
krypton	Kr	36	83.80	84
lanthanum	La	57	138.905 5	139
lawrencium	Lr	103		260
lead	Pb	82	207.2	207
lithium	Li	3	6.941	7
lutetium	Lu	71	174.967	175
magnesium	Mg	12	24.305	24.5
manganese	Mn	25	54.938 0	55
mendelevium	Md	101		258
mercury	Hg	80	200.59	200.5
molybdenum	Mo	42	95.94	96
neodymium	Nd	60	144.24	144
neon	Ne	10	20.179	20
neptunium	Np	93	237.048 2	237
nickel	Ni	28	58.70	58.5
niobium	Nb	41	92.906 4	93
nitrogen	N	7	14.006 7	14
nobelium	No	102		259
osmium	Os	76	190.2	190
oxygen	O	8	15.999 4	16
palladium	Pd	46	106.4	106
phosphorus	P	15	30.973 76	31
platinum	Pt	78	195.09	195
plutonium	Pu	94		244
polonium	Po	84		209
potassium	K	19	39.098 3	39
praseodymium	Pr	59	140.907 7	141
promethium	Pm	61		145
protactinium	Pa	91	231.035 9	231
radium	Ra	88	226.025 4	226
radon	Rn	86		222
rhenium	Re	75	186.207	186
rhodium	Rh	45	102.905 5	103
rubidium	Rb	37	85.467 8	85.5
ruthenium	Ru	44	101.07	101
samarium	Sm	62	150.4	150.5

element	symbol	atomic number	relative atomic mass accurate	approx.
scandium	Sc	21	44.9559	45
selenium	Se	34	78.96	79
silicon	Si	14	28.0855	28
silver	Ag	47	107.868	108
sodium	Na	11	22.98977	23
strontium	Sr	38	87.62	87.5
sulphur	S	16	32.06	32
tantalum	Ta	73	180.9479	181
technetium	Tc	43		97
tellurium	Te	52	127.60	127.5
terbium	Tb	65	158.9254	159
thallium	Tl	81	204.37	204.5
thorium	Th	90	232.0381	232
thulium	Tm	69	168.9342	169
tin	Sn	50	118.69	118.5
titanium	Ti	22	47.90	48
tungsten	W	74	183.85	184
uranium	U	92	238.029	238
vanadium	V	23	50.9415	51
xenon	Xe	54	131.30	131.5
ytterbium	Yb	70	173.04	173
yttrium	Y	39	88.9059	89
zinc	Zn	30	65.38	65.5
zirconium	Zr	40	91.22	91

Logarithms

	0	1	2	3	4	5	6	7	8	9	differences								
											1	2	3	4	5	6	7	8	9
10	0000	0043	0086	0128	0170	0212	0253	0294	0334	0374	4	8	12	17	21	25	29	33	37
11	0414	0453	0492	0531	0569	0607	0645	0682	0719	0755	4	8	11	15	19	23	26	30	34
12	0792	0828	0864	0899	0934	0969	1004	1038	1072	1106	3	7	10	14	17	21	24	28	31
13	1139	1173	1206	1239	1271	1303	1335	1367	1399	1430	3	6	10	13	16	19	23	26	29
14	1461	1492	1523	1553	1584	1614	1644	1673	1703	1732	3	6	9	12	15	18	21	24	27
15	1761	1790	1818	1847	1875	1903	1931	1959	1987	2014	3	6	8	11	14	17	20	22	25
16	2041	2068	2095	2122	2148	2175	2201	2227	2253	2279	3	5	8	11	13	16	18	21	24
17	2304	2330	2355	2380	2405	2430	2455	2480	2504	2529	2	5	7	10	12	15	17	20	22
18	2553	2577	2601	2625	2648	2672	2695	2718	2742	2765	2	5	7	9	12	14	16	19	21
19	2788	2810	2833	2856	2878	2900	2923	2945	2967	2989	2	4	7	9	11	13	16	18	20
20	3010	3032	3054	3075	3096	3118	3139	3160	3181	3201	2	4	6	8	11	13	15	17	19
21	3222	3243	3263	3284	3304	3324	3345	3365	3385	3404	2	4	6	8	10	12	14	16	18
22	3424	3444	3464	3483	3502	3522	3541	3560	3579	3598	2	4	6	8	10	12	14	15	17
23	3617	3636	3655	3674	3692	3711	3729	3747	3766	3784	2	4	6	7	9	11	13	15	17
24	3802	3820	3838	3856	3874	3892	3909	3927	3945	3962	2	4	5	7	9	11	12	14	16
25	3979	3997	4014	4031	4048	4065	4082	4099	4116	4133	2	3	5	7	9	10	12	14	15
26	4150	4166	4183	4200	4216	4232	4249	4265	4281	4298	2	3	5	7	8	10	11	13	15
27	4314	4330	4346	4362	4378	4393	4409	4425	4440	4456	2	3	5	6	8	9	11	13	14
28	4472	4487	4502	4518	4533	4548	4564	4579	4594	4609	2	3	5	6	8	9	11	12	14
29	4624	4639	4654	4669	4683	4698	4713	4728	4742	4757	1	3	4	6	7	9	10	12	13
30	4771	4786	4800	4814	4829	4843	4857	4871	4886	4900	1	3	4	6	7	9	10	11	13
31	4914	4928	4942	4955	4969	4983	4997	5011	5024	5038	1	3	4	6	7	8	10	11	12
32	5051	5065	5079	5092	5105	5119	5132	5145	5159	5172	1	3	4	5	7	8	9	11	12
33	5185	5198	5211	5224	5237	5250	5263	5276	5289	5302	1	3	4	5	6	8	9	10	12
34	5315	5328	5340	5353	5366	5378	5391	5403	5416	5428	1	3	4	5	6	8	9	10	11
35	5441	5453	5465	5478	5490	5502	5514	5527	5539	5551	1	2	4	5	6	7	9	10	11
36	5563	5575	5587	5599	5611	5623	5635	5647	5658	5670	1	2	4	5	6	7	8	10	11
37	5682	5694	5705	5717	5729	5740	5752	5763	5775	5786	1	2	3	5	6	7	8	9	10
38	5798	5809	5821	5832	5843	5855	5866	5877	5888	5899	1	2	3	5	6	7	8	9	10
39	5911	5922	5933	5944	5955	5966	5977	5988	5999	6010	1	2	3	4	5	7	8	9	10
40	6021	6031	6042	6053	6064	6075	6085	6096	6107	6117	1	2	3	4	5	6	8	9	10
41	6128	6138	6149	6160	6170	6180	6191	6201	6212	6222	1	2	3	4	5	6	7	8	9
42	6232	6243	6253	6263	6274	6284	6294	6304	6314	6325	1	2	3	4	5	6	7	8	9
43	6335	6345	6355	6365	6375	6385	6395	6405	6415	6425	1	2	3	4	5	6	7	8	9
44	6435	6444	6454	6464	6474	6484	6493	6503	6513	6522	1	2	3	4	5	6	7	8	9
45	6532	6542	6551	6561	6571	6580	6590	6599	6609	6618	1	2	3	4	5	6	7	8	9
46	6628	6637	6646	6656	6665	6675	6684	6693	6702	6712	1	2	3	4	5	6	7	7	8
47	6721	6730	6739	6749	6758	6767	6776	6785	6794	6803	1	2	3	4	5	5	6	7	8
48	6812	6821	6830	6839	6848	6857	6866	6875	6884	6893	1	2	3	4	4	5	6	7	8
49	6902	6911	6920	6928	6937	6946	6955	6964	6972	6981	1	2	3	4	4	5	6	7	8
50	6990	6998	7007	7016	7024	7033	7042	7050	7059	7067	1	2	3	3	4	5	6	7	8
51	7076	7084	7093	7101	7110	7118	7126	7135	7143	7152	1	2	3	3	4	5	6	7	8
52	7160	7168	7177	7185	7193	7202	7210	7218	7226	7235	1	2	2	3	4	5	6	7	7
53	7243	7251	7259	7267	7275	7284	7292	7300	7308	7316	1	2	2	3	4	5	6	6	7
54	7324	7332	7340	7348	7356	7364	7372	7380	7388	7396	1	2	2	3	4	5	6	6	7

	0	**1**	**2**	**3**	**4**	**5**	**6**	**7**	**8**	**9**	differences								
											1	2	3	4	5	6	7	8	9
55	7404	7412	7419	7427	7435	7443	7451	7459	7466	7474	1	2	2	3	4	5	5	6	7
56	7482	7490	7497	7505	7513	7520	7528	7536	7543	7551	1	2	2	3	4	5	5	6	7
57	7559	7566	7574	7582	7589	7597	7604	7612	7619	7627	1	2	2	3	4	5	5	6	7
58	7634	7642	7649	7657	7664	7672	7679	7686	7694	7701	1	1	2	3	4	4	5	6	7
59	7709	7716	7723	7731	7738	7745	7752	7760	7767	7774	1	1	2	3	4	4	5	6	7
60	7782	7789	7796	7803	7810	7818	7825	7832	7839	7846	1	1	2	3	4	4	5	6	6
61	7853	7860	7868	7875	7882	7889	7896	7903	7910	7917	1	1	2	3	4	4	5	6	6
62	7924	7931	7938	7945	7952	7959	7966	7973	7980	7987	1	1	2	3	3	4	5	6	6
63	7993	8000	8007	8014	8021	8028	8035	8041	8048	8055	1	1	2	3	3	4	5	5	6
64	8062	8069	8075	8082	8089	8096	8102	8109	8116	8122	1	1	2	3	3	4	5	5	6
65	8129	8136	8142	8149	8156	8162	8169	8176	8182	8189	1	1	2	3	3	4	5	5	6
66	8195	8202	8209	8215	8222	8228	8235	8241	8248	8254	1	1	2	3	3	4	5	5	6
67	8261	8267	8274	8280	8287	8293	8299	8306	8312	8319	1	1	2	3	3	4	5	5	6
68	8325	8331	8338	8344	8351	8357	8363	8370	8376	8382	1	1	2	3	3	4	4	5	6
69	8388	8395	8401	8407	8414	8420	8426	8432	8439	8445	1	1	2	2	3	4	4	5	6
70	8451	8457	8463	8470	8476	8482	8488	8494	8500	8506	1	1	2	2	3	4	4	5	6
71	8513	8519	8525	8531	8537	8543	8549	8555	8561	8567	1	1	2	2	3	4	4	5	5
72	8573	8579	8585	8591	8597	8603	8609	8615	8621	8627	1	1	2	2	3	4	4	5	5
73	8633	8639	8645	8651	8657	8663	8669	8675	8681	8686	1	1	2	2	3	4	4	5	5
74	8692	8698	8704	8710	8716	8722	8727	8733	8739	8745	1	1	2	2	3	4	4	5	5
75	8751	8756	8762	8768	8774	8779	8785	8791	8797	8802	1	1	2	2	3	3	4	5	5
76	8808	8814	8820	8825	8831	8837	8842	8848	8854	8859	1	1	2	2	3	3	4	5	5
77	8865	8871	8876	8882	8887	8893	8899	8904	8910	8915	1	1	2	2	3	3	4	4	5
78	8921	8927	8932	8938	8943	8949	8954	8960	8965	8971	1	1	2	2	3	3	4	4	5
79	8976	8982	8987	8993	8998	9004	9009	9015	9020	9025	1	1	2	2	3	3	4	4	5
80	9031	9036	9042	9047	9053	9058	9063	9069	9074	9079	1	1	2	2	3	3	4	4	5
81	9085	9090	9096	9101	9106	9112	9117	9122	9128	9133	1	1	2	2	3	3	4	4	5
82	9138	9143	9149	9154	9159	9165	9170	9175	9180	9186	1	1	2	2	3	3	4	4	5
83	9191	9196	9201	9206	9212	9217	9222	9227	9232	9238	1	1	2	2	3	3	4	4	5
84	9243	9248	9253	9258	9263	9269	9274	9279	9284	9289	1	1	2	2	3	3	4	4	5
85	9294	9299	9304	9309	9315	9320	9325	9330	9335	9340	1	1	2	2	3	3	4	4	5
86	9345	9350	9355	9360	9365	9370	9375	9380	9385	9390	1	1	2	2	3	3	4	4	5
87	9395	9400	9405	9410	9415	9420	9425	9430	9435	9440	0	1	1	2	2	3	3	4	4
88	9445	9450	9455	9460	9465	9469	9474	9479	9484	9489	0	1	1	2	2	3	3	4	4
89	9494	9499	9504	9509	9513	9518	9523	9528	9533	9538	0	1	1	2	2	3	3	4	4
90	9542	9547	9552	9557	9562	9566	9571	9576	9581	9586	0	1	1	2	2	3	3	4	4
91	9590	9595	9600	9605	9609	9614	9619	9624	9628	9633	0	1	1	2	2	3	3	4	4
92	9638	9643	9647	9652	9657	9661	9666	9671	9675	9680	0	1	1	2	2	3	3	4	4
93	9685	9689	9694	9699	9703	9708	9713	9717	9722	9727	0	1	1	2	2	3	3	4	4
94	9731	9736	9741	9745	9750	9754	9759	9763	9768	9773	0	1	1	2	2	3	3	4	4
95	9777	9782	9786	9791	9795	9800	9805	9809	9814	9818	0	1	1	2	2	3	3	4	4
96	9823	9827	9832	9836	9841	9845	9850	9854	9859	9863	0	1	1	2	2	3	3	4	4
97	9868	9872	9877	9881	9886	9890	9894	9899	9903	9908	0	1	1	2	2	3	3	4	4
98	9912	9917	9921	9926	9930	9934	9939	9943	9948	9952	0	1	1	2	2	3	3	4	4
99	9956	9961	9965	9969	9974	9978	9983	9987	9991	9996	0	1	1	2	2	3	3	3	4

Antilogarithms

	0	1	2	3	4	5	6	7	8	9	differences								
											1	2	3	4	5	6	7	8	9
.00	1000	1002	1005	1007	1009	1012	1014	1016	1019	1021	0	0	1	1	1	1	2	2	2
.01	1023	1026	1028	1030	1033	1035	1038	1040	1042	1045	0	0	1	1	1	1	2	2	2
.02	1047	1050	1052	1054	1057	1059	1062	1064	1067	1069	0	0	1	1	1	1	2	2	2
.03	1072	1074	1076	1079	1081	1084	1086	1089	1091	1094	0	0	1	1	1	1	2	2	2
.04	1096	1099	1102	1104	1107	1109	1112	1114	1117	1119	0	1	1	1	1	2	2	2	2
.05	1122	1125	1127	1130	1132	1135	1138	1140	1143	1146	0	1	1	1	1	2	2	2	2
.06	1148	1151	1153	1156	1159	1161	1164	1167	1169	1172	0	1	1	1	1	2	2	2	2
.07	1175	1178	1180	1183	1186	1189	1191	1194	1197	1199	0	1	1	1	1	2	2	2	2
.08	1202	1205	1208	1211	1213	1216	1219	1222	1225	1227	0	1	1	1	1	2	2	2	3
.09	1230	1233	1236	1239	1242	1245	1247	1250	1253	1256	0	1	1	1	1	2	2	2	3
.10	1259	1262	1265	1268	1271	1274	1276	1279	1282	1285	0	1	1	1	1	2	2	2	3
.11	1288	1291	1294	1297	1300	1303	1306	1309	1312	1315	0	1	1	1	2	2	2	2	3
.12	1318	1321	1324	1327	1330	1334	1337	1340	1343	1346	0	1	1	1	2	2	2	2	3
.13	1349	1352	1355	1358	1361	1365	1368	1371	1374	1377	0	1	1	1	2	2	2	3	3
.14	1380	1384	1387	1390	1393	1396	1400	1403	1406	1409	0	1	1	1	2	2	2	3	3
.15	1413	1416	1419	1422	1426	1429	1432	1435	1439	1442	0	1	1	1	2	2	2	3	3
.16	1445	1449	1452	1455	1459	1462	1466	1469	1472	1476	0	1	1	1	2	2	2	3	3
.17	1479	1483	1486	1489	1493	1496	1500	1503	1507	1510	0	1	1	1	2	2	2	3	3
.18	1514	1517	1521	1524	1528	1531	1535	1538	1542	1545	0	1	1	1	2	2	2	3	3
.19	1549	1552	1556	1560	1563	1567	1570	1574	1578	1581	0	1	1	1	2	2	3	3	3
.20	1585	1589	1592	1596	1600	1603	1607	1611	1614	1618	0	1	1	1	2	2	3	3	3
.21	1622	1626	1629	1633	1637	1641	1644	1648	1652	1656	0	1	1	2	2	2	3	3	3
.22	1660	1663	1667	1671	1675	1679	1683	1687	1690	1694	0	1	1	2	2	2	3	3	3
.23	1698	1702	1706	1710	1714	1718	1722	1726	1730	1734	0	1	1	2	2	2	3	3	4
.24	1738	1742	1746	1750	1754	1758	1762	1766	1770	1774	0	1	1	2	2	2	3	3	4
.25	1778	1782	1786	1791	1795	1799	1803	1807	1811	1816	0	1	1	2	2	2	3	3	4
.26	1820	1824	1828	1832	1837	1841	1845	1849	1854	1858	0	1	1	2	2	3	3	3	4
.27	1862	1866	1871	1875	1879	1884	1888	1892	1897	1901	0	1	1	2	2	3	3	3	4
.28	1905	1910	1914	1919	1923	1928	1932	1936	1941	1945	0	1	1	2	2	3	3	4	4
.29	1950	1954	1959	1963	1968	1972	1977	1982	1986	1991	0	1	1	2	2	3	3	4	4
.30	1995	2000	2004	2009	2014	2018	2023	2028	2032	2037	0	1	1	2	2	3	3	4	4
.31	2042	2046	2051	2056	2061	2065	2070	2075	2080	2084	0	1	1	2	2	3	3	4	4
.32	2089	2094	2099	2104	2109	2113	2118	2123	2128	2133	0	1	1	2	2	3	3	4	4
.33	2138	2143	2148	2153	2158	2163	2168	2173	2178	2183	0	1	1	2	2	3	3	4	4
.34	2188	2193	2198	2203	2208	2213	2218	2223	2228	2234	1	1	2	2	3	3	4	4	5
.35	2239	2244	2249	2254	2259	2265	2270	2275	2280	2286	1	1	2	2	3	3	4	4	5
.36	2291	2296	2301	2307	2312	2317	2323	2328	2333	2339	1	1	2	2	3	3	4	4	5
.37	2344	2350	2355	2360	2366	2371	2377	2382	2388	2393	1	1	2	2	3	3	4	4	5
.38	2399	2404	2410	2415	2421	2427	2432	2438	2443	2449	1	1	2	2	3	3	4	4	5
.39	2455	2460	2466	2472	2477	2483	2489	2495	2500	2506	1	1	2	2	3	3	4	5	5
.40	2512	2518	2523	2529	2535	2541	2547	2553	2559	2564	1	1	2	2	3	4	4	5	5
.41	2570	2576	2582	2588	2594	2600	2606	2612	2618	2624	1	1	2	2	3	4	4	5	5
.42	2630	2636	2642	2649	2655	2661	2667	2673	2679	2685	1	1	2	2	3	4	4	5	6
.43	2692	2698	2704	2710	2716	2723	2729	2735	2742	2748	1	1	2	3	3	4	4	5	6
.44	2754	2761	2767	2773	2780	2786	2793	2799	2805	2812	1	1	2	3	3	4	4	5	6
.45	2818	2825	2831	2838	2844	2851	2858	2864	2871	2877	1	1	2	3	3	4	5	5	6
.46	2884	2891	2897	2904	2911	2917	2924	2931	2938	2944	1	1	2	3	3	4	5	5	6
.47	2951	2958	2965	2972	2979	2985	2992	2999	3006	3013	1	1	2	3	3	4	5	5	6
.48	3020	3027	3034	3041	3048	3055	3062	3069	3076	3083	1	1	2	3	4	4	5	6	6
.49	3090	3097	3105	3112	3119	3126	3133	3141	3148	3155	1	1	2	3	4	4	5	6	6

	0	1	2	3	4	5	6	7	8	9	differences								
											1	2	3	4	5	6	7	8	9
.50	3162	3170	3177	3184	3192	3199	3206	3214	3221	3228	1	1	2	3	4	4	5	6	7
.51	3236	3243	3251	3258	3266	3273	3281	3289	3296	3304	1	2	2	3	4	5	5	6	7
.52	3311	3319	3327	3334	3342	3350	3357	3365	3373	3381	1	2	2	3	4	5	5	6	7
.53	3388	3396	3404	3412	3420	3428	3436	3443	3451	3459	1	2	2	3	4	5	6	6	7
.54	3467	3475	3483	3491	3499	3508	3516	3524	3532	3540	1	2	2	3	4	5	6	6	7
.55	3548	3556	3565	3573	3581	3589	3597	3606	3614	3622	1	2	2	3	4	5	6	7	7
.56	3631	3639	3648	3656	3664	3673	3681	3690	3698	3707	1	2	3	3	4	5	6	7	8
.57	3715	3724	3733	3741	3750	3758	3767	3776	3784	3793	1	2	3	3	4	5	6	7	8
.58	3802	3811	3819	3828	3837	3846	3855	3864	3873	3882	1	2	3	4	4	5	6	7	8
.59	3890	3899	3908	3917	3926	3936	3945	3954	3963	3972	1	2	3	4	5	5	6	7	8
.60	3981	3990	3999	4009	4018	4027	4036	4046	4055	4064	1	2	3	4	5	6	6	7	8
.61	4074	4083	4093	4102	4111	4121	4130	4140	4150	4159	1	2	3	4	5	6	7	8	9
.62	4169	4178	4188	4198	4207	4217	4227	4236	4246	4256	1	2	3	4	5	6	7	8	9
.63	4266	4276	4285	4295	4305	4315	4325	4335	4345	4355	1	2	3	4	5	6	7	8	9
.64	4365	4375	4385	4395	4406	4416	4426	4436	4446	4457	1	2	3	4	5	6	7	8	9
.65	4467	4477	4487	4498	4508	4519	4529	4539	4550	4560	1	2	3	4	5	6	7	8	9
.66	4571	4581	4592	4603	4613	4624	4634	4645	4656	4667	1	2	3	4	5	6	7	9	10
.67	4677	4688	4699	4710	4721	4732	4742	4753	4764	4775	1	2	3	4	5	7	8	9	10
.68	4786	4797	4808	4819	4831	4842	4853	4864	4875	4887	1	2	3	4	6	7	8	9	10
.69	4898	4909	4920	4932	4943	4955	4966	4977	4989	5000	1	2	3	5	6	7	8	9	10
.70	5012	5023	5035	5047	5058	5070	5082	5093	5105	5117	1	2	4	5	6	7	8	9	11
.71	5129	5140	5152	5164	5176	5188	5200	5212	5224	5236	1	2	4	5	6	7	8	10	11
.72	5248	5260	5272	5284	5297	5309	5321	5333	5346	5358	1	2	4	5	6	7	9	10	11
.73	5370	5383	5395	5408	5420	5433	5445	5458	5470	5483	1	3	4	5	6	8	9	10	11
.74	5495	5508	5521	5534	5546	5559	5572	5585	5598	5610	1	3	4	5	6	8	9	10	12
.75	5623	5636	5649	5662	5675	5689	5702	5715	5728	5741	1	3	4	5	7	8	9	10	12
.76	5754	5768	5781	5794	5808	5821	5834	5848	5861	5875	1	3	4	5	7	8	9	11	12
.77	5888	5902	5916	5929	5943	5957	5970	5984	5998	6012	1	3	4	5	7	8	10	11	12
.78	6026	6039	6053	6067	6081	6095	6109	6124	6138	6152	1	3	4	6	7	8	10	11	13
.79	6166	6180	6194	6209	6223	6237	6252	6266	6281	6295	1	3	4	6	7	9	10	11	13
.80	6310	6324	6339	6353	6368	6383	6397	6412	6427	6442	1	3	4	6	7	9	10	12	13
.81	6457	6471	6486	6501	6516	6531	6546	6561	6577	6592	2	3	5	6	8	9	11	12	14
.82	6607	6622	6637	6653	6668	6683	6699	6714	6730	6745	2	3	5	6	8	9	11	12	14
.83	6761	6776	6792	6808	6823	6839	6855	6871	6887	6902	2	3	5	6	8	9	11	13	14
.84	6918	6934	6950	6966	6982	6998	7015	7031	7047	7063	2	3	5	6	8	10	11	13	15
.85	7079	7096	7112	7129	7145	7161	7178	7194	7211	7228	2	3	5	7	8	10	12	13	15
.86	7244	7261	7278	7295	7311	7328	7345	7362	7379	7396	2	3	5	7	8	10	12	13	15
.87	7413	7430	7447	7464	7482	7499	7516	7534	7551	7568	2	3	5	7	9	10	12	14	16
.88	7586	7603	7621	7638	7656	7674	7691	7709	7727	7745	2	4	5	7	9	11	12	14	16
.89	7762	7780	7798	7816	7834	7852	7870	7889	7907	7925	2	4	5	7	9	11	13	14	16
.90	7943	7962	7980	7998	8017	8035	8054	8072	8091	8110	2	4	6	7	9	11	13	15	17
.91	8128	8147	8166	8185	8204	8222	8241	8260	8279	8299	2	4	6	8	9	11	13	15	17
.92	8318	8337	8356	8375	8395	8414	8433	8453	8472	8492	2	4	6	8	10	12	14	15	17
.93	8511	8531	8551	8570	8590	8610	8630	8650	8670	8690	2	4	6	8	10	12	14	16	18
.94	8710	8730	8750	8770	8790	8810	8831	8851	8872	8892	2	4	6	8	10	12	14	16	18
.95	8913	8933	8954	8974	8995	9016	9036	9057	9078	9099	2	4	6	8	10	12	15	17	19
.96	9120	9141	9162	9183	9204	9226	9247	9268	9290	9311	2	4	6	8	11	13	15	17	19
.97	9333	9354	9376	9397	9419	9441	9462	9484	9506	9528	2	4	7	9	11	13	15	17	20
.98	9550	9572	9594	9616	9638	9661	9683	9705	9727	9750	2	4	7	9	11	13	16	18	20
.99	9772	9795	9817	9840	9863	9886	9908	9931	9954	9977	2	5	7	9	11	14	16	18	20

Answers to questions

Chapter 1 (p. 11)   **1 a** $123.3\,dm^3$; **b** $116.1\,dm^3$ **3** $1.168\,g\,dm^{-3}$
4 $903.6\,cm^3$ **7** 16.22 **8** X_2H_6 **9** $1.555:1$ **15** $65.86(N_2)$; $15.2(CO_2)$;
$20.26(O_2)$ **16** 25.8%.

Chapter 2 (p. 21)   **13** 665.9×10^{12} **14** $104\,cm^3$ **21** 46 **22** $36.7\,cm^3$
23 MF_6; 238.2 **25** $24.64\,cm^3$; 230.9 **26** 35.24 per cent
27 53.94 per cent   **29** 2 per cent.

Chapter 3 (p. 36)   **6** 8.314; $393.5\,m\,s^{-1}$ **9** $461\,m\,s^{-1}$.

Chapter 5 (p. 48)   **8** 200×10^9 **9** 176×10^9.

Chapters 8 and 9 (p. 92)   **1** $13.56 \times 10^{-12}\,s^{-1}$; 36.13×10^9; 0.976;
$1.024\,g$ **2 a** $7.66 \times 10^9\,s$; **b** $169.8 \times 10^9\,s$ **3** $869.2 \times 10^{-6}\,g$; 1.3×10^9
year   **5** $800\,s^{-1}$ **15** $16.004\,335$ **19** 65.99 year.

Chapter 10 (p. 104)   **4 a** 1.24; **b** 6.20; 500 **6** $470\,MeV$ **7** $93.4\,MeV$
9 1.0039 **10** 0.51 **11** 11.01128 **12** 2.29×10^{12} **13** 728.3×10^6
14 $192.9\,MeV$ **16** $1.51 \times 10^{-29}\,kg$.

Chapter 15 (p. 174)   **4** 74 per cent.

Chapter 16 (p. 193)   **7 a** 0.16×10^{-18}; **b** 124×10^{-6}; **c** 10^{-2}; **d** 40
8 $520\,kJ\,mol^{-1}$; $418.6\,kJ\,mol^{-1}$; 241 nm.

Chapter 17 (p. 207)   **1** 0.2813; 71.88 per cent   **2** 25×10^{-3} **12** 0.19
14 0.0338 **16** 6 **17** 63.3 per cent N_2; 34.9 per cent O_2; 1.8 per cent Ar.

Chapter 19 (p. 229)   **9** $186.65\,kPa$ **10 a** $2799\,Pa$; $4533\,Pa$; **b** $7333\,Pa$;
c 38 per cent CH_3OH by volume   **11** Mol fraction of $A = 0.167$
12 $164.4\,cm^3$.

Chapter 20 (p. 237)   **1** 181.8 **4** 120.1; 229.9 **5** S_8; 1.9×10^{-3} **6** 145
10 126 **11** $390.4\,°C$; 134.6 **13 a** $7419\,g$; **b** $2581\,g$.

Chapter 21 (p. 244)   **1** 596 **2** $3.84\,g\,dm^{-3}$ **5** 34.2×10^3 **6** 68.1×10^3
7 240×10^3.

Chapter 22 (p. 258)   **1** $-548\,kJ$ **2 a** $-597.6\,kJ\,mol^{-1}$
b $-1255\,kJ\,mol^{-1}$ **3** $60.2\,kJ\,mol^{-1}$ **4** $-74.7\,kJ\,mol^{-1}$ **5** $4.247\,kg$
6 $-214.2\,kJ\,mol^{-1}$ **8** $37.61\,kJ\,mol^{-1}$ **10** $-2220\,kJ\,mol^{-1}$ **11** $2.9\,°C$

12 $-61.52\,\text{kJ}\,\text{mol}^{-1}$ **13** $-3242\,\text{kJ}\,\text{mol}^{-1}$ **14** $10.94\,\text{kJ}$; $10.73\,\text{kJ}$
15 $-460.2\,\text{kJ}$ **16** $-108.8\,\text{kJ}\,\text{mol}^{-1}$ **17 a** $90.4\,\text{kJ}\,\text{mol}^{-1}$;
$-40.6\,\text{kJ}\,\text{mol}^{-1}$ **b** $87.9\,\text{kJ}\,\text{mol}^{-1}$; $-43.1\,\text{kJ}\,\text{mol}^{-1}$ **18** $238.3\,\text{kJ}\,\text{mol}^{-1}$
26 a $-174.5\,\text{kJ}$; **b** $-11\,\text{kJ}$; **c** $-3120.2\,\text{kJ}$; **d** $-65.5\,\text{kJ}$; **e** $216.7\,\text{kJ}$
32 $151.8\,\text{kJ}\,\text{mol}^{-1}$.

Chapter 23 (p. 268) **16 a** $-25.6\,\text{kJ}$; **b** $-56.1\,\text{kJ}$; **c** $-81.8\,\text{kJ}$;
d $-152.4\,\text{kJ}$.

Chapter 24 (p. 283) **2** $12.14\,\text{J}\,\text{K}^{-1}\,\text{mol}^{-1}$ **4** $109\,\text{J}\,\text{K}^{-1}\,\text{mol}^{-1}$
18 a 9×10^{64}; **b** 2.982; **c** 55×10^{14}; **d** 0.335; **e** 6.46×10^{-22}
19 $404\,\text{K}$, $1109\,\text{K}$, $464\,\text{K}$, $1479\,\text{K}$.

Chapter 25 (p. 299) **1** 4; $99.57\,\text{g}$ **2 a** $0.9\,\text{mol}$; **b** $0.845\,\text{mol}$;
c $0.54\,\text{mol}$ **5** $2.925:1$ **12** $303\,\text{mol}^{-1}\,\text{dm}^3$ **13** 2.77 per cent
14 $8.044\times10^3\,\text{Pa}$ **15** 0.1089; $0.3838\,\text{atm}$ **16 a** $0.037\,\text{mol}\,\text{dm}^{-3}$;
b $0.063\,\text{mol}\,\text{dm}^{-3}$ **17 a** 50; **b** $5.9\,\text{atm}$; **c** $0.65\,\text{atm}$ **19 a** 625, 1156;
b $1:1.848$; **c** $84.42\,\text{mm}$ **20** $57.68\,\text{kJ}\,\text{mol}^{-1}$ **21** $171.6\,\text{kJ}$
22 $90.16\,\text{kJ}\,\text{mol}^{-1}$ **23** $505.4\,\text{kJ}$ **26 i** 0.25 per cent; **ii** 46.6 per cent.

Chapter 26 (p. 315) **2** 1 **3** $144.5\,\text{min}$ **5** $82.4\,\text{min}$ **6** $513\,\text{s}$ **7** 3 **8** 2
13 1.

Chapter 27 (p. 331) **1** $285.7\,\text{kJ}$ **3** 0.34; 0.50 **4** $53.59\,\text{kJ}$
17 a 160; **b** 119.6.

Chapter 29 (p. 352) **2 a** $1.524\,\text{cm}^{-1}$; **b** $0.021\,16\,\text{S}\,\text{cm}^{-1}$
3 a $0.5\,\text{cm}^{-1}$; **b** $0.068\,96\,\text{S}\,\text{cm}^{-1}$; $180.5\,\Omega$ **4** $625\times10^3\,\Omega$; $16\times10^{-6}\,\text{A}$
13 a 386.0; **b** 246.3 **15** 0.495 **16 i** $1:1$; **ii** $1:3$
20 $1.38\times10^{-5}\,\text{mol}\,\text{dm}^{-3}$ **23** $0.11x\,\text{cm}$.

Chapter 30 (p. 361) **1** $17.9\times10^{-6}\,\text{mol}\,\text{dm}^{-3}$
3 a $6.03\times10^{-3}\,\text{mol}\,\text{dm}^{-3}$; **b** $803\times10^{-6}\,\text{mol}\,\text{dm}^{-3}$;
c $134\times10^{-6}\,\text{mol}\,\text{dm}^{-3}$; **d** $828\times10^{-9}\,\text{mol}\,\text{dm}^{-3}$
6 $443\times10^{-6}\,\text{mol}\,\text{dm}^{-3}$ **7** 95.5 per cent **10 a** 0.0245; 0.190;
b $18\times10^{-6}\,\text{mol}\,\text{dm}^{-3}$; $10.85\times10^{-6}\,\text{mol}\,\text{dm}^{-3}$ **11** $-0.187\,°\text{C}$
12 $1.342\times10^{-3}\,\text{mol}\,\text{dm}^{-3}$ **14** $2.65\times10^{-5}\,\text{mol}\,\text{dm}^{-3}$.

Chapter 31 (p.379) **1** 20 per cent **2 a** $10^{-4}\,\text{mol}\,\text{dm}^{-3}$;
b $2.5\times10^{-4}\,\text{mol}\,\text{dm}^{-3}$ **3 a** 2.88; **b** 2; **c** 12; **d** 5.7 **8** 0.0132; 2.878
9 12; 3.162×10^{-6} **10 a** $0.0365\,\text{g}$; **b** 3.42; **c** 2.29×10^{-3}
11 $2.9\times10^{-3}\,\text{mol}\,\text{dm}^{-3}$; 2.54 **16** 3.89; nil **17** 4.57 **30** 8.37
31 0.0075 per cent **32** 8.87 **33** $0.73\,\text{g}$ **34** 9.43
37 $8.5\times10^{-6}\,\text{mol}\,\text{dm}^{-3}$; 5.07 **40 a** 11.2; **b** 10.2; **c** 3.25.

Chapter 33 (p. 401) **2** 12×10^{-3} **3** $1.7\times10^{-9}\,\text{mol}\,\text{dm}^{-3}$
4 $K=108x^5/214^5$ **5** $\sqrt{x(x+y)}$ **6 a** $10^{-23}\,\text{mol}\,\text{dm}^{-3}$; $10^{-21}\,\text{mol}\,\text{dm}^{-3}$;
b $25\times10^{-27}\,\text{mol}^2\,\text{dm}^{-6}$ **8 a** $1.332\times10^{-8}\,\text{mol}^2\,\text{dm}^{-6}$; **b** 4.038×10^{-4}.

Introduction to Physical Chemistry

Chapter 34 (p. 418)   **8** 0.38 V.

Chapter 35 (p. 432)   **21** -0.278 V; 2+   **22** $+0.53$ V; 9×10^8.

Chapter 36 (p. 448)   **1** $4xy/3$ **3** 90 per cent
5 a 50 cm^3; **b** 25 cm^3 **6 a** 14.36 A; **b** 14.21 A, 193 C   **8** 31.5 μm
14 0.56.

Chapter 37 (p. 462)   **4** 22×10^{-16} cm^2; 2.5 nm; 3.7 nm   **9** 6×10^2 m^2
10 10^{18}; 4.9 Pa; 4×10^{-6} °C   **11** 15×10^{18}; 1.2×10^3 m^2 **14** 718.6 Pa.

Index

The periodic table

s-block

1A	2A
1 H	
3 Li	4 Be
11 Na	12 Mg
19 K	20 Ca
37 Rb	38 Sr
55 Cs	56 Ba
87 Fr	88 Ra

d-block — transition elements →

3A	4A	5A	6A	7A		8		1B	2B
21 Sc	22 Ti	23 V	24 Cr	25 Mn	26 Fe	27 Co	28 Ni	29 Cu	30 Zn
39 Y	40 Zr	41 Nb	42 Mo	43 Tc	44 Ru	45 Rh	46 Pd	47 Ag	48 Cd
57* La	72 Hf	73 Ta	74 W	75 Re	76 Os	77 Ir	78 Pt	79 Au	80 Hg
89† Ac									

p-block

3B	4B	5B	6B	7B
5 B	6 C	7 N	8 O	9 F
13 Al	14 Si	15 P	16 S	17 Cl
31 Ga	32 Ge	33 As	34 Se	35 Br
49 In	50 Sn	51 Sb	52 Te	53 I
81 Tl	82 Pb	83 Bi	84 Po	85 At

0
2 He
10 Ne
18 Ar
36 Kr
54 Xe
86 Rn